CONVERSION FACTORS

MASS

$1.0\ \text{lb}_m = 453.59237\ \text{g}$
$1.0\ \text{slug} = 32.174\ \text{lb}_m$
$1.0\ \text{kg} = 2.2046\ \text{lb}_m$

LENGTH

$1.0\ \text{in.} = 2.54\ \text{cm} = 25.4\ \text{mm}$
$1.0\ \text{m} = 3.208\ \text{ft} = 39.37\ \text{in.}$
$1.0\ \text{cm} = 0.01\ \text{m} = 0.3937\ \text{in.} = 0.0323\ \text{ft}$
$1.0\ \text{mm} = 0.001\ \text{m} = 1 \times 10^{-3}\ \text{m}$
$1.0\ \mu\text{m} = 0.000001\ \text{m} = 1 \times 10^{-6}\ \text{m}$
$1.0\ \text{nm} = 0.000000001\ \text{m} = 1 \times 10^{-9}\ \text{m}$
$1.0\ \text{km} = 1000\ \text{m} = 0.612\ \text{miles}$
$1.0\ \text{miles} = 5280\ \text{ft}$

AREA

$1.0\ \text{m}^2 = 10.76\ \text{ft}^2$
$1.0\ \text{cm}^2 = 1 \times 10^{-4}\ \text{m}^2 = 0.155\ \text{in.}^2$

VOLUME

$1\ \text{l} = 1 \times 10^{-3}\ \text{m}^3$
$\qquad = 0.2642\ \text{gal}$
$1.0\ \text{gal} = 231.0\ \text{in.}^3 = 0.1337\ \text{ft}^3$
$1.0\ \text{gal} = 0.0037854\ \text{m}^3$
$1.0\ \text{ft}^3 = 0.0283\ \text{m}^3$

TIME

$1.0\ \text{min} = 60\ \text{s}$
$1.0\ \text{h} = 60\ \text{min}$
$1.0\ \text{day} = 8.64 \times 10^4\ \text{s}$

FORCE

$1.0\ \text{N} = 1\ \text{kg m/s}^2 = 1 \times 10^5\ \text{dyn}$
$1.0\ \text{lb} = 4.44822\ \text{N}$
$1.0\ \text{kg}_f = 9.806\ \text{N}$

PRESSURE OR STRESS

$1.0\ \text{Pa} = 1\ \text{N/m}^2$
$\qquad\qquad = 1.4504 \times 10^{-4}\ \text{lb/in.}^2$
$1.0\ \text{lb/in.}^2 = 6894.76\ \text{N/m}^2$
$1.0\ \text{atm} = 14.696\ \text{lb/in.}^2 = 760\ \text{Torr}$
$1.0\ \text{bar} = 14.505\ \text{lb/in.}^2$
$\qquad\qquad = 1 \times 10^5\ \text{N/m}^2$
$\qquad\qquad = 1 \times 10^8\ \text{dyn/cm}^2$
$1.0\ \text{in. Hg} = 3376.8\ \text{N/m}^2$
$1.0\ \text{in. H}_2\text{O} = 2.54\ \text{cm H}_2\text{O} = 248.8\ \text{N/m}^2$
$\qquad\qquad = 0.0362\ \text{lb/in.}^2$
$1.0\ \text{Torr} = 1\ \text{mm Hg}$
$1.0\ \mu\text{strain} = 10^{-6}\ \text{m/m}$

TEMPERATURE

$\text{K} = {}^\circ\text{C} + 273.15$
$1{}^\circ\text{C} = 1.8{}^\circ\text{F} = 1\ \text{K}$
${}^\circ\text{F} = 1.8{}^\circ\text{C} + 32$
${}^\circ\text{R} = {}^\circ\text{F} + 459.67$

VOLUME FLOW RATE

$1\ \text{gal/min} = 0.00223\ \text{ft}^3/\text{s} = 0.06309\ \text{l/s}$

ROTATION

$1\ \text{rev/s} = 2\pi\ \text{rad/s} = 60\ \text{rpm}$

FREQUENCY

$1\ \text{Hz} = 2\pi\ \text{rad/s}$

MOMENT OR TORQUE

$1.0\ \text{N m} = 0.7376\ \text{lb ft}$

Theory and Design for Mechanical Measurements

Third Edition

Richard S. Figliola
Clemson University

Donald E. Beasley
Clemson University

John Wiley & Sons, Inc.

Acquisitions Editor   *Joe Hayton*
Marketing Manager   *Katherine Hepburn*
Senior Production Editor   *Patricia McFadden*
Designer   *Maddy Lesure*
Illustration Editor   *Gene Aiello*
Cover Photo   *Lonnie Duka/Stone*

This book was set in *Times Ten* by *TechBooks* and printed and bound by *Hamilton Printing*. The cover was printed by *Lehigh Press, Inc.*

This book is printed on acid-free paper. ⊗

To order books or for customer service please, call 1(800)-CALL-WILEY (225-5945).

Library of Congress Cataloging in Publication Data:
Figliola, R. S.
Theory and design for mechanical measurements / Richard S. Figliola, Donald E. Beasley.—3rd ed.
p.   cm.
Includes bibliographical references and index.
ISBN 0-471-35083-4 (cloth/CD-ROM : alk. paper)
1. Mensuration.   I. Beasley, Donald E.   II. Title.

T50 .F54   2000
681′.2—dc21   00-036794

ISBN 0-471-35083-4

Printed in the United States of America

10 9 8 7 6 5

Preface

Our stimulus for producing the third edition of this text derives from the continuing need for engineers to design and conduct experiments with the associated challenges of instrumentation, data acquisition, and analysis. In the third edition, we have maintained the focus of the previous editions and have retained the following key objectives:

1. To provide a fundamental background in the theory of engineering measurements and measurement system performance.
2. To convey the principles and practice for the design of measurement systems and measurement test plans, including the role of statistics and uncertainty analysis in design.
3. To establish the fundamental principles and convey prevailing engineering practice for the measurement of those physical variables most important to engineering applications.

Implicit in these objectives is our assumption that certain aspects of measurements can be generalized, such as test plan design, signal reconstruction, or dynamic response. Other aspects are best treated in the context of the measurement of a specific physical quantity, such as strain or temperature. We feel that measurement system design provides an appropriate opportunity to integrate design into the curriculum. The development of a test plan, the selection of measurement techniques, and the analysis and presentation of test results fit well within the intent of design instruction.

The familiar reader will find the same overall structure as of previous editions. But on the basis of considerable reader input, we have revised the organization and discussion of uncertainty analysis, strain measurement, digital signal concepts, and filtering. End-of-chapter problems have been revised, with the significant addition of completely new exercises and a stress towards use of SI units. Measurement courses often introduce many new ideas, perhaps more than other courses in the traditional engineering curriculum. With this in mind, we have continued to edit the text to make the manuscript more easily readable and have avoided unnecessary expansion of the topical coverage.

Computer use is facilitated in the third edition by the inclusion of software that addresses the most fundamental issues in digital signal processing and instrument response. We intend this to be a useful learning tool for students, as well as a convenient instructional tool. The companion software is provided in the form of Matlab files, chosen because of the wide availability of this software in engineering programs. The software supports exploration of the concepts, as well as enhancing the examples and homework problems. The software provides the ability to interactively examine the effects of changes in such parameters as sampling rate for digital data acquisition, time constants for first-order systems, and many other parameters of interest. Instructions for installing and running the software can be found in the readme text file on the software disk. For those users who are familiar with Matlab, the file MainSwitch.m provides access to all of the program files. We have cited areas in the text where the companion software would enhance student learning. However, the text is totally self-supporting without use of this software.

Chapters 1 through 5 provide an introduction to measurement theory. Chapter 1, Basic Concepts of Measurement Methods, introduces the collective stages of the measurement system. The chapter stresses the importance of designing a test plan, as well as the concepts of calibration and standards. Chapter 2, Static and Dynamic Characteristics of Signals, addresses the makeup of signals, the basis for transmitting information throughout a measurement system. Both the static and dynamic components of signals are examined in terms of their time-invariant values and time-dependent values of amplitude and frequency. In Chapter 3, Measurement System Behavior, the response of measurement systems to such input signals is explored and the concepts of sensitivity, time response, and frequency response are presented.

An awareness of the measurement system design process and calibration procedures motivates the introduction of the statistical nature of physical variables and uncertainty analysis in Chapter 4, Probability and Statistics, and Chapter 5, Uncertainty Analysis. The importance of these two chapters cannot be overstressed, as practicing engineers are finding that their use goes beyond engineering measurements. Chapter 4 provides a sufficient introduction to probability and statistics to comprehend and report the behavior of measured variables. An associated course laboratory in the measurement of engineering variables would provide an excellent context for this. Chapter 5 provides a complete treatment of uncertainty analysis. It establishes a methodology for performing uncertainty analyses based on a variety of information. A working knowledge is developed through carefully selected example problems and further examples are integrated throughout the text. However, the concepts associated with uncertainty analysis are greatly enhanced through application, such as through associated laboratory exercises.

Chapter 6, Analog Electrical Devices and Measurements, describes basic analog electrical measurements and circuits. Many of these form important components of most practical measurement systems. Sampling concepts of data acquisition are retained in Chapter 7, Sampling and Digital Devices. In addition to developing the criteria for discrete sampling and the role of the Fourier transform in signal reconstruction, the basic components required for analog-to-digital or digital-to-digital communication are presented. This chapter also includes detailed discussions of computer-based data acquisition systems and methods for serial and parallel communication.

Chapters 8 through 12 describe the principles and practice for measuring important variables such as temperature, pressure, flow, displacement, force, power, and strain. Our goals in writing these chapters were both to provide an understanding of the physical principles used in such instrumentation and to provide sufficient, practical information to select components in assembling a measurement system. With each instrument discussed we have noted practical ranges of operation, required supporting equipment, and bias errors to be expected, as well as other helpful information.

The text is flexible and can be used in a variety of course structures at both the undergraduate and graduate levels. Chapters 1 through 5 and aspects of Chapters 6 and 7 and specific material from later chapters provide the material needed for a typically multihour undergraduate course. Chapters 4 through 12 are suitable for a more instrumentation-oriented laboratory course. Chapters 1, 4, and 5 could supplement any lab course. Chapters 6 through 12 have been written so that they can be used independently, but do make use of material from Chapters 3, 4, and 5. Because of the more mature student, a graduate course can cover the entire text in one semester.

We express our sincerest appreciation to the students, teachers, and engineers who have used our text. We are indebted to the many who have written us with their constructive comments and encouragement. We thank the Wiley editorial staff for their welcome assistance. The assistance of Dr. Timothy A. Conover in the development of the software companion disk is gratefully acknowledged. We are so very grateful to our wives and daughters, Suzanne and Elizabeth, and Leigh and Sarah for their patience, understanding, and valued help.

Richard S. Figliola
Donald E. Beasley
Clemson, South Carolina
March, 2000

Contents

? Spline Fitting ?

Chapter 1

Basic Concepts of Measurement Methods

1.1 INTRODUCTION

We make measurements everyday. For example, we routinely measure body weight on a scale or read the temperature of an outdoor thermometer. Most people put little thought into the selection of instruments for these very simple measurements. After all, the direct use of the data is clear to us, the types of instruments and techniques have long been established by custom, and the outcome of these measurements is not important enough to merit much attention to features such as accuracy or alternative methods. However, when the stakes become greater, the selection of measurement equipment and techniques and the interpretation of the measured data can demand considerable attention. Just contemplate on how you might verify that a new engine is built as designed and meets the performance specifications and emissions limits required.

But first things first. The primary objective in any measurement is to establish the value or the tendency of some variable. But this is based on the value or the tendency suggested by the measurement device. So just how does one establish the relationship between the real value of a variable and that actually measured? How can a measurement plan be devised so that the measurement provides the unambiguous information sought? How can a measurement system be used so that the engineer can easily interpret the measured data and be confident in their meaning? There are many considerations that have to be addressed to answer these basic but most important measurement questions.

At the onset, we want to stress that the subject of this text is real-life oriented. Specifying a measurement system and measurement procedures represents an open-ended design problem whose outcome will not have a unique solution. That means there may be several approaches to solving a measurement problem, and some will be better than others. This text emphasizes accepted procedures for analyzing a measurement problem to assist in the selection of equipment, methodology, and data analysis to meet the design objectives. Perhaps more than in any other technical field, the approach taken in measurement design and the outcome achieved will often depend on the attention and experience of the designer.

1.2 GENERAL MEASUREMENT SYSTEM

We will begin with a model of the generic measurement system. A *measurement*[1] assigns a specific value to a physical variable. That physical variable becomes the *measured variable*. A measurement system is a tool used for this quantification of the physical variable. As such, it is used to extend the abilities of the human senses, which, although they can detect and recognize different degrees of roughness, length, sound, color, and smell, are limited and are not very adept at assigning specific values to sensed variables. A general template for a measurement system is illustrated in Figure 1.1. Basically such a system consists of part or all of four general stages: (1) the sensor–transducer stage, (2) the signal-conditioning stage, (3) the output stage, and (4) the feedback-control stage. These stages form the bridge between the input to the measurement system and the system output, a quantity that is used to infer the value of the physical variable measured. The relationship between the input information, as acquired by the sensor, and the system output is established by a *calibration*.

The *sensor* is a physical element that uses some natural phenomenon to sense the variable being measured. The *transducer* converts this sensed information into a detectable signal form, which might be electrical, mechanical, optical, or otherwise. The goal is to convert the sensed information into a form that can be easily quantified.

For example, the liquid contained within the bulb on the common bulb thermometer of Figure 1.2 exchanges energy with its surroundings until the two are in thermal equilibrium. At that point they are at the same temperature. This energy exchange is the input signal to this measurement system. The phenomenon of thermal expansion of the liquid results in its movement up and down the stem, which in this case is the output signal from which we determine temperature. The liquid in the bulb acts as the sensor. By forcing the expanding liquid into a narrow capillary, this measurement system transforms thermal information into a mechanical displacement. Hence, the bulb's internal capillary design acts as a transducer.

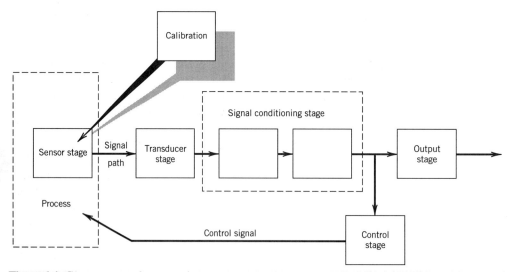

Figure 1.1 Components of a general measurement system.

---

[1]There are many terms introduced that are common in instrumentation. A glossary of terms is located in the back of the text for your reference.

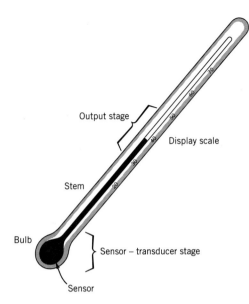

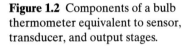

Figure 1.2 Components of a bulb thermometer equivalent to sensor, transducer, and output stages.

It is worth noting that the term "transducer" is also often used in reference to a packaged device, which may contain a sensor, transducer, and even some signal conditioning elements. The context in which the term is used usually prevents any ambiguity.

Sensor selection, placement, and installation are particularly important, because the input to the measurement system is the information sensed by the sensor. Accordingly, the interpretation of all information passed through and indicated by the system depends on that which is actually sensed by the sensor.

Signal conditioning equipment takes the transducer signal and modifies it to a desired form. This optional intermediate stage might be used to perform tasks such as increasing the magnitude of the signal through amplification, removing portions of the signal through some filtering technique, and/or providing mechanical or optical linkage between the transducer and the output stage, for example, converting a translational displacement of a sensor into a rotational displacement of a pointer. This stage can consist of one or more devices, which are often connected in series. For example, the diameter of the thermometer capillary relative to the bulb volume determines how far up the stem the liquid moves with increasing temperature. It "conditions" the signal by amplifying the liquid displacement.

The *output stage* provides an indication of the value of the measurement. The output equipment might be a simple readout display or a marked scale, or it might contain devices that can record the signal for later analysis. Examples of these devices are readout dials, recorders, and computer disk drives. The readout scale of the bulb thermometer in Figure 1.2 serves as the output stage of that measurement system.

In those measurement systems involved in process control, a fourth stage, the *feedback-control stage*, contains a controller that interprets the measured signal and makes a decision regarding the control of the process. This decision results in a change in a process parameter that affects the magnitude of the sensed variable. In simple controllers, this decision is based on the magnitude of the signal of the sensed variable, usually whether it exceeds some high or low set point; this value is set by the system operator. A simple measurement system with a control stage is a household furnace thermostat. The operator fixes the set point for temperature on the thermostat display, and the furnace is activated as the local temperature at the thermostat, as determined by the sensor within the device, rises or falls about the set point. In a more sophisticated controller, a signal from a measurement system can be used as an input to an "expert system" controller that, through an artificial intelligence scheme, determines the optimum set conditions for the process.

1.3 EXPERIMENTAL TEST PLAN

Suppose you wanted to answer the question, "What is the fuel use of my new car?" What might be your test plan? Two important variables to measure would be distance and fuel volume consumption. What other variables might influence your results? If your intent was to estimate the average fuel usage to expect over the course of ownership, then the driving route you choose would play a big role in the results and is a variable. Obviously, driving only on highways will impose a different trend on the results than driving in the city, so you might want to randomize your route by using various types of roads. If more than one driver uses the car, then the driver becomes a variable because each individual drives somewhat differently. Certainly weather and road conditions influence the results, and you might want to consider this in your plan. Thus we see that the utility of the measured data is very much affected by variables beyond the primary ones measured. In developing your test, the question you propose to answer will be a factor in developing your test plan, and you should be careful in defining that question so as to meet your objective. Imagine how your test would differ if you were interested instead in providing values used to advertise the expected average fuel use of a model of car. Also, you need to consider just how good an answer you need. Is 2.35 l/km or 1 mile/gal close enough? If not, then the test might require much tighter controls. Interestingly, this one example contains all the same elements of any sophisticated test. If you can conceptualize the factors influencing this test and how you will plan around them, then you are on track to handle almost any test. Before we move into the details of measurements, we focus here on some important concepts germane to all measurements and tests.

Experimental design involves developing a measurement test plan. A test plan will draw from the following three steps[2]:

1. *Parameter Design Plan.* This is the test objective and identification of process variables and parameters and a means for their control. Ask: "What question am I trying to answer? What has to be measured?" and "What variables will affect my results?"

2. *System and Tolerance Design Plan.* This is the selection of a measurement technique, equipment, and test procedure based on some preconceived tolerance limits for error.[3] Ask: "How will I do the measurement and how good do the results have to be?"

---

[2]These three strategies are similar to the bases for certain design methods [15] used in engineering systems design.

[3]The Tolerance Design Plan strategy used in this text draws on uncertainty (sensitivity) analyses. Sensitivity methods are common in design optimization.

3. *Data Reduction Design Plan.* Plan ahead on how to analyze, present, and use the anticipated data. Ask: "How will I interpret the resulting data? How will I use the data to answer my question?"

Going through all three steps in the test plan before any measurements are taken is a useful trick of a successful engineer. Often Step 3 will force you to reconsider Steps 1 and 2! In this section, concepts related to Step 1 are introduced.

Variables

"Identify the relevant process parameters and variables." This is the first step of experimental design measurement strategy. In addition to the targeted measured variable, there may be other variables pertinent to the measured process that affect the outcome. All known process variables should be listed and evaluated for any possible cause and effect relationships. If a change in one variable will not affect the value of another variable, the two are considered independent of each other. A variable that can be changed independently of other variables is known as an *independent variable.* A variable that is affected by changes in one or more other variables is known as a *dependent variable.*

The *control* of variables is important. A variable is controlled if it can be held at a constant value or at some prescribed condition during a measurement. Complete control of a variable would imply that it can be held to an exact prescribed value. Such complete control of a variable is not usually possible. We will use the adjective "controlled" to refer to a variable that can be held as prescribed, at least in a nominal sense. The cause and effect relationship between an independent variable and a dependent variable is found by applying a controlled value of the independent variable while measuring the dependent variable.

Variables can be described as being continuous or discrete in nature. In this text, we use the term "discrete variable" to refer to one whose possible values can be enumerated. For example, the outcome of the roll of a die represents a *discrete variable,* as the numbers on a die are discrete (1-2-3-4-5-6). Recall the driver or route driven in the fuel usage example earlier. Variables that describe an entity or specific item are discrete; examples include a test machine or instrument, a test specimen, or a test operator. Otherwise, the variable is considered to be *continuous.* For example, a variable whose value is determined from a pointer dial graduated between 1 and 6 would be continuous because the pointer could point to integers 1 through 6 or to any fractional value in between. Engineering variables such as displacement, pressure, strain, or temperature are continuous by their nature.

Variables that are not or cannot be controlled during measurement but that affect the value of the variable measured are called *extraneous variables.* Their influence can confuse the clear relation between cause and effect in a measurement. Would not the driving style affect the fuel consumption of a car? Then unless controlled, this influence will affect the result. Extraneous variables can introduce differences in repeated measurements of the same measured variable taken under seemingly identical operating conditions, or they can impose a false trend onto the behavior of that variable. The effects of extraneous variables take the form of signals superimposed over the measured signal, with such forms as noise and drift.

Consider a thermodynamics experiment to establish the boiling point of water. The apparatus for measuring the boiling point might yield the results shown in Figure 1.3 for three test runs conducted on separate days. Notice the different outcome for each test.

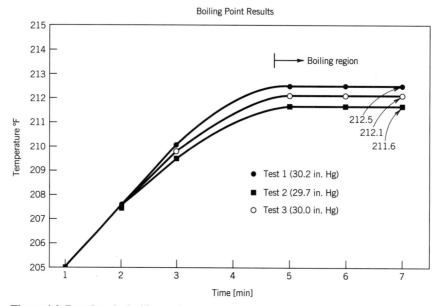

Figure 1.3 Results of a boiling point test for water.

Why should the data from three seemingly identical tests show such different results?

Suppose an evaluation of the measurement system accuracy accounts for only 0.1°F of the test data scatter. Thus another plausible contributing factor is the effect of an extraneous variable. Indeed, a close examination of the test data shows a measured variation in the barometric pressure. The pressure variation is consistent with the trend seen in the boiling point data. Because the local barometric pressure was not controlled (i.e., it was not held fixed among the tests), the pressure acted as an extraneous variable adding to the differences in outcomes between the test runs. Control important variables or be prepared to solve a puzzle!

Parameters

A *parameter* is defined in this text as a functional relationship between variables. A parameter that has an effect on the behavior of the measured variable is called a control parameter. Available methods for establishing control parameters based on known process variables include similarity and dimensional analysis techniques (e.g., [1–3]), and physical laws. A control parameter is completely controlled if it can be set and held at a constant value during a set of measurements.

As an example, the flow rate, Q, developed by a fan depends on rotational speed, n, and the diameter, d, of the fan. A control parameter for this group of three variables, found by similarity methods, is the fan flow coefficient, $C_1 = Q/nd^3$. For a given fan d is fixed (and therefore controlled), and if speed is somehow controlled, the fan flow rate associated with that speed can be measured and the flow coefficient can be determined. Parameters can also be affected by extraneous variables.

Noise and Interference

Just how extraneous variables affect measured data can be delineated into noise and interference. *Noise* is a random variation of the value of the measured signal as a consequence of the variation of the extraneous variables. Examples include variations in measured values arising from incomplete control of a variable, normal random variations in environmental conditions that affect the measured variable and/or the measurement system, or thermal noise (called Johnson noise) caused by the random temperature-induced motion of electrons within wiring. A completely controlled variable contains no noise.

Interference produces undesirable deterministic trends on the measured value because of extraneous variables. One form common to electrical instruments is that of a sinusoidal wave superimposed onto a measured signal path. Examples include well-defined deterministic variations in environmental conditions, local ac power line noise (60 or 50 Hz) or fluorescent lighting arc noise (120 or 100 Hz), and electromagnetic interference (EMI) and radio-frequency (rf) interference. Hum and acoustic feedback in public address and audio systems are ready examples of interference effects that are superimposed onto a desirable signal.

An undesirable situation arises if the period of the interference is longer than the period over which the measurement is made, because the interference will superimpose a false trend in the behavior of the measured variable.

Consider the effects of noise and interference on the signal, $y(t) = 2 + \sin 2\pi t$. As shown in Figure 1.4, noise adds to the scatter of the signal. Interference imposes a trend onto the signal. An important goal of a measurement plan should be to perform the measurement in a manner that will break up interference trends so that they appear as random variations in the data set. Although this will increase the scatter in the measured values of a data set, scatter can be handled by statistics. It is far more important to eliminate false trends in the data set, which will otherwise lead to a misinterpretation of the measured data. Randomization methods are available that can be easily incorporated into the measurement plan and will minimize or eliminate interference trends. Several are discussed in the paragraphs that follow.

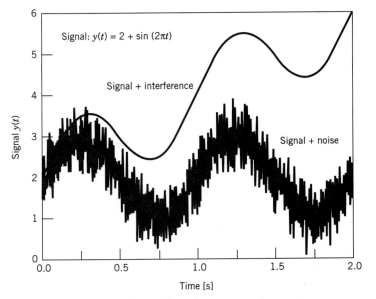

Figure 1.4 Effects of noise and interference superimposed on the signal $y(t) = 2 + \sin 2\pi t$.

Random Tests

Recall our car fuel use example. Let y be the fuel use, which depends on x_a, fuel volume consumption, and x_b, distance traveled. We determine y by varying these two variables (that is, we drive the car). However, the test result can be affected by extraneous variables such as route, driver, and weather and road conditions. For example, driving only on superhighways will impose a false (untypical) trend on our intended average fuel estimate, so we could drive on different types of roads to break up this trend. This approach introduces a random test strategy.

In general, consider the situation in which the dependent variable, y, is a function of several independent variables, x_a, x_b, ... However, the measurement of y can also be influenced by several extraneous variables, z_j, where $j = 1, 2, \ldots$, such that $y = f(x_a, x_b, \ldots; z_j)$. For the dependence of y on the independent variables to be found, they are varied. Although the influence of the z_j variables on these tests cannot be eliminated, the possibility of their introducing a false trend on y can be minimized by the use of a proper test strategy. A random test is one such strategy.

A *random test* is defined by a measurement matrix that sets a random order in the value of the independent variable applied. Trends normally introduced by the coupling of a relatively slow and uncontrolled variation in the extraneous variables with a sequential application in values of the independent variable applied will be broken up. This type of plan is effective for the local control of extraneous variables that change in a continuous manner. Consider Examples 1.1 and 1.2.

Discrete extraneous variables can also be treated a little differently. The use of different instruments, different test operators, and different test conditions are examples of discrete extraneous variables that can affect the outcome of a measurement. Randomizing a test matrix to minimize discrete influences can be done efficiently through the use of an experimental design using random blocks. A block consists of a data set of the measured variable in which the controlled variable is varied but the extraneous variable is fixed. The extraneous variable is varied between blocks. This enables some amount of local control over the discrete extraneous variable. In the fuel usage example, we might consider several blocks, each composed of a different driver (extraneous variable) driving similar routes, and averaging the results. Many strategies for randomized blocks exist, as do advanced statistical methods for data analysis (e.g., [4–6]). Consider Examples 1.3 and 1.4.

EXAMPLE 1.1

In the pressure calibration system shown in Figure 1.5, a sensor–transducer is exposed to a known pressure, p. The transducer, powered by an external supply, converts the sensed signal into a voltage that is measured by a voltmeter. The measurement approach is to control the applied pressure by the measured displacement of a piston that is used to compress a gas contained within the piston-cylinder chamber. The gas chosen closely obeys the ideal gas law. Hence, piston displacement, x, which sets the chamber volume, $V = (x \times \text{area})$, is easily related to chamber pressure. Identify the independent and dependent variables in the calibration and possible extraneous variables.

KNOWN

Pressure calibration system of Figure 1.5

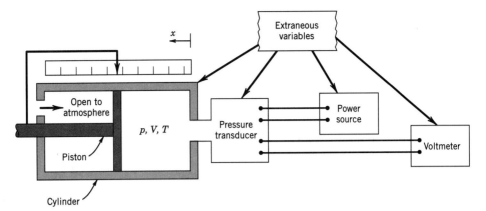

Figure 1.5 Pressure calibration rig.

FIND

Independent, dependent, and extraneous variables

SOLUTION

The control parameter for this problem can be formed from the ideal gas law: $pV/T =$ constant, where T is the gas temperature. An independent variable in the calibration is the piston displacement that sets the volume. This variable can be controlled by locking the piston into position. From the ideal gas law, gas pressure will also depend on temperature, and therefore temperature is also an independent variable. However, T and V are not in themselves independent according to the control parameter. Since volume is to be varied through variation of piston displacement, T and V can be controlled provided a mechanism is incorporated into the scheme to maintain a constant gas temperature within the chamber. This will also maintain chamber area constant, a relevant factor in controlling the volume. In that way, the applied variations in V will be the only effect on pressure, as desired. The dependent variable is the chamber gas pressure. The pressure sensor is exposed to the chamber gas pressure and, hence, this is the pressure sensed by it. Examples of likely extraneous variables would include noise effects due to the room temperature, z_1, and line voltage variations, z_2, which would affect the excitation voltage from the power supply and the performance of the voltmeter. Connecting wires between devices will act as an antenna and possibly will introduce interference, z_3, superimposed onto the electrical signal. Their magnitudes can be reduced by proper electrical shielding.

Hence, $p = f(V, T; z_1, z_2, z_3)$, where $V = f_1(x, T)$.

COMMENT

Even though we might try to keep the gas temperature constant, any slight variations in the gas temperature would affect the volume and pressure and, hence, will act as an additional extraneous variable!

EXAMPLE 1.2

Develop a test plan that will minimize the interference effects of the extraneous variables in Example 1.1.

KNOWN

$p = f(V, T; z_1, z_2, z_3)$, where $V = f_1(x; T)$. Control variable V is changed. Dependent variable p is measured.

FIND

Randomize the possible effects of extraneous variables

SOLUTION

Part of our test strategy is to vary volume, control gas temperature, and measure pressure. An important feature of all test plans is a strategy that minimizes the superposition of false trends onto the data set by the extraneous variables. Since z_1, z_2, and z_3 and any inability to hold the gas temperature constant are continuous extraneous variables, their influence on p can be randomized by a random test. This entails shuffling the order by which V is applied. Say that we pick six values of volume: V_1, V_2, V_3, V_4, V_5, and V_6, where the subscripts correspond to an increasing sequential order of the respective values of volume. Any random order will do fine. One possibility found by using the random function features of a handheld calculator is

$$V_2 \quad V_5 \quad V_1 \quad V_4 \quad V_6 \quad V_3$$

If we perform our measurements in a random order, interference trends will be broken up.

EXAMPLE 1.3

The manufacture of a particular composite material requires mixing a percentage by weight of binder with resin to produce a gel. The gel is used to impregnate a fiber to produce the composite material in a manual process called the lay-up. The strength, σ, of the finished material depends on the percent binder in the gel. However, the strength may also be lay-up operator dependent. Formulate a test matrix by which the strength to percent binder–gel ratio relationship under production conditions can be established.

KNOWN

$\sigma = f$ (binder; operator)

ASSUMPTION

Strength is affected only by binder and operator

FIND

Test matrix to randomize effects of operator

SOLUTION

The dependent variable, σ, is to be tested against the independent variable, percent binder–gel ratio. The operator is an extraneous variable in actual production. As a simple test, we could test the relationship between three binder–gel ratios, A, B, and

C, and measure strength. We could also choose three typical operators (z_1, z_2, and z_3) to produce N separate composite test samples for each of the three binder–gel ratios. This gives the three-block test pattern:

Block				
1	z_1:	A	B	C
2	z_2:	A	B	C
3	z_3:	A	B	C

In the analysis of the test, all of these data can be combined. The results of each block will include each operator's influence as a variation. We can assume that the order used within each block is unimportant. But if only the data from one operator are considered, the results may show a trend consistent with the lay-up technique of that operator. The test matrix here will randomize the influence of any one operator on the strength test results by introducing the influence of several operators.

EXAMPLE 1.4

Suppose following lay-up, the composite material of Example 1.3 is allowed to cure at a controlled but elevated temperature. We wish to develop a relationship between the binder–gel ratio and the cure temperature and strength. Develop a suitable test matrix.

KNOWN

$\sigma = f(\text{binder, temperature; operator})$

ASSUMPTION

Strength is affected only by binder, temperature, and operator

FIND

Test matrix to randomize effect of operator

SOLUTION

We develop a simple matrix to test for the dependence of composite strength on the independent variables of binder–gel ratio and cure temperature. We could proceed as in Example 1.3 and set up three randomized blocks for ratio and three for temperature for a total of 18 separate tests. Suppose instead we choose three temperatures, T_1, T_2, and T_3, along with three binder–gel ratios, A, B, and C, and three operators, z_1, z_2, and z_3, and set up a 3×3 test matrix representing a single randomized block. If we organize the block such that no operator runs the same test combination more than once, we randomize the influence of any one operator on a particular binder–gel ratio, temperature test.

	z_1	z_2	z_3
A	T_1	T_2	T_3
B	T_2	T_3	T_1
C	T_3	T_1	T_2

COMMENT

The suggested test matrix not only randomizes the extraneous variable but has reduced the number of tests by one-half over the direct use of three blocks for ratio and for temperature. However, either approach is fine. The above matrix is referred to as a Latin square [4–6].

If we wanted to include our ability to control the independent variables in the test data variations, we could duplicate the Latin-square test several times to build up a significant data base. Such a duplication is referred to as a replication.

Replication and Repetition

In general, the estimated value of a measured variable improves with the number of measurements. For example, a bearing manufacturer would obtain a better estimate of the mean diameter and the variation in the diameters of a batch of bearings by measuring many bearings rather than just a few. Repeated measurements made during any single test run or on a single batch are called *repetitions*. Repetition allows for quantifying the variation in a measured variable as it occurs during any one test or batch while the operating conditions are held under nominal control. However, repetition will not permit an assessment of how exact the operating conditions can be set.

If the bearing manufacturer was interested in how closely the bearing mean diameter was controlled in day-in and day-out operations with a particular machine or test operator, duplicate tests run on different days would be needed. An independent duplication of a set of measurements using similar operating conditions is referred to as a *replication*. It allows for quantifying the variation in a measured variable as it occurs between different tests, each having the same nominal values of operating conditions.

Finally, if the bearing manufacturer was interested in how closely the bearing mean diameter was controlled when different machines or different machine operators were used, duplicate tests using these different configurations might hold the answer. Here, replication provides a means to randomize the interference effects of the different bearing machines or operators mentioned. These act as extraneous variables on the diameter of the bearings produced.

Replication allows an assessment of the control on setting the operating conditions, that is, the ability to reset the conditions to some desired value. Ultimately, replication estimates our control over the procedure used.

EXAMPLE 1.5

Consider a room furnace thermostat. Set to some temperature, we can make repeated measurements (repetition) of room temperature and come to a conclusion about the average value and the variation in room temperature at that particular thermostat setting. Repetition permits us to estimate the variation in this measured variable. This repetition permits an assessment of how well we can maintain (control) room temperature at some thermostat setting.

Now suppose we change the set temperature to some arbitrary value but sometime later return it to the original setting. We now duplicate the repeated measurements (another repetition). The two sets of test data are replications of each other.

We might find that the average temperature in the second test differs from the first. The different averages suggest something about our ability to set and control the temperature in the room. Replication permits the assessment of how well we can duplicate a set of conditions.

Concomitant Methods

A good strategy is to incorporate the use of *concomitant methods* in a measurement plan. The goal is to obtain two or more estimates for the result, each based on a different method, which can be compared as a check for agreement. This may affect the experimental design in that additional variables may have to be measured, or the different method could be an analysis that estimates an expected value of the measurement. As an example, suppose we needed to establish the volume of a cylindrical rod of known material. We could simply measure the diameter and length of the rod to compute this. Alternatively, we could measure the weight of the rod and compute volume based on the specific weight of the material. The second method complements the first and provides an important check on the adequacy of the first estimate.

1.4   CALIBRATION

The relationship between the value of the input to the measurement system and the system's indicated output value is established during a calibration of the measurement system. A *calibration* is the act of applying a known value of input to a measurement system for the purpose of observing the system output. The known value used for the calibration is called the *standard*.

Static Calibration

The most common type of calibration is known as a *static calibration*. In this procedure, a known value is input to the system under calibration and the system output is recorded. The term "static" refers to a calibration procedure in which the values of the variables involved remain constant; that is, they do not change with time. In static calibrations, only the magnitudes of the known input and the measured output are important.

By application of a range of known values for the input and observation of the system output, a direct calibration curve can be developed for the measurement system. On such a curve the input, x, is plotted on the abscissa against the measured output, y, on the ordinate, such as indicated in Figure 1.6. In a calibration the input value should be a controlled independent variable, whereas the measured output value becomes the dependent variable of the calibration.

The static calibration curve describes the static input–output relationship for a measurement system and forms the logic by which the indicated output can be interpreted during an actual measurement. For example, the calibration curve is the basis for fixing the output display scale on a measurement system, such as that of Figure 1.2. Alternatively, a calibration curve can be used as part of developing a functional relationship, an equation known as a correlation, between input and output. A correlation will have the form $y = f(x)$ and is determined by applying physical reasoning and curve-fitting techniques to the calibration curve. The correlation can

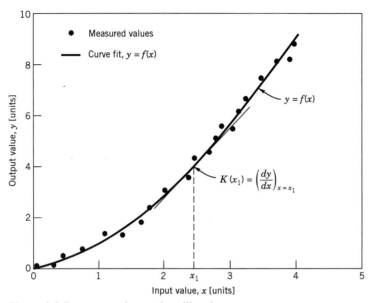

Figure 1.6 Representative static calibration curve.

then be used in later measurements to ascertain the unknown input value based on the output value, the value indicated by the measurement system.

Dynamic Calibration

When the variables of interest are time dependent and time-based information is sought, we need dynamic information. In a broad sense, dynamic variables are time dependent in both their amplitude and frequency content. A *dynamic calibration* determines the relationship between an input of known dynamic behavior and the measurement system output. Usually, such calibrations involve either a sinusoidal signal or a step change as the known input signal. The dynamic nature of signals and measurement systems is explored fully in Chapter 3.

Static Sensitivity

The slope of a static calibration curve yields the *static sensitivity*[4] of the measurement system. As depicted graphically in the calibration curve of Figure 1.6, the static sensitivity, K, at any particular static input value, say x_1, is evaluated by

$$K = K(x_1) = \left(\frac{dy}{dx} \right)_{x=x_1} \tag{1.1}$$

where K is a function of x. The static sensitivity is a measure relating the change in the indicated output associated with a given change in a static input. Since calibration curves can be linear or nonlinear depending on the measurement system and on the variable being measured, K may or may not be constant over a range of input values.

---

[4]Some texts refer to this as the static gain.

Range

The proper procedure for calibration is to apply known inputs ranging from the minimum to the maximum values for which the measurement system is to be used. These limits define the operating *range* of the system. The input operating range is defined as extending from x_{min} to x_{max}. This range defines its *input span*, expressed as

$$r_i = x_{max} - x_{min} \tag{1.2}$$

Similarly, the output operating range is specified from y_{min} to y_{max}. The *output span*, or *full-scale operating range (FSO)*, is expressed as

$$r_o = y_{max} - y_{min} \tag{1.3}$$

It is important to avoid extrapolation beyond the range of known calibration during measurement since the behavior of the measurement system is uncharted in these regions. As such, the range of calibration should be carefully selected.

Accuracy

The accuracy of a system can be estimated during calibration. If the input value of calibration is known exactly, then it can be called the true value. The *accuracy* of a measurement system refers to its ability to indicate a true value exactly. Accuracy is related to absolute error. *Absolute error*, ϵ, is defined as the difference between the true value applied to a measurement system and the indicated value of the system:

$$\epsilon = \text{true value} - \text{indicated value} \tag{1.4}$$

from which the percent accuracy is found by

$$A = \left(1 - \frac{|\epsilon|}{\text{true value}} \right) \times 100 \tag{1.5}$$

By definition, accuracy can be determined only when the true value is known, such as during a calibration.

An alternative form of calibration curve is the deviation plot, such as shown in Figure 1.7. Such a curve plots the difference or deviation between a true or expected value, y', and the indicated value, y, versus the indicated value. Deviation curves are extremely useful when the differences between the true and the indicated value are too small to suggest possible trends on direct calibration plots. They are often required in situations that require errors to be reduced to the minimums possible. In Figure 1.7, the voltage output during calibration from a temperature-sensing thermocouple is compared with the nominal expected values obtained from reference tables.

Precision and Bias Errors

The repeatability or precision of a measurement system refers to the ability of the system to indicate a particular value upon repeated but independent applications of a specific value of input. The *precision error* is a measure of the random variation found during repeated measurements. An estimate of a measurement system precision does not require a calibration, per se. However, note that a system that repeatedly indicates the same wrong value upon repeated application of a particular input would be considered to be very precise regardless of its accuracy.

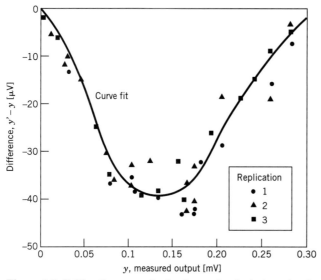

Figure 1.7 Calibration curve in the form of a deviation plot for a temperature sensor.

If the indicated measured value differs from the true value, the measured value is said to contain a bias error. The average error in a series of repeated calibration measurements defines the bias error. The *bias error* is the difference between the average value and the true value. Both precision and bias errors affect the measure of a system's accuracy.

The concepts of accuracy and of bias and precision errors in measurements can be illustrated by the throw of darts. Consider the dart board of Figure 1.8, where the goal is to throw the darts into the bull's-eye. For this analogy, the bull's-eye represents the true value and each throw represents a measurement value. In Figure 1.8(*a*), the thrower displays good precision (i.e., low precision error) in that each throw repeatedly hits the same spot on the board, but the thrower is not accurate in that the dart misses the bull's-eye each time. This thrower is precise, but we see that low precision error alone is not a measure of accuracy. The error in each throw can be computed from the distance between the bull's-eye and each dart. The average value of the error yields the bias error. This thrower has a bias to the left of the target. If the bias could be reduced, then this thrower's accuracy would improve. In Figure 1.8(*b*), the thrower displays high accuracy and high repeatability, hitting

(*a*) High repeatability gives low precision error but no direct indication of accuracy

(*b*) High accuracy means low precision and bias errors

(*b*) Bias and precision errors lead to poor accuracy

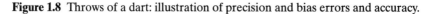

Figure 1.8 Throws of a dart: illustration of precision and bias errors and accuracy.

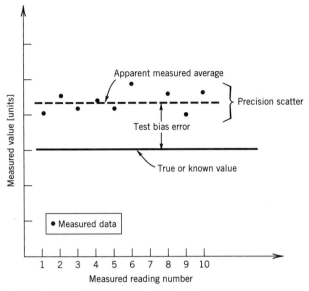

Figure 1.9 Effects of precision and bias errors on calibration readings.

the bull's-eye on each throw. Both throw scatter and bias error are near zero. High accuracy must imply both low precision and bias errors. In Figure 1.8(*c*), the thrower displays neither high precision nor accuracy, with the errant throws scattered around the board. Each throw contains a different amount of error. Whereas the bias error is the average of the errors in each throw, the precision error is related to the varying amount of error in the throws. The accuracy of this thrower's technique appears to be biased and lacking precision. The precision and bias errors of the thrower can be computed by using the statistical methods that are discussed in Chapters 4 and 5. Both precision and bias errors quantify the error in any set of measurements and are used to estimate accuracy.

Suppose a measurement system was used to measure a variable whose value was kept constant and known exactly, as in a calibration. For example, 10 independent measurements are made with the results, as shown in Figure 1.9. The variations in the measurements, the observed scatter in the data, would be related to the system precision error associated with the measurement of the variable. That is, the scatter is mainly due to (1) the measurement system and (2) the method of its use, since the value of the variable is essentially constant. However, the offset between the apparent average of the readings and the true value would provide a measure of the bias error to be expected from this measurement system.

In any measurement other than a calibration, the error cannot be known exactly since the true value is not known. However, from the results of a calibration, the operator might feel confident that the error is within certain bounds, a plus or minus range of the indicated reading. Since the magnitude of the error in any measurement can only be estimated, one refers to an estimate of the error in the measurement as the *uncertainty* present in the measured value. Uncertainty results from errors that are present in the measurement system, its calibration, and measurement technique, and is manifested by measurement system bias and precision errors. A method of estimating the uncertainty in a measurement is treated in detail in Chapter 5.

Table 1.1 Manufacturer's Specifications: Typical Pressure
Transducer

Operation	
Input range	0–1000 cm H_2O
Excitation	±15 V dc
Output range	0–5 V
Performance	
Linearity error	±0.5% FSO
Hysteresis error	Less than ±0.15% FSO
Sensitivity error	±0.25% of reading
Thermal sensitivity error	±0.02% /°C of reading
Thermal zero drift	0.02% /°C FSO
Temperature range	0–50 °C

The precision and bias errors of a measurement system are the result of several interacting errors inherent to the measurement system, the calibration procedure, and the standard used to provide the known value. These errors can be delineated and quantified as elemental errors through the use of particular calibration procedures and data reduction techniques. An example is given for a typical pressure transducer in Table 1.1.

Sequential Test

A *sequential test* applies a sequential variation in the input value over the desired input range. This may be accomplished by increasing the input value (upscale direction) or by decreasing the input value (downscale direction) over the full input range.

Hysteresis

The sequential test is an effective diagnostic technique for identifying and quantifying a hysteresis error in a measurement system. *Hysteresis error* refers to differences in the values found between going upscale and downscale in a sequential test. The effect of hysteresis in a calibration curve is illustrated in Figure 1.10(a). For a particular input value, the hysteresis error is found from the difference in the upscale and downscale output value by $e_h = (y)_{\text{upscale}} - (y)_{\text{downscale}}$. Hysteresis is usually specified for a measurement system in terms of the maximum hysteresis error found in the calibration, $e_{h_{\max}}$, as a percentage of full-scale output range:

$$\% e_{h_{\max}} = \frac{e_{h_{\max}}}{r_o} \times 100 \tag{1.6}$$

such as that indicated in Table 1.1. Hysteresis occurs when the output of a measurement system is dependent on the previous value indicated by the system. Such dependencies can be brought about through some realistic system limitations such as friction or viscous damping in moving parts or residual charge in electrical components. Some hysteresis is normal for any system and affects the precision of the system.

Random Test

A *random test* applies a randomly selected sequence of values of a known input over the intended calibration range. The random application of input tends to minimize

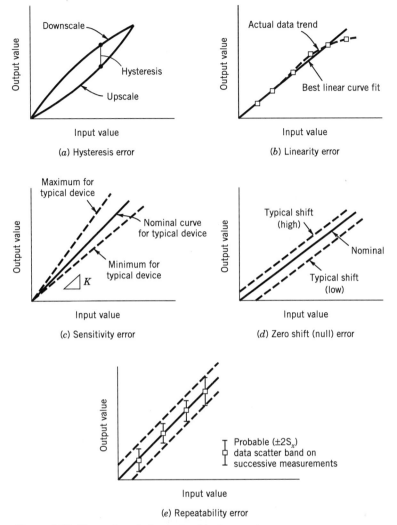

Figure 1.10 Examples of elements of instrument error.

the impact of interference. It breaks up hysteresis effects and observation errors. It ensures that each application of input value is independent of the previous. This reduces the calibration bias error. Generally, such a random variation in input value will more closely simulate the actual measurement situation.

A random test provides an important diagnostic test for the delineation of several measurement system performance characteristics based on a set of random calibration test data. In particular, the linearity error, sensitivity error, zero error, and instrument repeatability error, as illustrated in Figures 1.10(b)–1.10(e), can be quantified from a static random test calibration.

Linearity Error

Many instruments are designed to achieve a linear relation between an applied static input and indicated output value. Such a linear static calibration curve would have the general form

$$y_L(x) = a_0 + a_1 x \qquad (1.7)$$

where the curve fit $y_L(x)$ provides a predicted output value based on a linear relation between x and y. However, in real systems, truly linear behavior is only approximately achieved. As a result, measurement device specifications usually provide a statement as to the expected linearity of the static calibration curve for the device. The relation between $y_L(x)$ and measured value $y(x)$ is a measure of the nonlinear behavior of a system:

$$e_L(x) = y(x) - y_L(x) \qquad (1.8)$$

where $e_L(x)$ is the *linearity error* that arises in describing the actual system behavior by equation (1.7). Such behavior is illustrated in Figure 1.10(*b*), in which a linear curve has been fit through a calibration data set. For a measurement system that is essentially linear in behavior, the extent of possible nonlinearity in a measurement device is often specified in terms of the maximum expected linearity error for the calibration as a percentage of full-scale output range:

$$\% e_{L_{max}} = \frac{e_{L_{max}}}{r_o} \times 100 \qquad (1.9)$$

This is how the linearity error for the pressure transducer in Table 1.1 was estimated. Statistical methods of quantifying such data scatter about a line are discussed in Chapter 4.

Sensitivity and Zero Errors

The scatter in the data measured during a calibration affects the precision in the estimate of the slope of the calibration curve. As shown for the linear calibration curve in Figure 1.10(*c*), if we fix the zero intercept at zero (a zero output from the system for zero input), then the scatter in the data leads to a precision error in estimating the slope of the calibration curve. The *sensitivity error*, e_K, is a statistical measure of the precision error in the estimate of the slope of the calibration curve (as discussed in Chapter 4). The static sensitivity of a device is sometimes temperature dependent and this is often specified. In Table 1.1, the sensitivity error reflects calibration results at a constant reference ambient temperature, whereas the thermal sensitivity error was found by calibration at different temperatures.

If the zero intercept is not fixed but the sensitivity is constant, then drifting of the zero intercept introduces a vertical shift of the calibration curve, as shown in Figure 1.10(*d*). This shift of the zero intercept of the calibration curve is known as the *zero error*, e_z, of the measurement system. The zero error can usually be reduced by periodically adjusting the output from the measurement system under a zero input condition. However, some random variation in the zero intercept is common, particularly with electronic and digital equipment subjected to temperature variations (e.g., thermal zero drift in Table 1.1).

Instrument Repeatability

The ability of a measurement system to indicate the same value upon repeated but independent application of the same input is known as the instrument *repeatability*. Specific claims of repeatability are based on multiple calibration tests (replication) performed within a given lab on the particular unit. Repeatability, as shown in

Figure 1.10(*e*), is based on a statistical measure (developed in Chapter 4) called the standard deviation, S_x, a measure of the variation in the output for a given input. The value claimed is usually in terms of the maximum expected error as a percentage of full-scale output range:

$$\%(e_R)_{\text{max}} = \frac{2(S_x)}{r_o} \times 100 \tag{1.10}$$

The instrument repeatability reflects only the error found under controlled calibration conditions. It does not include the additional errors introduced during measurement that are due to variations in the measured variable or that are due to procedure.

Reproducibility

The term "reproducibility," when reported in instrument specifications, refers to the results of separate repeatability tests. Manufacturer claims of instrument reproducibility must be based on multiple repeatability tests (replication) performed in different labs on a single unit.

Instrument Precision

The term "instrument precision," when reported in instrument specifications, refers to the results of separate repeatability tests. Manufacturer claims of instrument precision must be based on multiple repeatability tests (replication) performed on different units of the same manufacture, preferably performed in different labs.

Overall Instrument Error

An estimate of the *overall instrument error* is made based on all known errors. This error is often misleadingly referred to as the instrument accuracy in some instrument specifications. An estimate is computed from the square root of the sum of the squares of all known errors. For M known errors, the instrument error, e_c, is estimated by

$$e_c = \left[e_1^2 + e_2^2 + \cdots + e_M^2\right]^{1/2} \tag{1.11}$$

For example, for an instrument having known hysteresis, linearity, sensitivity, and repeatability errors, the instrument error is estimated by

$$e_c = \left[e_h^2 + e_L^2 + e_K^2 + e_R^2\right]^{1/2} \tag{1.12}$$

1.5   STANDARDS

When a measurement system is calibrated it is compared with some standard whose value is presumably known. This standard may be a piece of equipment, an object having a well-defined physical attribute to be used as a comparison, or a well-accepted technique known to produce a reliable value. Let us explore how certain standards come to be and how these standards are the foundation of all measurements.

A *dimension* defines a physical variable that is used to describe some aspect of a physical system. A *unit* defines a quantitative measure of a dimension. For example, mass, length, and time describe basic dimensions with which we associate the units of kilogram, meter, and second. A *primary standard* defines the value of a unit. It provides the means to describe the unit with a unique number that can be understood throughout the world. The primary standard then assigns a unique value to a unit by definition! As such it must define the unit exactly. In 1960, the 11th General Conference on Weights and Measures, the international agency responsible for maintaining precise uniform standards of measurements, formally adopted the International System of Units (SI) as the international standard of units. The system has been adopted worldwide, although other unit systems are commonly used in the consumer market. These other unit systems are now treated as conversions from SI. Examples of these include the U.S. system of units found in the United States and the meter-kg-s (mks) system common to most of the world.

Primary standards are necessary because the value assigned to a unit is actually arbitrary. Whether a meter is the length of a king's arm or the distance light travels in a fraction of a second depends only on how we choose to define it. To avoid confusion, units are defined by international agreement through the use of primary standards. Once agreed upon, the primary standard forms the exact definition of the unit until it is changed by some later agreement. Important features sought in any standard should include the following:

- Global availability
- Continued reliability
- Stability

with minimal sensitivity to external environmental sources. Next we examine some basic dimensions and the primary standards that form the definition of the units that describe them [10].

Basic Dimensions and Their Units

Mass

The dimension of mass is defined in SI by the kilogram. Originally, the unit of the kilogram was defined by the mass of one liter of water at room temperature, but today an equivalent yet more consistent definition defines the kilogram exactly as the mass of a particular platinum-iridium bar that is maintained under very specific environmental conditions at the International Bureau of Weights and Measures located in Sevres, France. This particular bar forms the primary standard for the kilogram.

In the United States, the U.S. engineering standard unit system remains widely used. In the U.S. engineering unit system, mass is defined by the pound-mass, lb_m, which can be derived from the definition of the kilogram:

$$1\,lb_m = 0.45359237\,kg \tag{1.13}$$

Equivalent standards for the kilogram and the pound-mass are maintained by the U.S. National Institute of Standards and Technology (NIST, formerly the National Bureau of Standards, or NBS) in Gaithersburg, Maryland.

Time and Frequency

The dimension of time is defined by the unit of a second in both SI and U.S. engineering units. One second is defined [7] as the time elapsed during 9,192,631,770 periods of the radiation emitted between two excitation levels of the fundamental state of cesium-133. Despite this seemingly unusual definition, this primary standard can be reliably reproduced at suitably equipped laboratories throughout the world to within 2 parts in ten trillion.

The Bureau Internationale de l'Heure (BIH) in Paris maintains the primary standard for clock time. Periodically, adjustments to clocks around the world are made relative to the BIH clock so as to keep time synchronous.

The standard for cyclical frequency is based on the time standard. The standard unit is the hertz (1 Hz = 1 cycle/s). The cyclical frequency is related to the circular frequency (radians/second) by

$$1\,\text{Hz} = \frac{2\pi\,\text{rad}}{1\,\text{s}} \tag{1.14}$$

Both time and frequency standard signals are broadcast worldwide over designated radio stations for use in navigation and as a source of a standard for these dimensions.

Length

The meter is the SI unit for length. New primary standards are defined when the ability to determine the new standard becomes more accurate than the existing standard. In 1982, a new primary standard was adopted to define the unit of a meter. The primary standard for the dimension of length is now based on the dimension of time. One meter is exactly defined as the length traveled by light in 3.335641×10^{-9} s, a number derived from the known velocity of light in vacuum.

The U.S. engineering standard unit of the foot and related unit of the inch can be derived from the meter.

$$1\,\text{ft} = 0.30480060\,\text{m}$$
$$1\,\text{in.} = 0.02540005\,\text{m} \tag{1.15}$$

In the United States, primary standards for length are maintained at NIST.

Temperature

The kelvin, K, is the SI unit of thermodynamic temperature and is the fraction 1/273.16 of the thermodynamic temperature of the triple point of water. A temperature scale was devised by William Thomson, Lord Kelvin (1824–1907), and forms the basis for the absolute practical temperature scale in common use. This scale is based on polynomial interpolation between the equilibrium phase change points of a number of common pure substances from the triple point of equilibrium hydrogen (13.81 K) to the freezing point of pure gold (1337.58 K). Above 1337.58 K, the scale is based on Planck's law of radiant emissions. The details of the standard have been modified over the years but are governed by the International Temperature Scale—1990 [8], which superseded the International Practical Temperature Scale—1968 [9].

The U.S. engineering unit system uses the absolute scale of Rankine (°R). This and the common scales of Celsius (°C) and Fahrenheit (°F) are related to the Kelvin scale by the following:

$$(°\text{C}) = (\text{K}) - 273.15$$
$$(°\text{F}) = (°\text{R}) - 459.67$$
$$(°\text{F}) = 1.8 \times (°\text{C}) + 32.0 \tag{1.16}$$

Current

The SI electrical unit for current is the ampere. An ampere is defined as the constant current that, if maintained in two straight parallel conductors of infinite length and of negligible circular cross section and placed 1 m apart in vacuum, would produce a force equal to 2×10^{-7} N per meter of length between these conductors.

Derived Units

Although the above discussion does not cover all of the basic dimensions, it does cover those most commonly referred to in this text. Most other dimensions and their associated units can usually be defined in terms of the basic dimensions and units already described [10].

Force

According to Newton's law, force is proportional to mass times acceleration:

$$\text{force} = \frac{\text{mass} \times \text{acceleration}}{g_c}$$

where g_c is the proportionality constant.

In the SI system of units, force is defined by a derived unit called the newton (N)

$$1\,\text{N} = 1\frac{\text{kg m}}{\text{s}^2} \tag{1.17}$$

So for this system the value of g_c must be 1.0 (kg m)/(s^2 N). Note that the resulting expression for Newton's second law does not explicitly require the inclusion of g_c.

However, in U.S. engineering units, the units of force and mass are related through the following definition: 1 lb$_m$ exerts a force of 1 lb in a standard gravitational field. With this definition,

$$1\,\text{lb} = \frac{(1\,\text{lb}_m)(32.174\,\text{ft/s}^2)}{g_c} \tag{1.18}$$

and g_c must take on the value of 32.174 (lb$_m$ ft)/(lb s^2). In the U.S. system, the pound is a defined quantity and g_c must be derived through Newton's law. Similarly, in the mks system, which uses the kilogram-force, the value for g_c takes on a value of 9.806 (kg m)/(s^2 kg$_f$).

Many engineers have some difficulty with using g_c in the non-SI systems. Actually, whenever force and mass appear in the same expression, just remember to relate them by using g_c through Newton's law:

$$g_c = \frac{mg}{F} = 1\frac{\text{kg m/s}^2}{\text{N}} = 32.2\frac{\text{lb}_m\,\text{ft/s}^2}{\text{lb}} = 9.806\frac{\text{kg m/s}^2}{\text{kg}_f} \tag{1.19}$$

Other Units

Energy is defined as force times distance with the standard SI unit of the joule (J):

$$1\,\text{J} = 1\frac{\text{kg m}^2}{\text{s}^2}$$

Power is defined as energy per unit time in terms of the watt (W):

$$1\,W = 1\frac{J}{s}$$

Stress and pressure are defined as force per unit area in terms of the pascal (Pa):

$$1\,Pa = 1\,N/m^2$$

Electrical Dimensions

The units for fundamental dimensions of electrical potential and resistance are based on the definitions of the absolute volt (V) and ohm (Ω), respectively. One ohm absolute is defined by 0.9995 times the resistance to current flow of a column of mercury that is 1.063 m in length and has a mass of 0.0144521 kg at 273.15 K.

On a practical level, working standards for resistance take the form of certified standard resistors or resistance boxes, and they are used as standards for comparison in the calibration of resistance-measuring devices. The practical potential standard makes use of a standard cell consisting of a saturated solution of cadmium sulfate. The potential difference of two conductors connected across such a solution is set at 1.0183 V at 293 K. The standard cell maintains a constant electromotive force over very long periods of time, provided that it is not subjected to a current drain exceeding 100 mA for more than a few minutes. The standard cell is typically used as a standard for comparison for voltage measurement devices.

A chart for converting between units is included inside the text cover. Table 1.2 lists some standard units used in the SI and U.S. engineering systems.

Table 1.2 Standard Dimensions and Units

	Unit	
Dimension	SI	U.S.
Primary		
Length	meter (m)	foot (ft)
Mass	kilogram (kg)	pound-mass (lb_m)
Time	second (s)	second (s)
Temperature	kelvin (K)	rankine ($^\circ$R)
Current	ampere (A)	ampere (A)
Derived		
Force	newton (N)	pound-force (lb)
Voltage	volt (V)	volt (V)
Resistance	ohm (Ω)	ohm (Ω)
Capacitance	farad (F)	farad (F)
Inductance	henry (H)	henry (H)
Pressure, stress	pascal (Pa)	pound/foot^2 (psf)
Energy	joule (J)	British thermal unit (BTU)
Power	watt (W)	foot-pound (ft-lb)

Table 1.3 Hierarchy of Standards

Primary standard
Interlaboratory transfer standard
Local standard
Working instrument

Hierarchy of Standards

When we use a known value as the input to a measurement system during a calibration, it becomes the standard on which the calibration is based. Obviously, the actual primary standards are impractical as standards for normal calibration use, but they serve as a reference for exactness. It would not be reasonable to travel to France to calibrate an ordinary laboratory scale by using the primary standard for mass (nor would it be permitted!). So for practical reasons, there exists a hierarchy of secondary standards that attempt to duplicate the primary standards. Secondary standards provide reasonable approximations to the primary standards but can be accessed more readily for calibration purposes. However, an amount of uncertainty must be accepted with the use of standards that are replicas of primary standards. At the top of the standards hierarchy, just below the primary standard, are the national reference standards maintained by designated standards laboratories throughout the world. In the United States, NIST maintains primary and secondary standards and recommends standard procedures for the calibration of measurement systems.

Each subsequent level of the hierarchy is derived by calibration against the standard at the previous higher level. Table 1.3 lists an example of such a lineage for standards from a primary standard at a national lab down to a working standard, such as might be used in a typical laboratory to calibrate a measurement system. As one moves down through this lineage, the degree of exactness by which a standard approximates the primary standard from which it is derived deteriorates. That is, increasing elements of error are introduced into the standard as one goes from one generation of standard to the next. As a common example, an institution might maintain its own local standard, which it uses to calibrate the everyday measurement devices found in the individual laboratories throughout the institution. Calibration of the local standard might be against an NIST transfer standard. NIST will periodically calibrate its own transfer standard. This is illustrated for a temperature standard traceability hierarchy in Table 1.4. Therefore, the uncertainty in the approximation of the known value increases as one moves down the hierarchy. It follows, then, that since the calibration determines the relationship between the input value and the output value, the accuracy of the calibration will depend in part on the accuracy of the standard. But if typical working standards contain some error, how is accuracy

Table 1.4 Temperature Standard Traceability (Example)

	Standard	Error[a]
Level	Method	[°C]
Primary	Fixed thermodynamic points	0
Transfer	Platinum resistance thermometer	±0.005
Local	Platinum resistance thermometer	±0.01
Working instrument	Glass bulb thermometer	±0.1

[a] Typical instrument bias and precision errors.

ever determined? At best, accuracy can only be estimated, and the confidence in that estimate will depend on the quality of the standard and calibration techniques used.

Test Standards

The term "standard" can also be applied in other ways. *Test standards* refer to well-defined test procedures, technical terminology, methods to construct test specimens or test devices, and/or methods for data reduction. The goal of a test standard is to provide consistency in the conduct and reporting of a certain type of measurement between test facilities. Similarly, *test codes* refer to procedures for the manufacture, installation, calibration, performance specification, and safe operation of equipment.

Diverse examples of test standards and codes are illustrated in readily available documents ([e.g., 11–14]) from professional societies, such as the American Society of Mechanical Engineers (ASME) and the American Society of Testing and Materials (ASTM). For example, ASME Power Test Code 19.5 provides detailed designs and operation procedures for flow meters, whereas ASTM Test Standard F558-88 provides detailed procedures for evaluating vacuum cleaner cleaning effectiveness and controls the language for product performance claims. Test standards and codes are legal instruments, which must be observed by engineers.

1.6   PRESENTING DATA

Since we use several plotting formats throughout this text to present data, it is best to introduce these formats here. Data presentation conveys significant information about the relationship between variables. Software is readily available to assist in providing high-quality plots, or plots can be generated manually by using graph paper. Several forms of plotting formats (or graph paper) are discussed as follows.

Rectangular Coordinate Format

In a rectangular grid format, both the ordinate and the abscissa have uniformly sized divisions providing a linear scale. This is the most common format used for constructing plots and establishing the form of the relationship between the independent and dependent variable.

Semilog Coordinate Format

In a semilog format, one coordinate has a linear scale and one coordinate has a logarithmic scale. Plotting values on a logarithmic scale is the same as performing the logarithmic operation on those values; for example, plotting y vs. x with a logarithmic x axis is the same as plotting y vs. $\log x$ on linear axes. Logarithmic scales are advantageous when one of the variables spans more than 1 order of magnitude. In particular, the semilog format may be convenient when the data approximately follow a relationship of the forms $y = ae^x$ or $y = a \times 10^x$, as a linear curve will result in each case. A natural logarithmic operation can also be conveyed on a logarithmic scale as the relation, $\ln y = 2.3 \log y$, is just a scaling operation between these two scales.

Full-Log Coordinate Format

The full-log or log–log format has logarithmic scales for both axes. Such a format is preferred when both variables contain data values that span more than 1 order of magnitude. With data that follow a trend of the form $y = ax^n$, a linear curve will be obtained in the log–log format.

1.7   SUMMARY

During a measurement the input signal is not known but is inferred from the value of the output signal from the measurement system. However, the output signal can be affected by many variables that will introduce variation and trends. Careful test planning is required to reduce such effects. Only during calibration is the input value actually known, and this information is used to develop the relation between input and output necessary to interpret the output signal during later measurements. An important step in the design of a measurement system is the inclusion of a means for a reproducible calibration that closely simulates the type of signal to be input during actual measurements.

REFERENCES

1. Bridgeman, P. W., *Dimensional Analysis*, 2d ed., Yale University Press, New Haven, CN, 1931; paperback Y-82, 1963.
2. Duncan, W. J., *Physical Similarity and Dimensional Analysis*, Arnold, London, 1953.
3. Massey, B. S., *Units, Dimensions and Physical Similarity*, Van Nostrand Reinhold, New York, 1971.
4. Lipsen, C., and Sheth, N. J., *Statistical Design and Analysis of Engineering Experimentation*, McGraw-Hill, New York, 1973.
5. Peterson, R. G., *Design and Analysis of Experiments*, Marcel Dekker, New York, 1985.
6. Mead, R., *The Design of Experiments: Statistical Principles for Practical Application*, Cambridge University Press, New York, 1988.
7. *NBS Technical News Bulletin* 52:1, January 1968.
8. International Temperature Scale—1990, *Metrologia* 27:3, 1990.
9. International Practical Temperature Scale—1968 (English version), *Metrologia* 4, October 1968. Also see the amended version EPT-76, *Metrologia* 12, 1976.
10. Taylor, B., *Guide for the Use of the International System of Units*, NIST Special Publication 811, 1995.
11. *ASME Power Test Codes*, American Society of Mechanical Engineers, New York.
12. *ASHRAE Handbook of Fundamentals*, American Society of Heating, Refrigeration and Air Conditioning Engineers, New York, 1998.
13. *American National Standard*, American National Standards Institute, New York.
14. *ASTM Standards*, American Society for Testing and Materials, Philadelphia, PA.
15. Peace, G. S., *Taguchi Methods*, Addison-Wesley, Reading, MA, 1993.

NOMENCLATURE

e_c	instrument error	x	independent variable; input value
e_h	hysteresis error		
e_K	sensitivity error	y	dependent variable; output value
e_L	linearity error	y_L	linear polynomial
e_R	repeatability error	A	accuracy
e_z	zero error	K	static sensitivity
p	pressure [$m\,l^{-1}\,t^{-2}$]	T	temperature [°]
r_i	input span	V	volume [l^3]
r_o	output span	ϵ	absolute error

Table 1.5 Calibration Data

X [cm]	Y [V]	X [cm]	Y [V]
0.5	0.4	10.0	15.8
1.0	1.0	20.0	36.4
2.0	2.3	50.0	110.1
5.0	6.9	100.0	253.2

PROBLEMS

1.1 Discuss what is meant by a primary standard and by a secondary standard. In general, what is meant by the term "standard" in the context of a measurement?

1.2 When should a calibration be performed? What is the purpose of a calibration?

1.3 Suppose you found an old dial thermometer in a stockroom. Discuss several methods by which you might estimate its precision? Its accuracy?

1.4 Under which situations would a dynamic calibration of a measurement system be required?

1.5 How would the resolution of the display scale of an instrument affect its accuracy? Explain in terms of precision and bias errors.

1.6 How would the hysteresis of an instrument affect its accuracy? Explain in terms of precision and bias errors.

1.7 Select three different types of measurement systems with which you have experience and identify which attributes of the system comprise the measurement system stages of Figure 1.1.

1.8 Identify the measurement system stages for the following systems: (a) room thermostat, (b) automobile speedometer, (c) portable CD stereo system, (d) antilock braking system (automobile), and (e) audio speaker. (Refer back to Figure 1.1 and use other resources, such as a library or internet search, as needed to learn more about each system.)

1.9 Specify the range of the calibration data of Table 1.5.

1.10 For the calibration data of Table 1.5, plot the results by using rectangular and log–log scales. Discuss the apparent advantages of either presentation.

1.11 For the calibration data of Table 1.5, determine the static sensitivity of the system at (a) $X = 5$; (b) $X = 10$; (c) $X = 20$. For which input values is the system more sensitive?

1.12 Consider the voltmeter calibration data in Table 1.6. Plot the data by using a suitable scale. Specify the percent maximum hysteresis based on full-scale range. Input X is based on a standard known to be accurate to better than 0.05 mV.

1.13 Three clocks are compared with a time standard at three successive hours. The data are given in Table 1.7. Arrange these clocks in order of estimated accuracy. Justify your choices.

Table 1.6 Voltmeter Calibration Data

Increasing Input [mV]		Decreasing Input [mV]	
X	Y	X	Y
0.0	0.1	5.0	5.0
1.0	1.1	4.0	4.2
2.0	2.1	3.0	3.2
3.0	3.0	2.0	2.2
4.0	4.1	1.0	1.2
5.0	5.0	0.0	0.2

Table 1.7 Clock Calibration Data

Clock	Standard Time		
	1:00:00	2:00:00	3:00:00
	Indicated Time		
A	1:02:23	2:02:24	3:02:25
B	1:00:05	2:00:05	3:00:05
C	1:00:01	1:59:58	3:00:01

1.14 Each of the following equations can be represented as a straight line on a x–y plot by choosing the appropriate axis scales. Plot them in both a rectangular coordinate format and then in an appropriate format to yield a straight line. Explain how the plot operation yields the straight line. Variable y has units of volts. Variable x has units of meters (use a range of $0.01 \leq x \leq 10.0$). Note: this is easily done by using a spreadsheet program in which you can compare the use of different axis scales.

a. $y = x^2$
b. $y = 1.1x$
c. $y = 2x^{0.5}$
d. $y = 10x^4$
e. $y = 10e^{-2x}$

1.15 Plot $y = 10e^{-5x}$ V on in semilog format (use three cycles). Determine the slope of the equation at $x = 0$, $x = 2$, and $x = 20$.

1.16 Plot the following data on an appropriate set of axes. Estimate the static sensitivity K at each X.

Y [V]	X [m]
2.9	0.5
3.5	1.0
4.7	2.0
9.0	5.0

1.17 The following data have the form $y = ax^b$. Plot the data in an appropriate format to estimate the coefficients a and b. Estimate the static sensitivity K at each value of X. How is K affected by X?

Y [V]	X [m]
0.14	0.5
2.51	2.0
15.30	5.0
63.71	10.0

1.18 For the calibration data given, plot the calibration curve on an appropriate set of axes. Estimate the static sensitivity of the system at each X. Then plot K against X. Comment on the behavior of the static sensitivity with static input magnitude for this system.

Y [cm]	X [kPa]
4.76	0.05
4.52	0.1
3.03	0.5
1.84	1.0

1.19 A bulb thermometer hangs outside a window and is used to measure the outside temperature. Comment on the extraneous variables that might affect the difference between the actual outside temperature and the indicated temperature from the thermometer.

1.20 A synchronous electric motor test stand permits either the variation of input voltage or output shaft load and the subsequent measurement of motor efficiency, winding temperature, and input current. Comment on the independent, dependent, and extraneous variables for a motor test.

1.21 The transducer specified in Table 1.1 is chosen to measure a nominal pressure of 500 cm H_2O. The ambient temperature is expected to vary between 18°C and 25°C during tests. Estimate the magnitude of each elemental error affecting the measured pressure.

1.22 A force measurement system (weight scale) has the following specifications:

Range:	0–1000 N
Linearity error:	0.10% FSO
Hysteresis error:	0.10% FSO
Repeatability error:	0.15% FSO
Zero drift:	0.20% FSO

Estimate the overall instrument error for this system based on available information.

1.23 An engineer ponders a test plan, thinking: "What strategy should I include to estimate any variation over time of the measured variable? What strategy should I include to estimate my control of the independent variable during the test? What does the engineer mean?"

1.24 If the outcome of a test were suspected to be dependent on the ability to control the test operating conditions, what strategy should be incorporated into the test plan to estimate this effect?

1.25 State the purpose of using randomization methods during a test.

1.26 Provide an example of repetition and replication in a test plan from your own experience.

1.27 Develop a test plan that might be used to estimate the average temperature that could be maintained in a heated room as a function of the heater thermostat setting.

1.28 Develop a test plan that might be used to evaluate the fuel efficiency of a production model automobile. Explain your reasoning.

1.29 A race engine shop has just completed two engines of the same design. How might you determine which engine performs better (i) on a test stand (engine dynamometer) and (ii) at the race track? Describe some measurements that you feel might be useful and how you might use that information. Discuss possible differences between the two tests and how these might influence the results (e.g., you can control room conditions on a test stand but not at a track).

1.30 A large batch of carefully made machine shafts can be manufactured on one of four lathes by one of 12 quality machinists. Set up a test matrix to estimate the tolerances that can be held within a production batch. Explain your reasoning.

1.31 Suggest an approach(es) to estimate the linearity error and the hysteresis error of a measurement system.

1.32 Suggest a test matrix to evaluate the wear performance of four different brands of aftermarket passenger car tires of the same size, load, and speed ratings on a fleet of eight cars of the same make. If the cars were not of the same make, what would change?

1.33 The relation between the flow rate through a pipeline of area, A, and the pressure drop, dp, across an orifice-type flow meter inserted in that line (Figure 1.11) is given by

$$Q = CA\sqrt{\frac{2dp}{\rho}}$$

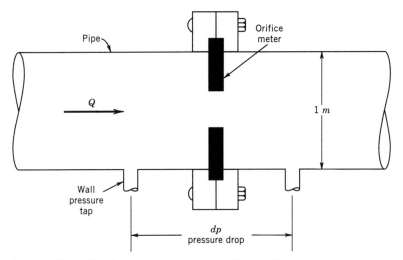

Figure 1.11 Orifice flow meter setup for Problem 1.33.

For a pipe diameter of 1 m and a flow range of $20°C$ H_2O between 2 and 10 m³/min and $C = 0.75$, plot the expected form of the calibration curve for flow rate versus pressure drop over the flow range. Is the static sensitivity a constant? Incidentally, such a calibration would be governed by ANSI/ASME Test Standard PTC 19.5.

1.34 For the orifice meter calibration in Problem 1.32, would the term "linearity error" have a meaning for this system? Explain. Also, list the dependent and independent variables in the calibration.

1.35 A piston engine manufacturer uses four different subcontractors to plate the pistons for a make of engine. Plating thickness is important in quality control (performance and part life). Devise a test matrix to assess how well the manufacturer can control plating under its current system.

1.36 A simple thermocouple circuit is formed by using two wires of different alloys: One end of the wires is twisted together to form the measuring junction, and the other ends are connected to a voltmeter and form the reference junction. A voltage is set up by the difference in temperature between the two junctions. For a given pair of alloy materials and reference junction temperatures, the temperature of the measuring junction is inferred from the measured voltage difference. For a measurement, what variables have to be controlled? What are the dependent and independent variables?

1.37 A linear variable displacement transducer (LVDT) senses displacement and outputs a voltage, which is linear to the input. Figure 1.12 shows an LVDT setup for static calibration using a micrometer to apply the known displacement and a voltmeter for the output. A strictly regulated dc voltage powers the transducer. What are the independent and dependent variables in this calibration? Can you suggest any extraneous variables? What would be involved in a replication?

1.38 For the LVDT calibration of the previous problem, what would be involved in determining the repeatability of the instrument? The reproducibility? What effects are different in the two tests? Explain.

1.39 A manufacturer wants to quantify the expected average fuel mileage of a product line of automobile. They decide that they can put one or more cars on a chassis dynamometer, which runs the wheels at desired speeds and loads, to assess this. Alternatively, they can use drivers and drive the cars over some selected course instead. (i) Discuss the merits of either approach, considering the control of variables and identifying extraneous variables. (ii) Can you recognize that somewhat different tests might provide answers to different questions? For example, discuss the difference in meanings possible from the results of operating one car on the

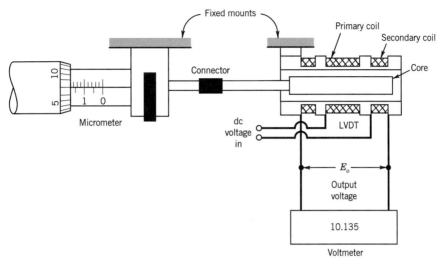

Figure 1.12 LVDT setup for Problem 1.37.

dynamometer and comparing it to one driver on a course. Cite other examples. (iii) Are these two test methods examples of concomitant methods?

1.40 You estimate your car's fuel use by comparing fuel volume used over a known distance. Your brother, who drives the same model car, disagrees with your claimed results based on his own experience. How might you justify the differences based on the concepts of control of variables, interference and noise effects, and test matrix used?

1.41 In discussing concomitant methods, we cited an example of computing the volume of a cylindrical rod based on its average dimensions versus known weight and material properties. Although we should not expect too different of an answer with either technique, identify where noise and interference effects will affect the result of either method.

1.42 When a strain gauge is stretched under uniaxial tension, its resistance varies with the imposed strain. A resistance bridge circuit is used to convert the resistance change into a voltage. Suppose a known tensile load were applied to the system shown in Figure 1.13 and the output measured on a voltmeter. What are the independent and dependent variables in this calibration? Can you suggest any extraneous variables? What would be involved in a replication?

1.43 For the strain gauge calibration of the previous problem, what would be involved in determining the repeatability of the instrument? The reproducibility? A statement on instrument precision? What effects are different in the tests? Explain.

1.44 A major tennis manufacturer is undertaking a test program for shoes, tennis balls, and tennis strings. Develop a test plan for the situations described in the list that follows. In each case provide details for how the test should be conducted, and describe expected difficulties in interpreting the results. The tests are to be conducted during college tennis matches. Four teams of six players each are to test the products under match (as opposed to practice) conditions. A tennis team consists of six players, and a match consists of six singles matches and three doubles matches.

- Tennis shoes: Two different sole designs and materials are to be wear tested. The life of the soles is known to be strongly affected by court surface and playing style.
- Tennis strings: A new tennis string material is to be tested for durability. String failure occurs as a result of breakage or loss of tension. String life is a function of the racquet, the player's style, and string tension.

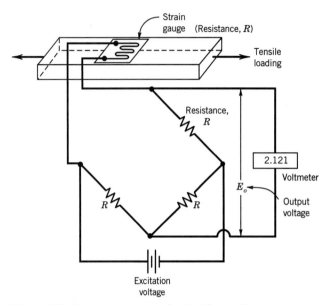

Figure 1.13 Strain gauge setup for Problem 1.42.

- Tennis balls: Two tennis balls are to be play tested for durability. The condition of a tennis ball can be described by the coefficient of restitution, and the total weight (since the cover material actually is lost during play). A player's style and the court surface are the primary determining factors on tennis ball wear.

1.45 Research the following test codes (these are available in most libraries). Write a short report describing the coverage of the code and its intent: (a) ASTM F 558-88 (Air Performance of Vacuum Cleaners), (b) ANSI Z21.86 (Gas Fired Space Heating Appliances), (c) ISO 10770-1:1998 (Test Methods for Hydraulic Control Valves), (d) ANSI/ASME PTC19.1-1998 (Measurement Uncertainty: Instruments and Apparatus), (e) ISO 7401:1988 (Road Vehicles: Lateral Response Test Methods), and (f) your local municipal building code or housing ordinance.

Chapter 2

Static and Dynamic Characteristics of Signals

2.1 INTRODUCTION

A measurement system takes an *input* quantity and transforms it into an *output* quantity that can be observed or recorded, such as the movement of a pointer on a dial or the magnitude of a digital display. Our goal in this chapter is to understand the characteristics of both the input signals to a measurement system and the resulting output signals. The shape and form of a signal is often referred to as its *waveform*. The waveform contains information about the *magnitude* and *amplitude*, which indicate the size of the input quantity, and the *frequency*, which indicates the way the signal changes in time. An understanding of waveforms is required for the selection of measurement systems and the interpretation of measured signals.

2.2 INPUT–OUTPUT SIGNAL CONCEPTS

Two important tasks that engineers face in the measurement of physical variables are (1) selecting a measurement system and (2) interpreting the output from a measurement system. A simple example of selecting a measurement system might be the selection of a tire gauge for measuring the air pressure in a bicycle tire or in a car tire, as shown in Figure 2.1. The gauge for the car tire would have to be able to indicate pressures up to 275 kPa (40 lb/in.^{2}), but the bicycle tire gauge would be required to indicate higher pressures, maybe up to 700 kPa (100 lb/in.^{2}). This idea of the *range* of an instrument, its lower to upper measurement limits, is basic to all measurement systems and demonstrates that some basic understanding of the nature of the input signal, in this case the magnitude, is necessary in evaluating or selecting a measurement system for a particular application.

A much more difficult task is the evaluation of the output of a measurement system when the time or spatial behavior of the input is not known. The pressure in a tire does not change while we are trying to measure it, but what if we wanted to measure pressure in a cylinder in an automobile engine? Would the tire gauge or another gauge based on its operating principle work? We know that the pressure in the cylinder varies with time. If our task was to select a measurement system to determine this time-varying pressure, information about the pressure variations in the cylinder would be necessary. From thermodynamics and the speed range of the engine it may be possible to estimate the magnitude of pressures to be expected, and the rate with which they change. From that we can select an appropriate measurement

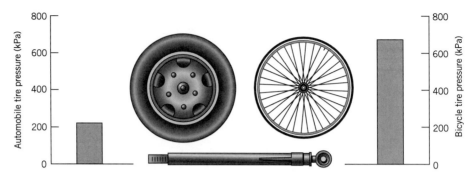

Figure 2.1 Measurement system selection based on input signal range.

system. But to do this, we need to develop a way to express this idea of the magnitude and rate of change of a variable.

Many measurement systems exhibit similar responses under a variety of conditions, which suggests that the performance and capabilities of measuring systems may be described in a generalized way. To examine further the generalized behavior of measurement systems, we first examine the possible forms of the input and output signals. We will associate the term "signal" with the "transmission of information." A *signal* is the physical information about a measured variable being transmitted between a process and the measurement system, between the stages of a measurement system, or the output from a measurement system.

Classification of Waveforms

Signals may be classified as either analog, discrete time, or digital. *Analog* describes a signal that is continuous in time. Since physical variables tend to be continuous in nature, an analog signal provides a ready representation of their time-dependent behavior. In addition, the magnitude of the signal is continuous and thus can have any value within the operating range. An analog signal is shown in Figure 2.2(a); a similar continuous signal would result from a recording of the pointer rotation with time for the output display shown in Figure 2.2(b). Contrast this continuous signal with the signal shown in Figure 2.3(a). This format represents a *discrete time signal*, for which information about the magnitude of the signal is available only at discrete

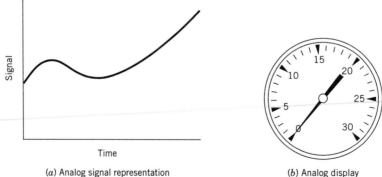

Figure 2.2 Analog signal concepts.

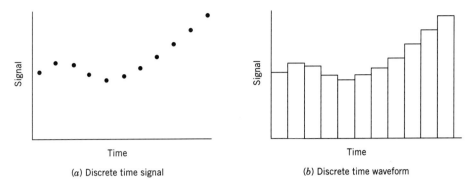

Figure 2.3 Discrete time signal concepts.

points in time. A discrete time signal usually results from measuring a continuous variable at finite time intervals.

Because information in the signal shown in Figure 2.3(a) is available only at discrete times, some assumption must be made about the behavior of the measured variable during the times when it is not being sampled. One approach is to assume the signal is constant between samples. The waveform that results from this assumption is shown in Figure 2.3(b). Clearly, as the time between samples is reduced, the difference between the discrete variable and the continuous signal it represents decreases.

Digital signals are particularly useful when data acquisition and processing are performed by using a digital computer. A digital signal has two important characteristics. First, a digital signal exists at discrete values in time, like a discrete time signal. Second, the magnitude of a digital signal is discrete, determined by a process known as quantization at each discrete point in time. Quantization assigns a single number to represent a range of magnitudes of a continuous signal.

Figure 2.4(a) shows digital and analog forms of the same signal where the magnitude of the digital signal can have only certain discrete values. Thus, a digital signal provides a quantized magnitude at discrete times. The waveform that would result from assuming that the signal is constant between sampled points in time is shown in Figure 2.4(b). As an example of quantization, consider a digital watch that displays time in hours and minutes. For the entire duration of 1 min, a single numerical value is displayed until it is updated at the next discrete time step. As such, the continuous physical variable time is quantized in its conversion to a digital display.

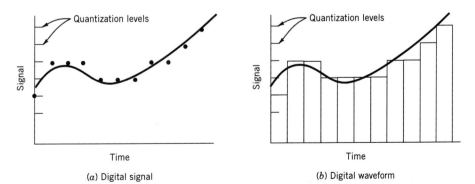

Figure 2.4 Digital signal representation and waveform.

Sampling of an analog signal to produce a digital signal can be accomplished by using an analog-to-digital (A/D) converter, a type of solid-state device that converts an analog voltage signal to a binary number system representation. The limited resolution of the binary number that corresponds to a range of voltages creates the quantization levels and ranges.

For example, a compact disk player is built around technology that relies on the conversion of a continuously available signal, such as music from a microphone, into a digital form. The digital information is stored on a compact disk and later read in digital form by a laser playback system [1]. However, to serve as input to a traditional stereo amplifier, and since speakers and the human ear are analog devices, the digital information is converted back into a continuous voltage signal for playback.

Signal Waveforms

In addition to the classification of signals as analog, discrete time, or digital, some description of the waveform associated with a signal is useful. Signals may be characterized as either static or dynamic. A *static signal* does not vary with time. The diameter of a shaft is an example. However, this idea of a completely static signal requires some discussion. Many physical variables change slowly enough in time, compared to the process with which they interact, that for all practical purposes these signals may be considered static in time. For example, the voltage across the terminals of a battery is approximately constant over its useful life. Or consider measuring temperature by using an outdoor thermometer; since the outdoor temperature does not change significantly in a matter of minutes, this input signal may be considered as static when compared to our time period of interest. A mathematical representation of a static signal is given by a constant, as indicated in Table 2.1. In contrast, often we are interested in how the measured variable changes with time. This leads us to consider time-varying signals further.

A *dynamic signal* is defined as a time-dependent signal. In general, dynamic signal waveforms, $y(t)$, may be classified as shown in Table 2.1. A *deterministic signal* is one that varies in time in a predictable manner, such as a sine wave, a step function, or a ramp function, as shown in Figure 2.5. A signal is *steady periodic* if the variation of the magnitude of the signal repeats at regular intervals in time. Examples of steady periodic behaviors would include the motion of an ideal pendulum, or the

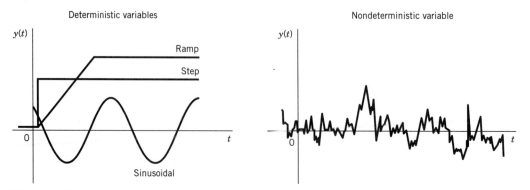

Figure 2.5 Examples of dynamic signals.

Table 2.1 Classification of Waveforms

I. Static
$$y(t) = A_0$$

II. Dynamic
    Periodic waveforms
        Simple periodic waveform
$$y(t) = A_0 + C\sin(\omega t + \phi)$$
        Complex periodic waveform
$$y(t) = A_0 + \sum_{n=1}^{\infty} C_n \sin(n\omega t + \phi_n)$$
    Aperiodic waveforms
        Step[a]
$$\begin{aligned} y(t) &= A_0 U(t) \\ &= A_0 \quad \text{for } t > 0 \end{aligned}$$
        Ramp
$$y(t) = Kt \quad \text{for } 0 < t < t_f$$
        Pulse[b]
$$y(t) = A_0 U(t) - A_0 U(t - t_1)$$

III. Nondeterministic waveform
$$y(t) \approx A_0 + \sum_{n=1}^{\infty} C_n \sin(\omega_n t + \phi_n)$$

[a] $U(t)$ represents the unit step function, which is zero for $t < 0$ and 1 for $t \geq 0$.
[b] t_1 represents the pulse width.

temperature variations in the cylinder of an internal combustion engine, under steady operating conditions. Periodic waveforms may be classified as simple or complex. A simple periodic waveform contains only one frequency. A *complex periodic waveform* contains multiple frequencies and is represented as a superposition of multiple simple periodic waveforms. *Aperiodic* is used to describe deterministic signals that do not repeat at regular intervals, such as a step function.

Also described in Figure 2.5 is a *nondeterministic signal*, which has no discernible pattern of repetition. A nondeterministic signal cannot be prescribed before it occurs, although certain characteristics of the signal may be known in advance. As an example, consider the transmission of data files from one computer to another. Signal characteristics such as the rate of data transmission and the possible range of signal magnitude are known for any signal in this system. However, it would not be possible to predict future signal characteristics based on existing information in such a signal. Such a signal is properly characterized as being nondeterministic. Nondeterministic signals are generally described by their statistical characteristics or some approximate series.

2.3 SIGNAL ANALYSIS

In this section, we consider concepts related to the characterization of signals. A measurement system produces a signal that may be analog, discrete time, or digital. An analog signal is continuous with time and has a magnitude that is analogous to the magnitude of the physical variable being measured. Consider the analog signal shown in Figure 2.6(a), which is continuous over the recorded time period from t_1 to

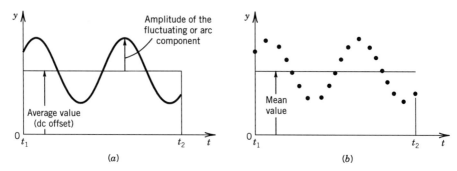

Figure 2.6 Analog and discrete representations of a dynamic signal.

t_2. The average or mean value[1] of this signal is found by

$$\bar{y} \equiv \frac{\int_{t_1}^{t_2} y(t)\, dt}{\int_{t_1}^{t_2} dt} \tag{2.1}$$

The mean value provides a measure of the static portion of a signal over the time, $t_2 - t_1$. It is sometimes called the *dc component* or dc offset of the signal.

The mean value does not provide any indication of the amount of variation in the dynamic portion of the signal. The characterization of the dynamic portion, or *ac component*, of the signal may be illustrated by considering the average power dissipated in an electrical resistor through which a fluctuating current flows. The power dissipated in a resistor caused by the flow of a current is

$$P = I^2 R$$

where

P = power dissipated
I = current
R = resistance

If the current varies in time, the total electrical energy dissipated in the resistor over the time t_1 to t_2 would be

$$\int_{t_1}^{t_2} P\, dt = \int_{t_1}^{t_2} [I(t)]^2 R\, dt \tag{2.2}$$

The current $I(t)$ would, in general, include both a dc component and a fluctuating ac component. Consider finding the magnitude of a constant effective current, I_e, that would produce the same total energy dissipation in the resistor as the time-varying current, $I(t)$, over the time period t_1 to t_2. Assuming that the resistance, R, is constant, this current would be determined by equating $(I_e)^2 R(t_2 - t_1)$ with equation (2.2) to yield

$$I_e = \sqrt{\frac{1}{t_2 - t_1} \int_{t_1}^{t_2} [I(t)]^2\, dt} \tag{2.3}$$

This value is called the root-mean-square value or rms value of the current; clearly it is the average value of the square of the current over the measured time period.

---

[1] Strictly speaking, for a continuous signal the mean value and the average value are the same. This is not true for discrete time signals.

Based on this reasoning, the *rms value* of any continuous analog variable $y(t)$ over the time $t_2 - t_1$ is expressed as

$$y_{rms} = \sqrt{\frac{1}{t_2 - t_1} \int_{t_1}^{t_2} y^2 \, dt} \tag{2.4}$$

A time-dependent analog signal, $y(t)$, can be represented by a discrete set of N numbers over the time period from t_1 to t_2 through the conversion

$$y(t) \longrightarrow \{y(r \, \delta t)\} \quad r = 1, \ldots, N$$

which uses the sampling convolution

$$\{y(r \, \delta t)\} = y(t)\delta(t - r \, \delta t)$$
$$= \{y_i\} \quad i = 1, 2, \ldots, N$$

Here, $\delta(t - r \, \delta t)$ is the delayed unit impulse function, δt is the sample time increment between each number, and $N \, \delta t = t_2 - t_1$ gives the total sample period over which the measurement of $y(t)$ takes place. The effect of discrete sampling on the original analog signal is demonstrated in Figure 2.6(*b*), in which the analog signal has been replaced by $\{y(r \, \delta t)\}$, which represents N values of a discrete time signal for $y(t)$.

For either a discrete time signal or a digital signal, the mean value can be estimated by

$$\bar{y} = \frac{1}{N} \sum_{i=1}^{N} y_i \tag{2.5}$$

where each y_i is a discrete number in the data set of $\{y(n \, \delta t)\}$. The mean approximates the static component of the signal over the time interval t_1 to t_2. The rms value can be found as

$$y_{rms} = \sqrt{\frac{1}{N} \sum_{i=1}^{N} y_i^2} \tag{2.6}$$

The rms value takes on additional physical significance when either the signal contains no dc component or the dc component has been subtracted from the signal. The rms value of a signal having a zero mean is a statistical measure of the magnitude of the fluctuations in the signal, relative to the mean value.

Effects of Signal-Averaging Period

The choice of a time period, t_f, where $t_f = t_2 - t_1$, for signal analysis depends on the intended purpose of the analysis. For simple periodic functions, averaging the signal over a time exactly equal to the period of the function results in a true representation of the long-term average value of the signal. This is also true for the rms value. Averaging a simple periodic signal over a time period that is not exactly the period of the function can produce misleading results. However, as the averaging time period becomes long relative to the signal period, the resulting values will accurately represent the signal. Complex waveforms and nondeterministic signals have a range of frequencies present; as such, no single averaging period will produce an exact representation of the signal. In such cases, the signal should be analyzed over a period of time that is long compared with the longest period contained within the signal waveform. This can be tested by comparing the resulting mean and rms values determined over increasingly longer periods.

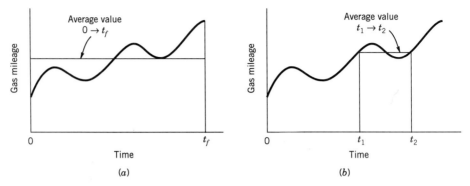

Figure 2.7 Effect of time period on mean value for a nondeterministic signal.

Consider the signal corresponding to the time-dependent fuel consumption of an automobile shown in Figure 2.7. Certainly, the mean value determined for the t_f used in Figure 2.7(a) will differ from that used in Figure 2.7(b). However, a mean value over the smaller time interval of Figure 2.7(b) might yield information on performance under very specific conditions relevant to that portion of the signal, such as during the climbing of a hill. There are no rules to assist in the selection of t_f beyond the development of a measurement plan based on a careful evaluation of the measurement objective and the intended use of the results.

DC Offset

When the ac component of the signal is of primary interest, the dc component can be removed. This procedure often allows a change of scale in the display or plotting of a signal such that fluctuations that were a small percentage of the dc signal can be more clearly observed without the superposition of the large dc component. The enhancement of fluctuations through the subtraction of the average value is illustrated in Figure 2.8. Figure 2.8(a) contains the entire complex waveform consisting of both static and dynamic parts. When the dc component is removed, the dynamic portion can be amplified for enhancement, as in Figure 2.8(b).

EXAMPLE 2.1

Suppose the current passing through a resistor can be described by

$$I(t) = 10 \sin t$$

where I represents the time-dependent current in amperes. Establish the mean and rms values of current over a time from 0 to t_f, with $t_f = \pi$ and then with $t_f = 2\pi$. How do the results relate to the power dissipated in the resistor?

KNOWN

$$I(t) = 10 \sin t \text{ A}$$

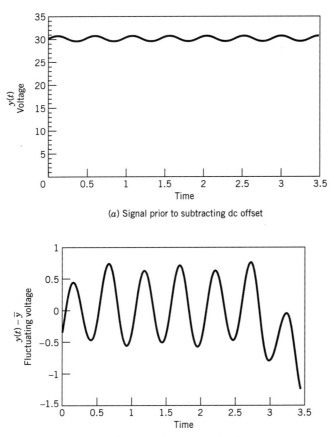

(a) Signal prior to subtracting dc offset

(b) Fluctuating component of signal

Figure 2.8 Effect of subtracting the dc offset for a dynamic signal.

FIND

$\overline{I}$ and I_{rms} with $t_f = \pi$ and 2π

SOLUTION

The average value for a time from 0 to t_f is found from equation (2.1) as

$$\overline{I} = \frac{\int_0^{t_f} I(t)\, dt}{\int_0^{t_f} dt} = \frac{\int_0^{t_f} 10 \sin t\, dt}{t_f}$$

Evaluation of this integral yields

$$\overline{I} = \frac{1}{t_f}[-10\cos t]_0^{t_f}$$

With $t_f = \pi$, the average value, $\overline{I}$, is $20/\pi$ A. For the case $t_f = 2\pi$, the evaluation of the integral yields an average value of zero.

The rms value for the time period 0 to t_f is given by the application of equation (2.4), which yields

$$I_{rms} = \sqrt{\frac{1}{t_f} \int_0^{t_f} I(t)^2 \, dt} = \sqrt{\frac{1}{t_f} \int_0^{t_f} (10 \sin t)^2 \, dt}$$

This integral is evaluated as

$$I_{rms} = \sqrt{\frac{100}{t_f} \left(-\frac{1}{2} \cos t \sin t + \frac{t}{2} \right) \Big|_0^{t_f}}$$

For $t_f = \pi$, the rms value is $\sqrt{50}$ A. Evaluation of the integral for the rms value with $t_f = 2\pi$ also yields $\sqrt{50}$ A.

COMMENT

Although the average value over the period 2π is zero, the power dissipated in the resistor must be the same over both the positive and negative half-cycles of the sine function. Thus, the rms value of current is the same for the time period of π and 2π and is indicative of the power dissipated.

2.4 SIGNAL AMPLITUDE AND FREQUENCY

A key factor in measurement system behavior is the nature of the input signal to the system. A means is needed to classify waveforms for both the input signal and the resulting output signal relative to their magnitude and frequency. It would be very helpful if the behavior of measurement systems could be defined in terms of their response to a limited number and type of input signals. This is, in fact, exactly the case: A very complex signal, even one that is nondeterministic in nature, can be approximated as an infinite series of sine and cosine functions, as suggested in Table 2.1. The method of expressing such a complex signal as a series of sines and cosines is called *Fourier analysis*. Nature provides some experiences that support our contention that complex signals can be represented by the addition of a number of simpler periodic functions. For example, rich musical tones can be generated by combining a number of different pure tones, and an excellent physical analogy for Fourier analysis is provided by the separation of white light through a prism. Figure 2.9 illustrates the transformation of a complex waveform, represented by white light, into its simpler components, represented by the colors in the spectrum. In this example, the colors in the spectrum are represented as simple periodic functions that combine to form white light. Fourier analysis is roughly the mathematical equivalent of a prism, and it yields a representation of a complex signal in terms of simple periodic functions.

The representation of complex and nondeterministic waveforms by simple periodic functions allows measurement system response to be reasonably well defined by examining the ouput resulting from a few specific input waveforms, one of which is a simple periodic. As represented in Table 2.1, a simple periodic waveform has a single, well-defined amplitude and a single frequency. Before a generalized response of measurement systems can be determined, an understanding of the method of representing complex signals in terms of simpler functions is necessary.

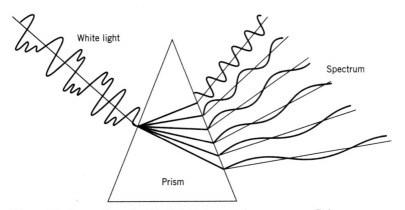

Figure 2.9 Separation of white light into its color spectrum. Color corresponds to a particular frequency or wavelength; light intensity corresponds to varying amplitudes.

Periodic Signals

The fundamental concepts of frequency and amplitude can be understood through the observation and analysis of periodic motions. Although sines and cosines are by definition geometric quantities related to the lengths of the sides of a right triangle, for our purposes sines and cosines are best thought of as mathematical functions that describe specific physical behaviors of systems. These behaviors are described by differential equations, which have sines and cosines as their solutions. As an example, consider a mechanical vibration of a mass attached to a linear spring, as shown in Figure 2.10. For a linear spring, the spring force, F, and displacement, y, are related by $F = ky$, where k is the constant of proportionality, called the spring constant. Application of Newton's second law to this system yields a governing equation for the displacement, y, as

$$m\frac{d^2y}{dt^2} + ky = 0 \qquad (2.7)$$

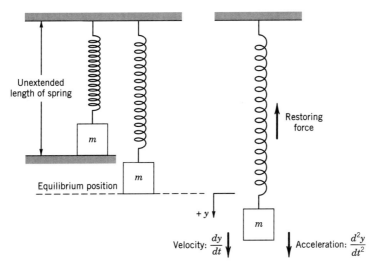

Figure 2.10 Spring-mass system.

This linear, second-order differential equation with constant coefficients describes the motion of the idealized spring mass system when there is no external force applied. The general form of the solution to this equation is

$$y = A\cos \omega t + B \sin \omega t \tag{2.8}$$

where $\omega = \sqrt{k/m}$. Physically we know that if the mass is displaced from the equilibrium point and released, it will oscillate about the equilibrium point. The time required for the mass to finish one complete cycle of the motion is called the *period*.

Frequency is related to the period and is defined as the number of complete cycles of the motion per unit time. This frequency, f, is measured in cycles per second (Hz; 1 cycle/s $= 1$ Hz). The term ω is also a frequency, but instead of having units of cycles per second, it has units of radians per second. This frequency, ω, is called the circular frequency since it relates directly to cycles on the unit circle, as illustrated in Figure 2.11. The relationship between ω, f, and the period, T, is

$$T = \frac{2\pi}{\omega} = \frac{1}{f} \tag{2.9}$$

In equation (2.8), the sine and cosine terms can be combined if a phase angle is introduced such that

$$y = C \cos(\omega t - \phi) \tag{2.10a}$$

or

$$y = C \sin(\omega t + \phi^*) \tag{2.10b}$$

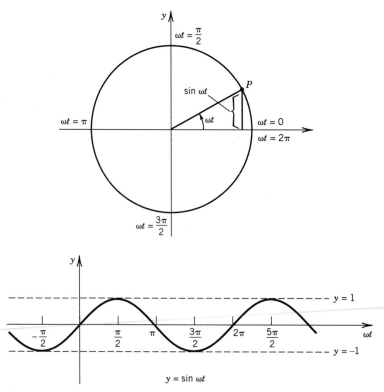

Figure 2.11 Relationship between cycles on the unit circle and circular frequency.

The values of C, ϕ, and ϕ^* are found from the following trigonometric identities:

$$A \cos \omega t + B \sin \omega t = \sqrt{A^2 + B^2} \cos(\omega t - \phi)$$

$$A \cos \omega t + B \sin \omega t = \sqrt{A^2 + B^2} \sin(\omega t + \phi^*) \qquad (2.11)$$

$$\phi = \tan^{-1} \frac{B}{A} \quad \phi^* = \tan^{-1} \frac{A}{B} \quad \phi^* = \frac{\pi}{2} - \phi$$

The size of the maximum and minimum displacements from the equilibrium position, or the value C, is the amplitude of the oscillation. The concepts of amplitude and frequency are essential for the description of time-dependent signals.

Frequency Analysis

Many signals that result from the measurement of dynamic variables are nondeterministic in nature and have a continuously varying rate of change. These signals, having *complex waveforms*, present difficulties in the selection of a measurement system and in the interpretation of an output signal. However, it is possible to separate a complex signal, or any signal for that matter, into a number of sine and cosine functions. In other words, any complex signal can be thought of as being made up of sines and cosines of differing periods and amplitudes, which are added together in an infinite trignometric series. This representation of a signal as a series of sines and cosines is called a *Fourier series*. Once a signal is broken down into a series of periodic functions, the importance of each frequency can be easily determined. This information about frequency content allows the proper choice of a measurement system, and a precise interpretation of output signals.

In theory, Fourier analysis allows essentially all mathematical functions of practical interest to be represented by an infinite series of sines and cosines.[2]

The following definitions relate to Fourier analysis:

1. A function $y(t)$ is a *periodic function* if there is some positive number T such that

$$y(t + T) = y(t)$$

The *period* for $y(t)$ is T. If both $y_1(t)$ and $y_2(t)$ have period T, then

$$ay_1 + by_2$$

also has a period of T (a and b are constants).

2. A *trigonometric series* is given by

$$A_0 + A_1 \cos t + B_1 \sin t + A_2 \cos 2t + B_2 \sin 2t + \cdots$$

where A_n and B_n are the coefficients of the series.

EXAMPLE 2.2

As a physical (instead of mathematical) example of the frequency content of a signal, consider stringed musical instruments, such as guitars and violins. When a string is caused to vibrate by plucking or bowing, the sound is a result of the motion of the string and the resonance of the instrument. (The concept of resonance will be explored in Chapter 3.) The musical pitch for such an instrument is the lowest frequency

---

[2]A periodic function may be represented as a Fourier series if the function is piecewise continuous over the limits of integration and the function has a left- and right-hand derivative at each point in the interval. The sum of the resulting series is equal to the function at each point in the interval, except points where the function is discontinuous.

of the string vibrations. Our ability to recognize differences in musical instruments is primarily a result of the higher frequencies present in the sound, which are usually integer multiples of the fundamental frequency. These higher frequencies are called *harmonics*.

The motion of a vibrating string is best thought of as being composed of several basic motions, which together create a musical tone. Figure 2.12 illustrates the vibration modes associated with a string plucked at its center. The string vibrates

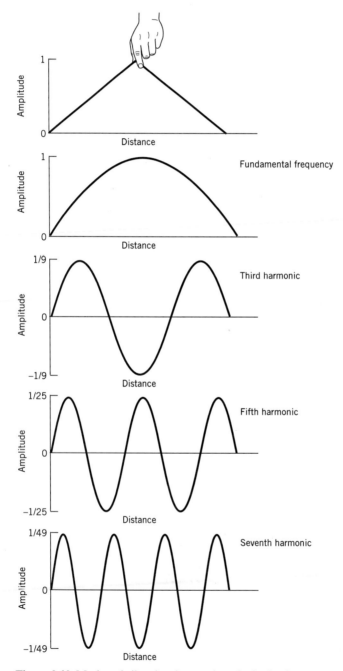

Figure 2.12 Modes of vibration for a string plucked at its center.

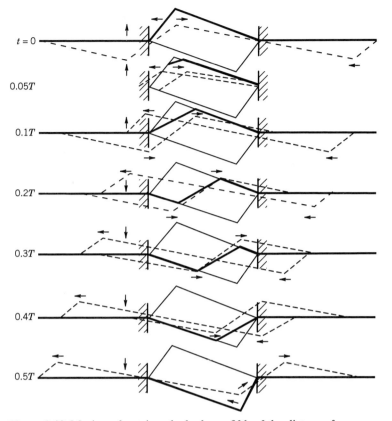

Figure 2.13 Motion of a string plucked one-fifth of the distance from a fixed end. [From N. H. Fletcher and T. D. Rossing, *The Physics of Musical Instruments*. Copyright © 1991 by Springer-Verlag, New York. Reprinted by permission.]

with a fundamental frequency and odd-numbered harmonics, each having a specific phase relationship with the fundamental. The relative strength of each harmonic is graphically illustrated through its amplitude in Figure 2.12. Figure 2.13 shows the motion, which is caused by plucking a string one-fifth of the distance from a fixed end. The resulting frequencies are illustrated in Figure 2.14. Notice that the fifth harmonic is missing from the resulting sound.

Musical sound from a vibrating string is analogous to a measurement system input or output, which contains many frequency components. Fourier analysis and frequency spectra provide insightful and practical means of reducing such complex signals into a combination of simple waveforms. One goal in this chapter is to understand the frequency and amplitude analysis of complex signals.

Fourier Series and Coefficients

A periodic function $y(t)$ with a period $T = 2\pi$ is to be represented by a trigonometric series, such that for any t,

$$y(t) = A_0 + \sum_{n=1}^{\infty} (A_n \cos nt + B_n \sin nt) \qquad (2.12)$$

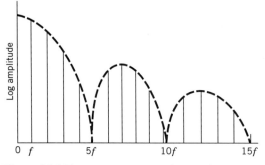

Figure 2.14 Frequency spectrum for a string plucked one-fifth of the distance from a fixed end. [From N. H. Fletcher and T. D. Rossing, *The Physics of Musical Instruments*. Copyright © 1991 by Springer-Verlag, New York. Used by permission.]

With $y(t)$ known, the coefficients A_n and B_n are to be determined. For A_0 to be determined, equation (2.12) is integrated from $-\pi$ to $+\pi$:

$$\int_{-\pi}^{\pi} y(t)\, dt = A_0 \int_{-\pi}^{\pi} dt + \sum_{n=1}^{\infty} \left(A_n \int_{-\pi}^{\pi} \cos nt\, dt + B_n \int_{-\pi}^{\pi} \sin nt\, dt \right) \quad (2.13)$$

Since

$$\int_{-\pi}^{\pi} \cos nt\, dt = 0 \quad \text{and} \quad \int_{-\pi}^{\pi} \sin nt\, dt = 0$$

equation (2.13) yields

$$A_0 = \frac{1}{2\pi} \int_{-\pi}^{\pi} y(t)\, dt \quad (2.14)$$

The coefficient A_m may be determined by multiplying equation (2.12) by $\cos mt$ and integrating from $-\pi$ to π. The resulting expression for A_m is

$$A_m = \frac{1}{\pi} \int_{-\pi}^{\pi} y(t)\, \cos mt\, dt \quad (2.15)$$

Similarly, multiplying equation (2.12) by $\sin mt$ and integrating from $-\pi$ to π yields B_m. Thus, for a function $y(t)$ with a period 2π, the coefficients of the trigonometric series representing $y(t)$ are given by the Euler formulas:

$$A_0 = \frac{1}{2\pi} \int_{-\pi}^{\pi} y(t)\, dt$$

$$A_n = \frac{1}{\pi} \int_{-\pi}^{\pi} y(t)\, \cos nt\, dt \quad (2.16)$$

$$B_n = \frac{1}{\pi} \int_{-\pi}^{\pi} y(t) \sin nt\, dt$$

The trigonometric series corresponding to $y(t)$ is called the Fourier series for $y(t)$, and the coefficients A_n and B_n are called the Fourier coefficients of $y(t)$. Functions represented by a Fourier series normally do not have a period of 2π. However, the transition to an arbitrary period can be affected by a change of scale, yielding a new set of Euler formulas, described as follows.

Fourier Coefficients for Functions Having Arbitrary Periods

The coefficients of a trigonometric series representing a function of frequency ω are given by the Euler formulas,

$$A_0 = \frac{1}{T} \int_{-T/2}^{T/2} y(t)\, dt$$

$$A_n = \frac{2}{T} \int_{-T/2}^{T/2} y(t) \cos n\omega t\, dt \qquad (2.17)$$

$$B_n = \frac{2}{T} \int_{-T/2}^{T/2} y(t) \sin n\omega t\, dt$$

where $n = 1, 2, 3, \ldots$, and $T = 2\pi/\omega$ is the period of $y(t)$. The trigonometric series that results from these coefficients is a Fourier series and may be written as

$$y(t) = A_0 + \sum_{n=1}^{\infty} (A_n \cos n\omega t + B_n \sin n\omega t) \qquad (2.18)$$

In the series for $y(t)$ in equation (2.18), when $n = 1$ the corresponding terms in the Fourier series are called *fundamental* and have the lowest frequency in the series. The *fundamental frequency* for this Fourier series is $\omega = 2\pi/T$. Frequencies corresponding to $n = 2, 3, 4, \ldots$ are known as *harmonics*, with, for example, $n = 2$ representing the second harmonic.

A series of sines and cosines may be written as a series of either sines or cosines through the introduction of a phase angle, so that the Fourier series in equation (2.18),

$$y(t) = A_0 + \sum_{n=1}^{\infty} (A_n \cos n\omega t + B_n \sin n\omega t)$$

may be written as

$$y(t) = A_0 + \sum_{n=1}^{\infty} C_n \cos(n\omega t - \phi_n) \qquad (2.19)$$

or

$$y(t) = A_0 + \sum_{n=1}^{\infty} C_n \sin(n\omega t + \phi_n^*) \qquad (2.20)$$

where

$$C_n = \sqrt{A_n^2 + B_n^2}$$

$$\tan \phi_n = \frac{B_n}{A_n} \qquad \tan \phi_n^* = \frac{A_n}{B_n} \qquad (2.21)$$

Even and Odd Functions

A function $g(t)$ is even if it is symmetric about the origin, which may be stated, for all t,

$$g(-t) = g(t)$$

A function $h(t)$ is odd if, for all t,

$$h(-t) = -h(t)$$

For example, $\cos nt$ is even, whereas $\sin nt$ is odd. A particular function or waveform may be even, odd, or neither even nor odd.

Fourier Cosine Series

If $y(t)$ is even, its Fourier series will contain only cosine terms:

$$y(t) = \sum_{n=1}^{\infty} A_n \cos \frac{2\pi nt}{T} = \sum_{n=1}^{\infty} A_n \cos n\omega t \qquad (2.22)$$

Fourier Sine Series

If $y(t)$ is odd, its Fourier series will contain only sine terms

$$y(t) = \sum_{n=1}^{\infty} B_n \sin \frac{2n\pi t}{T} = \sum_{n=1}^{\infty} B_n \sin n\omega t \qquad (2.23)$$

Functions that are neither even nor odd result in a Fourier series that contains both sine and cosine terms.

EXAMPLE 2.3

Determine the Fourier series that represents the function shown in Figure 2.15.

KNOWN

$T = 10\,(-5\,\text{to}\,+5)$
$A_0 = 0$

FIND

The Fourier coefficients $A_1, A_2, \ldots$ and $B_1, B_2, \ldots$

SOLUTION

Since the function shown in Figure 2.15 is odd, the Fourier series will contain only sine terms [see equation (2.23)]:

$$y(t) = \sum_{n=1}^{\infty} B_n \sin \frac{2n\pi t}{T}$$

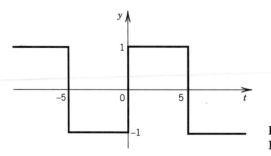

Figure 2.15 Function represented by a Fourier series in Example 2.3.

where

$$B_n = \frac{2}{10} \left[\int_{-5}^{0} (-1) \sin\left(\frac{2n\pi t}{10}\right) dt + \int_{0}^{5} (1) \sin\left(\frac{2n\pi t}{10}\right) dt \right]$$

$$B_n = \frac{2}{10} \left\{ \left[\frac{10}{2n\pi} \cos\left(\frac{2n\pi t}{10}\right) \right]_{-5}^{0} + \left[\frac{-10}{2n\pi} \cos\left(\frac{2n\pi t}{10}\right) \right]_{0}^{5} \right\}$$

$$B_n = \frac{2}{10} \left\{ \frac{10}{2n\pi} [1 - \cos(-n\pi) - \cos(n\pi) + 1] \right\}$$

The reader can verify that all $A_n = 0$ for this function. For even values of n, B_n is identically zero and for odd values of n,

$$B_n = \frac{4}{n\pi}$$

The resulting Fourier series is then

$$y(t) = \frac{4}{\pi} \sin\frac{2\pi}{10}t + \frac{4}{3\pi} \sin\frac{6\pi}{10}t + \frac{4}{5\pi} \sin\frac{10\pi}{10}t + \cdots$$

Note that the fundamental frequency is $\omega = 2\pi/10$ rad/s and the subsequent terms are the odd-numbered harmonics of ω.

COMMENT

Consider the function given by

$$y(t) = 1.0 \quad 0 < t < 5$$

We may represent $y(t)$ by a Fourier series if we extend the function beyond the specified range (0–5), either as an even periodic extension or as an odd periodic extension. (Because we are interested only in the Fourier series representing the function over the range $0 < t < 5$, we can impose any behavior outside of that domain that helps to generate the Fourier coefficients!) Let's choose an odd periodic extension of $y(t)$; the resulting function remains identical to the function shown in Figure 2.15.

EXAMPLE 2.4

Find the Fourier coefficients of the periodic function

$$y(t) = -5 \quad \text{when} \ -\pi < t < 0$$
$$= +5 \quad \text{when} \quad 0 < t < \pi$$

and $y(t + 2\pi) = y(t)$. Plot the resulting first four partial sums for the Fourier series.

KNOWN

Function $y(t)$ over the range $-\pi$ to π

FIND

Coefficients A_n and B_n

SOLUTION

The function as stated is periodic, having a period of $T = 2\pi$ (i.e., $\omega = 1$ rad/s), and is identical in form to the odd function examined in Example 2.3. Since this function is also odd, the Fourier series contains only sine terms, and

$$B_n = \frac{1}{\pi} \int_{-\pi}^{\pi} y(t) \sin n\omega t\, dt = \frac{1}{\pi} \left[\int_{-\pi}^{\pi} (-5) \sin nt\, dt + \int_{-\pi}^{\pi} (5) \sin nt\ dt \right]$$

which yields upon integration

$$\frac{1}{\pi} \left\{ \left[(5) \frac{\cos nt}{n} \right]_{-\pi}^{0} - \left[(5) \frac{\cos nt}{n} \right]_{0}^{\pi} \right\}$$

Thus,

$$B_n = \frac{10}{n\pi} (1 - \cos n\pi)$$

which is zero for even n, and $20/n\pi$ for odd values of n. The Fourier series can then be written[3]

$$y(t) = \frac{20}{\pi} \left(\sin t + \frac{1}{3} \sin 3t + \frac{1}{5} \sin 5t + \cdots \right)$$

Figure 2.16 shows the first four partial sums of this function, as they compare to the function they represent. Note that at the points of discontinuity of the function, the Fourier series takes on the arithmetic mean of the two function values.

COMMENT

As the number of terms in the partial sum increases, the Fourier series approximation of the square wave function becomes even better. For each additional term in the series, the number of "humps" in the approximate function in each half-cycle corresponds to the number of terms included in the partial sum.

The program file *FourCoef* on the companion software disk illustrates the behavior of partial sums for several waveforms.

EXAMPLE 2.5

As an example of interpreting the frequency content of a given signal, consider the output voltage from a rectifier. A rectifier functions to "flip" the negative half of an

---

[3] If we assume that the sum of this series most accurately represents the function y at $t = \pi/2$. then

$$y \left(\frac{\pi}{2} \right) = 5 = \frac{20}{\pi} \left(1 - \frac{1}{3} + \frac{1}{5} - + \cdots \right)$$

or

$$\frac{\pi}{4} = 1 - \frac{1}{3} + \frac{1}{5} - \frac{1}{7} + - \cdots$$

This series approximation of π was first obtained by Gottfried Wilhelm Leibniz (1646–1716) in 1673 from geometrical reasoning.

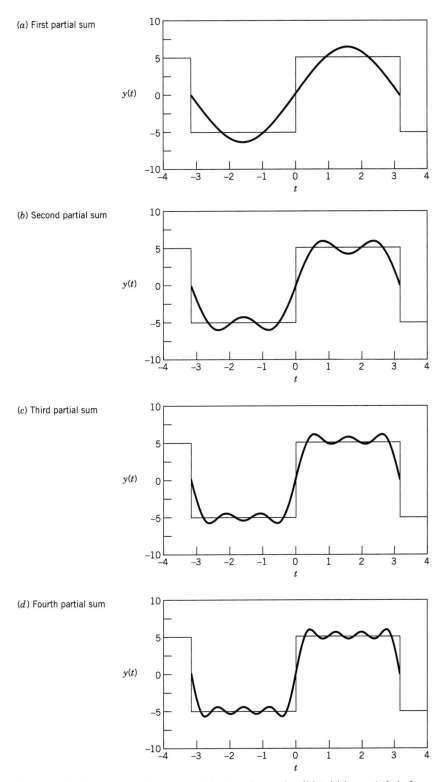

Figure 2.16 First four partial sums of the Fourier series $(20/\pi)(\sin t + 1/3\sin 3t + 1/5\sin 5t + \cdots)$ in comparison with the exact waveform.

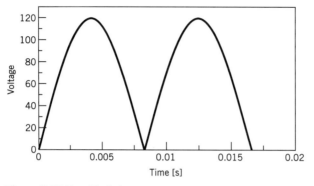

Figure 2.17 Rectified sine wave.

alternating current (ac) into the positive half-plane, resulting in a signal that appears as shown in Figure 2.17. For the ac signal the voltage is given by

$$E(t) = 120 \sin 120\pi t$$

The period of the signal is 1/60 s, and the frequency is 60 Hz.

KNOWN

The rectified signal can be expressed as

$$E(t) = |120 \sin 120\pi t|$$

FIND

The frequency content of this signal as determined from a Fourier series analysis

SOLUTION

The frequency content of this signal can be determined by expanding the function in a Fourier series. The coefficients may be determined by using the Euler formulas, keeping in mind that the rectified sine wave is an even function. The coefficient A_0 is determined from

$$A_0 = 2\left[\frac{1}{T}\int_0^{T/2} y(t)\,dt\right] = \frac{2}{1/60}\int_0^{1/120} 120 \sin 120\pi t\,dt$$

with the result that

$$A_0 = \frac{2 \times 60 \times 2}{\pi} = 76.4$$

The remaining coefficients in the Fourier series may be expressed as

$$A_n = \frac{4}{T}\int_0^{T/2} y(t) \cos\frac{2n\pi t}{T}\,dt$$

$$= \frac{4}{1/60}\int_0^{1/120} 120 \sin 120\pi t \cos 120 n\pi t\,dt$$

For values of n that are odd, the coefficient A_n is identically zero. For values of n that

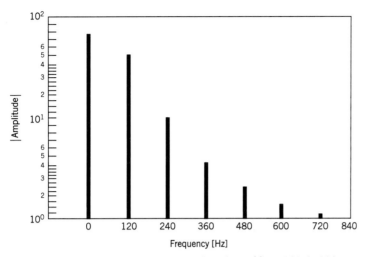

Figure 2.18 Frequency content of the function $y(t) = |120 \sin 120\pi t|$.

are even, the result is

$$A_n = \frac{120}{\pi} \left(\frac{-2}{n-1} + \frac{2}{n+1} \right) \tag{2.24}$$

The Fourier series for the function $|120 \sin 120\pi t|$ is

$$76.4 - 50.93 \cos 240\pi t$$
$$- 10.10 \cos 480\pi t - 4.37 \cos 720\pi t \dots$$

Figure 2.18 shows the frequency content of the rectified signal, based on a Fourier series expansion.

COMMENT

The frequency content of a signal is determined by examining the amplitude of the various frequency components that are present in the signal. For a periodic mathematical function, these frequency contributions can be illustrated by expanding the function in a Fourier series and plotting the amplitudes of the contributing sine and cosine terms.

 On the companion software disk, several programs exist (*FourCoef, FunSpect, DataSpect*) to help the reader explore the concept of the superposition of simple periodic signals to create a more complex signal. The software allows you to create your own functions and to explore their frequency and amplitude content.

2.5   FOURIER TRANSFORM AND THE FREQUENCY SPECTRUM

The previous discussion of Fourier analysis demonstrates that an arbitrary, but known, function can be expressed as a series of sines and cosines known as a Fourier series. The coefficients of the Fourier series specify the amplitudes of the sines and cosines, each having a specific frequency. Unfortunately, in most practical measurement applications the input signal may not be known in functional form. Therefore, although the theory of Fourier analysis demonstrates that any function can be expressed as a Fourier series, the analysis presented so far has not provided a specific technique for analyzing measured signals. Such a technique for the decomposition of a measured dynamic signal in terms of amplitude and frequency is now described.

Recall that the dynamic portion of a signal of arbitrary period can be described from equation (2.17).

$$A_n = \frac{2}{T} \int_{-T/2}^{T/2} y(t) \cos n\omega t \, dt$$

$$B_n = \frac{2}{T} \int_{-T/2}^{T/2} y(t) \sin n\omega t \, dt$$

(2.25)

where the amplitudes A_n and B_n correspond to the nth frequency of a Fourier series.

If we consider the period of the function to approach infinity, we can eliminate the constraint on Fourier analysis that the signal be a periodic waveform. In the limit as T approaches infinity the Fourier series becomes an integral. The spacing between frequency components becomes infinitesimal. This means that coefficients A_n and B_n become continuous functions of frequency and can be expressed as $A(\omega)$ and $B(\omega)$, where

$$A(\omega) = \int_{-\infty}^{\infty} y(t) \cos \omega t \, dt$$

$$B(\omega) = \int_{-\infty}^{\infty} y(t) \sin \omega t \, dt$$

(2.26)

The Fourier coefficients of $A(\omega)$ and $B(\omega)$ are known as the components of the Fourier transform of $y(t)$. To develop the Fourier transform, consider the complex number defined as

$$Y(\omega) \equiv A(\omega) - iB(\omega) \tag{2.27}$$

where $i = \sqrt{-1}$. Then from equations (2.26) it follows directly that

$$Y(\omega) = \int_{-\infty}^{\infty} y(t) \, (\cos \omega t - i \, \sin \omega t) \, dt \tag{2.28}$$

Introducing the identity

$$e^{-i\theta} = \cos \theta - i \, \sin \theta$$

leads to

$$Y(\omega) = \int_{-\infty}^{\infty} y(t) e^{i\omega t} \, dt \tag{2.29}$$

Recalling from equation (2.9) that the cyclical frequency, f, in hertz, is related to the circular frequency and its period by

$$f = \omega/2\pi = 1/T$$

we can rewrite equation (2.29) as

$$Y(f) = \int_{-\infty}^{\infty} y(t) e^{-i2\pi f t} \, dt \tag{2.30}$$

Equation (2.29) or (2.30) provides the two-sided *Fourier transform* of $y(t)$. The significance of the Fourier transform, $Y(f)$, is that it describes the signal as a continuous function of frequency. If $y(t)$ is known or measured, then its Fourier transform will provide the amplitude-frequency properties of the signal that otherwise are not readily apparent in its time-based form. This property is analogous to the optical properties displayed by the prism in Figure 2.9.

If $Y(f)$ is known or measured, we can recover the signal $y(t)$ from

$$y(t) = \frac{1}{2\pi} \int_{-\infty}^{\infty} Y(f)e^{i2\pi ft}\, df \tag{2.31}$$

Equation (2.31) describes the *inverse Fourier transform* of $Y(f)$. It suggests that given the amplitude-frequency properties of a signal we can reconstruct the original signal $y(t)$.

The Fourier transform is a complex number having a magnitude and a phase

$$Y(f) = |Y(f)|e^{i\phi(f)} = A(f) - iB(f) \tag{2.32}$$

The magnitude of $Y(f)$ is given by

$$|Y(f)| = \sqrt{\mathrm{Re}[Y(f)]^2 + \mathrm{Im}[Y(f)]^2} \tag{2.33}$$

and the phase by

$$\phi(f) = \tan^{-1}\frac{\mathrm{Im}[Y(f)]}{\mathrm{Re}[Y(f)]} \tag{2.34}$$

As noted earlier, the Fourier coefficients are related to cosine and sine terms. Then the amplitude of $y(t)$ can be expressed by its amplitude spectrum

$$C(f) = \sqrt{A(f)^2 + B(f)^2} \tag{2.35}$$

and its phase shift by

$$\phi(f) = \tan^{-1}\frac{B(f)}{A(f)} \tag{2.36}$$

Thus the introduction of the Fourier transform has provided a method to decompose a measured signal $y(t)$ into its amplitude-frequency components. Later we will see how important this method is, particularly when digital sampling is used to measure and interpret an analog signal.

A variation of the amplitude spectrum is the power spectrum, which is given by $C(f)^2/2$. Further details concerning the properties of the Fourier transform and spectrum functions can be found in [2, 3]. An excellent historical account and discussion of the wide-ranging applications is found in [4].

Discrete Fourier Transform

As a practical matter, it is likely that if $y(t)$ is measured and recorded then it will be stored in the form of a discrete time series. A data-acquisition system with digital computer is the most common approach for this. Data are acquired over a finite period of time rather than the mathematically convenient infinite period of time. A discrete data set containing N values representing a time interval from 0 to t_f will accurately represent the signal, provided that the measuring period has been properly chosen and is sufficiently long. We will deal with the details for such period selection in a discussion on sampling concepts in Chapter 7. The preceding analysis will now be extended to accommodate a discrete series.

Consider the time-dependent portion of the signal $y(t)$ that is measured N times at equally spaced time intervals δt. In this case, the continuous signal $y(t)$ will be replaced by the discrete time signal given by $y(r\delta t)$ for $r = 1, 2, \ldots, N$. In effect, the relationship between $y(t)$ and $\{y(r\delta t)\}$ is described by a set of impulses of an amplitude determined by the value of $y(t)$ at each time step $r\delta t$. This transformation from a continuous to discrete time signal is described by

$$\{y(r\delta t)\} = y(t)\boldsymbol{\delta}(t - r\delta t) \quad r = 1, 2, \ldots, N \tag{2.37}$$

where $\boldsymbol{\delta}(t - r\delta t)$ is the delayed unit impulse function and $\{y(r\delta t)\}$ refers to the discrete data set given by $y(r\delta t)$ for $r = 1, 2, \ldots, N$.

An approximation to the Fourier transform integral of equation (2.30) for use on a discrete data set is the discrete Fourier transform (DFT). The DFT is given by

$$Y(f_k) = \frac{2}{N} \sum_{r=1}^{N} y(r\delta t)e^{-i2\pi rk/N} \quad k = 1, 2, \ldots, \frac{N}{2} \tag{2.38}$$

where $f_k = k\delta f$ with $k = 1, 2, \ldots, N/2$ and with $\delta f = 1/N\delta t$. Here, δf is the frequency resolution of the DFT with each $Y(f_k)$ spaced in frequency increments of δf. In developing equation (2.38) from (2.30), t was replaced by $r\delta t$ and f was replaced by $k/N\delta t$. The factor $2/N$ scales the transform when it is obtained from a data set of finite length.

The DFT as expressed by equation (2.38) yields $N/2$ discrete values of the Fourier transform of $\{y(r\delta t)\}$. This is the so-called one-sided or half-transform, as it assumes that the data set is one sided, extending from $t = 0$ to t_f, and it returns only positive-valued frequencies.

Equation (2.38) performs the numerical integration required by the Fourier integral. Equations (2.35), (2.36), and (2.38) demonstrate that the application of the DFT on the discrete series of data, $y(r\delta t)$, permits the decomposition of the discrete data in terms of frequency and amplitude content. Hence, with the use of this method, a measured discrete signal of unknown functional form can be reconstructed as a Fourier series by using Fourier transform techniques.

Software for computing the Fourier transform of a discrete signal is provided on the companion software disk to this text. The time required to compute directly the DFT algorithm described increases at a rate that is proportional to N^2. This makes it inefficient for use with data sets of large N. A fast algorithm for computing the DFT, known as the *fast Fourier transform* (FFT), was developed by Cooley and Tukey [5]. This method is widely available, is the basis of most Fourier analysis software packages, and is the version used on the companion software disk. The FFT algorithm is discussed in most advanced texts on signal analysis (e.g., [6, 7]). The accuracy of discrete Fourier analysis depends on the frequency content[4] of $y(t)$ and on the Fourier transform frequency resolution. An extensive discussion of these interrelated parameters is given in Chapter 7.

EXAMPLE 2.6

Convert the continuous signal described by $y(t) = 10 \sin 2\pi t$ V into a discrete set of eight numbers, using a time increment of 0.125 s.

---

[4]The value of $1/\delta t$ must be more than twice the highest frequency contained in $y(t)$.

Table 2.2 Discrete Data Set for $y(t) = 10 \sin 2\pi t$

r	$y(r\,\delta t)$	r	$y(r\,\delta t)$
1	7.071	5	−7.071
2	10.000	6	−10.000
3	7.071	7	−7.071
4	0.000	8	0.000

KNOWN

The signal has the form $y(t) = C_1 \sin 2\pi f_1 t$,

where

$$f_1 = \omega_1/2\pi = 1 \text{ Hz}$$
$$C(f_1 = 1 \text{ Hz}) = 10 \text{ V}$$
$$\phi(f_1) = 0$$
$$\delta t = 0.125 \text{ s}$$
$$N = 8$$

FIND

$\{y(r\delta t)\}$

SOLUTION

The measurement of $y(t)$ every 0.125 s produces the discrete data set $\{y(r\delta t)\}$ given in Table 2.2. The measurement produces one complete period of the signal with a signal duration of $N\delta t = 1$ s. The signal and its discrete representation as a series of impulses in a time domain are plotted in Figure 2.19. See Example 2.7 for more on how to create this discrete series by using common software.

EXAMPLE 2.7

Estimate the amplitude spectrum of the discrete data set in Example 2.6.

KNOWN

Discrete data set $\{y(r\delta t)\}$ of Table 2.2
$\delta t = 0.125$ s
$N = 8$

FIND

$C(f)$

SOLUTION

 We can either use the Fourier transform (file *FunSpect*) software provided on the companion software disk or a spreadsheet program to find $C(f)$ (see the Comment for details). For $N = 8$, a one-sided Fourier transform algorithm will return $N/2$ values for $C(f)$, with each successive amplitude corresponding to a frequency spaced

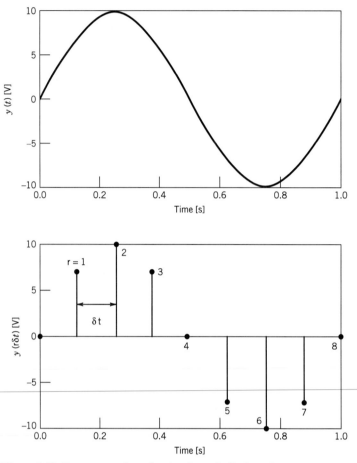

Figure 2.19 Representation of a simple periodic function as a discrete signal.

at intervals of $1/N\delta t = 0.125$ Hz. The amplitude spectrum is shown in Figure 2.20 and has a spike of 10 V centered at a frequency of 1 Hz.

COMMENT

The discrete series and the amplitude spectrum are easily reproduced by using either the program file called *FunSpect* on the companion software disk (which uses

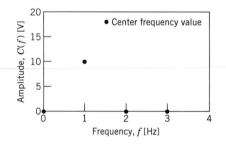

Figure 2.20 Amplitude as a function of frequency for a discrete representation of $10\sin 2\pi t$ resulting from a discrete Fourier transform algorithm.

Matlab) or by using spreadsheet software. The Fourier analysis capability of this software can also be used on an existing data set, such as with the program file called *DataSpect*. The following discussion of Fourier Analysis using a spreadsheet program makes specific reference to procedures and commands from Microsoft® Excel*; similar functions are available in other spreadsheet programs. When a spreadsheet is used, the $N = 8$ data point sequence $\{y(r\delta t)\}$ is created as in Column 3. Under Tools/Data Analysis, select Fourier Analysis. At the prompt, enter the cells containing $\{y(r\delta t)\}$ and define the cell destination. The analysis executes the DFT of equation (2.38) by using the FFT algorithm and it returns N complex numbers, the Fourier coefficients $Y(f)$ of equation (2.32). The analysis returns the two-sided transform [as in equation (2.30)], which contains N coefficients. However, a finite data set is one sided, so we are interested in only the first $N/2$ Fourier coefficients, shown in Column 4 (the second $N/2$ coefficients just mirror the first $N/2$). To find the coefficient magnitude as given by equation (2.33), compute or use the function IMABS $(=\sqrt{A^2 + B^2})$ on each of the first $N/2$ coefficients, and then scale each coefficient magnitude by dividing by $N/2$. The $N/2$ scaled coefficients (Column 5) now represent the discrete amplitudes corresponding to $N/2$ discrete frequencies (Column 6) extending from $f = 0$ to $(N/2 - 1)/N\delta t$ Hz, with each frequency separated by $1/N\delta t$.

Spreadsheet

			Column		
1	2	3	4	5	6
r	$t(s)$	$y(r\delta t)$	$Y(f) = A - Bi$	$C(f)$	$f(\text{Hz})$
1	0.125	7.07	0	0	0
2	0.25	10	$28.28 - 28.28i$	10	1
3	0.375	7.07	0	0	2
4	0.5	0	0	0	3
5	0.625	−7.07			
6	0.75	−10			
7	0.875	−7.07			
8	1	0			

The software on the companion disk performs these same operations in Matlab[5] as follows:

$t = 1/8 : 1/8 : 1;$ defines time from 0.125 s to 1 s in increments of 0.125 s

$y = 10^* \sin(2^* \text{pi}^* t);$ creates the discrete time series of $N = 8$

$y\text{coef} = \text{fft}(y);$ performs the Fourier analysis; returns N coefficients

$c = \text{coef}/4$ divides by $N/2$ to scale and determine the magnitudes

When signal frequency content is not known prior to conversion to a discrete signal, it is necessary to experiment with the parameters of frequency resolution, N and δt, to obtain an unambiguous representation of the signal. Techniques for this will be explored later in Chapter 7.

---

*Microsoft® and Excel are either registered trademarks or trademarks of Microsoft Corporation in the United States and/or other countries

[5]Matlab is a registered trademark of Mathworks, Inc.

2.6 SUMMARY

This chapter has provided a fundamental basis for the description of signals. The capabilities of a measurement system can be properly specified only if the nature of the input signal is known. The descriptions of general classes of input signals will be seen, in Chapter 3, to allow universal descriptions of measurement system dynamic behavior.

Any signal can be represented by a static magnitude and a series of varying frequencies and amplitudes. As such, measurement system selection and design must consider the frequency content of the input signals the system is intended to measure. Fourier analysis was introduced to allow a precise definition of the frequencies and the phase relationships among various frequency components within a particular signal. In Chapter 7, it will be shown as a tool for the accurate interpretation of discrete signals.

In conclusion, signal characteristics form an important basis for the selection of measurement systems and the interpretation of measurement system output. In Chapter 3 these ideas are combined with the concept of a generalized set of measurement system behaviors. The combination of generalized measurement system behavior and generalized descriptions of input waveforms provides for an understanding of a wide range of instruments and measurement systems.

REFERENCES

1. Monforte, J., The digital reproduction of sound, *Scientific American*, 251(6):78, 1984.
2. Bracewell, R. N., *The Fourier Transform and Its Applications*, 2d ed., McGraw-Hill, New York, 1986.
3. Champeney, D. C., *Fourier Transforms and Their Physical Applications*, Academic, London, 1973.
4. Bracewell, R. N., The Fourier transform, *Scientific American*, 260(6):86, 1989.
5. Cooley, J. W., and Tukey, J. W., *Math Computation* 19:207, April 1965 (see also Special Issue on the fast Fourier transform, *IEEE Transactions on Audio and Electroacoustics* AU-2, June 1967).
6. Bendat, J. S., and Piersol, A. G., *Measurement and Analysis of Random Data*, Wiley, New York, 1966 (see also Bendat, J. S., and Piersol, A. G., *Engineering Applications of Correlation and Spectral Analysis*, Wiley, New York, 1980).
7. Cochran, W. T., et al., What is the fast Fourier transform?, *Proceedings of the IEEE* 55(10):1664, 1967.

Suggested Reading

Halliday, D., and Resnick, R., *Fundamentals of Physics*, 5th ed., Wiley, New York, 1999.
Kreyszig, E., *Advanced Engineering Mathematics*, 8th ed., Wiley, New York, 1998.

NOMENCLATURE

f	frequency, in Hz $[t^{-1}]$	N	total number of discrete data points
k	spring constant $[m\,t^{-2}]$; integer	T	period $[t]$
m	mass $[m]$	U	unit step function
t	time $[t]$	Y	Fourier transform
y	dependent variable	α	angle [rad]
y_m	discrete data points	β	angle [rad]
A	amplitude	δf	frequency resolution $[t^{-1}]$
B	amplitude	δt	sample time increment $[t]$
C	amplitude	ϕ	phase angle [rad]
F	force $[m\,l\,t^{-2}]$	ω	circular frequency, in rad/s $[t^{-1}]$

PROBLEMS

2.1 Define the term signal as it relates to measurement systems. Provide two examples of static and dynamic input signals to measurement systems.

2.2 List the important characteristics of signals and define each.

2.3 Determine the average and rms values for the function

$$y(t) = 30 + 2\cos 6\pi t$$

over the time periods (a) 0–0.1 s, (b) 0.4–0.5 s, (c) 0–0.33 s, and (d) 0–20 s. Comment on the nature and meaning of the results in terms of an analysis of dynamic signals.

2.4 The following values are obtained by sampling two time-varying signals once every 0.4 s:

t	$y_1(t)$	$y_2(t)$	t	$y_1(t)$	$y_2(t)$
0.0	0	0			
0.4	11.76	15.29	2.4	−11.76	−15.29
0.8	19.02	24.73	2.8	−19.02	−24.73
1.2	19.02	24.73	3.2	−19.02	−24.73
1.6	11.76	15.29	3.6	−11.76	−15.29
2.0	0	0	4.0	0	0

Determine the mean and the rms values for this discrete data. Discuss the significance of the rms value in distinguishing these signals.

2.5 A *moving average* is an averaging technique that can be applied to an analog, discrete time, or digital signal. A moving average is based on the concept of windowing, as illustrated in Figure 2.21. That portion of the signal that lies inside the window is averaged and the average values plotted as a function of time as the window moves across the signal. A 10-point moving average of the signal in Figure 2.21 is plotted in Figure 2.22.

a. Discuss the effects of employing a moving average on the signal depicted in Figure 2.21.
b. Develop a computer-based algorithm for computing a moving average; determine the effect of the width of the averaging window on the signal described by

$$y(t) = \sin 5t + \cos 11t$$

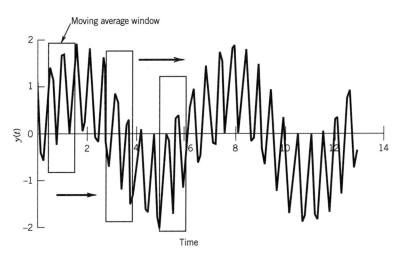

Figure 2.21 Moving average and windowing.

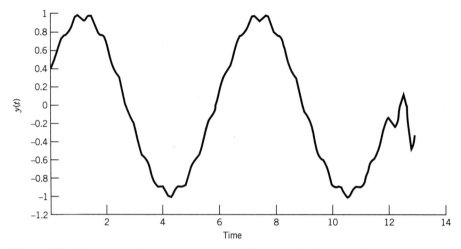

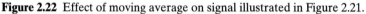

Figure 2.22 Effect of moving average on signal illustrated in Figure 2.21.

This signal should be represented as a discrete time signal by computing the value of the function at equally spaced time intervals. An appropriate time interval for this signal would be 0.05 s. Examine the signal with averaging windows of 4 and 30 points.

2.6 Determine the value of the spring constant that would result in a spring-mass system that would execute one complete cycle of oscillation every 2.7 s, for a mass of 0.5 kg. What natural frequency does this system exhibit in radians/second?

2.7 A spring with $k = 5000$ N/cm supports a mass of 1 kg. Determine the natural frequency of this system in radians/second and hertz.

2.8 For the following sine and cosine functions, determine the period, the frequency in hertz, and the circular frequency in radians/second (note: t represents time in seconds).

    a. $\sin 2\pi t/5$
    b. $5\cos 20t$
    c. $\sin 3n\pi t$ for $n = 1$ to ∞

2.9 Express the following function in terms of a cosine term only:

$$y(t) = 5\sin 4t + 3\cos 4t$$

2.10 Express the function

$$y(t) = 4\sin 2\pi t + 15\cos 2\pi t$$

in terms of (a) a cosine term only and (b) a sine term only.

2.11 Express the Fourier series given by

$$y(t) = \sum_{n=1}^{\infty} \frac{2\pi n}{6}\sin n\pi t + \frac{4\pi n}{6}\cos n\pi t$$

using only cosine terms.

2.12 If T is a period of $y(x)$, show that nT for $n = 2, 3, \ldots$ is a period of that function.

2.13 The nth partial sum of a Fourier series is defined as

$$A_0 + A_1\cos\omega_1 t + B_1\sin\omega_1 t + \cdots + A_n\cos\omega_n t + B_n\sin\omega_n t$$

For the third partial sum of the Fourier series given by

$$y(t) = \sum_{n=1}^{\infty} \frac{3n}{2}\sin nt + \frac{5n}{3}\cos nt$$

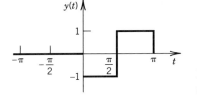

Figure 2.23 Function to be expanded in a Fourier series in Problem 2.15.

  a. What is the fundamental frequency and the associated period?
  b. Express this partial sum as cosine terms only.

2.14 For the Fourier series given by

$$y(t) = 4 + \sum_{n=1}^{\infty} \frac{2n\pi}{10} \cos \frac{n\pi}{4}t + \frac{120n\pi}{30} \sin \frac{n\pi}{4}t$$

where t is time in seconds:

  a. What is the fundamental frequency in hertz and radians/second?
  b. What is the period T associated with the fundamental frequency?
  c. Express this Fourier series as an infinite series containing sine terms only.

2.15 Find the Fourier series of the function shown in Figure 2.23, assuming the function has a period of 2π. Plot an accurate graph of the first three partial sums of the resulting Fourier series.

2.16 Determine the Fourier series for the function

$$y(t) = t \quad -5 < t < 5$$

by expanding the function as an odd periodic function with a period of 10 units, as shown in Figure 2.24. Plot the first, second, and third partial sums of this Fourier series.

2.17 a. Show that $y(t) = t^2(-\pi < t < \pi)$, $y(t + 2\pi) = y(t)$ has the Fourier series

$$y(t) = \frac{\pi^2}{3} - 4\left(\cos t - \frac{1}{4}\cos 2t + \frac{1}{9}\cos 3t - + \cdots \right)$$

  b. By setting $t = \pi$ in this series, show that a series approximation for π, first discovered by Euler, results as

$$\sum_{n=1}^{\infty} \frac{1}{n^2} = 1 + \frac{1}{4} + \frac{1}{9} + \frac{1}{16} + \cdots = \frac{\pi^2}{6}$$

2.18 Determine the Fourier series that represents the function $y(t)$ where

$$y(t) = t \quad \text{for } 0 < t < 1$$

and

$$y(t) = 2 - t \quad \text{for } 1 < t < 2$$

Clearly explain your choice for extending the function to make it periodic.

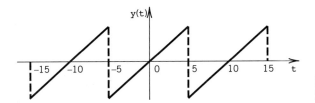

Figure 2.24 Sketch for Problem 2.16.

2.19 Classify the following signals as static or dynamic, and identify any that may be classified as periodic:

    a. $\sin 10t$ V
    b. $5 + 2\cos 2t$ m
    c. $5t$ s
    d. 2 V

2.20 A particle executes linear harmonic motion around the point $x = 0$. At time zero the particle is at the point $x = 0$ and has a velocity of 5 cm/s. The frequency of the motion is 1.0 Hz. Determine: (a) the period, (b) the amplitude of the motion, (c) the displacement as a function of time, and (d) the maximum speed.

2.21 Define the following characteristics of signals: (a) frequency content, (b) amplitude, (c) magnitude, and (d) period.

2.22 Construct an amplitude spectrum plot for the Fourier series in Problem 2.16 for $y(t) = t$. Discuss the significance of this spectrum for measurement or interpretation of this signal. Hint: The plot can be done by inspection or by using software such as *DataSpect*.

2.23 Construct an amplitude spectrum plot for the Fourier series in Problem 2.17 for $y(t) = t^2$. Discuss the significance of this spectrum for selecting a measurement system. Hint: The plot can be done by inspection or by using software.

2.24 Sketch representative waveforms of the following signals, and represent them as mathematical functions (if possible):

    a. The output signal from the thermostat on a refrigerator.
    b. The electrical signal to a spark plug in a car engine.
    c. The input to a cruise control fron an automobile.
    d. A pure musical tone (e.g., 440 Hz is an A).
    e. The note produced by a guitar string.
    f. AM and FM radio signals.

2.25 Represent the function

$$e(t) = 5\sin 31.4t + 2\sin 44t$$

as a discrete set of $N = 128$ numbers separated by a time increment of $(1/N)$. Use an appropriate algorithm to construct an amplitude spectrum from this data set. (Hint: A spreadsheet or program file *DataSpect* will handle this).

2.26 Repeat Problem 2.25 by using a data set of 256 numbers at $\delta t = (1/N)$ and $\delta t = (1/2N)$ s. Compare and discuss the results.

2.27 A particular strain sensor is mounted to an aircraft wing that is subjected to periodic wind gusts. The strain measurement system indicates a periodic strain that ranges from 3250×10^{-6} in./in. to 4150×10^{-6} in./in. at a frequency of 1 Hz. Determine:

    a. The average value of this signal.
    b. The amplitude and the frequency of this output signal when expressed as a simple periodic function.
    c. A Fourier series that represents this signal.

    Construct an amplitude spectrum plot for the output signal.

2.28 For a dynamic calibration involving a force measurement system, a known force is applied to a sensor. The force varies between 100 and 170 N at a frequency of 10 rad/s. State the average (static) value of the input signal, its amplitude, and its frequency. Assuming that the signal may be represented by a simple periodic waveform, express the signal as a Fourier series.

2.29 A displacement sensor is placed on a dynamic calibration rig known as a shaker. This device produces a known periodic displacement that serves as the input to the sensor. If the known displacement is set to vary between 2 and 5 mm at a frequency of 100 Hz, express the input signal as a Fourier series. Plot the signal in the time domain, and construct an amplitude spectrum plot.

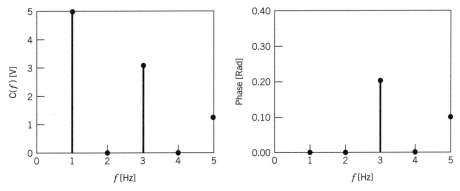

Figure 2.25 Spectrum for Problem 2.31.

2.30 Classify the following signals as completely as possible, using concepts discussed in this chapter: (a) clock face having hands, (b) Morse code, (c) musical score, as input to a musician, (d) flashing neon sign, (e) telephone conversation, and (f) fax transmission.

2.31 Describe the signal defined by the amplitude spectrum and its phase shift of Figure 2.25 in terms of its Fourier series. What is the frequency resolution of these plots? What was the sample time increment used?

2.32 For the even-functioned triangle wave signal defined by

$$y(t) = (4C/T)t + C \quad - T/2 \le t \le 0$$
$$y(t) = (-4C/T)t + C \quad 0 \le t \le T/2$$

where C is an amplitude and T is the period:

a. Show that this signal can be represented by the Fourier series

$$y(t) = \sum_{n=1}^{\infty} \frac{4C(1 - \cos n\pi)}{(\pi n)^2} \cos \frac{2\pi nt}{T}$$

b. Expand the first three nonzero terms of the series. Identify the terms containing the fundamental frequency and its harmonics.

c. Plot each of the first three terms on a separate graph. Then plot the sum of the three. For this, set $C = 1$ V and $T = 1$ s. Note: the program file *FourCoef* on the software companion disk is useful in this problem.

d. Sketch the amplitude spectrum for the first three terms of this series, first separately for each term and then combined.

2.33 Figure 2.16 illustrates how including higher-frequency terms in a Fourier series refines the accuracy of the series representation of the original function. However, measured signals often display an amplitude spectrum for which amplitudes decrease with frequency. Using Figure 2.16 as a resource, discuss the effect of high frequency, low amplitude noise on a measured signal. What signal characteristics would be unaffected? Consider both periodic and random noise.

Chapter 3

Measurement System Behavior

3.1 INTRODUCTION

Each measurement system will respond differently to different types of input signals. Therefore, a particular system may not be suitable for the measurement of certain input signals. However, a measurement system will always output information, regardless of how well (or poorly!) this information might reflect the actual input signal. Accurately interpreting the information about an input signal based on the information contained in the output signal requires understanding how the system responds to a variety of input signals. In this chapter, the concept of simulating the measurement system behavior through mathematical modeling is introduced. From this modeling, the important aspects of how a measurement system responds and how that pertains to system design and specification will be introduced and investigated.

Throughout this chapter, we will use the term "measurement system" in a generic sense. We refer either to the response of the measurement system as a whole or to the response of any component or instrument that makes up that system. Both are important and are interpreted in similar ways. Each individual stage of the measurement system will have its own response to a given input. The overall system response will be affected by the response of each stage that makes up the complete system.

3.2 GENERAL MODEL FOR A MEASUREMENT SYSTEM

As pointed out in Chapter 2, all input and output signals can be broadly classified as being static, dynamic, or some combination of the two. For a static signal, only the signal magnitude is needed to reconstruct the input signal based on the indicated output signal. Consider the measurement of the length of a board by using a ruler. Once the ruler is positioned, the indication (output) of the magnitude of length is immediately displayed since the board length (input) does not change over the time required to make the measurement. Thus, the board length represents a static input signal that is interpreted through the static magnitude output indicated by the ruler.

Dynamic Measurements

For dynamic signals, the signal amplitude, frequency, and/or general waveform become necessary to reconstruct the input signal. Since dynamic signals vary with time, the measurement system must be able to respond fast enough literally to keep

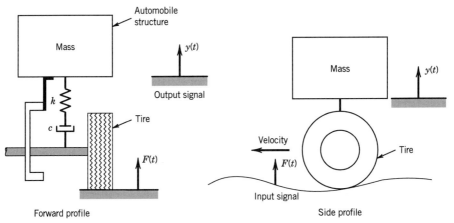

Figure 3.1 Model of an automobile suspension with input and output signals.

up with the input signal. Consider the time response of a typical bulb thermometer when measuring body temperature. The thermometer, initially at room temperature or so, is placed under the tongue. Even after several seconds, the thermometer does not indicate the expected value of body temperature and its display continually changes. What has happened? Surely body temperature is not changing. Use of the magnitude of the output signal after only several seconds might lead to false conclusions about one's health! Experience shows that within a few minutes, the correct body temperature will be indicated; so we wait. In this example, body temperature itself is a constant during the measurement, but the input signal to the thermometer is suddenly changed from room temperature to body temperature; that is, it undergoes a step change. This is a dynamic event as the thermometer (the measurement system) sees it! The thermometer must gain energy from its new environment to reach thermal equilibrium, and this takes a finite amount of time. The ability of any measuring system to follow dynamic signals is a characteristic of the design of the measuring system components. The importance of measurement system response is apparent even when the most common of measurement systems is used.

Now consider the task of assessing the ride quality of an automobile suspension system. A simplified view for one wheel of this system is shown in Figure 3.1. As a tire moves along the road, the road surface provides the time-dependent input signal, $F(t)$, to the suspension at the tire contact point. The motion sensed by the passengers, $y(t)$, is a basis for the ride quality and can be described by a waveform, which will depend on the input from the road and the behavior of the suspension. An engineer must anticipate the form of input signals and design the suspension to attain a desirable output signal.

Measurement systems play a key role in documenting ride quality. However, just as the road and car interact to provide ride quality, the input signal and the measurement system interact to form an output signal. During an actual measurement, we must deduce the input signal based on the output signal and the measurement system characteristics. Therefore, it is important to understand how a measurement system responds to different forms of input signals.

The behavior of measurement systems to a few special inputs will, for the most part, define the input–output signal relationships necessary to enable the correct interpretation of the measured signal. We will show that only a few measurement system characteristics are needed to define the system response, so that dynamic

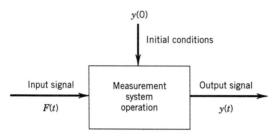

Figure 3.2 Measurement system operation on an input signal, $F(t)$, provides the output signal, $y(t)$.

calibrations can be restricted to a few specific tests. The results of these tests can be used to judge the suitability of a particular system for a particular measurement.

With the previous discussion in mind, consider that the primary task of a measurement system is to sense an input signal and to translate that information into a readily understandable and quantifiable output form. Then one can assume that a measurement system performs some mathematical operation on a sensed input. In fact, a general measurement system can be represented by a differential equation, which describes the operation that a measurement system performs on the input signal.

This concept of an input signal being operated on by the measurement system that provides an output signal is illustrated in Figure 3.2, with the mathematical operation that the measurement system performs represented within the box. For an input signal, $F(t)$, the system performs some operation that yields the output signal, $y(t)$. Then we must use $y(t)$ to infer $F(t)$. Therefore, at least a qualitative understanding of the operation that the measurement system performs is imperative to interpret the input signal correctly. We will propose a general mathematical model for a measurement system. Then, by representing a typical input signal as some function that acts as an input to the model, we can study just how the measurement system would behave by solving the model equation. In essence, we will perform the analytical equivalent of a system calibration. This information can then be used to determine those input signals for which a particular measurement system is best suited.

Measurement System Model

Consider, then, the following model of a measurement system, which consists of a general nth-order linear ordinary differential equation in terms of a general output signal, represented by variable $y(t)$, and subject to a general input signal, represented by the forcing function, $F(t)$:

$$a_n \frac{d^n y}{dt^n} + a_{n-1} \frac{d^{n-1} y}{dt^{n-1}} + \cdots + a_1 \frac{dy}{dt} + a_0 y = F(t) \tag{3.1}$$

The coefficients $a_0, a_1, a_2, \ldots, a_n$ represent physical system parameters whose properties and values will depend on the measurement system itself. The forcing function can be generalized into the mth-order form

$$F(t) = b_m \frac{d^m x}{dt} + b_{m-1} \frac{d^{m-1} x}{dt^{m-1}} + \cdots + b_0 x \quad m \le n$$

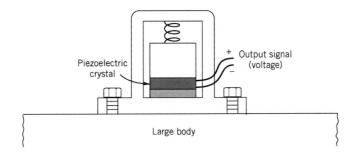

(a) Piezoelectric accelerometer attached to large body

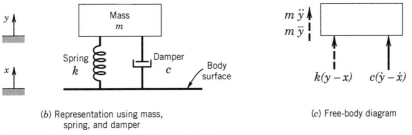

(b) Representation using mass, spring, and damper

(c) Free-body diagram

Figure 3.3 Accelerometer of Example 3.1.

where $b_0, b_1, \ldots, b_m$ also represent physical system parameters. Real measurement systems can be modeled in this manner by considering their governing system equations. These equations are generated by the application of pertinent fundamental physical laws of nature to the measurement system. Our discussion will be limited to measurement system concepts, but a general treatment of systems can be found in textbooks dedicated to that topic (e.g., [1–3]).

EXAMPLE 3.1

As an illustration, consider the seismic accelerometer depicted in Figure 3.3(a). Various configurations of this instrument are used in seismic and vibration engineering to determine the motion of large bodies to which the accelerometer is attached. Basically, as the small accelerometer mass reacts to motion it places the piezoelectric crystal into compression or tension, which causes a surface charge to develop on the crystal. The charge is proportional to the motion. As the large body moves, the mass of the accelerometer will move with an inertial response. The stiffness of the spring, k, will provide a restoring force to move the accelerometer mass back to equilibrium while internal frictional damping, c, will oppose any displacement away from equilibrium. A model of this measurement device in terms of ideal elements of stiffness, mass, and damping is given in Figure 3.3(b) and the corresponding free-body diagram in Figure 3.3(c). Let y denote the position of the small mass within the accelerometer and x denote the displacement of the body. Solving Newton's second law for the free body yields the second-order linear, ordinary differential equation,

$$m\frac{d^2 y}{dt^2} + c\frac{dy}{dt} + ky = c\frac{dx}{dt} + kx$$

Since the displacement y is the pertinent output from the accelerometer that is due to displacement x, the equation has been written such that all output terms, that is, all the y terms, are on the left side. All other terms are considered to be input signals and are placed on the right side. Comparing this to the general form for a second-order equation ($n = 2$) from equation (3.1),

$$a_2 \frac{d^2 y}{dt^2} + a_1 \frac{dy}{dt} + a_0 y = b_1 \frac{dx}{dt} + b_0 x$$

we observe that $a_2 = m$, $a_1 = b_1 = c$, $a_0 = b_0 = k$, and that the forces developed as a result of the velocity and displacement of the body become the inputs to the accelerometer. If we could anticipate the waveform of x [e.g., $x(t) = x_0 \sin \omega t$], we could simulate the measurement system response to this input by solving the equation for $y(t)$.

Fortunately, many measurement systems can be modeled by zero-, first-, or second-order linear, ordinary differential equations. More complex systems can usually be simplified to a lower order. Our intention here is to attempt to understand how systems behave and how response is closely related to the design features of a measurement system; it is not to simulate the exact system behavior. The exact input–output relationship is **always** found from calibration, but modeling will guide us in narrowing down our choice to specific instruments and determining the type, range, and specifics of calibration. Next, we will examine several special cases of equation (3.1) that identify the most important concepts of measurement system behavior.

3.3 SPECIAL CASES OF THE GENERAL SYSTEM MODEL

Zero-Order Systems

A basic model for measurement systems used to measure static signals is the zero-order system model. It is the simplest model for a measurement system and is represented by a zero-order differential equation:

$$a_0 y = F(t)$$

with $F(t) = b_0 x$. Dividing through by a_0 gives

$$y = KF(t) \tag{3.2}$$

where $K = 1/a_0$. K is called the *static sensitivity* of the system. This system property was introduced in Chapter 1 as the relation between the static output change associated with a change in input. In zero-order behavior, the system output responds to the input signal instantly. If an input signal of magnitude $F(t) = A$ were applied, the instrument would indicate KA, as specified by equation (3.2). The scale of the measuring device would be calibrated to indicate A directly.

For real systems, the zero-order system concept is used to model the measurement system response to static inputs. In fact, the zero-order concept appropriately models any system during a static calibration. When dynamic input signals are involved, the output signal can be considered valid only when it is at static equilibrium. This is because most real measurement systems possess inertial or storage

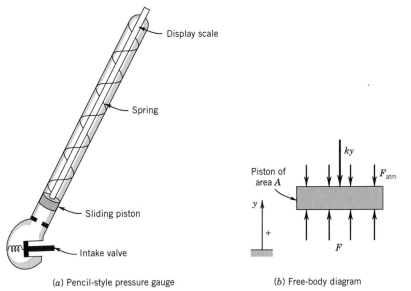

(a) Pencil-style pressure gauge (b) Free-body diagram

Figure 3.4 Pressure gauge of Example 3.2.

capabilities that require higher-order differential equations to correctly model their time-dependent behavior to dynamic input signals.

Determination of K. The static sensitivity is found from the static calibration of the measurement system. It is the slope of the calibration curve.

EXAMPLE 3.2

A pencil-type pressure gauge commonly used to measure tire pressure can be modeled at static equilibrium by considering the force balance on the gauge sensor, a piston that slides up and down a cylinder. In Figure 3.4(*a*), the piston motion is restrained by an internal spring of stiffness, k, so that at static equilibrium the absolute pressure force, F, bearing on the piston equals the force exerted on the piston by the spring, F_s, plus the atmospheric pressure force, F_{atm}. In this manner, the transduction of pressure into displacement occurs. Considering the piston free body at static equilibrium in Figure 3.4(*b*), the static force balance, $\Sigma \mathbf{F} = 0$, yields

$$ky = F - F_{atm}$$

where y is measured relative to some static reference position marked as zero on the output display. Pressure is simply the force acting inward over the piston surface area, A. Thus dividing through by area provides the zero-order response equation between output displacement and input pressure

$$y = (A/k)(p - p_{atm})$$

The term $p - p_{atm}$ represents the pressure relative to atmospheric pressure. It is the pressure indicated by this gauge. A direct comparison with equation (3.2) implies that the input pressure is translated into piston displacement through the static sensitivity, $K = A/k$. The system operates on pressure so as to bring about a relative displacement of the piston, the magnitude of which is used to indicate the pressure.

The spring stiffness and piston area affect the magnitude of this displacement, factors considered in its design. The exact static input–output relationship is found through calibration of the gauge. Since such elements as piston inertia and frictional dissipation were not considered, this model would not be appropriate for studying the dynamic response of the piston.

First-Order Systems

Measurement systems that contain storage elements cannot respond instantaneously to changes in input. The bulb thermometer discussed in Section 3.2 is a good example. The bulb exchanges energy with its environment until the two are at the same temperature, storing energy during the exchange. The temperature of the bulb sensor will change with time until this equilibrium is reached, which accounts physically for its less than immediate response. The rate at which temperature changes with time can be modeled with a first-order derivative, and the thermometer behavior can be modeled as a first-order equation. In general, systems with a storage or dissipative capability but negligible inertial forces may be modeled by using a first-order differential equation of the form

$$a_1 \dot{y} + a_0 y = F(t) \tag{3.3}$$

with $\dot{y} = dy/dt$. Dividing through by a_0 gives

$$\tau \dot{y} + y = KF(t) \tag{3.4}$$

where $\tau = a_1/a_0$; τ is called the *time constant* of the system. Regardless of the physical dimensions of a_0 and a_1, their ratio will always have the dimensions of time. The time constant provides a measure of the speed of system response, and as such is an important specification in measuring dynamic input signals. To explore this concept more fully, consider the response of the general first-order system to the following two forms of an input signal: the step function and the simple periodic function.

Step Function Input

The step function, $AU(t)$, is defined as

$$AU(t) = 0 \quad t \leq 0^-$$

$$AU(t) = A \quad t \geq 0^+$$

where A is the amplitude of the step function and $U(t)$ is defined as in Figure 3.5 as the unit step function. Physically, this function describes a sudden change in the input signal from a constant value of one magnitude to a constant value of some other magnitude, such as a sudden change in temperature, pressure, or loading. Application of a step function input to a measurement system provides information about the speed at which a system responds to a change in input signal. To illustrate this, let us apply a step function as an input to the general first-order system.

Setting $F(t) = AU(t)$ in equation (3.4) gives

$$\tau \dot{y} + y = KAU(t) = KF(t)$$

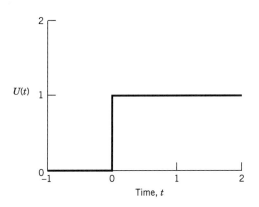

Figure 3.5 The unit step function, $U(t)$.

with an arbitrary initial condition, $y(0) = y_0$. The solution of this differential equation yields for $t \geq 0^+$

$$\underbrace{y(t)}_{\text{time response}} = \underbrace{KA}_{\text{steady response}} + \underbrace{(y_0 - KA)e^{-t/\tau}}_{\text{transient response}} \qquad (3.5)$$

The solution of the differential equation, $y(t)$, is the *time response* (or simply the *response*) of the system. Equation (3.5) describes the behavior of the system to a step change in input. This means that $y(t)$ is in fact the output indicated by the display stage of the system. It should represent the time variation of the output display of the measurement system if an actual step change were applied to the system. We have simply used mathematics to simulate this response.

The second term on the right side of equation (3.5) is known as the *transient response* of $y(t)$ since, as $t \rightarrow \infty$, the magnitude of this term eventually reduces to zero. The first term is known as the steady response since, as $t \rightarrow \infty$, the response of $y(t)$ approaches this steady value. The *steady response* is that portion of the output signal that remains after the transient response has decayed to zero.

For illustrative purposes only, let $y_0 < A$ so that the time response becomes as shown in Figure 3.6. Over time, the indicated output value rises from its initial value, at the instant the change in input is applied, to an eventual constant value, $y_\infty = KA$, at steady response. Compare this general time response to the recognized behavior of the bulb thermometer when measuring body temperature as discussed earlier. We

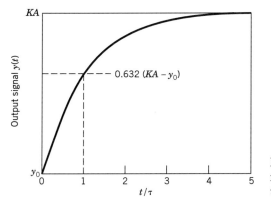

Figure 3.6 First-order system time response to a step function input: the time response, $y(t)$.

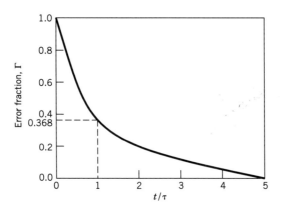

Figure 3.7 First-order system time response to a step function input: the error fraction, Γ.

see a qualitative similarity. In fact, the measurement of body temperature represents a real step function input to the thermometer, itself a real first-order measuring system.

Suppose we rewrite response equation (3.5) in a form

$$\Gamma(t) = \frac{y(t) - y_\infty}{y_0 - y_\infty} = e^{-t/\tau} \tag{3.6}$$

where the term Γ is the *error fraction* of the output signal. Equation (3.6) is plotted in Figure 3.7, where the time axis has been nondimensionalized by the time constant. It can be seen that the error fraction decreases from a value of 1 and approaches a value of 0 with increasing t/τ. At the instant just after the step change in input is introduced, $\Gamma = 1.0$ so that the indicated output from the measurement system is 100% in error. This implies that the system has responded 0% to the input change. From Figure 3.7, it is apparent that as time increases the system responds with decreasing error in its indicated value. Let the percent response of the system to a step change be given as $(1 - \Gamma) \times 100$. Then by $t = \tau$, where $\Gamma = 0.368$, the system will have responded to 63.2% of the step change. Further, when $t = 2.3\tau$ the system will have responded to 90% of the step change; by $t = 5\tau$, 99.3%. Values for the percent response and the corresponding error as functions of t/τ are summarized in Table 3.1. The time required for a system to respond to a value that is 90% of the step input, $y_\infty - y_0$, is called the *rise time* of the system, a term often cited in instrument specifications.

Table 3.1 First-Order System Response and Error Fraction

t/τ	% Response	Γ	% Error
0	0.0	1.0	100.0
1	63.2	0.368	36.8
2	86.5	0.135	13.5
2.3	90.0	0.100	10.0
3	95.0	0.050	5.0
5	99.3	0.007	0.7
∞	100.0	0.0	0.0

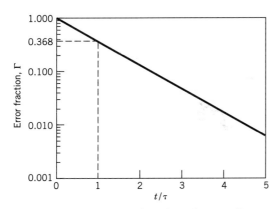

Figure 3.8 The error fraction plotted on semilog coordinates.

From this behavior, one can see that the time constant provides a measure of the speed of system response to a change in input value. A smaller time constant indicates a shorter time between the instant that an input is applied and when the system reaches an essentially steady output. The time constant is the time required for a first-order system to achieve 63.2% of the step change magnitude, $y_\infty - y_0$. The time constant is a system property.

Determination of τ**.** According to the development just discussed, the time constant of a first-order measurement system can be experimentally determined by recording the system's response to a step function input of a known magnitude. In practice, it is best to record that response from $t = 0$ until a steady response is achieved. The data can then be plotted as error fraction versus time on a semilog plot, such as in Figure 3.8. This type of plot is equivalent to the transformation

$$\ln \Gamma = 2.3 \log \Gamma = -(1/\tau)t \qquad (3.7)$$

which is of the linear form, $Y = mX + B$. A curve fit through the data will provide a good estimate of the slope, m, of the resulting plot. From equation (3.7), we see that $m = -1/\tau$, which yields the estimate for τ.

This method offers several advantages over attempting to compute τ directly from the time required to achieve 63.2% of the step change magnitude. First, real systems may deviate somewhat from perfect first-order behavior. In itself, this is fine, since our intention is to study the system behavior, not to simulate it exactly. On rectangular grid plots, such as Figure 3.7, deviations from perfect first-order behavior will not be easy to detect, but on a semilog plot such deviations are readily apparent as clear trends away from a linear curve. Modest deviations do not pose any problem. However, strong deviations provide a good indication that the system is not behaving as expected, thus requiring a closer examination of either the system operation, the step function experiment, or the assumed form of the system model. Second, the acquisition of data during the step function experiment is prone to some error (precision error) in each data point. The use of a data curve fit to determine τ utilizes all of the data over time so as to minimize the influence of an error in any one data point. Third, the method eliminates the need to accurately determine the $\Gamma = 1.0$ and 0.368 points, which are difficult to establish in practice.

EXAMPLE 3.3

Suppose a bulb thermometer originally indicating 20°C is suddenly exposed to a fluid temperature of 37°C. Develop a simple model to simulate the thermometer output response.

KNOWN

$$T(0) = 20°C$$
$$T_\infty = 37°C$$
$$F(t) = [T_\infty - T(0)]U(t)$$

ASSUMPTIONS

To keep things simple, assume the following:

No installation effects (neglect conduction and radiation effects)
Sensor mass is mass of liquid in bulb only
Uniform temperature within bulb (lumped mass)
Thermometer scale is calibrated to indicate temperature

FIND

$T(t)$

SOLUTION

Consider the energy balance developed in Figure 3.9. According to the first law of thermodynamics, the rate at which energy is exchanged between the sensor and its environment through convection, $\dot{Q}$, must be balanced by the storage of energy within the thermometer, dE/dt. This conservation of energy is written as

$$\frac{dE}{dt} = \dot{Q}$$

Energy storage in the bulb is manifested by a change in bulb temperature so that for a constant mass bulb, $dE(t)/dt = mc_v\, dT(t)/dt$. Energy exchange by convection between the bulb at $T(t)$ and an environment at T_∞ has the form $\dot{Q} = hA\Delta T$. The

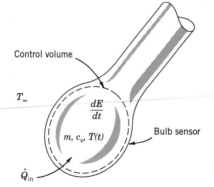

Figure 3.9 Thermometer and energy balance of Example 3.3.

first law can be written as

$$mc_v \frac{dT(t)}{dt} = hA_s[T_\infty - T(t)]$$

This equation can be written in the form

$$mc_v \frac{dT(t)}{dt} + hA_s[T(t) - T(0)] = hA_s[T_\infty - T(0)]U(t)$$
$$= hA_s F(t)$$

with initial condition $T(0)$

where

m = mass of liquid within thermometer
c_v = specific heat of liquid within thermometer
h = convection heat transfer coefficient between bulb and environment
A_s = thermometer surface area

The term hA_s controls the rate at which energy can be transferred between a fluid and a body; it is analogous to an electrical conductance. By comparison with equation (3.3), $a_0 = hA_s$, $a_1 = mc_v$, and $b_0 = hA_s$. Rewriting for $t \geq 0^+$ and simplifying yields

$$\frac{mc_v}{hA_s} \frac{dT(t)}{dt} + T(t) = T_\infty$$

From equation (3.4), this implies that the time constant and static sensitivity are

$$\tau = \frac{mc_v}{hA_s} \qquad K = \frac{hA_s}{hA_s} = 1$$

A direct comparison with equation (3.5) yields the thermometer response:

$$T(t) = T_\infty + [T(0) - T_\infty]e^{-t/\tau}$$
$$= 37 - 17e^{-t/\tau}\,[°C]$$

COMMENT

An interactive example of a first-order system based on the thermometer problem (Examples 3.3–3.5) is contained on the companion software disk in program file *FirstOrd*. The user can choose input functions and note the system response.

It is clear that the time constant, τ, of the thermometer can be reduced by decreasing its mass-to-area ratio or by increasing h (for example, this can be accomplished by stirring, shaking, or other methods that increase the velocity of the fluid around the sensor). Without modeling, such information could be ascertained only by trial and error, a time-consuming and costly method with no assurance of success. Also, it is significant that we found that the response of the temperature measurement system in this case will depend on the environmental conditions of the measurement that control h, since the magnitude of h affects the magnitude of τ. If h is not controlled during response tests (i.e., if it is an extraneous variable), ambiguous results are possible. For example, the curve of Figure 3.8 will become nonlinear and/or replications will not yield the same values for τ.

Review of this example should make it apparent that the results of a well-executed step calibration may not be indicative of an instrument's performance during a measurement if the measurement conditions will differ from those existing during the step calibration.

EXAMPLE 3.4

For the thermometer of Example 3.3 subjected to a step change in input, calculate the 90% rise time in terms of t/τ.

KNOWN

Same as Example 3.3

ASSUMPTIONS

Same as Example 3.3

FIND

90% response time in terms of t/τ

SOLUTION

The percent response of the system is given by $(1 - \Gamma) \times 100$ with the error fraction, Γ, defined by equation (3.6). From equation (3.5), we note that at $t = \tau$, the thermometer will indicate $T(\tau) = 30.75°\text{C}$, which represents only 63.2% of the step change from 20 to 37°C. The 90% rise time represents the time required for Γ to drop to a value of 0.10. Then

$$\Gamma = 0.10 = e^{-t/\tau}$$

or

$$t/\tau = 2.3$$

COMMENT

In general, a time equivalent to 2.3τ is required to achieve 90% of the applied step input for a first-order system.

EXAMPLE 3.5

A particular thermometer is subjected to a step change, such as in Example 3.3, in an experimental exercise to determine its time constant. The temperature data are recorded with time and presented in Figure 3.10. Determine the time constant for this thermometer. In the experiment, the heat transfer coefficient, h, is estimated to be 6 W/m^2 °C from heat transfer correlations.

KNOWN

Data of Figure 3.10
$h = 6$ W/m^2 °C

ASSUMPTIONS

First-order behavior
Constant properties
Model of Example 3.3

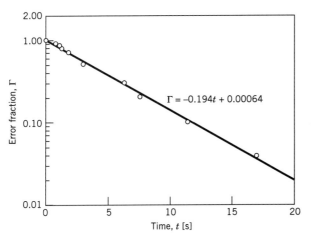

Figure 3.10 Temperature–time history of Example 3.5.

FIND

τ

SOLUTION

According to equation (3.7), the time constant should be the negative reciprocal of the slope of a curve drawn through the data of Figure 3.10. Aside from the first few data points, the data appear to follow a linear trend, indicating a nearly first-order behavior and validating our model assumption. One method of estimating τ would be simply to draw a curve through the linear portion of the data and compute a slope, m:

$$m = -\frac{1}{\tau} = \frac{2.3(\log \Gamma_2 - \log \Gamma_1)}{t_2 - t_1}$$

But a better method would be to use a linear curve fit. A first-order least-squares fit[1] yields the equation

$$2.3 \log \Gamma = (-0.194)t + 0.00064$$

With $m = -0.194/\text{s} = -1/\tau$, the time constant can be estimated to be 5.15 s for a constant $h = 6$ W/m^2 °C. $\tau = \dfrac{1}{(0.194\frac{1}{s})} = 5.15 \text{ sec}$

COMMENT

Based on the physical assumptions made, a first-order model results from a simple energy balance on this system. First-order behavior implies that the data should fit the straight line of a semilog plot. If the experimental data were to deviate significantly from first-order behavior, one would want to examine the conduct of the experiment and/or the rigor of the model assumptions relative to the real system.

---

[1] The least-squares approach to curve fitting is discussed in detail in Chapter 4.

Sine Function Input

Periodic signals are commonly encountered in many engineering processes, such as vibrating structures, vehicle suspension dynamics, reciprocating pump flows, and environmental temperatures. When periodic inputs are applied to a first-order system, the frequency of the input has an important influence on the measuring system time response, affecting the output signal. This behavior can be studied effectively by application of a simple periodic waveform as an input to the system. Consider the first-order measuring system to which an input of the form $F(t) = A \sin \omega t$ is applied for $t \geq 0^+$:

$$\tau \dot{y} + y = KA \sin \omega t$$

with initial conditions $y(0) = y_0$. The general solution to this differential equation yields the measuring system output signal, the time response to the applied input, $y(t)$:

$$y(t) = Ce^{-t/\tau} + \frac{KA}{[1 + (\omega\tau)^2]^{1/2}} \sin(\omega t - \tan^{-1} \omega\tau) \qquad (3.8)$$

where the value for C will depend on the exact value of y_0.

So what has happened? The output signal, $y(t)$, consists of a transient and a steady response. The first term on the right side is the transient response. As t increases, this term decays to zero and no longer influences the output signal. Transient response is important only during the initial period following the application of the new input. We already have information about the system transient response from the step function study, so let us focus our attention on the second term, the steady response. This term persists for as long as the periodic input is maintained. From equation (3.8), it is seen that the frequency of the steady response term remains the same as the input signal frequency, but note that the amplitude of the steady response depends on the value of the applied frequency, ω. Also, the phase angle of the periodic function has changed.

Equation (3.8) can be rewritten in a general form

$$y(t) = Ce^{-t/\tau} + B(\omega) \sin[\omega t + \phi(\omega)]$$
$$B(\omega) = \frac{KA}{[1 + (\omega\tau)^2]^{1/2}} \qquad (3.9)$$
$$\phi(\omega) = -\tan^{-1} \omega\tau$$

where $B(\omega)$ represents the amplitude of the steady response and the angle $\phi(\omega)$ is known as the *phase shift*. A relative illustration between the input signal and the system output response is given in Figure 3.11 for an arbitrary frequency and system time constant. From equation (3.9), it is seen that both B and ϕ are frequency dependent. Hence, the exact form of the output response will depend on the value of the frequency of the input signal. The steady response of any system to which a periodic input of frequency, ω, is applied is known as the *frequency response* of that system. The frequency dependence of a system affects the magnitude of amplitude B and brings about a time delay between when the input is applied and when the measuring system responds to the input signal. This time delay, β_1, is seen in the phase shift, $\phi(\omega)$, of the steady response. For a phase shift given in radians, the time delay in units of time is

$$\beta_1 = \frac{\phi(\omega)}{\omega}$$

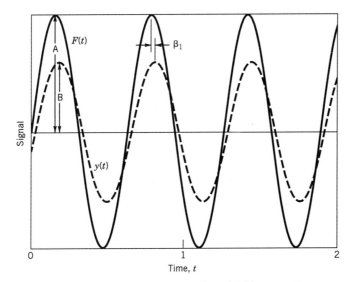

Figure 3.11 Relationship between a sinusoidal input and output: amplitude, frequency, and time lag.

that is, we can write

$$\sin(\omega t + \phi) = \sin\left[\omega\left(t + \frac{\phi(\omega)}{\omega}\right)\right] = \sin[\omega(t + \beta_1)]$$

The value for β_1 will be negative, indicating a time delay between the output and input signals.

Since equation (3.9) applies to all first-order measuring systems, the magnitude and phase shift by which the output signal differs from the input signal are predictable. Here we define the magnitude ratio, $M(\omega)$, for a first-order system subjected to a simple periodic input as

$$M(\omega) = \frac{B}{KA} = \frac{1}{[1 + (\omega\tau)^2]^{1/2}} \tag{3.10}$$

The magnitude ratio for a first-order system is plotted in Figure 3.12 and the

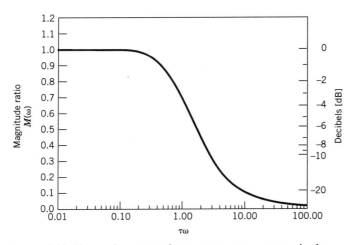

Figure 3.12 First-order system frequency response: magnitude ratio.

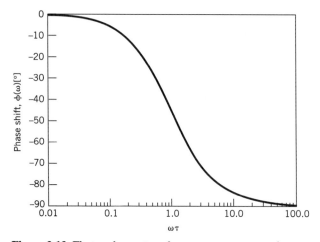

Figure 3.13 First-order system frequency response: phase shift.

corresponding phase shift is plotted in Figure 3.13. The effects of both system time constant and input signal frequency on frequency response are apparent in both figures. This behavior can be interpreted in the following manner. For those values of $\omega\tau$ for which the system responds with values of $M(\omega)$ near unity, the measurement system will transfer all or nearly all of the input signal amplitude to the output with very little time delay; that is, B will be close to KA in magnitude and $\phi(\omega)$ will be near $0°$. At large values of $\omega\tau$ the measurement system will essentially remove, that is, filter out, the frequency information of the input signal by responding with very small amplitudes, shown by small $M(\omega)$, and larger time delays.

Any combination of ω and τ will produce the same results. If one wants to measure signals with high-frequency content and transfer the information about the signal to the output stage of the measurement system, then a system having a small τ will be necessary. In contrast, systems of large τ may be adequate to measure signals of very low-frequency content.

The *dynamic error*, $\delta(\omega)$, of a system can be defined as $\delta(\omega) = M(\omega) - 1$ and represents a measure of the inability of a system to adequately reconstruct the amplitude of the input signal for a particular input frequency. Measurement systems with a magnitude ratio at or near unity over the anticipated frequency band of the input signal are preferred to minimize $\delta(\omega)$. Perfect reproduction of the input signal is not possible, so some dynamic error is inevitable. For a first-order system, the *frequency bandwidth* is defined traditionally as the frequency band over which $M(\omega) \geq 0.707$; in terms of the decibel, defined as

$$dB = 20\log M(\omega) \tag{3.11}$$

it is the band of frequencies within which $M(\omega)$ does not drop by more than -3 dB.

The functions $M(\omega)$ and $\phi(\omega)$ represent the frequency response of the measurement system to periodic inputs. These equations and universal curves are intended to be used for guidance in the selection of measurement systems and system components. They are not meant to be used to correct measured data. This can be seen by considering Figures 3.12 and 3.13. As $M(\omega)$ falls away from unity and as $\phi(\omega)$ falls away from zero, these curves take on rather steep slopes. In these regions of the curves, small errors in the prediction of τ and deviations of the real systems from ideal first-order behavior can lead to errors of unpredictable magnitude.

Determination of Frequency Response.   The frequency response of a measurement system is found by a dynamic calibration. In this case, the calibration would entail applying a simple periodic waveform of known amplitude and frequency to the system sensor stage and measuring the corresponding output stage amplitude and phase shift. In practice, however, developing a method to produce a periodic input signal in the form of a physical variable may demand considerable ingenuity and effort. Hence, in many situations an engineer will elect to rely on modeling to infer system frequency response behavior.

EXAMPLE 3.6

A temperature sensor is to be selected to measure temperature within a reaction vessel. It is suspected that the temperature will behave as a simple periodic waveform with a frequency somewhere between 1 and 5 Hz. Several size sensors are available, each with a known time constant. Based on the time constant, select a suitable sensor, assuming that a dynamic error of $\pm 2\%$ is acceptable.

KNOWN

$$1 \le f \le 5 \text{ Hz}$$
$$|\delta(\omega)| \le 0.02$$

ASSUMPTIONS

First-order system
$F(t) = A \sin \omega t$

FIND

time constant, τ

SOLUTION

A $|\delta(\omega)| \le 0.02$ implies that a magnitude ratio of between $0.98 \le M \le 1.02$ is sought. From Figure 3.12, it is seen that first-order systems never exceed $M = 1$. So the constraint becomes $0.98 \le M \le 1$. Then,

$$0.98 \le M(\omega) = \frac{1}{[1 + (\omega\tau)^2]^{1/2}}$$

From Figure 3.12, this constraint is maintained over the range $0 \le \omega\tau \le 0.2$. It is also seen in this figure that for a system of fixed time constant, the smallest value of $M(\omega)$ will occur at the largest frequency. Thus setting $\omega = 2\pi(5)$ rad/s and solving for $M(\omega) = 0.98$ yields $\tau \le 6.4$ ms. Accordingly, a sensor with a time constant of 6.4 ms or smaller will be suitable.

Second-Order Systems

Systems that possess inertia contain a second-derivative term in their model equation (see Example 3.1). A system that is modeled by a second-order differential equation is called a second-order system. Examples of second-order measurement systems include accelerometers and diaphragm pressure transducers (including microphones).

In general, a second-order measurement system subjected to an arbitrary input, $F(t)$, can be described by an equation of the form

$$a_2 \ddot{y} + a_1 \dot{y} + a_0 y = F(t) \tag{3.12}$$

where a_0, a_1, and a_2 are physical parameters used to describe the system and $\ddot{y} = d^2y/dt^2$. This equation can be rewritten as

$$\frac{1}{\omega_n^2} \ddot{y} + \frac{2\zeta}{\omega_n} \dot{y} + y = KF(t) \tag{3.13}$$

where

$$\omega_n = \sqrt{\frac{a_0}{a_2}} \equiv \text{natural frequency of the system}$$

$$\zeta = \frac{a_1}{2(a_0 a_2)^{1/2}} \equiv \text{damping ratio of the system}$$

Consider the homogeneous solution to equation (3.13). Its form will depend on the roots of the characteristic equation to (3.13)

$$\frac{1}{\omega_n^2}\lambda^2 + \frac{2\zeta}{\omega_n}\lambda + 1 = 0$$

This quadratic equation has two roots:

$$\lambda_{1,2} = -\zeta\omega_n \pm \omega_n\sqrt{\zeta^2 - 1}$$

Depending on the value for ζ, three forms of homogeneous solution are possible:

$0 < \zeta < 1$ (underdamped system solution)

$$y_h(t) = Ce^{-\zeta\omega_n t} \sin\left(\omega_n\sqrt{1 - \zeta^2}\, t + \Theta\right) \tag{3.14a}$$

$\zeta = 1$ (critically damped system solution)

$$y_h(t) = C_1 e^{\lambda_1 t} + C_2 t e^{\lambda_2 t} \tag{3.14b}$$

$\zeta > 1$ (overdamped system solution)

$$y_h(t) = C_1 e^{\lambda_1 t} + C_2 e^{\lambda_2 t} \tag{3.14c}$$

The homogeneous solution determines the transient response of a system. It is seen that for systems having $0 \leq \zeta < 1$, the transient response will be oscillatory, whereas for $\zeta \geq 1$, the transient response will not oscillate. The critically damped solution, $\zeta = 1$, denotes the demarcation between oscillatory and nonoscillatory behavior in the transient response. The damping ratio is a measure of system damping, a property of a system that enables it to dissipate energy internally.

Step Function Input

Again, the step function input is applied to determine the general behavior and speed at which the system will respond to a change in input. The response of a second-order measurement system to a step function input is found from the solution

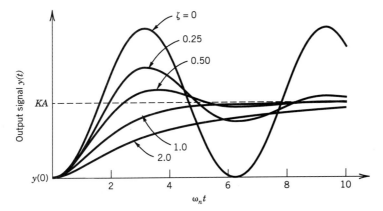

Figure 3.14 Second-order system time response to a step function input.

of equation (3.13), with $F(t) = AU(t)$, to be

$$y(t) = KA - KAe^{-\zeta \omega_n t}$$
$$\times \left[\frac{\zeta}{(1 - \zeta^2)^{1/2}} \sin \left(\omega_n \sqrt{1 - \zeta^2}\, t \right) + \cos \left(\omega_n \sqrt{1 - \zeta^2}\, t \right) \right] \quad 0 \le \zeta \le 1$$

$$(3.15a)$$

$$y(t) = KA - KA(1 + \omega_n t)e^{-\omega_n t} \quad \zeta = 1 \qquad (3.15b)$$

$$y(t) = KA - KA$$
$$\times \left[\frac{\zeta + \sqrt{\zeta^2 - 1}}{2\sqrt{\zeta^2 - 1}} e^{(-\zeta + \sqrt{\zeta^2 - 1})\omega_n t} + \frac{\zeta - \sqrt{\zeta^2 - 1}}{2\sqrt{\zeta^2 - 1}} e^{(-\zeta - \sqrt{\zeta^2 - 1})\omega_n t} \right] \quad \zeta > 1$$

$$(3.15c)$$

where the initial conditions, $y(0)$ and $\dot{y}(0)$, have been set to zero for convenience.

Equations $(3.15a)$–$(3.15c)$ are plotted in Figure 3.14 for several values of ζ. The interesting feature is the transient response. For underdamped systems, the transient response is oscillatory about the steady value and occurs with a period

$$T_d = \frac{2\pi}{\omega_d} \qquad (3.16)$$

$$\omega_d = \omega_n \sqrt{1 - \zeta^2} \qquad (3.17)$$

where ω_d is called the *ringing frequency*. In instruments, this oscillatory behavior is called "ringing." The ringing phenomenon and the associated ringing frequency are properties of the measurement system and are independent of the input signal. It is the free oscillation frequency of a system displaced from its equilibrium.

The duration of the transient response is controlled by the $\zeta \omega_n$ term. The system settles to KA more quickly when it is designed with a larger $\zeta \omega_n$. Nevertheless, for all systems with $\zeta > 0$, the response will eventually indicate the steady value of $y_\infty = KA$ as $t \to \infty$.

The time required for a second-order system to first achieve a value within 90% of the step input, $KA - y_0$, is defined as its *rise time*. Rise time can be decreased by decreasing the damping ratio, as seen in Figure 3.14. However, the severe ringing associated with very lightly damped systems can delay the time to achieve a steady value relative to systems of higher damping. This is demonstrated by comparing the response at $\zeta = 0.25$ with the response at $\zeta = 1$ in Figure 3.14. With this in mind, the time required for a measurement system's oscillations to settle to within $\pm 10\%$ of the steady value, KA, is defined as the *settling time* or response time. The settling time is an approximate measure of the time to steady response. A damping ratio of ~ 0.7 appears to offer a good compromise between ringing and settling time. If an error fraction (Γ) of a few percent is acceptable, then a system with $\zeta = 0.7$ will achieve steady response in about one-half the time of a system having $\zeta = 1$. For this reason, most measurement systems intended to measure sudden changes in input signal will typically be designed such that parameters $a_0, a_1,$ and a_2 provide a damping ratio between 0.6 and 0.8.

Determination of Ringing Frequency and Rise and Settling Times.   The experimental determination of the ringing frequency associated with under-damped systems is performed by applying a step input to the second-order measurement system and recording the response with time. This type of calibration will also yield information concerning the time to steady response of the system, which includes rise and settling times. Example 3.8 describes such a test. Typically, measurement systems suitable for dynamic signal measurements are rated according to their 90% rise time and settling time.

EXAMPLE 3.7

Determine the physical parameters that affect the natural frequency and damping ratio of the accelerometer of Example 3.1.

KNOWN

Accelerometer shown in Figure 3.3

ASSUMPTIONS

Second-order system as modeled in Example 3.1

FIND

ω_n and ζ

SOLUTION

A comparison between the governing equation for the accelerometer in Example 3.1 and equation (3.13) suggests

$$\omega_n = \sqrt{\frac{k}{m}} \qquad \zeta = \frac{c}{2(km)^{1/2}}$$

Accordingly, the physical parameters of mass, spring stiffness, and frictional damping control the natural frequency and damping ratio of this measurement system.

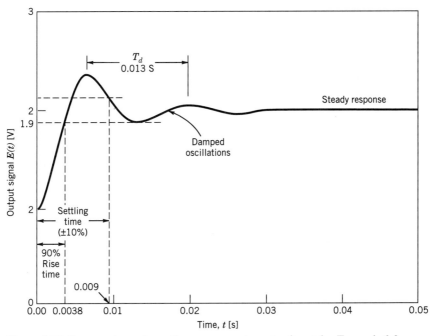

Figure 3.15 Pressure transducer time response to a step input for Example 3.8.

EXAMPLE 3.8

The curve given in Figure 3.15 was reproduced from an oscilloscope trace of the voltage signal output of a diaphragm pressure transducer subjected to a step change in input. During a static calibration using a pressure standard, it is found that the pressure–voltage relationship was linear over the range 1 atmosphere (atm) to 4 atm and that the static sensitivity of the measurement system was 1 V/atm. For the step test, it is known that the initial pressure was atmospheric pressure, p_a, and the final pressure was $2p_a$. Estimate the 90% rise time, the 90% settling time, and the ringing frequency of the measurement system.

KNOWN

$p(0) = 1$ atm
$p_\infty = 2$ atm
$K = 1$ V/atm

ASSUMPTIONS

Oscilloscope does not affect transducer performance
Second-order system

FIND

90% rise and settling times; ω_d

SOLUTION

The ringing behavior of the system noted on the oscilloscope trace supports the assumption that the transducer can be described as having a second-order behavior.

From the given information,

$$E(0) = Kp(0) = 1 \text{ V}$$
$$E_\infty = Kp_\infty = 2 \text{ V}$$

so that the step change observed on the oscilloscope trace should appear as a magnitude of 1 V. The 90% rise time will occur when the output first achieves a value of 1.9 V. The 90% settling time will occur when the output from the measurement system settles between $1.9 \text{ V} \le E(t) \le 2.1 \text{ V}$. From Figure 3.15, the 90% rise occurs in ~4 ms and the 90% settling time in ~9 ms. The period of the ringing behavior, T_d, is judged to be ~13 ms for an $\omega_d \approx 485$ rad/s.

Sine Function Input

The response of a second-order system to a sinusoidal input of the form, $F(t) = A \sin \omega t$, is given by

$$y(t) = y_h + \frac{KA \sin[\omega t + \phi(\omega)]}{\{[1 - (\omega/\omega_n)^2]^2 + (2\zeta\omega/\omega_n)^2\}^{1/2}} \tag{3.18}$$

with frequency-dependent phase shift

$$\phi(\omega) = -\tan^{-1} \frac{2\zeta\omega/\omega_n}{1 - (\omega/\omega_n)^2} \tag{3.19}$$

The exact form for y_h is found from equations (3.14a)–(3.14c) and depends on the value of ζ. The steady response, the second term on the right side, has the general form

$$y_{\text{steady}}(t) = B(\omega) \sin[\omega t + \phi(\omega)] \tag{3.20}$$

with amplitude $B(\omega)$. A comparison of equations (3.18) and (3.20) indicates that the amplitude of the steady response of a second-order system subjected to a sinusoidal input is dependent on the value of ω. Thus, the amplitude of the output signal from a second-order measurement system is frequency dependent. In general, we can define the magnitude ratio, $M(\omega)$, for a second-order system as

$$M(\omega) = \frac{B}{KA} = \frac{1}{\{[1 - (\omega/\omega_n)^2]^2 + (2\zeta\omega/\omega_n)^2\}^{1/2}} \tag{3.21}$$

The magnitude ratio for a second-order system is plotted in Figure 3.16 for several values of damping ratio. A corresponding plot of the phase-shift dependency on input frequency and damping ratio is shown in Figure 3.17. For an ideal measurement system, $M(\omega)$ would equal unity and $\phi(\omega)$ would equal zero for all values of frequency. Instead, real systems behave in a nearly ideal manner over only a portion of the frequency curve. In general, $M(\omega)$ will approach zero and $\phi(\omega)$ will approach $-\pi$ as ω/ω_n becomes large. Keep in mind that ω_n is a property of the measurement system, whereas ω is a property of the input signal.

System Characteristics

Several tendencies are apparent in Figures 3.16 and 3.17. For a system of zero damping, $\zeta = 0$, $M(\omega)$ will approach infinity and $\phi(\omega)$ jumps to $-\pi$ in the vicinity of

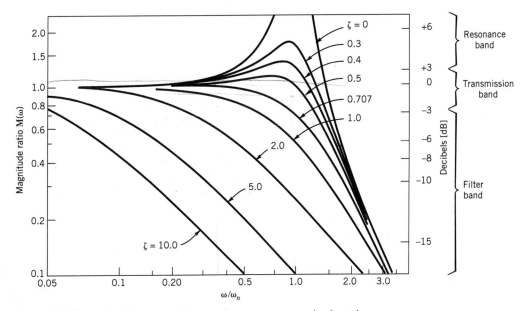

Figure 3.16 Second-order system frequency response: magnitude ratio.

$\omega = \omega_n$. This behavior is characteristic of system resonance. Real systems possess some amount of damping, which modifies the abruptness and magnitude of resonance, but underdamped systems may still achieve resonance. This region on Figures 3.16 and 3.17 is called the *resonance band* of the system, referring to the range

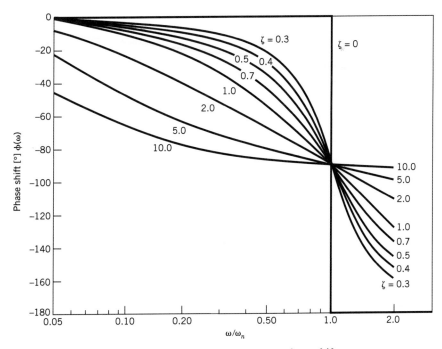

Figure 3.17 Second-order system frequency response: phase shift.

of frequencies over which the system is in resonance. Resonance in under-damped systems occurs at the *resonance frequency*, $\omega_R = \omega_n\sqrt{1 - 2\zeta^2}$. The resonance frequency is a property of the measurement system. Resonance is excited by a periodic input signal. The resonance frequency differs from the ringing frequency of free oscillation. Resonance behavior results in values of $M(\omega) > 1$ and considerable phase shift. For most applications, operation at frequencies within the resonance band is undesirable and could even be damaging to some delicate sensors. Resonance behavior is very nonlinear and results in distortion of the output signal. In contrast systems having $\zeta > 0.707$ do not resonate.

At low values of ω/ω_n, $M(\omega)$ remains near unity and $\phi(\omega)$ near zero. This indicates that information concerning the input signal of frequency, ω, will be passed through to the output signal with little alteration in the amplitude or phase shift. This region on the frequency response curves is called the *transmission band*. The actual extent of the frequency range for near unity gain depends on the system damping ratio. The transmission band of a system is either specified by its frequency bandwidth, typically defined for a second-order system as $3 \text{ dB} \geq M(\omega) \geq -3 \text{ dB}$, or specified explicitly by some other range of dynamic error. Operation of a measurement system within its transmission band is preferred when it is desirable to measure the dynamic content of the input signal.

At large values of ω/ω_n, $M(\omega)$ approaches zero. In this region, the measurement system will attenuate the amplitude information in the input signal. A large phase shift occurs. This region is known as the *filter band*, typically defined as the frequency range over which $M(\omega) \leq -3 \text{ dB}$. Most readers are familiar with the use of a filter to remove undesirable features from desirable product. Operation within this band is useful to remove input signal information at frequencies within the filter band.

EXAMPLE 3.9

Determine the frequency response of a pressure transducer that has a damping ratio of 0.5 and a ringing frequency (found by a step test) of 1200 Hz.

KNOWN

$\zeta = 0.5$

$\omega_d = 2\pi(1200 \text{ Hz}) = 7540 \text{ rad/s}$

ASSUMPTIONS

Second-order system

FIND

$M(\omega)$ and $\phi(\omega)$

SOLUTION

The frequency response of a measurement system is given by $M(\omega)$ and $\phi(\omega)$ as defined in equations (3.19) and (3.21). Since $\omega_d = \omega_n\sqrt{1 - \zeta^2}$, the natural frequency of the pressure transducer is found to be $\omega_n = 8706 \text{ rad/s}$. The frequency response at selected frequencies is computed from equations (3.19) and (3.21):

ω[rad/s]	$M(\omega)$	$\phi(\omega)$[°]
500	1.00	−3.3
2,600	1.04	−18.2
3,500	1.07	−25.6
6,155	1.15	−54.7
7,540	1.11	−73.9
8,706	1.00	−90.0
50,000	0.05	−170.2

COMMENT

Note the resonance behavior in the transducer response which peaks at $\omega_R = 6155$ rad/s. Resonance effects can be minimized by operating a measurement system at input frequencies less than ~30% of the system's natural frequency. The response of second-order systems can be explored by using software on the companion disk see program file *SecondOrd*).

EXAMPLE 3.10

An accelerometer is to be selected to measure a time-dependent motion. In particular, input signal frequencies below 100 Hz are of prime interest. Select a set of acceptable parameter specifications for the instrument, assuming a dynamic error of ±5%.

KNOWN

$f \leq 100$ Hz (i.e., $\omega \leq 628$ rad/s)

ASSUMPTIONS

Second-order system
Dynamic error of ±5% acceptable

FIND

Select ω_n and ζ

SOLUTION

To meet a ±5% dynamic error constraint, we will want $0.95 \leq M(\omega) \leq 1.05$ over the frequency range $0 \leq \omega \leq 628$ rad/s. This problem is certainly open ended in that a number of instruments with different ω_n will do the task. So as one solution, let us set $\zeta = 0.7$ and then solve for the required ω_n by using equation (3.21):

$$1.05 \geq M(\omega) = \frac{1}{\{[1 - (\omega/\omega_n)^2]^2 + [2\zeta(\omega/\omega_n)]^2\}^{1/2}}$$

$$0.95 \leq M(\omega) = \frac{1}{\{[1 - (\omega/\omega_n)^2]^2 + [2\zeta(\omega/\omega_n)]^2\}^{1/2}}$$

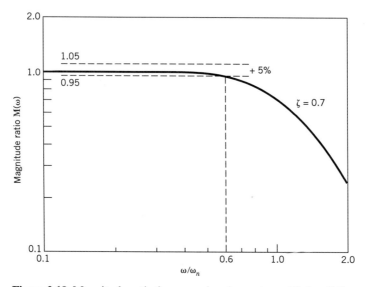

Figure 3.18 Magnitude ratio for second-order system with $\zeta = 0.7$ for Example 3.10.

With $\omega = 628$ rad/s, these equations yield $\omega_n \geq 1047$ rad/s. Alternatively, we could plot equation (3.21) with $\zeta = 0.7$, as shown in Figure 3.18. In Figure 3.18, we find that $0.95 \leq M(\omega) \leq 1.05$ over the frequency range $0 \leq \omega/\omega_n \leq 0.6$. Again, this puts $\omega_n \geq 1047$ rad/s as being acceptable. So as one solution, an instrument having $\zeta = 0.7$ and $\omega_n \geq 1047$ rad/s will meet the problem constraints.

3.4 TRANSFER FUNCTIONS

Consider the schematic representation in Figure 3.19. The measurement system operates on the input signal, $F(t)$, by some function, $G(s)$, so as to indicate the output signal, $y(t)$. This operation can be explored by taking the Laplace transform of both sides of the differential equation, (3.4), which describes the general first-order measurement system. One obtains

$$Y(s) = K\frac{1}{\tau s + 1}F(s) + \frac{y_0}{\tau s + 1}$$

where $y_0 = y(0)$. This can be rewritten as

$$Y(s) = KG(s)F(s) + Q(s)G(s) \tag{3.22}$$

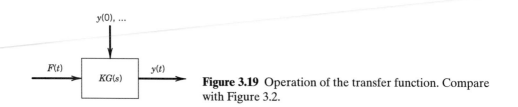

Figure 3.19 Operation of the transfer function. Compare with Figure 3.2.

where $G(s)$ is the transfer function of the first-order system given by

$$G(s) = \frac{1}{\tau s + 1} \tag{3.23}$$

and $Q(s)$ is the system initial state function. Because it includes $KF(t)$, the first term on the right side of equation (3.22) contains the information that describes the steady response of the measurement system to the input signal, whereas the second term describes its transient response. So the transfer function $G(s)$ describes the complete time response of the measurement system. As indicated in Figure 3.19, the transfer function defines the mathematical operation that the measurement system performs on $F(t)$ to yield the time response of the system.

The system frequency response, which has been shown to be given by $M(\omega)$ and $\phi(\omega)$, can be found by finding the value of $G(s)$ at $s = i\omega$. This yields the complex number,

$$\begin{aligned} G(s = i\omega) &= \frac{1}{\tau i\omega + 1} \\ &= M(\omega)e^{i\phi(\omega)} \end{aligned} \tag{3.24}$$

where $G(i\omega)$ is a vector on the real–imaginary plane having a magnitude, $M(\omega)$, and inclined at an angle, $\phi(\omega)$, relative to the real axis as indicated in Figure 3.20. For the first-order system, the magnitude of $G(i\omega)$ is simply equation (3.10) and the phase-shift angle is given in equation (3.9).

For a second-order system, the governing equation is defined by equation (3.13), with initial conditions $y(0) = y_0$ and $\dot{y}(0) = \dot{y}_0$. The Laplace transform yields

$$Y(s) = \frac{1}{(1/\omega_n^2)s^2 + (2\zeta/\omega_n)s + 1}KF(s) + \frac{s\dot{y}_0 + y_0}{(1/\omega_n^2)s^2 + (2\zeta/\omega_n)s + 1} \tag{3.25}$$

which can be represented by

$$Y(s) = KG(s)F(s) + Q(s)G(s) \tag{3.26}$$

By inspection, the transfer function is given by

$$G(s) = \frac{1}{(1/\omega_n^2)s^2 + (2\zeta/\omega_n)s + 1} \tag{3.27}$$

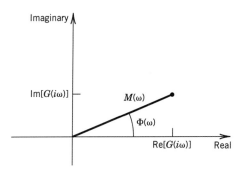

Figure 3.20 Complex plane approach to describing frequency response.

Solving for $G(s)$ at $s = i\omega$, one obtains for a second-order system

$$G(s = i\omega) = \frac{1}{(i\omega)^2/\omega_n^2 + 2\zeta i\omega/\omega_n + 1}$$
$$= M(\omega)e^{i\phi(\omega)} \tag{3.28}$$

which yields the same magnitude ratio and phase shift relations as given by equations (3.21) and (3.19).

3.5   PHASE LINEARITY

It can be noted from Figures 3.16 and 3.17 that systems having a damping ratio near 0.7 possess the broadest frequency range over which $M(\omega)$ will remain at or near unity and that over this same frequency range the phase shift will essentially vary in a linear manner with frequency. Although it is not possible to design a measurement system without accepting some amount of phase shift, it is desirable to design a system such that the phase shift varies linearly with frequency. This is because measurement system operation in the regions of Figure 3.17 where phase shift varies nonlinearly with frequency will often bring about a significant distortion in the waveform of the output signal relative to the input signal. Distortion refers to a notable change in the shape of the waveform, as opposed to simply an amplitude alteration or relative phase shift. To minimize distortion, many measurement systems are designed with $0.6 \leq \zeta \leq 0.8$.

Signal distortion can be illustrated by considering a general function, $u(t)$:

$$u(t) = \sum_{n=1}^{\infty} \sin n\omega t = \sin \omega t + \sin 2\omega t + \cdots \tag{3.29}$$

Suppose during measurement a phase shift of this signal were to occur such that the phase shift remained linearly proportional to the frequency; that is, the measured signal, $v(t)$, could be represented by

$$v(t) = \sin(\omega t - \phi) + \sin(2\omega t - 2\phi) + \cdots \tag{3.30}$$

Or, by setting

$$\theta = (\omega t - \phi) \tag{3.31}$$

we can write

$$v(t) = \sin \theta + \sin 2\theta + \cdots \tag{3.32}$$

Inspection of $v(t)$ shows it to be equivalent to the original signal, $u(t)$. If the phase shift were not linearly related to the frequency, an output signal different in form from the input signal would result. This is demonstrated in Example 3.11.

EXAMPLE 3.11

Consider the effect of variations in phase shift with frequency on a measured signal by examination of the signal defined by the function

$$u(t) = \sin t + \sin 5t$$

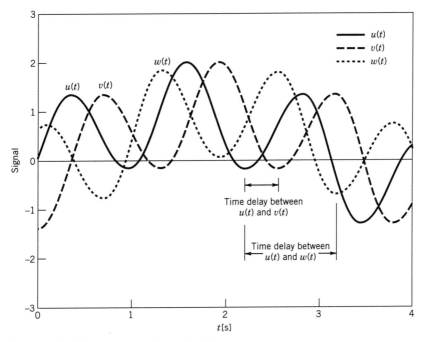

Figure 3.21 Waveforms for Example 3.11.

Suppose this signal is measured in such a way that a phase shift that is linearly proportional to the frequency occurs in the form

$$v(t) = \sin(t - 0.35) + \sin[5t - 5(0.35)]$$

Both $u(t)$ and $v(t)$ are plotted in Figure 3.21. It is apparent that the two waveforms have remained identical except that $v(t)$ lags $u(t)$ by some time increment. However, suppose $u(t)$ were measured in such a way that the relation between phase shift and frequency was nonlinear, such as in the signal output form

$$w(t) = \sin(t - 0.35) + \sin(5t - 5)$$

The $w(t)$ signal is also plotted in Figure 3.21, which can be seen to produce a differently behaving waveform than $u(t)$. This different behavior is *signal distortion*. In a comparison of $u(t)$, $v(t)$, and $w(t)$, it should be apparent that distortion can be brought about by the nonlinear relation of phase shift with frequency.

3.6   MULTIPLE-FUNCTION INPUTS

So far we have discussed the response of a measurement system to a signal containing only a single frequency. What about the response of a measurement system to multiple input frequencies? Or to an input that consists of both a static and a dynamic part, such as a periodic pressure, strain, or temperature signal? When models are used that are linear, ordinary differential equations subjected to inputs that are linear in terms of the dependent variable, the principle of superposition of linear systems will apply to the solution of these equations. The *principle of superposition* states that a linear combination of input signals applied to a linear measurement system produces an output signal that is simply the linear addition of the separate output signals that would result if each input term had been applied separately. Because the form of the transient response is not affected by the input function, we can focus on the steady response. In general, we can write that if the forcing function of a form

$$F(t) = A_0 + \sum_{i=1}^{\infty} (A_i \sin \omega_i t) \tag{3.33}$$

is applied to a system, then the combined steady response will have the form

$$KA_0 + \sum_{n=1}^{\infty} B(\omega_i) \sin[\omega_i t + \phi(\omega_i)] \tag{3.34}$$

where $B(\omega_i) = KA_i M(\omega_i)$. The development of the superposition principle can be found in basic texts on dynamic systems (e.g., [4]).

EXAMPLE 3.12

A second-order instrument having a $K = 1$ unit/unit, $\zeta = 2$, and $\omega_n = 628$ rad/s is to be used to measure an input signal of the form

$$F(t) = 5 + 10 \sin 25t + 20 \sin 400t$$

Predict its steady output signal.

KNOWN

Second-order system
$K = 1$ unit/unit
$\zeta = 2.0$
$\omega_n = 628$ rad/s
$F(t) = 5 + 10 \sin 25t + 20 \sin 400t$

ASSUMPTIONS

Linear system (superposition holds)

FIND

$y(t)$

SOLUTION

Since $F(t)$ has a form consisting of a linear addition of multiple input functions, the steady response signal will have the form of equation (3.33).

$$y(t) = 5K + 10KM(25 \text{ rad/s}) \sin[25t + \phi(25 \text{ rad/s})]$$
$$+ 20KM(400 \text{ rad/s}) \sin[400t + \phi(400 \text{ rad/s})]$$

Using equations (3.19) and (3.21), or, alternatively, Figures 3.16 and 3.17, with $\omega_n =$

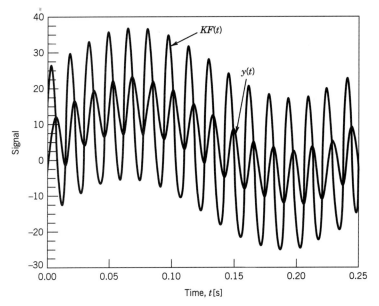

Figure 3.22 Input and output signals for Example 3.12.

628 rad/s and $\zeta = 2.0$, we find that

$$M(25 \text{ rad/s}) = 0.99 \qquad \phi(25 \text{ rad/s}) = -9.1°$$
$$M(400 \text{ rad/s}) = 0.39 \qquad \phi(400 \text{ rad/s}) = -77°$$

So that the output signal will have the form

$$y(t) = 5 + 9.9\sin(25t - 9.1°) + 7.8\sin(400t - 77°)$$

The output signal is plotted against the input signal in Figure 3.22. The amplitude spectra for both the input signal and the output signal are shown in Figures 3.23(*a*) and 3.23(*b*). Spectra can also be generated by using software on the companion disk using program files *FunSpect* and *DataSpect*.

COMMENT

It is seen that information concerning the average value and concerning the 25 rad/s component of the input signal will be passed along to the output signal. However, amplitude information concerning the 400 rad/s component of the input signal has been severely reduced (down 61%) in the output signal. If information concerning the 400 rad/s component is important, then it would be better to select an instrument that has an improved frequency response in the 400 rad/s range.

3.7 COUPLED SYSTEMS

When a measurement system consists of more than one instrument, the measurement system behavior can become more complicated. As instruments in each stage of the system are connected, the output from one stage becomes the input to the next stage and so forth. Such measurement systems will have an output response to the original input signal that is some combination of the individual instrument responses to the

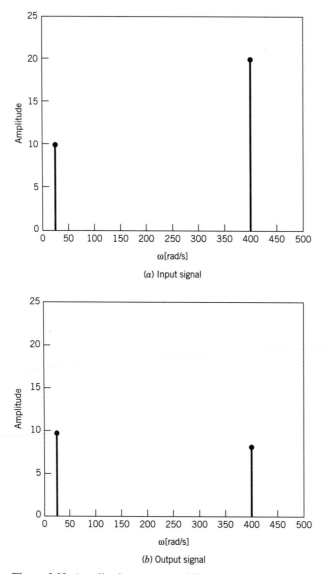

Figure 3.23 Amplitude spectrum of Example 3.12.

input. However, the system concepts of zero-, first-, and second-order systems studied previously can be used for a case-by-case study of the coupled measurement system. This is done by considering the input to each stage of the measurement system as the output of the previous stage.

This concept is easily illustrated by considering a first-order sensor that may be connected to a second-order output device (for example, a temperature sensor–transducer connected to a strip chart recorder). Suppose the input to the sensor is a simple periodic waveform, $F(t) = A \sin \omega t$. The transducer will respond with the output signal of the form of equation (3.8):

$$y_t(t) = C e^{-t/\tau} + \frac{K_t A}{[1 + (\omega\tau)^2]^{1/2}} \sin(\omega t + \phi_t)$$

$$\phi_t = -\tan^{-1}(\omega\tau)$$

(3.35)

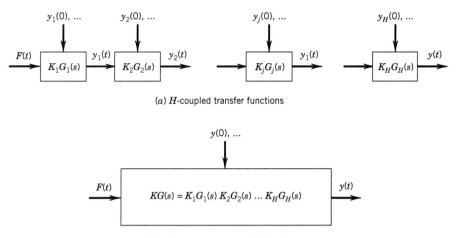

(a) H-coupled transfer functions

(b) Equivalent system transfer function

Figure 3.24 Coupled systems, describing the system transfer function.

where subscript t refers to the transducer. However, the transducer output signal now becomes the input signal, $F_2(t) = y_t$, to the second-order device. The output from the second-order device, $y_s(t)$, will be a second-order response to the input $F_2(t)$,

$$y_s(t) = y_h(t) + \frac{K_t K_s A \sin(\omega t + \phi_t + \phi_s)}{[1 + (\omega\tau)^2]^{1/2}\{[1 - (\omega/\omega_n)^2]^2 + (2\zeta\omega/\omega_n)^2\}^{1/2}}$$

$$\phi_s = -\tan^{-1} \frac{2\zeta\omega/\omega_n}{1 - (\omega/\omega_n)^2} \tag{3.36}$$

where subscript s refers to the strip chart recorder and $y_h(t)$ is the transient response. The output signal displayed on the recorder, $y_s(t)$, is the measurement system response to the original input signal to the transducer, $F(t) = A \sin \omega t$. The amplitude of the recorder output signal is the product of the static sensitivities and magnitude ratios of a first- and second-order system. The phase shift is the sum of the phase shifts of the two systems.

From equation (3.36) we can make a general observation. Consider the schematic representation in Figure 3.24, which depicts a measurement system consisting of H interconnected devices, $j = 1, 2, \ldots, H$, each device described by a linear system model. The overall transfer function of the combined system, $G(s)$, will be the product of the transfer functions of each of the individual devices, $G_j(s)$, such that

$$KG(s) = K_1 G_1(s) K_2 G_2(s) \cdots K_H G_H(s) \tag{3.37}$$

At $s = i\omega$, equation (3.37) gives

$$KG(i\omega) = (K_1 K_2 \cdots K_H) \times [M_1(\omega) M_2(\omega) \cdots M_H(\omega)] e^{i[\phi_1(\omega) + \phi_2(\omega) + \cdots + \phi_H(\omega)]} \tag{3.38}$$

According to equation (3.38), given an input signal to device 1, the system steady output signal at device H will be described by the system frequency response $G(s = i\omega) = M(\omega)e^{i\phi(\omega)}$, with an overall system static sensitivity described by

$$K = K_1 K_2 K_3 \ldots K_H \tag{3.39}$$

The overall system magnitude ratio will be the product

$$M(\omega) = M_1(\omega) M_2(\omega) \ldots M_H(\omega) \tag{3.40}$$

and the overall system phase shift will be the sum

$$\phi(\omega) = \phi_1(\omega) + \phi_2(\omega) + \cdots + \phi_H(\omega) \tag{3.41}$$

This will hold true provided that there do not exist any significant loading effects, a situation discussed in Chapter 6. The response of coupled systems can be further explored by using the program file *MultiSys* found on the companion disk.

3.8   SUMMARY

The response of a measurement system to a time-dependent input depends on several factors, including the inherent speed of response of that system and the frequency response of that system. Modeling has enabled us to develop and to illustrate these concepts. Those system design parameters that affect system response are exposed through modeling, and this assists in instrument selection. Modeling has also suggested the methods by which measurement system specifications such as the time constant, response time, frequency response, damping ratio, and resonance frequency can be determined both analytically and experimentally. An interpretation of these system properties and their effect on system performance was determined.

The speed of response of a system to a change in input is estimated by use of the step function input. The system parameters of the time constant, for first-order systems, and the natural frequency and damping ratio, for second-order systems, are used as indicators of system response speed. The magnitude ratio and phase shift define the frequency response of any system and are found by an input of a periodic waveform to a system. Figures 3.12, 3.13, 3.16, and 3.17 are universal frequency response curves for first- and second-order systems, respectively. These curves can be found in most engineering and mathematical handbooks and can be applied to any first- or second-order system, as appropriate.

REFERENCES

1. Close, C. M., and Frederick, D. K., *Modeling and Analysis of Dynamic Systems*, 2d ed., Wiley, New York, 1994.
2. Doebelin, E. O., *System Modeling and Response. Theoretical and Experimental Approaches*, Wiley, New York, 1980.
3. Ogata, K., *System Dynamics*, 3d ed., Prentice-Hall, Upper Saddle River, NJ, 1992.
4. Palm, W. J., III, *Modeling, Analysis and Control of Dynamic Systems*, 2d ed., Wiley, New York, 2000.

Suggested Reading

Raven, F., *Automatic Control Engineering*, 5th ed., McGraw-Hill, New York, 1994.

NOMENCLATURE

$a_0, a_1, \ldots, a_n$	physical coefficients	$p(t)$	pressure $[m\,l^{-1}\,t^{-2}]$
$b_0, b_1, \ldots, b_m$	physical coefficients	t	time $[t]$
c	damping coefficient $[m\,t\,l]$	$x(t)$	independent variable
f	cyclical frequency [Hz]	$y(t)$	dependent variable
k	spring constant or stiffness $[m\,t^{-2}]$	y^n	nth time derivative of $y(t)$
h	heat transfer coefficient $[m\,l\,t^{-3}/°C]$	$y^n(0)$	initial condition of y^n
m	mass $[m]$	A	input signal amplitude or magnitude

B	output signal amplitude	$U(t)$	unit step function
C	constant	β_1	time lag [t]
$E(t)$	energy [$m\,l\,t^{-3}$] or voltage [V]	$\delta(\omega)$	dynamic error
F	force [$m\,l\,t^{-2}$]	τ	time constant [t]
$F(t)$	forcing function	$\phi(\omega)$	phase shift
$G(s)$	transfer function	ω	circular frequency [t^{-1}]
K	static sensitivity	ω_n	natural frequency [t^{-1}]
$M(\omega)$	magnitude ratio, B/KA	ω_d	ringing frequency [t^{-1}]
$T(t)$	temperature [°]	ω_R	resonance frequency [t^{-1}]
T_d	ringing period [t]	ζ	damping ratio
		Γ	error fraction

Subscripts

0	initial value	∞	final or steady value
h	homogeneous solution		

PROBLEMS

Although not required, software found on the companion disk can be useful for many of these problems.

3.1 From a static calibration, a measurement system is found to have a static sensitivity of 2 V/kg. If an input range of 1 kg to 10 kg is to be measured, determine the expected range of values for the output signal. What would be the significance of increasing the static sensitivity?

3.2 Determine the 75%, 90%, and 95% response time for each of the systems given (assume zero initial conditions):

  a. $0.4\dot{T} + T = 4U(t)$
  b. $\ddot{y} + 2\dot{y} + 4y = U(t)$
  c. $2\ddot{P} + 8\dot{P} + 8P = 2U(t)$
  d. $5\dot{y} + 5y = U(t)$

3.3 A special sensor is designed to sense the percent vapor present in a liquid–vapor mixture. If during a static calibration the sensor indicates 80 units when in contact with 100% liquid, 0 units with 100% vapor, and 40 units with a 50–50% mixture, determine the static sensitivity of the sensor.

3.4 A system is modeled by the equation

$$0.5\dot{y} + y = F(t)$$

Initially, the output signal is steady at 75 V. The input signal is suddenly increased to 100 V.

  a. Determine the response equation.
  b. On the same graph, plot both the input signal and the system time response until a steady response is reached.

3.5 Suppose a thermometer similar to that of Example 3.3 is known to have a time constant of 30 s in a particular application. Plot its time response to a step change from 32°F to 120°F. Determine its 90% rise time.

3.6 Referring back to Example 3.3, a student establishes the time constant of a temperature sensor by first holding it immersed in hot water and then suddenly removing it and holding it immersed in cold water. Several other students perform the same test with similar sensors. Overall, their results are inconsistent with each other, with estimated time constants differing by as much as a factor of 1.2. Offer any suggestions as to why this might happen. Hint: Try this yourself and think about the control of test conditions.

3.7 A thermocouple, which responds as a first-order instrument, has a time constant of 20 ms. Determine its 90% rise time.

3.8 During a step function calibration, a first-order instrument is exposed to a step change of 100 units. If after 1.2 s the instrument indicates 80 units, estimate the instrument time constant. Estimate the error in the indicated value after 1.5 s. $y(0) = 0$ units; $K = 1$ unit/unit.

3.9 Estimate the dynamic error that would result from measuring a 2-Hz periodic waveform, using a first-order system with time constant of 0.7 s.

3.10 A first-order instrument with time constant of 1 s is used to measure a signal that can be represented by $F(t) = 10 \cos 2.5t$. Write the expected indicated steady response output signal. What is the expected time lag between input and output signal? $y(0) = 0$; $K = 1$.

3.11 A first-order instrument with a time constant of 2 s is to be used to measure a periodic input. If a dynamic error of $\pm 2\%$ can be tolerated, determine the maximum frequency of a periodic input that can be measured. What is the associated time lag (in seconds) at this frequency?

3.12 Determine the frequency response $[M(\omega)$ and $\phi(\omega)]$ for an instrument having a time constant of 10 ms. Estimate the instrument's frequency range to keep its dynamic error within 10%.

3.13 A temperature measuring device with a time constant of 0.15 s outputs a voltage that is linearly proportional to temperature. The device is used to measure an input signal of the form $T(t) = 115 + 12 \sin 2t \, °C$. Plot the input signal and the predicted output signal with time, assuming first-order behavior and a static sensitivity of 5 mV/°C. Determine the dynamic error and time lag in the steady response. $T(0) = 115°C$.

3.14 A first-order sensor is to be installed into a reactor vessel to monitor temperature. If a sudden rise in temperature greater than 100°C should occur, shutdown of the reactor will have to begin within 5 s after reaching 100°C. Determine the maximum allowable time constant for the sensor.

3.15 A single-loop LR circuit having a resistance of 1 MΩ is to be used as a low-pass filter between an input signal and a voltage measurement device. To attenuate undesirable frequencies above 1000 Hz by at least 50%, select a suitable inductor size if the time constant for this circuit is given by L/R.

3.16 A measuring system has a natural frequency of 0.5 rad/s, a damping ratio of 0.5, and a static sensitivity of 0.5 m/V. Estimate its 90% rise time and settling time if $F(t) = 2U(t)$ and the initial condition is zero. Plot the response $y(t)$ and its indicate transient and steady responses.

3.17 Plot the frequency response, based on equations (3.17) and (3.19), for an instrument having a damping ratio of 0.6. Determine the frequency range over which the dynamic error remains within 5%. Repeat for a damping ratio of 0.9 and 2.0.

3.18 The output from a temperature system indicates a steady, time-varying signal having an amplitude that varies between 30°C and 40°C with a single frequency of 10 Hz. Express the output signal as a waveform equation, $y(t)$. If the dynamic error is to be less than 1%, what must the system time constant be?

3.19 A cantilever beam instrumented with strain gauges is used as a force scale. A step test on the beam provides a measured damped oscillatory signal with time. If the signal behavior is second order, show how a data reduction design plan could use the information in this signal to determine the natural frequency and damping ratio of the cantilever beam. (Hint: consider the shape of the decay of the peak values in the oscillation).

3.20 A step test of a transducer brings on a damped oscillation decaying to a steady value. If the period of oscillation is 0.577 ms, what is the transducer ringing frequency?

3.21 If an instrument has a known damping ratio of 0.8, ringing frequency of 1000 Hz, and static sensitivity of 1.5 V/V, determine its expected steady response to an input signal that oscillates sinusoidally between 12 and 24 V at a frequency of 300 Hz.

3.22 An application demands that a sinusoidal pressure variation of 250 Hz be measured with no more than 2% dynamic error. In selecting a suitable pressure transducer from a vendor catalog, you note that a desirable line of transducers has a fixed natural frequency of 600 Hz but that you have a choice of transducer damping ratios of between 0.5 and 1.5 in increments of 0.05. Select a suitable transducer.

3.23 A compact disk player is to be isolated from room vibrations by placing it on an isolation pad. The isolation pad can be considered as a board of mass, m, a foam mat of stiffness, k, and a damping coefficient, c. For expected vibrations in the frequency range of between 2 and 40 Hz, select reasonable values for m, k, and c such that the room vibrations are attenuated by at least 50%. Assume that the only degree of freedom is in the vertical direction.

3.24 A single-loop RCL electrical circuit can be modeled as a second-order system in terms of current. Show that the differential equation for such a circuit subjected to a forcing function potential $E(t)$ is given by

$$L\frac{d^2 I}{dt^2} + R\frac{dI}{dt} + \frac{I}{C} = E(t)$$

Determine the natural frequency and damping ratio for this system. For a forcing potential, $E(t) = 1 + 0.5 \sin 2000t \, \text{V}$, determine the system steady response when $L = 2H$, $C = 1 \, \mu\text{F}$, and $R = 10,000 \, \Omega$. Plot the steady output signal and input signal versus time, $I(0) = dI(0)/dt = 0$.

3.25 A transducer that behaves as a second-order instrument has a damping ratio of 0.7 and a natural frequency of 1000 Hz. It is to be used to measure a signal containing frequencies as large as 750 Hz. If a dynamic error of $\pm 10\%$ can be tolerated, is this transducer a good choice?

3.26 A strain gauge measurement system is mounted on an airplane wing to measure wing oscillation and strain during wind gusts. The strain system has a 90% rise time of 100 ms, a ringing frequency of 1200 Hz, and a damping ratio of 0.8. Estimate the dynamic error in measuring a 1-Hz oscillation. Also estimate the time lag. Explain in words the meaning of this information.

3.27 An instrument having a resonance frequency of 1414 rad/s with a damping ratio of 0.5 is used to measure a signal of $\sim$6000 Hz. Estimate the expected dynamic error and phase shift.

3.28 Select one set of appropriate values for damping ratio and natural frequency for a second-order instrument used to measure frequencies up to 100 rad/s with no more than $\pm 10\%$ dynamic error. A catalog offers models with damping ratios of 0.4, 1, and 2 and natural frequencies of 200 and 500 rad/s. Explain your reasoning.

3.29 A signal of the form $F(t) = \sin t + 0.3 \sin 20t \, \text{N}$ is input to a first-order system that has a time constant of 0.2 s and a static sensitivity of 1 V/N. Determine the steady response (steady output signal) from the system. Discuss the transfer of information from the input to the output. Can the input signal be resolved based on the output?

3.30 A force transducer having a damping ratio of 0.5 and a natural frequency of 4000 Hz is available for use to measure a periodic signal of 2000 Hz. Show that the transducer fails a $\pm 10\%$ dynamic error constraint. Estimate its resonance frequency.

3.31 An accelerometer, whose frequency response is defined by equations (3.19) and (3.21), has a damping ratio of 0.4 and a natural frequency of 18,000 Hz. It is used to sense the relative displacement of a beam to which it is attached. If an impact to the beam imparts a vibration at 4500 Hz, estimate the dynamic error and phase shift in the accelerometer output. Estimate its resonance frequency.

3.32 Derive the equation form for the magnitude ratio and phase shift of the seismic accelerometer of Example 3.1. Does its frequency response differ from that predicted by equations 3.19 and 3.21? For what type of measurement would you suppose this instrument would be best suited?

3.33 Suppose the pressure transducer of Example 3.9 had a damping ratio of 0.6. Plot its frequency response $M(\omega)$ and $\phi(\omega)$. At which frequency is $M(\omega)$ a maximum?

3.34 The output stage of a first-order transducer is to be connected to a second-order display stage device. The transducer has a known time constant of 1.4 ms and static sensitivity of 2 V/°C, whereas the display device has values of sensitivity, damping ratio, and natural frequency of 1 V/V, 0.9, and 5000 Hz, respectively. Determine the steady response of this measurement system to an input signal of the form $T(t) = 10 + 50 \sin 628t \, °\text{C}$.

3.35 The displacement of a solid body is to be monitored by a seismic transducer (second-order system) with the signal output displayed on a recorder (second-order system). The displacement is expected to vary sinusoidally between 2 and 5 mm at a rate of 85 Hz. Select appropriate design specifications for the measurement system for no more than a 5% dynamic error (i.e., specify an acceptable range for natural frequency and damping ratio for each device).

3.36 A force measurement system has a resonance frequency of 82.5 rad/s, a damping ratio of 0.4, and static sensitivity of 2 V/N. If an input of the form

$$F(t) = 3 + \sin 8t + \sin 165t \text{ N}$$

is to be applied, write the expected form of the output signal in volts. Comment on the suitability of the instrument for this measurement. Optional: use the accompanying software or similar software to estimate and plot the amplitude spectrum.

3.37 A signal suspected to be of the nominal form

$$y(t) = 5 \sin 1000t \text{ mV}$$

is input to a first-order instrument having a time constant of 100 ms and $K = 1$ V/V. It is then to be passed through a second-order amplifier having $K = 100$ V/V, a natural frequency of 15,000 Hz, and a damping ratio of 0.8. What is the expected form of the output signal, $y(t)$? Estimate the dynamic error and phase lag in the output. Is this system a good choice? If not, do you have any suggestions?

3.38 A typical modern dc audio amplifier has a frequency bandwidth of 0–20,000 Hz ± 1 dB. Explain the meaning of this specification and its relation to music reproduction.

3.39 The displacement of a rail vehicle chassis as it rolls down a track is measured by using a transducer ($K = 10$ mV/mm, $\omega_n = 10,000$ rad/s, $\zeta = 0.6$) and a recorder ($K = 1$ mm/mV, $\omega_n = 700$ rad/s, $\zeta = 0.7$). The resulting amplitude spectrum of the output signal consists of a spike of 90 mm at 2 Hz and a spike of 50 mm at 40 Hz. Are the measurement system specifications suitable for the displacement signal? (If not, suggest changes.) Estimate the actual displacement of the chassis. State any assumptions.

3.40 It is expected that the amplitude spectrum of the time-varying displacement signal from a vibrating U-shaped tube will have a spike at 85, 147, 220, and 452 Hz. Each spike is related to an important vibrational mode of the tube. Displacement transducers available for this application have a range of natural frequencies from 500 Hz through 1000 Hz, with a fixed damping ratio of ~0.5. The transducer output is to be monitored on a DFT-based spectrum measurement device that has a frequency bandwidth extending from 0.1 Hz to 250 kHz. Within the stated range of availability, select a suitable transducer for this measurement.

3.41 A sensor mounted to a cantilever beam indicates beam motion with time. When the beam is deflected and released (step test), the sensor signal indicates that the beam oscillates as an underdamped second-order system with a ringing frequency of 10 rad/s. The maximum displacement amplitudes are measured at three different times corresponding to the 1st, 16th, and 32nd cycles and found to be 17, 9, and 5 mV, respectively. Estimate the damping ratio and natural frequency of the beam based on this measured signal, $K = 1$ mm/mV.

Chapter 4

Probability and Statistics

4.1 INTRODUCTION

Suppose we had a large box containing thousands of similar round bearings. To get an idea of the size of the bearings, we might measure two dozen. The resulting diameter values form a data set, which we use to infer something about the size of the bearings in the box as a whole, such as average size and size variation. But how close are the values from our data set to the actual average size and variation of all the bearings in the box? If we took another data set, should we expect the values to be exactly the same? These are questions that have answers in probability and statistics.

Engineering measurements taken repeatedly under seemingly identical conditions will normally show variations in measured values. Sources that contribute to these variations include the following:

Measurement System

- Resolution
- Repeatability

Measurement Procedure and Technique

- Repeatability

Measured Variable

- Temporal variation
- Spatial variation

For a given set of measurements, we want to be able to quantify (1) a single representative value that best characterizes the average of the data set, as well as (2) a representative value that provides a measure of the variation in the measured data set. Furthermore, we will want to know how well this single average value represents the true average value of the variable measured. This is done by establishing (3) an interval about the representative average value in which the true value is expected to lie. Statistics and probability provide the tools needed to arrive at these three estimates.

This chapter presents an introduction to the concepts of probability and statistics at a level sufficient to provide information for a large class of engineering judgments. Such information also allows for the reduction of raw data into results. Some of these materials, such as the concepts of mean value and standard deviation, should already be familiar to the engineering student and are presented along with their relations with probability to provide for a correct interpretation of data.

4.2   STATISTICAL MEASUREMENT THEORY

A *sample* of data refers to a set of data obtained during repeated measurements of a variable under fixed operating conditions. This measured variable is also known as the *measurand*. Fixed operating conditions imply that the external conditions that control the process from which the measured value is obtained are held at fixed values while the sample is obtained. In actual engineering practice, the ability to control the operating conditions at truly fixed conditions may be impossible, and the term "fixed operating conditions" should be considered in a nominal sense. That is, the process conditions are maintained as closely as possible.

In this chapter, we will consider only those situations in which the bias error is negligible.[1] Recall from Chapter 1 that this is the case when the average error in a data set is zero.

We begin by considering the measurement problem of estimating the *true mean value*, x', based on the information derived from the repeated measurement of x. The true value is the value we want to estimate from the measurement. A sample of the variable x under controlled, fixed operating conditions renders a finite number of data points. We use this data to infer x'. We can imagine that if the number of data points is very small, then our estimation of x' from the data set could be heavily influenced by the value of any one data point. If this data point showed a large variation from x' relative to the other data, then the estimate of the true value could show a large error. If the data set were larger, then the influence of any one data point would be offset by the larger influence of the other data. As $N \to \infty$, all the possible variations in x would be included in the data set. From a practical view, only finite-sized data sets may be possible, in which case the measured data can provide only some estimate of the true value.

From a statistical analysis of the data set and an analysis of sources of error that influence these data, we can estimate x' as

$$x' = \bar{x} \pm u_x \quad (P\%) \tag{4.1}$$

where $\bar{x}$ represents the most probable estimate of x' based on the available data and u_x the confidence interval or uncertainty in that estimate at some probability level, $P\%$. The confidence interval (or uncertainty) is based both on estimates of the precision error and on the bias error in the measurement of x. In this chapter, we seek methods that will estimate x' and the precision error in x caused only by the variation in the data set.

Probability Density Functions

Regardless of the care taken in obtaining a set of data from independent measurements under identical conditions, random scatter in the data values will normally occur. Therefore, the measurand is also known as a *random variable*. If the variable is continuous in time or space, then it is said to be a continuous random variable. A variable represented only by discrete values is called a discrete random variable. During repeated measurements of a variable under fixed operating conditions, each data point may tend to assume one particular value or lie within some interval about this value more often than not, when all the data are compared. This tendency toward

---

[1] Bias error will not vary with repeated measurements and does not affect the statistics of the measurement. Bias is considered in Chapter 5.

Table 4.1 Sample of Random Variable x

i	x_i	i	x_i
1	0.98	11	1.02
2	1.07	12	1.26
3	0.86	13	1.08
4	1.16	14	1.02
5	0.96	15	0.94
6	0.68	16	1.11
7	1.34	17	0.99
8	1.04	18	0.78
9	1.21	19	1.06
10	0.86	20	0.96

one central value about which all the other values are scattered is known as a *central tendency* of a random variable.[2] Probability deals with the concept that a particular interval of values for a variable will be measured with some frequency relative to any other interval.

The central value and those values scattered about it can be determined from the probability density of the measured variable. The frequency with which the measured variable assumes a particular value or interval of values is described by its probability density. Consider a sample of x shown in Table 4.1, which consists of N individual measurements, x_i, where $i = 1, 2, \ldots, N$, each measurement taken at random but under identical test operating conditions. The measured values of this variable are plotted on a single axis in Figure 4.1.

In Figure 4.1, there exists a region on the axis where the data points tend to clump; this region contains the central value. Such behavior is typical of most engineering measurands. We might expect that the true mean value of x is contained somewhere in this clump.

This description for variable x can be extended. Suppose we replot the data of Table 4.1. The abscissa will be divided between the maximum and minimum measured values of x into K small intervals. Let the number of times, n_j, that a measured value assumes a value within an interval defined by $x - \delta x \leq x < x + \delta x$ be plotted on the ordinate. For small N, K should be conveniently chosen but such that $n_j \geq 5$ for at least one interval. For $N > 40$, an estimate [1] of the number of intervals K required for a viable statistical analysis is found from

$$K = 1.87(N - 1)^{0.40} + 1 \tag{4.2}$$

The resulting plot is called a *histogram* of the variable. The histogram is just another way of viewing both the tendency and the density of a variable. If the ordinate were nondimensionalized by dividing n_j by the total number of measurements of the variable, N, a *frequency distribution* of the variable would result. For any value of

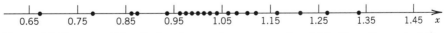

Figure 4.1 Concept of density in reference to a measured variable (from Example 4.1).

---

[2]Not all random variables display a central tendency; e.g., the value of a fair roll of a die could show values from 1 through 6 with equal probability of 1/6.

the variable, the frequency, f_j, at which that value of the variable occurred is found from its frequency distribution. This concept is illustrated in Example 4.1.

EXAMPLE 4.1

Compute the histogram and frequency distribution for the data of Table 4.1.

KNOWN

Data of Table 4.1

$N = 20$

ASSUMPTIONS

Fixed operating conditions

FIND

Histogram and frequency distribution

SOLUTION

To develop the histogram, compute a reasonable number of intervals for this data set. For $N = 20$, a convenient estimate of K is found from equation (4.2) to be

$$K = 1.87(N-1)^{0.40} + 1 = 7$$

Next, determine the maximum and minimum values of the data set and divide this range into K intervals. For a minimum of 0.68 and a maximum of 1.34, a $\delta x = 0.05$ is chosen. The intervals are shown below. $interval = \frac{1.34 - 0.68}{7}$

j	Interval	n_j	$f_j = n_j/N$
1	$0.65 \leq x_i < 0.75$	1	0.05
2	$0.75 \leq x_i < 0.85$	1	0.05
3	$0.85 \leq x_i < 0.95$	3	0.15
4	$0.95 \leq x_i < 1.05$	7	0.35
5	$1.05 \leq x_i < 1.15$	4	0.20
6	$1.15 \leq x_i < 1.25$	2	0.10
7	$1.25 \leq x_i < 1.35$	2	0.10

Since at least one interval has an $n_j \geq 5$, the interval number is adequate. The results are plotted in Figure 4.2. The plot displays a definite central tendency at the maximum frequency of occurrence within the interval 0.95 to 1.05.

COMMENT

The sum of the number of occurrences,

$$\sum_{j=1}^{K} n_j$$

must equal the total number of measurements, N. Likewise, the area under the percent frequency distribution curve will always equal the total frequency of occurrence

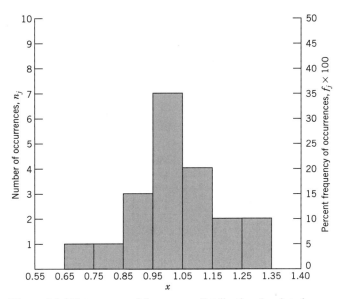

Figure 4.2 Histogram and frequency distribution for data in Table 4.1.

of 100%, that is

$$100 \times \sum_{j=1}^{K} f_j = 100\%$$

The *probability density function*, $p(x)$, results from the frequency distribution, in the limit as $N \to \infty$ and $\delta x \to 0$, by

$$p(x) = \lim_{N \to \infty, \, \delta x \to 0} \frac{n_j}{N(2\delta x)} \qquad (4.3)$$

The probability density function defines the probability that a measured variable might assume a particular value on any individual measurement. It also provides the central tendency of the variable. This central tendency is the desired representative value which gives the best estimate of the true mean value.

The actual shape that the probability density function will assume depends on the variable it represents and the circumstances affecting the process from which the variable is obtained. There are a number of standard probability density functions that suggest how a variable will be distributed. The specific values of the variable and the width of the distribution depend on the actual process, but the overall shape of the plot will most likely fit some standard distribution. A number of standard distributions that engineering data are likely to follow along with specific comments regarding the types of processes from which these are likely to be found are given in Table 4.2. Generally, experimentally determined histograms are used to determine which standard distribution the measured variable tends to follow. In turn, the standard distribution is used to interpret the data. Of course, the list in Table 4.2 is not all inclusive, and the reader is referred to more complete treatments of this subject [2, 4].

Table 4.2 Standard Statistical Distributions and Relations to Measurements

Distribution	Applications	Mathematical Representation
Normal	Most physical properties that are continuous or regular in time or space. Variations due to precision error.	$p(x) = \dfrac{1}{\sigma(2\pi)^{1/2}} \exp\left[-\dfrac{1}{2} \dfrac{(x-x')^2}{\sigma^2} \right]$
Log normal	Failure or durability projections; events whose outcomes tend to be skewed toward the extremity of the distribution.	$p(x) = \dfrac{1}{x\sigma(2\pi)^{1/2}} \exp\left[-\dfrac{1}{2} \ln\dfrac{(x-x')^2}{\sigma^2} \right]$
Poisson	Events randomly occurring in time; $p(x)$ refers to probability of observing x events in time t. Here λ refers to x'.	$p(x) = \dfrac{e^{-\lambda}\lambda^x}{x!}$
Weibull	Fatigue tests; similar to log normal applications.	See [4]
Binomial	Situations describing the number of occurrences, n, of a particular outcome during N independent tests where the probability of any outcome, P, is the same.	$p(n) = \left[\dfrac{N!}{(N-n)!n!} \right] P^n (1-P)^{N-n}$

Popcorn

Dice

Table 4.2 (*Continued*)

Shape

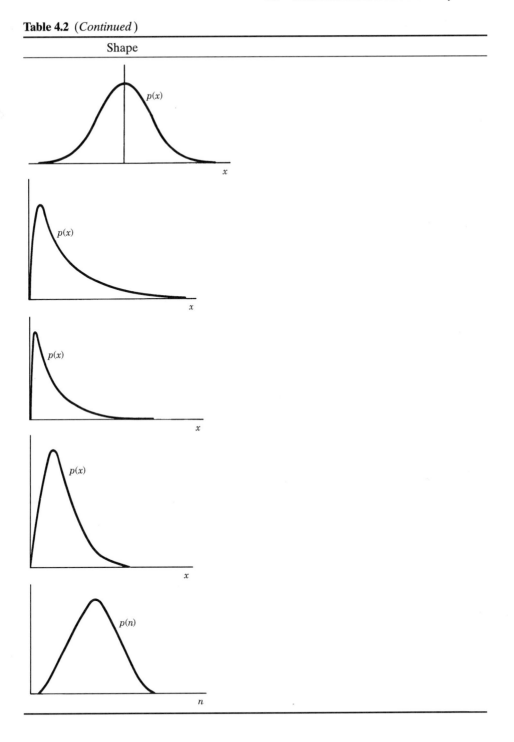

Regardless of the type of distribution assumed by a variable, a variable that shows a central tendency can be described and quantified through its mean value and variance. The mean value or central tendency of a continuous random variable, $x(t)$, having a probability density function $p(x)$, is given by

$$x' = \lim_{T \to \infty} \frac{1}{T} \int_0^T x(t)\, dt \tag{4.4a}$$

which for any continuous random variable x is equivalent to

$$x' = \int_{-\infty}^{\infty} x p(x)\, dx \tag{4.4b}$$

If the measured variable is described by discrete data, the mean value of measured variable, x_i, where $i = 1, 2, \ldots, N$, is given by

$$x' = \lim_{N \to \infty} \frac{1}{N} \sum_{i=1}^{N} x_i \tag{4.5}$$

Physically, the width of the density function reflects the data variation. For a continuous random variable, the variance is given by

$$\sigma^2 = \lim_{T \to \infty} \frac{1}{T} \int_0^T [x(t) - x']^2\, dt \tag{4.6a}$$

which is equivalent to

$$\sigma^2 = \int_{-\infty}^{\infty} (x - x')^2 p(x)\, dx \tag{4.6b}$$

or for discrete data, the variance is given by

$$\sigma^2 = \lim_{N \to \infty} \frac{1}{N} \sum_{i=1}^{N} (x_i - x')^2 \tag{4.7}$$

The standard deviation, σ, a commonly used statistical parameter, is defined as the square root of the variance.

A fundamental difficulty arises in the definitions given by equations 4.3–4.7 in that they assume an infinite number of measurements. But what if the data set is finite? Real data set sizes may range from as few as one to some large but finite number. For now we will study "infinite statistics" to introduce the connection between probability and statistics. Then we will turn our attention to the practical treatment of finite data sets.

4.3　INFINITE STATISTICS

A common distribution found in measurements is the normal or Gaussian distribution.[3] Many measurands common to engineering measurements are described by this distribution, which predicts that the scatter seen in a measured data set will be distributed symmetrically about some central tendency. The measurands of length, temperature, pressure, and velocity will likely display such a distribution, provided

---

[3] Actually, this distribution was independently suggested in the 18th century by Gauss, Laplace, and DiMoivre. However, fate is fickle and Gauss retains the honor.

the measurements are made repeatedly while the operating conditions are held fixed. Scatter brought about by process unsteadiness or by sources of precision error are "normally" distributed. The shape of the normal distribution is the familiar bell curve.

The probability density function for a random variable, x, having a normal distribution is defined as

$$p(x) = \frac{1}{\sigma(2\pi)^{1/2}} \exp\left[-\frac{1}{2}\frac{(x-x')^2}{\sigma^2}\right] \tag{4.8}$$

where x' is defined as the true mean value of x and σ^2 is the true variance of x. Hence, the exact form of $p(x)$ depends on the specific values for x' and σ. Note that a maximum in $p(x)$ will occur at $x = x'$, the true mean value. This implies that in the absence of bias the central tendency of a random variable having a normal distribution is toward its true mean value. The value most expected from any single measurement, that is, the most probable value, would be the true mean value.

How can we predict the probability that any future measurement will fall within some stated interval? The probability, $P(x)$, that random variable, x, will assume a value within the interval $x' \pm \delta x$ is given by the area under $p(x)$. This area is found by integration over the interval. Thus, this probability is given by

$$P(x' - \delta x \le x \le x' + \delta x) = \int_{x'-\delta x}^{x'+\delta x} p(x)\,dx \tag{4.9}$$

Integration of equation (4.9) is made easier through the following transformations. Begin by defining the terms $\beta = (x - x')/\sigma$ as the standardized normal variate for any value x and $z_1 = (x_1 - x')/\sigma$ as the variable that specifies an interval on $p(x)$. It follows that

$$dx = \sigma\,d\beta \tag{4.10}$$

so that equation (4.9) becomes

$$P(-z_1 \le \beta \le z_1) = \frac{1}{(2\pi)^{1/2}} \int_{-z_1}^{z_1} e^{-\beta^2/2}\,d\beta \tag{4.11}$$

Since for a normal distribution, $p(x)$ is symmetrical about x', one can write

$$\frac{1}{(2\pi)^{1/2}} \int_{-z_1}^{z_1} e^{-\beta^2/2} d\beta = 2\left[\frac{1}{(2\pi)^{1/2}} \int_{0}^{z_1} e^{-\beta^2/2} d\beta\right] \tag{4.12}$$

Known as the *normal error function*, the term in brackets in equation (4.12) provides one-half of the probability sought from equation (4.9) or its equivalent, equation (4.11). This half value is tabulated in Table 4.3 for the interval defined by z_1 in Figure 4.3.

It should now be clear that the statistical terms defined by equations (4.4)–(4.7) are actually statements of probability. The area under the portion of the probability density function curve, $p(x)$, defined by the interval $x' - z_1\sigma \le x \le x' + z_1\sigma$ provides the probability that a measurement will assume a value within that interval. Direct integration of $p(x)$ for a normal distribution between the limits $x' \pm z_1\sigma$ yields that for $z_1 = 1.0$, 68.26% of the area under $p(x)$ lies within $\pm 1.0\sigma$ of x'. This means that there is a 68.26% chance that a measurement of x will have a value within the interval $x' \pm 1.0\sigma$. As the interval defined by z_1 is increased, the probability of occurrence

Table 4.3 Probability Values for Normal Error Function

One-Sided Integral Solutions for $p(z_1) = \dfrac{1}{(2\pi)^{1/2}} \displaystyle\int_0^{z_1} e^{-\beta^2/2}\,d\beta$

$z_1 = \dfrac{x_1 - x'}{\sigma}$	1.90 0.00	1.91 0.01	1.92 0.02	1.93 0.03	1.94 0.04	1.95 0.05	1.96 0.06	1.97 0.07	1.98 0.08	1.99... 0.09
0.0	0.0000	0.0040	0.0080	0.0120	0.0160	0.0199	0.0239	0.0279	0.0319	0.0359
0.1	0.0398	0.0438	0.0478	0.0517	0.0557	0.0596	0.0636	0.0675	0.0714	0.0753
0.2	0.0793	0.0832	0.0871	0.0910	0.0948	0.0987	0.1026	0.1064	0.1103	0.1141
0.3	0.1179	0.1217	0.1255	0.1293	0.1331	0.1368	0.1406	0.1443	0.1480	0.1517
0.4	0.1554	0.1591	0.1628	0.1664	0.1700	0.1736	0.1772	0.1808	0.1844	0.1879
0.5	0.1915	0.1950	0.1985	0.2019	0.2054	0.2088	0.2123	0.2157	0.2190	0.2224
0.6	0.2257	0.2291	0.2324	0.2357	0.2389	0.2422	0.2454	0.2486	0.2517	0.2549
0.7	0.2580	0.2611	0.2642	0.2673	0.2704	0.2734	0.2764	0.2794	0.2823	0.2852
0.8	0.2881	0.2910	0.2939	0.2967	0.2995	0.3023	0.3051	0.3078	0.3106	0.3133
0.9	0.3159	0.3186	0.3212	0.3238	0.3264	0.3289	0.3315	0.3340	0.3365	0.3389
1.0	0.3413	0.3438	0.3461	0.3485	0.3508	0.3531	0.3554	0.3577	0.3599	0.3621
1.1	0.3643	0.3665	0.3686	0.3708	0.3729	0.3749	0.3770	0.3790	0.3810	0.3830
1.2	0.3849	0.3869	0.3888	0.3907	0.3925	0.3944	0.3962	0.3980	0.3997	0.4015
1.3	0.4032	0.4049	0.4066	0.4082	0.4099	0.4115	0.4131	0.4147	0.4162	0.4177
1.4	0.4192	0.4207	0.4222	0.4236	0.4251	0.4265	0.4279	0.4292	0.4306	0.4319
1.5	0.4332	0.4345	0.4357	0.4370	0.4382	0.4394	0.4406	0.4418	0.4429	0.4441
1.6	0.4452	0.4463	0.4474	0.4484	0.4495	0.4505	0.4515	0.4525	0.4535	0.4545
1.7	0.4554	0.4564	0.4573	0.4582	0.4591	0.4599	0.4608	0.4616	0.4625	0.4633
1.8	0.4641	0.4649	0.4656	0.4664	0.4671	0.4678	0.4686	0.4693	0.4699	0.4706
1.9	0.4713	0.4719	0.4726	0.4732	0.4738	0.4744	0.4750	0.4758	0.4761	0.4767
2.0	0.4772	0.4778	0.4783	0.4788	0.4793	0.4799	0.4803	0.4808	0.4812	0.4817
2.1	0.4821	0.4826	0.4830	0.4834	0.4838	0.4842	0.4846	0.4850	0.4854	0.4857
2.2	0.4861	0.4864	0.4868	0.4871	0.4875	0.4878	0.4881	0.4884	0.4887	0.4890
2.3	0.4893	0.4896	0.4898	0.4901	0.4904	0.4906	0.4909	0.4911	0.4913	0.4916
2.4	0.4918	0.4920	0.4922	0.4925	0.4927	0.4929	0.4931	0.4932	0.4934	0.4936
2.5	0.4938	0.4940	0.4941	0.4943	0.4945	0.4946	0.4948	0.4949	0.4951	0.4952
2.6	0.4953	0.4955	0.4956	0.4957	0.4959	0.4960	0.4961	0.4962	0.4963	0.4964
2.7	0.4965	0.4966	0.4967	0.4968	0.4969	0.4970	0.4971	0.4972	0.4973	0.4974
2.8	0.4974	0.4975	0.4976	0.4977	0.4977	0.4978	0.4979	0.4979	0.4980	0.4981
2.9	0.4981	0.4982	0.4982	0.4983	0.4984	0.4984	0.4985	0.4985	0.4986	0.4986
3.0	0.49865	0.4987	0.4987	0.4988	0.4988	0.4988	0.4989	0.4989	0.4989	0.4990

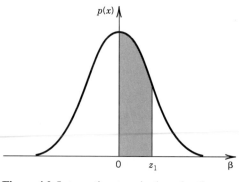

Figure 4.3 Integration terminology for the normal error function.

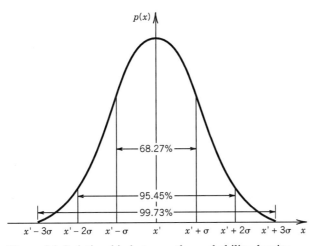

Figure 4.4 Relationship between the probability density function and its statistical parameters x' and σ for a normal distribution.

increases. For

$$z_1 = 2.0, \qquad 95.45\% \text{ of the area under } p(x) \text{ lies within } \pm z_1\sigma \text{ of } x'.$$
$$z_1 = 3.0, \qquad 99.73\% \text{ of the area under } p(x) \text{ lies within } \pm z_1\sigma \text{ of } x'.$$

This concept is illustrated in Figure 4.4.

EXAMPLE 4.2

Using the probability values in Table 4.3, show that the probability that a measurement will yield a value within $x' \pm \sigma$ is 0.6826, or 68.26%.

KNOWN

Table 4.3
$z_1 = 1$

ASSUMPTIONS

Data follow a normal distribution

FIND

$$P(x' - \sigma \leq x \leq x' + \sigma)$$

SOLUTION

To estimate the probability that a single measurement will yield a value within some interval, we need to evaluate the integral

$$\frac{1}{(2\pi)^{1/2}} \int_0^{z_1=1.0} e^{-\beta^2/2} \, d\beta$$

over the interval defined by z_1. Table 4.3 lists the solutions for this integral. Using

Table 4.3 for $z_1 = 1.0$, we find $P(z_1) = 0.3413$. However, since $z_1 \overset{z_1}{=} (x_1 - x')/\sigma$, then the probability that any measurement of x will produce a value within the interval $0 \leq x \leq x_1$ or $0 \leq x \leq x' + \sigma$ is 34.13%. Since the normal distribution is symmetric about x', the probability that x will fall within the interval defined between $-z_1\sigma$ and $+z_1\sigma$ for $z_1 = 1.0$ is $(2)(0.3413) = 0.6826$, or 68.26%. Accordingly, if a measurement of x were made, the probability that the value indicated by the measurement system would lie within the interval $x' - \sigma \leq x \leq x' + \sigma$ would be 68.26%.

COMMENT

Similarly, for $z_1 = 1.96$, the probability would be 95.0%.

It follows directly that the representative value that characterizes the variation of a measured data set is the standard deviation. The probability that the ith measured value of x will have a value between $x' \pm z_1\sigma$ is $2P(z_1) \times 100 = P\%$. This is written

$$x_i = x' \pm z_1\sigma \quad (P\%) \tag{4.13}$$

Thus, simple statistical analyses can provide useful quantification of a measured variable in terms of probability. This in turn can be useful in engineering situations in which the probable outcome of a measured variable has to be predicted or specified.

EXAMPLE 4.3

It is known that the statistics of a well-defined voltage signal are given by $x' = 8.5$ V and $\sigma^2 = 2.25$ V^2. If a single measurement of the voltage signal is made, determine the probability that the measured value will be between 10.0 and 11.5 V.

KNOWN

$x' = 8.5$ V
$\sigma^2 = 2.25$ V^2
$\sigma = 1.50$ V

ASSUMPTIONS

Signal has a normal distribution about x'

FIND

$P(10.0 \leq x \leq 11.5)$

SOLUTION

The standard deviation of the variable is $\sqrt{\sigma^2} = 1.5$ V. The probability that a value will fall between $8.5 \leq x \leq 10.0$ is given in Table 4.3 for $z_1 = (10.0 - 8.5)/1.5 = 1$. This yields $P(8.5 \leq x \leq 10.0) = P(z_1) = 0.3413$. Similarly, for the interval defined by $8.5 \leq x \leq 11.5$, $P(8.5 \leq x \leq 11.5) = P(z_1 = 2) = 0.4772$. The probability that x will fall into the interval $10.0 \leq x \leq 11.5$ is given by the area under $p(x)$ over this interval. This area is given by

$$P(10.0 \leq x \leq 11.5) = P(8.5 \leq x \leq 11.5) - P(8.5 \leq x \leq 10.0)$$
$$= 0.1359$$

There is a 13.59% probability that the measurement will yield a value between 10.0 and 11.5 V.

COMMENT

In general, the probability that a measured value will lie within an interval defined by any two values of z_1, such as z_a and z_b, is found by integration of $p(x)$ between z_a and z_b. For a normal probability density function, this probability is identical to $P(z_b) - P(z_a)$.

4.4   FINITE STATISTICS

If we recall the box of bearings discussed in Section 4.1, some two dozen bearings were measured, having been randomly selected from a population containing thousands. Can we use the resulting statistics from this sample to characterize the mean size and variance of the bearings within the box? Within the constraints imposed by probability, it is possible to estimate the true mean and true variance of the box of bearings from such a finite sample size. The method is now discussed.

Suppose we examine the case in which a measurement of some random variable $x(t)$ is to be performed. Assume for the moment that N discrete measurements (N repetitions) of this variable $x(t)$ have been made, each measurement represented by x_i, where $i = 1, 2, \ldots.N$, and N is finite. When N is less than infinity (or the entire population), all of the characteristics of a measured value may not be contained in the N data points. As such, statistical values obtained from finite-sized data sets should be regarded only as estimates of the true statistics of the measurand. Such statistics will be called finite statistics. Whereas *the true behavior of a variable is described by its infinite statistics, finite statistics describe only the behavior of the finite data set.*

Finite-sized data sets can provide the statistical estimates known as the *sample mean* value and the *sample variance*, defined by

$$\bar{x} = \frac{1}{N} \sum_{i=1}^{N} x_i \tag{4.14a}$$

$$S_x^2 = \frac{1}{N-1} \sum_{i=1}^{N} (x_i - \bar{x})^2 \tag{4.14b}$$

where $(x_i - \bar{x})$ is called the deviation of x_i. The *sample standard deviation*, S_x, is defined as $\sqrt{S_x^2}$. The sample mean value provides a most probable estimate of the true mean value, x'. The sample variance represents a measure of the *precision* of a measurement. These equations are robust and provide reasonable statistical estimates, regardless of the probability density function of the measurand.

In general, the number of degrees of freedom in a data set is the number of independent measurements available for estimating a statistical value. However, in N measurements having a central tendency, the data will be scattered about a mean value. The freedom of any data point in the measurement to assume any value, therefore, becomes restricted by this mean value. Hence, the degrees of freedom, ν, in the measure of data scatter is reduced by one to $N-1$, as seen in equation (4.14b).

The predictive utility of infinite statistics can be extended to data sets of finite sample size with only some modification. When sample sizes are finite, the z variable

Table 4.4 Student-t Distribution

$\nu = N-1$	t_{50}	t_{90}	t_{95}	t_{99}
1	1.000	6.314	12.706	63.657
2	0.816	2.920	4.303	9.925
3	0.765	2.353	3.182	5.841
4	0.741	2.132	~~2.770~~ 2.776	4.604
5	0.727	2.015	2.571	4.032
6	0.718	1.943	2.447	3.707
7	0.711	1.895	2.365	3.499
8	0.706	1.860	2.306	3.355
9	0.703	1.833	2.262	3.250
10	0.700	1.812	2.228	3.169
11	0.697	1.796	2.201	3.106
12	0.695	1.782	2.179	3.055
13	0.694	1.771	2.160	3.012
14	0.692	1.761	2.145	2.977
15	0.691	1.753	2.131	2.947
16	0.690	1.746	2.120	2.921
17	0.689	1.740	2.110	2.898
18	0.688	1.734	2.101	2.878
19	0.688	1.729	2.093	2.861
20	0.687	1.725	2.086	2.845
21	0.686	1.721	2.080	2.831
30	0.683	1.697	2.042	2.750
40	0.681	1.684	2.021	2.704
50	0.680	1.679	2.010	2.679
60	0.679	1.671	2.000	2.660
∞	0.674	1.645	1.960	2.576

described does not provide a reliable weight estimate of the true probability. However, the sample variance can be weighted in such a manner so as to compensate for the difference between the finite statistical estimates and the infinite statistics for a measured variable. For a normal distribution of x about some sample mean value. $\bar{x}$, one can state that statistically

$$x_i = \bar{x} \pm t_{\nu,P} S_x \quad (P\%) \tag{4.15}$$

where the variable $t_{\nu,P}$ is obtained from a new weighting function used for finite data sets and that replaces the z variable. This new variable is referred to as the t estimator. The interval $\pm t_{\nu,P} S_x$ represents a precision interval, given at probability $P\%$, within which one should expect any measured value to fall.

The value for the t estimator is a function of the probability, P, and the degrees of freedom, ν, in the standard deviation. These t values can be obtained from Table 4.4, which is a tabulation of the *Student-t distribution* as developed by William S. Gosset[4] (1876–1937), who recognized that the use of the z variable with S_x in place of σ was not reliable. Careful inspection of the t chart shows that the t value inflates the size of the interval required to attain a percent probability, $P\%$, to describe x. That is,

[4]At the time, Gosset was employed as a brewer and statistician by a well-known Irish brewery. You might pause to reflect on his multifarious contributions.

it has the effect of increasing the magnitude of $t_{v,P} S_x$ relative to $z_1 \sigma$ ($P\%$) when N is finite. As N approaches infinity, t approaches those values given by the z variable just as S_x must approach σ. It should be understood that for very small sample sizes ($N \leq 10$), sample statistics can be misleading. In those situations, other information regarding the measurement may be required, including additional measurements.

Standard Deviation of the Means

If we were to measure another two dozen of the bearings discussed in Section 4.1, we would expect the statistics from this new sample of randomly selected bearings to differ somewhat from the previous sample. This is simply due to the combined effects of a finite sample size and random variation in bearing size from manufacturing tolerances. So if this is expected behavior, how can we quantify how good the estimate is of the true mean based on a sample mean? That method is now discussed.

Suppose we were to measure a variable N times under fixed operating conditions. If we replicated this procedure M times, somewhat different estimates of the sample mean value and sample variance would be obtained for each of the M data sets. Why? The chance occurrence of events in any finite sample will affect the estimate of sample statistics. In fact, in experimental situations it is quite straightforward to demonstrate the variation of the sample mean value among different finite-sized samples of the same variable, even under identical operating conditions.[5] After M replications of the N measurements, a set of mean values that arc themselves normally distributed about some central value would be obtained. In fact, regardless of the probability density function of the measurand, the mean values obtained from M replications will follow a normal distribution.[6] This process is visualized in Figure 4.5. The amount of variation possible in the sample means would depend on two values: the sample variance, S_x^2, and sample size, N. The discrepancy tends to increase with variance and decrease with $N^{1/2}$.

This tendency for finite sample sets to have somewhat different statistics should not be surprising, for that is precisely the problem inherent to a finite data set. The

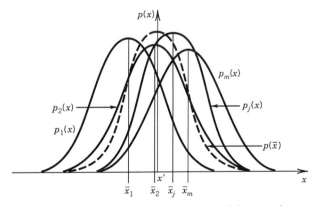

Figure 4.5 Normal distribution tendency of the sample means about a true value and in the absence of bias.

---

[5]This provides a compelling argument for providing replication in the measurement test plan.

[6]This is a consequence of the *central limit theorem* [2, 4].

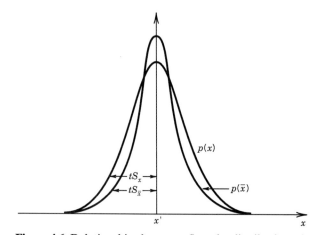

Figure 4.6 Relationships between S_x and a distribution of x and between $S_{\bar{x}}$ and the true value x'.

variation in the sample statistics will be characterized by a normal distribution of the sample mean values about the true mean. The variance of the distribution of mean values that could be expected can be estimated from a single finite data set through the *standard deviation of the means*, $S_{\bar{x}}$.

$$S_{\bar{x}} = \frac{S_x}{N^{1/2}} \tag{4.16}$$

An illustration of the relation between the standard deviation of a data set and the standard deviation of the means is given in Figure 4.6. The standard deviation of the means is a property of a finite data set. It reflects an estimate of how the sample mean values may be distributed about a true mean value.

So how good is the estimate of the true mean of a variable based on a finite-sized sample? The standard deviation of the means represents a measure of the precision in a sample mean. The range over which the possible values of the true mean value might lie at some probability level, P, based on the information from a sample data set is given as

$$\bar{x} \pm t_{v,P} S_{\bar{x}} \quad (P\%)$$

where $\pm t_{v,P} S_{\bar{x}}$ represents a *precision interval*, at the assigned probability, $P\%$ within which one should expect the true value of x to fall. As such, *the precision interval is a quantified measure of the precision error in the estimate of the true value of variable x*. This estimate of the true mean value based on a finite data set is stated as

$$x' = \bar{x} \pm t_{v,P} S_{\bar{x}} \quad (P\%) \tag{4.17}$$

EXAMPLE 4.4

Consider the data of Table 4.1. (a) Compute the sample statistics for this data set. (b) Estimate the interval of values over which 95% of the measurements of the measurand should be expected to lie. (c) Estimate the true mean value of the measurand at 95% probability based on this finite data set.

KNOWN

Table 4.1

$N = 20$

ASSUMPTIONS

Data set follows a normal distribution

FIND

$\bar{x}, \bar{x} \pm t S_x$, and $\bar{x} \pm t S_{\bar{x}}$

SOLUTION

The sample mean value is computed for the $N = 20$ values by the relation

$$\bar{x} = \frac{1}{20} \sum_{i=1}^{20} x_i = 1.02$$

This, in turn, is used to compute the sample standard deviation

$$S_x = \sqrt{\frac{1}{19} \sum_{i=1}^{20} (x_i - 1.02)^2} = 0.16$$

The degrees of freedom in the standard deviation are $v = N - 1 = 19$. From Table 4.4 at 95% probability, $t_{19,95}$ is 2.093. Then, the interval of values in which 95% of the measurements of x should lie is given by equation (4.15):

$$x_1 = \bar{x} \pm (2.093 \times 0.16) = 1.02 \pm 0.33 \quad (95\%)$$

Accordingly, if a 21st data point were to be taken, there is a 95% probability that its value would lie between 0.69 and 1.35.

The true mean value is estimated by the sample mean value. However, the precision interval for this estimate is $\pm t_{19,95} S_{\bar{x}}$, where

$$S_{\bar{x}} = \frac{S_{\bar{x}}}{N^{1/2}} = \frac{0.16}{(20)^{1/2}} = 0.04$$

Then, from equation (4.17),

$$x' = \bar{x} \pm t_{19,95} S_{\bar{x}} = 1.02 \pm 0.08 \quad (95\%) \tag{4.18}$$

Pooled Statistics

As discussed in Chapter 1, a good test plan uses replication, as well as repetition, in the measurement. Since replications are independent estimates of the same measured value, their data represent separate data samples that can be combined to provide a better statistical estimate of a measured variable than are obtained from a single sample. Samples that are grouped in a manner so as to determine a common set of statistics are said to be pooled.

Consider M replicates of a measurement of variable x, each of N repeated readings so as to yield the data set x_{ij}, where $i = 1, 2, \ldots, N$ and $j = 1, 2, \ldots, M$. It is assumed that bias error remains negligible in each replication. The *pooled mean* of x is defined by

$$\langle \bar{x} \rangle = \frac{1}{MN} \sum_{j=1}^{M} \sum_{i=1}^{N} x_{ij} \tag{4.19}$$

The *pooled standard deviation* of x is defined by

$$\langle S_x \rangle = \sqrt{\frac{1}{M(N-1)} \sum_{j=1}^{M} \sum_{i=1}^{N} (x_{ij} - \bar{x}_j)^2} = \sqrt{\frac{1}{M} \sum_{j=1}^{M} S_{x_j}^2} \tag{4.20}$$

with degrees of freedom, $v = M(N-1)$. The *pooled standard deviation of the means* of x is defined by

$$\langle S_{\bar{x}} \rangle = \frac{\langle S_x \rangle}{(MN)^{1/2}} \tag{4.21}$$

If the number of measurements of x are not the same between replications, then it is appropriate to weight each replication by its particular degrees of freedom. The pooled mean is then defined by its weighted mean

$$\langle \bar{x} \rangle = \frac{\sum_{j=1}^{M} N_j \bar{x}_j}{\sum_{j=1}^{M} N_j} \tag{4.22}$$

where subscript j refers to a particular data set. The pooled standard deviation is given by

$$\langle S_x \rangle = \sqrt{\frac{v_1 S_{x_1}^2 + v_2 S_{x_2}^2 + \cdots + v_M S_{x_M}^2}{v_1 + v_2 + \cdots + v_M}} \tag{4.23}$$

with degrees of freedom $v = \sum_{j=1}^{M} (v_j) = \sum_{j=1}^{M} (N_j - 1)$ and the pooled standard deviation of the means by

$$\langle S_{\bar{x}} \rangle = \frac{\langle S_x \rangle}{\left(\sum_{j=1}^{M} N_j \right)^{1/2}} \tag{4.24}$$

4.5   CHI-SQUARED DISTRIBUTION

In the previous section, we discussed how finite-sized data sets of the same measured variable would have somewhat different statistics. We used this argument to develop the concept of the standard deviation of the means as a precision indicator in the mean value. Similarly, we can estimate the precision by which S_x^2 predicts σ^2. If we plotted the sample standard deviation for many data sets, each having N data points, we would generate the probability density function, $p(\chi^2)$. The $p(\chi^2)$ follows the so-called *chi-squared* (χ^2) *distribution* depicted in Figure 4.7.

For the normal distribution, the χ^2 statistic is defined by (see [1, 2, or 4])

$$\chi^2 = v S_x^2 / \sigma^2 \tag{4.25}$$

with degrees of freedom $v = N - 1$.

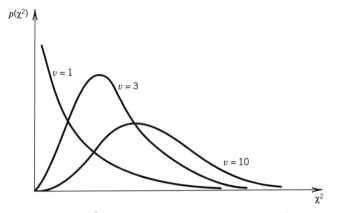

Figure 4.7 The χ^2 distribution with its dependency on degrees of freedom.

Precision Interval in a Sample Variance

The precision interval for the sample variance can be formulated by the probability statement

$$P\left(\chi^2_{1-\alpha/2} \leq \chi^2 \leq \chi^2_{\alpha/2}\right) = 1 - \alpha \tag{4.26}$$

with a probability of $P(\chi^2) = 1 - \alpha$. The term α is called the level of significance. Combining equations (4.25) with (4.26) gives

$$P\left[\nu S_x^2 / \chi^2_{\alpha/2} \leq \sigma^2 \leq \nu S_x^2 / \chi^2_{1-\alpha/2}\right] = 1 - \alpha \tag{4.27}$$

For example, the 95% precision interval by which S_x^2 estimates σ^2 is given by

$$\nu S_x^2 / \chi^2_{.025} < \sigma^2 < \nu S_x^2 / \chi^2_{.975} \quad (95\%) \tag{4.28}$$

Note that this interval is bounded by the 2.5% and 97.5% levels of significance.

The χ^2 distribution estimates the discrepancy expected as a result of random chance. Values for χ^2_α are tabulated in Table 4.5 as a function of the degrees of freedom. The $P(\chi^2)$ value equals the area under $p(\chi^2)$ as measured from the left, and the α value is the area as measured from the right, as noted in Figure 4.8. The total area under $p(\chi^2)$ is equal to unity.

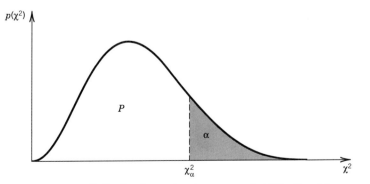

Figure 4.8 The χ^2 distribution as it relates to probability P and to the level of significance, $\alpha(= 1 - P)$.

Table 4.5 Values for χ_α^2

ν	$\chi_{0.99}^2$	$\chi_{0.975}^2$	$\chi_{0.95}^2$	$\chi_{0.90}^2$	$\chi_{0.50}^2$	$\chi_{0.05}^2$	$\chi_{0.025}^2$	$\chi_{0.01}^2$
1	0.000	0.000	0.000	0.016	0.455	3.84	5.02	6.63
2	0.020	0.051	0.103	0.211	1.39	5.99	7.38	9.21
3	0.115	0.216	0.352	0.584	2.37	7.81	9.35	11.3
4	0.297	0.484	0.711	1.06	3.36	9.49	11.1	13.3
5	0.554	0.831	1.15	1.61	4.35	11.1	12.8	15.1
6	0.872	1.24	1.64	2.20	5.35	12.6	14.4	16.8
7	1.24	1.69	2.17	2.83	6.35	14.1	16.0	18.5
8	1.65	2.18	2.73	3.49	7.34	15.5	17.5	20.1
9	2.09	2.70	3.33	4.17	8.34	16.9	19.0	21.7
10	2.56	3.25	3.94	4.78	9.34	18.3	20.5	23.2
11	3.05	3.82	4.57	5.58	10.3	19.7	21.9	24.7
12	3.57	4.40	5.23	6.30	11.3	21.0	23.3	26.2
13	4.11	5.01	5.89	7.04	12.3	22.4	24.7	27.7
14	4.66	5.63	6.57	7.79	13.3	23.7	26.1	29.1
15	5.23	6.26	7.26	8.55	14.3	25.0	27.5	30.6
16	5.81	6.91	7.96	9.31	15.3	26.3	28.8	32.0
17	6.41	7.56	8.67	10.1	16.3	27.6	30.2	33.4
18	7.01	8.23	9.39	10.9	17.3	28.9	31.5	34.8
19	7.63	8.91	10.1	11.7	18.3	30.1	32.9	36.2
20	8.26	9.59	10.9	12.4	19.3	31.4	34.2	37.6
30	15.0	16.8	18.5	20.6	29.3	43.8	47.0	50.9
60	37.5	40.5	43.2	46.5	59.3	79.1	83.3	88.4

EXAMPLE 4.5

Ten steel tension specimens are tested from a large batch, and a sample variance of $(200 \text{ kN/m}^2)^2$ is found. State the true variance expected at 95% confidence.

KNOWN

$S_x^2 = 40{,}000 \ (\text{kN/m}^2)^2$
$N = 10$

FIND

Precision interval for σ^2

SOLUTION

With $\nu = N - 1 = 9$, we find from Table 4.5, $\chi^2 = 19.0$ at $\alpha = 0.025$ written $\chi_{.025}^2$ and $\chi^2 = 2.7$ at $\alpha = 0.975$. Thus, from equation (4.27),

$$(9)(40000)/19.0 \leq \sigma^2 \leq (9)(40000)/2.7 \quad (95\%)$$

or, the precision interval for the variance is

$$18.947 \leq \sigma^2 \leq 133{,}333 \ (\text{kN/m}^2)^2 \quad (95\%)$$

or

$$(138)^2 \leq \sigma^2 \leq (365)^2 \leq (\text{kN/m}^2)^2 \quad (95\%)$$

This is the precision interval about σ^2 that is due to random chance. As N becomes larger, the precision interval will narrow.

EXAMPLE 4.6

A manufacturer knows from experience that the variance in the diameter of the roller bearings used in their bearings is 3.15 μm^2. Rejecting bearings drives up the unit cost. However, they reject any batch of roller bearings if the sample variance of 20 pieces selected at random exceeds 5 μm^2. Assuming a normal distribution, what is the probability that any given batch will be rejected even though its true variance is actually within the tolerance limits?

KNOWN

$\sigma^2 = 3.15\,\mu m^2$
$S_x^2 = 5\,\mu m^2$ based on $N = 20$

ASSUMPTIONS

Bearing size is normally distributed

FIND

χ_α^2

SOLUTION

This problem could be reposed as follows: What is the probability that S_x^2 based on 20 measurements will not predict σ^2 for the entire batch?
   For $v = N - 1 = 19$, and with equation (4.25), the χ^2 value is

$$\chi_\alpha^2(v) = vS_x^2/\sigma^2 = 30.16$$

An inspection of Table 4.5 shows $\chi_{.05}^2(19) = 30.1$, so we can take $\alpha \approx 0.05$. Taking χ_α^2 as a measure of discrepancy caused by random chance, we interpret this result as a 5% chance that a batch actually within tolerance will be rejected. So there is a probability, $P = 1 - \alpha$, of 95% that S_x^2 predicts the σ^2 for the batch. Rejecting this batch is a good decision.

Goodness-of-Fit Test

Just how well does a set of measurements follow an assumed distribution function? For example, in Example 4.4, we assumed that the data of Table 4.1 followed a normal distribution based only on the rough form of its histogram (Figure 4.2). A more rigorous approach would apply the chi-squared test using the chi-squared distribution. The chi-squared test provides a measure of the discrepancy between the measured variation of a data set and the variation predicted by the assumed distribution function.
   To begin, construct a histogram of K intervals from a data set of N measurements. This establishes the measured number of occurrences, n_j, that the measured value lies within the jth interval. Then calculate the degrees of freedom in the variance for

the data set, $v = N - m$, where m is the number of restrictions imposed. From v, estimate the predicted number of occurrences, n'_j, to be expected from the distribution function. For this test, the χ^2 value is calculated from the entire histogram by

$$\chi^2 = \frac{\sum\limits_j (n_j - n'_j)^2}{n'_j} \qquad j = 1, 2, \ldots, K \qquad (4.29)$$

For a given degree of freedom, the better a data set fits the assumed distribution function, the lower its χ^2 value will be, whereas the higher the χ^2 value, the more dubious is the fit.

The χ^2_α table, given in Table 4.5, can be interpreted as a measure of the discrepancy expected as a result of random chance. For example, a value for α of 0.95 implies that 95% of the discrepancy between the histogram and the assumed distribution is due to random variation only. With $P(\chi^2) = 1 - \alpha$, this leaves only 5% of the discrepancy caused by a systematic tendency, such as a different distribution. In general, a $P(\chi^2) < 0.05$ confers a very strong measure of a good fit to the assumed distribution, an unequivocal result. In contrast, values within the range

$$5\% \leq P(\chi^2) \leq 95\%$$

suggest that the data *could* be described by the distribution function assumed (an equivocal or ambiguous result). In that case, the use of the assumed distribution may be as reasonable as some other. Lastly, values of $P(\chi^2) > 95\%$ provide a strong measure against the assumed distribution. A different distribution should be attempted.

As with all finite data sets, statistics can only be used to suggest what is probable. Conclusions are left to the user. For example, data that fit a distribution too well could be suspected of being "constructed" or "fixed," such as might be the outcome from a loaded pair of dice (or other games of chance).

EXAMPLE 4.7

Test the hypothesis that the variable x as given by the measured data of Table 4.1 is described by a normal distribution.

KNOWN

Table 4.1 and Histogram of Figure 4.2
From Example 4.4: $\bar{x} = 1.02$ $S_x = 0.16$ $N = 20$

FIND

Apply the chi-squared test to the data set.

SOLUTION

Figure 4.2 (Example 4.1) provides a histogram for the data of Table 4.1 giving the values for n_j for $K = 7$ intervals. To evaluate the hypothesis, we must find the predicted number of occurrences, n'_j, for each of the seven intervals based on a normal distribution. To do this: Substitute $x' = \bar{x}$ and $\sigma = S_x$ and compute the probabilities based on the z values.

Table 4.6 Chi-Squared Test for Example 4.7

j	n_j	n'_j	$(n_j - n'_j)^2/n'_j$
1	1	0.72	0.11
2	1	1.96	0.47
3	3	3.74	0.15
4	7	8.50	0.27
5	4	3.96	0
6	2	2.66	0.16
7	2	1.12	0.69
			$\chi_\alpha^2 = 1.85$

For example, consider the second interval ($j = 2$). With the use of Table 4.3, the predicted probabilities are

$$P(0.75 \leq x < 0.85) = P(0.75 \leq x_i < x') - P(0.85 \leq x_i < x')$$
$$= P(z_a) - P(z_b)$$
$$= P(1.6875) - P(1.0625)$$
$$= 0.454 - 0.356 = 0.098$$

So for a normal distribution, we should expect the measured value to lie within the second interval for 9.8% of the measurements. With

$$n'_2 = N \times P(0.75 \leq x_i < 0.85) = 20 \times 0.098 = 1.96$$

that is, 1.96 occurrences are expected out of 20 measurements in this second interval. The measured data set shows $n_2 = 1$.

The results are summarized in Table 4.6 with x^2 based on equation (4.29). Because two calculated statistical values ($\bar{x}$ and S_x) are used in the computations, the degrees of freedom in χ^2 are restricted by $m = 2$:

$$\nu = K - m = 7 - 2 = 5$$

and from Table 4.5, for $\chi_\alpha^2(\nu) = 1.85$, $\alpha \approx 0.84$ or $P(\chi^2) \approx 0.16$. Although there is a high probability that the discrepancy between the histogram and the normal distribution is only due to a random variation of a finite data set, we should consider this result as being equivocal. The hypothesis that x is described by a normal distribution is neither proven nor disproven.

4.6   REGRESSION ANALYSIS

The discussion so far has dealt solely with the statistical determination of a measurand under fixed operating conditions. The measured variable is usually a function of one or more independent variables that are controlled during the measurement. When the measured variable is sampled, these variables are controlled, to the extent possible, as are all the other operating conditions. Following the sample, one of these variables is changed and a sample is made under the new operating conditions. This is the procedure used to document the relationship between the measured variable and an independent process variable. A regression analysis can often be used to establish a functional relationship between the dependent variable and the independent

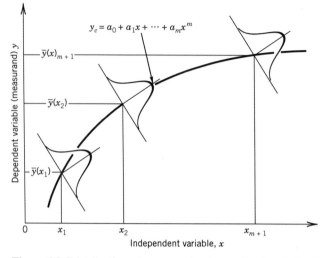

Figure 4.9 Distribution of measured value y about each fixed value of independent variable x. Curve y_c represents a possible functional relationship.

variable, which will hold on the average. It is used in those situations in which the relationship anticipated is either polynomial in form or can be approximated by a Fourier series. This discussion pertains directly to polynomial fits. More information on Fourier series fits can be found in [3].

A regression analysis assumes that the variation found in the dependent measured variable, the measurand, follows a normal distribution about each fixed value of the independent variable. Such behavior is illustrated in Figure 4.9 by considering the dependent variable $y_{i,j}$ consisting of N measurements, $i = 1, 2, \ldots, N$, of y at each of n values of independent variable, $x_j, j = 1, 2, \ldots, n$. This type of behavior is most common during calibrations, where the input to the measuring system, x, is held nominally fixed while the measurement of y occurs. It is also frequent in many types of measurements in which the dependent variable is assumed to be measured under fixed operating conditions. Repeated measurements of y will yield a normal distribution with variance $S_y^2(x_j)$, about some mean value, $\bar{y}(x_j)$.

Most spreadsheet and higher-level software packages provide the computer algorithms necessary to perform a regression analysis. Regardless of the package used, the engineer remains responsible for selecting the appropriate analysis for a given data set and interpreting the results of such analyses. The following discussion presents the concepts of regression analysis and its limitations.

Least-Squares Regression Analysis

The regression analysis for a single variable of the form $y = f(x)$ provides an mth-order polynomial fit of the data in the form

$$y_c = a_0 + a_1 x + a_2 x^2 + \cdots + a_m x^m \tag{4.30}$$

where y_c refers to the value of the dependent variable obtained directly from the polynomial equation for a given value of x. For n different values of the independent variable included in the analysis, the highest order, m, of the polynomial that can be determined is restricted to $m \leq n - 1$. The values of the m coefficients $a_0, a_1, \ldots, a_m$ are determined by the analysis. The most common form of regression analysis for engineering applications is the *method of least-squares*. The least-squares technique attempts to minimize the sum of the squares of the deviations between the actual data and the polynomial fit of a stated order by adjusting the values of the coefficients as necessary.

An mth-order polynomial relationship is to be found for a set of N data points of the form (x, y) in which x and y are the independent and dependent variables, respectively. Consider the situation in which N values of y exist, y_i, where $i = 1, 2, \ldots, N$, over n values of x. The task is to find the $m + 1$ coefficients, $a_0, a_1, \ldots, a_m$, of the polynomial of equation (4.30). Define the deviation between any dependent variable y_i and the polynomial as $y_i - y_{ci}$, where y_{ci} is the value of the polynomial evaluated at the data point (x_i, y_i). The sum of the squares of this deviation for all values of $y_i, i = 1, 2, \ldots, N$ is

$$D = \sum_{i=1}^{N}(y_i - y_{ci})^2 \tag{4.31}$$

The goal is to reduce D to a minimum for a given order of polynomial.

Combining equations (4.30) and (4.31), one can write

$$D = \sum_{i=1}^{N}[y_i - (a_0 + a_1 x + \cdots + a_m x^m)]^2 \tag{4.32}$$

Now the total differential of D is dependent on the $m + 1$ coefficients through

$$dD = \frac{\partial D}{\partial a_0}da_0 + \frac{\partial D}{\partial a_1}da_1 + \cdots + \frac{\partial D}{\partial a_m}da_m$$

To minimize the sum of the squares of the deviations, one wants dD to be zero. This is accomplished by setting each of the partial derivatives equal to zero:

$$
\begin{aligned}
\frac{\partial D}{\partial a_0} &= 0 = \frac{\partial}{\partial a_0}\left\{\sum_{i=1}^{N}[y_i - (a_0 + a_1 x + \cdots + a_m x^m)]^2\right\} \\
\frac{\partial D}{\partial a_1} &= 0 = \frac{\partial}{\partial a_1}\left\{\sum_{i=1}^{N}[y_i - (a_0 + a_1 x + \cdots + a_m x^m)]^2\right\} \\
&\vdots \qquad\qquad\qquad\qquad \vdots \\
\frac{\partial D}{\partial a_m} &= 0 = \frac{\partial}{\partial a_m}\left\{\sum_{i=1}^{N}[y_i - (a_0 + a_1 x + \cdots + a_m x^m)]^2\right\}
\end{aligned}
\tag{4.33}
$$

This yields $m + 1$ equations, which are solved simultaneously to yield the unknown regression coefficients, $a_0, a_1, \ldots, a_m$.

In general, the polynomial found by using a regression analysis will not fit every data point (x_i, y_i) exactly. Associated with any order of polynomial curve fit through a given set of data points, there will exist some deviation between the data point and the polynomial. One can compute a standard deviation based on the deviation of each data point and the fit by

$$S_{yx} = \sqrt{\frac{\sum\limits_{i=1}^{N}(y_i - y_{ci})^2}{\nu}} \tag{4.34}$$

where v is the degrees of freedom of the fit, $v = N - (m+1)$. S_{yx} is referred to as the *standard error of the fit* and is a measure of the precision with which a polynomial describes the behavior of the data set.

The best order of polynomial fit to a particular data set is the lowest order of fit that reduces S_{yx} to an acceptable value *and* maintains a logical physical sense between dependent and independent variables. This latter point is important. If the underlying physics of a problem implies that a certain order relationship should exist between dependent and independent variables, there is no sense in forcing the data to fit any other order of polynomial regardless of the value of S_{yx}.

In engineering, the independent variable is often a known and controlled value. This is particularly true during calibration. In such cases, we can assume that the variance of the curve fit line is not due to the independent value. We can state the curve fit with its precision interval as

$$y_c \pm t_{v,P} S_{yx} \quad (P\%) \tag{4.35}$$

where y_c is defined by equation (4.30).

In contrast, if we consider variability in both the independent and dependent variables, then the curve fit and its precision interval are estimated by [1, 4]

$$y_c \pm t_{v,P} S_{yx} \left[\frac{1}{N} + \frac{(x_i - \bar{x})^2}{\displaystyle\sum_{i=1}^{N}(x_i - \bar{x})^2} \right]^{1/2} \quad (P\%) \tag{4.36}$$

This is the case if, say, we wished to compare the dietary fat intake of males with their blood cholesterol levels. If a random sample of the male population were to be taken, we would have no control over the independent variable, in this case, the number of individuals having a certain fat intake level. Because there would generally be fewer data points at the extreme ends of the range, the precision error at the extremes of the curve fit would increase. This is reflected in equation (4.36).

There is no rule that can be used to estimate which order fit will yield an acceptable value of S_{yx} without trial and error. This is the attractive feature of having a least-squares software package available. The choice of the actual order of fit used is always a compromise between the precision needed and the convenience of using a low-order polynomial.

EXAMPLE 4.8

The following data are suspected to follow a linear relationship. Find an appropriate equation of the first-order form.

x[cm]	y[V]
1.0	1.2
2.0	1.9
3.0	3.2
4.0	4.1
5.0	5.3

KNOWN

Independent variable, x

Dependent measured variable, y

$N = 5$

ASSUMPTIONS

Linear relation

FIND

$y_c = a_0 + a_1 x$

SOLUTION

We seek a polynomial of the form $y_c = a_0 + a_1 x$, which minimizes the term

$$D = \sum_{i=1}^{N} (y_i - y_{ci})^2$$

setting derivatives to zero:

$$\frac{\partial D}{\partial a_0} = 0 = -2 \sum_{i=1}^{N} [y_i - (a_0 + a_1 x_i)]$$

$$\frac{\partial D}{\partial a_1} = 0 = -2 \sum_{i=1}^{N} [y_i - (a_0 + a_1 x_i)] x_i$$

Solving simultaneously for the coefficients a_0 and a_1 yields

$$a_0 = \frac{\Sigma x_i \, \Sigma x_i y_i - \Sigma x_i^2 \, \Sigma y_i}{(\Sigma x_i)^2 - N \Sigma x_i^2}$$

$$a_1 = \frac{\Sigma x_i \, \Sigma y_i - N \Sigma x_i y_i}{(\Sigma x_i)^2 - N \Sigma x_i^2}$$

(4.37)

From the data set, one finds from equation (4.37) $a_0 = 0.02$ and $a_1 = 1.04$. Hence,

$$y_c = 0.02 + 1.04x \quad \text{V}$$

COMMENT

Although the polynomial described by y_c is the linear curve fit for this data set, we still have no idea of how good this curve fits this data set or even if a first-order fit is appropriate. This is studied next.

Linear Polynomials

For linear polynomials a correlation coefficient, r, can be found by

$$r = \sqrt{1 - \frac{S_{yx}^2}{S_y^2}}$$

(4.38)

where

$$S_y^2 = \frac{1}{N-1} \sum_{i=1}^{N} (y_i - \bar{y})^2$$

The correlation coefficient represents a quantitative measure of the linear association between x and y. It is bounded by ± 1, which represents perfect correlation; the sign indicates that y increases or decreases with x. For $\pm 0.9 < r \leq \pm 1$, a linear regression can be considered as a reliable relation between y and x. Alternatively, the value r^2 is often reported, which is indicative of how well the variance in y is accounted for by the fit. It is a ratio of the variation assumed by the linear fit to the actual measured variations in the data. However, the correlation coefficient and the r^2 value are only indicators of the hypothesis that y and x are linearly related. They are not effective precision indicators of y_c. The S_{yx} value is used for that purpose.

The precision estimate of the slope of the fit can be estimated by

$$S_{a_1} = S_{yx} \sqrt{\frac{N}{N \sum_{i=1}^{N} x_i^2 - \left(\sum_{i=1}^{N} x_i \right)^2}} \tag{4.39}$$

For example, S_{a_1} would provide a measure of the static sensitivity error of a measurement system based on a linear fit of the calibration data.

The precision estimate of the zero intercept can be estimated by

$$S_{a_0} = S_{yx} \sqrt{\frac{N \sum_{i=1}^{N} x_i^2}{N \left[N \sum_{i=1}^{N} x_i^2 - \left(\sum_{i=1}^{N} x_i \right)^2 \right]}} \tag{4.40}$$

An error in a_0 would offset a calibration curve from its y intercept. The derivation and further discussion on equations (4.38)–(4.40) can be found in [1, 2, 4].

EXAMPLE 4.9

Compute the correlation coefficient and the standard error of the fit for the data in Example 4.8.

KNOWN

$y_c = 0.02 + 1.04x$ V

ASSUMPTIONS

Errors are normally distributed

FIND

r and S_{yx}

SOLUTION

Direct application of equation (4.38) with the data set yields for the correlation coefficient $r = 0.996$. An equivalent estimator is r^2. Here $r^2 = 0.99$, which indicates that

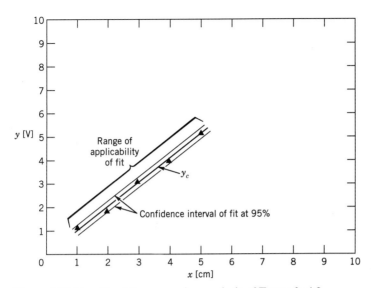

Figure 4.10 Results of the regression analysis of Example 4.9.

99% of the variance in y is accounted for by the fit, whereas only 1% is unaccountable. These values suggest that a linear fit is a reliable relation between x and y. The precision error between the data and this fit can be quantified through S_{yx}. Using equation (4.36), $S_{yx} = 0.16$ with degrees of freedom, $\nu = N - (m+1) = 3$.

COMMENT

The t estimator, $t_{3,95} = 3.18$, establishes a precision or confidence interval about the fit of $\pm(t_{3,95}S_{yx}) = \pm 0.50$. Accordingly, the polynomial fit can be stated at 95% confidence as

$$y_c = 1.04x + 0.02 \pm 0.50 \, \text{V} \quad (95\%)$$

This curve is plotted in Figure 4.10 with its 95% confidence interval. The regression polynomial with its precision interval is the only acceptable way to report a curve fit to a data set.

EXAMPLE 4.10

A velocity probe provides a voltage output that is related to velocity, U, by the form $E = a + bU^m$. A calibration is run and the data $(N = 5)$ are recorded as follows. Estimate an appropriate curve fit.

U[ft/s]	E_i[V]
0.0	3.19
10.0	3.99
20.0	4.30
30.0	4.48
40.0	4.65

KNOWN

$N = 5$

ASSUMPTIONS

Data related by $E = a + bU^m$

FIND

a and b

SOLUTION

The equation can be transformed into

$$\log(E - a) = \log b + m \log U$$

which has the linear form

$$Y = B + mX$$

The value of a is evaluated at $U = 0$ ft/s to be 3.19 V. The values for Y and X are computed as follows with the corresponding deviations from the resulting fit:

U	Y	X	$E_i - E_{c_i}$
0.0	—	—	0.0
10.0	−0.097	1.0	−0.01
20.0	0.045	1.30	0.02
30.0	0.111	1.48	0.0
40.0	0.164	1.60	−0.01

From equation (4.37), the values for Y and X are substituted with the results $a_0 = B = -0.525$ and $a_1 = m = 0.43$. The standard error of the fit is found from equation (4.25) to be $S_{yx} = 0.01$. From Table 4.4, $t_{3,95} = 3.18$ so that the precision interval $\pm t_{v,P} S_{yx}$ is found to be ± 0.03 V. This implies that

$$Y = -0.525 + 0.43X \pm 0.03 \text{ V} \quad (95\%)$$

The polynomial is transformed back to the form $E = a + bU^m$:

$$E_c = 3.19 + 0.30U^{0.43} \text{ V}$$

The curve fit with its 95% precision interval is denoted in Figure 4.11.

A regression analysis of multiple variables of the form $y = f(x_1, x_2, \ldots)$ is also possible. It will not be discussed here, but the concepts generated for the single-variable analysis are carried through for a multiple-variable analysis. The interested reader is referred to references [2, 4].

4.7 DATA OUTLIER DETECTION

It is not uncommon for a measured data set to contain one or more spurious data points that appear to be unrelated to the expected tendency of the data set. Data that lie outside the probability of normal variation can bias the sample mean value estimate and increase the precision estimates of the measured data set. Statistical

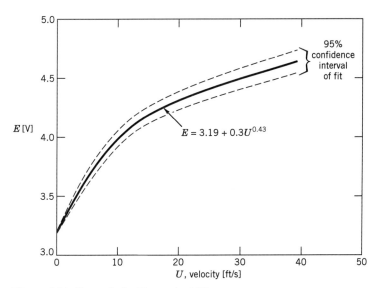

Figure 4.11 Curve fit for Example 4.10.

techniques can be used to detect extraneous data points, which are known as *outliers* because their measured values fall outside the probable range of values for a measurand. Once they are detected, a decision as to whether to remove the data points from the data set can be made. One is strongly cautioned to be certain that these extraordinary deviations from central tendency are truly spurious, and not a unique physical tendency of the measurand, prior to their removal. Data set histograms are particularly effective for visual detection of suspected outliers. If outliers are removed from a data set, the data statistics should be recomputed by using the remaining data.

Several methods exist for the detection of outliers. The specific methods presented here were chosen for their ease of use. The objective of the methods is to detect those data points that fall outside the normal range of variation expected in a data set based on the variance of that data set. This range is defined by some multiple of the standard deviation of the data set. Since a multiple of the standard deviation is related to the area under the probability density function, the basis for the outlier detection lies with the probability of occurrence for the value in question.

The simplest method for outlier detection would be to calculate the data set statistics and to label all data points that lie outside the range of 99.8% probability of occurrence, $\bar{x} \pm t_{v,99.8}S_x$, as outliers. This *three-sigma*[7] test works well with small data sets of 10 or more points and is easily programmed for use with computer-based data reduction schemes.

A more rigorous variation of this test, particularly well suited to large data sets, is the *modified three-sigma* test for outliers. The sample mean and standard deviations are estimated and a modified z variable is computed for each data point by

$$z_0 = \left| \frac{x_i - \bar{x}}{S_x} \right| \tag{4.41}$$

The probability that x lies outside the one-sided range defined by 0 and z_0 is $0.5 - P(z_0)$, where $P(z_0)$ is found from the one-sided z chart of Table 4.3. For N data points, if $N[0.5 - P(z_0)] \leq 0.1$, the data point can be considered as an outlier. Other

---

[7]So named since, as $v \to \infty$, $t \to 3$ for 99.8% probability.

methods of outlier detection can be found in statistics texts and measurement codes (e.g., [2, 5]).

EXAMPLE 4.11

Consider the data given here for 10 measurements of tire pressure taken with an inexpensive handheld gauge. Compute the statistics of the data set; then test for outliers by using the modified three-sigma test.

n_i	x[psi]	n_i	x[psi]
1	28	6	24
2	31	7	29
3	27	8	28
4	28	9	18
5	29	10	27

KNOWN

$N = 10$

ASSUMPTIONS

Each measurement obtained under fixed conditions

FIND

$\bar{x}$ and S_x; apply outlier detection tests

SOLUTION

Based on the 10 data points, the sample mean and sample variance can be found from equations (4.14a) and (4.14b) to be $\bar{x} = 27$ psi, $S_x = 3.8$. However, one would not expect tire pressure to vary much between two readings beyond the precision capabilities of the measurement system and technique, so data point 9 looks suspect as an outlier. Apply the modified three-sigma test to this data point.

For $x = 18$, $z_0 = 2.34$ and, from Table 4.3, $P(z_0) = 0.4904$, so we find $0.5 - P(z_0) = 0.0096$. Hence, $N[0.5 - P(z_0)] = 0.096$, a value that is less than 0.1. Data point 9 can be considered an outlier; that is, the magnitude of its deviation from the mean value is greater than that expected by normal variation. Since this data point clearly stands out as unusual in light of the measurement undertaken, it can be eliminated as a bad point. The remaining data variations reflect the precision of the measurement (variable, procedure, and instrument). The data set statistics become $N = 9, \bar{x} = 28$ psi, $S_{\bar{x}} = 2.0$, and $S_{\bar{x}} = 0.7$. When bias errors are neglected, the best estimate of the tire pressure can be stated as

$$x' = \bar{x} \pm t_{8,95} S_x = 28 \pm 1.6 \, \text{psi} \quad (95\%)$$

4.8 NUMBER OF MEASUREMENTS REQUIRED

Statistics can be used to assist in the design and planning of a test program. For example, how many measurements, N, are required to estimate the true mean value,

x', with acceptable precision? To answer this question, begin with equation (4.17), which expresses the true value based on a sample mean and its precision interval:

$$x' = \bar{x} \pm t_{v,95} \frac{S_x}{N^{1/2}}$$

We can express the precision interval in equation (4.17) as CI, that is,

$$\text{CI} = \pm t_{v,95} \frac{S_x}{N^{1/2}} \quad (95\%) \tag{4.42}$$

To evaluate equation (4.42), we must assign a value to S_x. The value of S_x should be a conservative estimate based on previous test data, prior experience, or manufacturer's information.

The precision interval is two sided about the mean, defining a range from $-t_{v,95}S_x/N^{1/2}$ to $+t_{v,95}S_x/N^{1/2}$. We introduce the one-sided precision value d as

$$d = \frac{\text{CI}}{2} = \frac{t_{v,95}S_x}{N^{1/2}} \tag{4.43}$$

Then, it follows that the required number of measurements is estimated by

$$N \approx \left(\frac{t_{v,95}S_x}{d} \right) \quad (95\%) \tag{4.44}$$

The approximation serves as a reminder that this expression is based on an assumed value for S_x. The accuracy of equation (4.43) will depend on how well the assumed value for S_x approximates σ.

The obvious deficiency in this method is that an estimate for the sample variance is needed. One way around this is to make a preliminary small number of measurements, N_1, to obtain an estimate of the sample variance, S_1, to be expected. Then S_1 is used to estimate the number of measurements required. The total number of measurements, N_T, will be estimated by

$$N_T \approx \left(\frac{t_{N_1-1,95}S_1}{d} \right)^2 \quad (95\%) \tag{4.45}$$

This establishes that $N_T - N_1$ additional measurements will be required.

EXAMPLE 4.12

Determine the number of measurements required to reduce the precision interval of the mean value of a variable to within 1 unit if the variance of the variable is estimated to be ~64 units.

KNOWN

CI $= 1$ unit $P = 95\%$

$d = 1/2$ $\sigma^2 = 64$ units

ASSUMPTIONS

$\sigma^2 \approx S_x^2$

FIND

N required

SOLUTION

Because equation (4.44) has two unknowns, begin this problem by guessing at some value for N. Then, using this guessed value, compute the t variable at the probability level desired. An updated value for N can then be found from the formulation

$$N = \left(\frac{t_{v,95}S_x}{d}\right)^2 \quad (95\%)$$

Then use trial and error iteration to converge on a value for N.

Suppose we begin by guessing that $N = 500$. Then

$$v = 499 \qquad t_{499,95} = 1.96 \Rightarrow N = 983$$

So, now guess $N = 983$. Then

$$v = 982 \qquad t_{982,95} = 1.96 \Rightarrow N = 983$$

We have converged on $N = 983$. Thus, at least 983 measurements must be made to achieve the desired precision interval in the measured variable. An analysis of the results after 983 measurements should be made to ensure that the variance level used was representative of the actual data set.

COMMENT

Since the precision interval is reduced as $N^{1/2}$, the procedure of increasing N to decrease the precision interval becomes one of diminishing returns.

EXAMPLE 4.13

From 51 measurements of a variable, S_1 is found to be 160. For a 95% precision interval of 60 units in the mean value, estimate the total number of measurements required.

KNOWN

$S_1 = 160$ units $N_1 = 51$
$d = CI/2 = 30$ $t_{50,95} = 2.01$

ASSUMPTIONS

$\sigma \approx S_1$

FIND

N_T

SOLUTION

The total number of measurements required is estimated by

$$N_T \approx \left(\frac{t_{N_1-1,95}S_1}{d}\right)^2 \quad (95\%)$$

$$= \left(\frac{2.01 \times 160}{30}\right)^2 = 115$$

Table 4.7 Summary Table for a Sample of N Data Points

Sample mean	$\bar{x} = \dfrac{1}{N} \displaystyle\sum_{i=1}^{N} x_i$
Sample standard deviation	$S_x = \sqrt{\dfrac{1}{N-1} \displaystyle\sum_{i=1}^{N} (x_i - \bar{x})^2}$
Standard deviation of the means	$S_{\bar{x}} = [S_x/(N)^{1/2}]$
Precision interval for a single data point, x_1	$\pm t_{v,P} S_x \quad (P\%)$
Precision interval for a mean value, $\bar{x}$	$\pm t_{v,P} S_{\bar{x}} \quad (P\%)$
Precision interval for a curve fit between x and y	$\pm t_{v,P} S_{yx} \quad (P\%)$

Thus, as a first guess, a total of 115 measurements are estimated to be necessary. This means that an additional 64 measurements should be taken. The N_T data must then be reanalyzed to be certain that the constraint is met.

4.9 SUMMARY

The normal scatter of data about some central mean value is brought about through several contributing factors, including the process variable's own temporal unsteadiness and spatial distribution under nominally fixed operating conditions, as well as precision errors in the measurement system and in the measurement procedure. Data scatter introduces an additional element, an uncertain vagueness, into the measurement scheme, which requires statistical methods to quantify. Statistics, then, becomes a powerful tool used to interpret and present data. In this chapter, we developed the most basic methods used to understand and quantify finite data sets. Methods to estimate the true mean value based on a limited number of data points and the precision in such estimates were presented along with treatment of data curve fitting. A summary table of these statistical estimators is given as Table 4.7.

REFERENCES

1. Kendal, M. G., and A. Stuart. *Advanced Theory of Statistics*, Vol. 2, Griffin, London, 1961. See also Bendat, J., and Piersol, A., *Random Data: Analysis and Measurement*, 2d ed. Wiley, New York, 1986.
2. Lipson, C., and Sheth, N. J., *Statistical Design and Analysis of Engineering Experiments*, McGraw-Hill, New York. 1973.
3. James, M. L., Smith, G. M., and Wolford, J. C., *Applied Numerical Methods For Digital Computation*, 2d ed., Harper & Row, New York, 1977.
4. Miller, I., and Freund, J. E., *Probability and Statistics for Engineers*, 3d ed., Prentice-Hall, Englewood Cliffs, NJ., 1985.
5. Measurement Uncertainty PTC 19.1, *ASME Power Test Codes*, American Society of Mechanical Engineers, New York, 1998.

NOMENCLATURE

$a_0, a_1, \ldots, a_m$	polynomial regression coefficients	$p(x)$	probability density function of x
f_j	frequency of occurrence	r	correlation coefficient
		t	Student-t variable

x	variable	S_x^2	sample variance of x
x'	true mean value of the	$\langle S_x \rangle$	pooled sample standard
	measurand x		deviation of x
$\bar{x}$	sample mean value of x	$\langle S_{\bar{x}} \rangle$	pooled standard deviation
$\langle \bar{x} \rangle$	pooled sample mean of x		of the means of x
y	variable	$\langle S_x^2 \rangle$	pooled sample variance of x
z_1	z variable	S_{yx}	standard error of the fit
N	total number of measurements		between y and x
$P(x)$	probability of measuring	β	normalized standard variate
	a value of x	σ	true standard deviation
$P\%$	percent probability		of the measurand
S_x	sample standard deviation x	σ^2	true variance of the measurand
$S_{\bar{x}}$	sample standard deviation of the	χ^2	chi-squared value
	means of x	ν	degrees of freedom

PROBLEMS

4.1 A large data set ($N > 1000$) has a mean value of 9.2 units and a standard deviation of 1.1 units. Determine the range of values in which 50% of the data set should be found, assuming a normal probability.

4.2 For a very large data set ($N > 10,000$), a mean value of 204 units with a standard deviation of 18 units has been determined. Determine the range of values in which 90% of the data fall.

4.3 At a fixed operating setting, the pressure in a line downstream of a reciprocating compressor is found to have a mean value of 121.6 psi with a standard deviation of 14 psi based on a very large data set obtained from continuous monitoring. What is the probability that the line pressure will exceed 150 psi during any measurement?

4.4 Consider the toss of four pennies. There are 2^4 possible outcomes of a single toss. Develop a histogram of the number of heads that can appear on any toss. Does it look like a Gaussian distribution? Should this be expected? What is the probability that three heads will appear on any toss?

4.5 As a game, slide a matchbook across a table, trying to make it stop at the same predefined point on each attempt. Measure the distance from the starting point to its stopping point. Repeat this 10, 20, and 50 times. Plot the frequency distributions for each set. Would you expect them to look like a Gaussian distribution? What statistical outcomes would distinguish a better player in this game?
    Problems 6 through 18 refer to the data in Table 4.8. Assume that the data have been recorded from a single process under a fixed operating condition.

4.6 Develop a histogram for the data listed in column 1.

4.7 Develop a frequency distribution plot for the data given in column 2.

4.8 Develop a histogram for the data in column 3.

4.9 For the data in column 1, determine the mean value and standard deviation. State the degrees of freedom in each.

4.10 For the data in column 3, determine the mean value and estimate the standard deviation of the means. State the degrees of freedom in each.

4.11 For the data in column 1, estimate the range of values in which you would expect 95% of all possible measured values for this operating condition to lie.

4.12 For the data in column 1, determine the best estimate of the mean value at a 95% confidence level. How does this differ from Problem 4.11?

4.13 For the data in column 2, estimate the range of values in which you would expect 90% of all possible values measured under this operating condition to lie.

Table 4.8 Sample Data For Exercise Problems

d [mm]	d [mm]	d [mm]
4.90	4.32	4.60
5.02	4.98	4.71
4.90	4.89	4.78
5.10	5.01	4.32
5.02	4.18	4.98
4.74	5.00	4.85
4.78	4.66	5.12
4.94	4.74	5.09
4.98	4.92	4.90
4.50	5.01	5.20
4.63	4.72	4.56
4.70	4.63	4.49
4.26	4.79	4.65
5.22	4.98	4.35
4.92	4.44	4.94
4.55	4.44	4.65
4.27	4.42	4.93
4.45	4.56	4.41
4.65	4.30	4.66
4.85	5.22	4.59

4.14 For the data in column 2, determine the best estimate for the mean value at a 90% confidence level.

4.15 For the data in column 3, estimate the range of values in which you would expect 95% of all possible measured values for this operating condition to fall.

4.16 For the data in column 1, determine the best estimate of the mean value at a 95% confidence level.

4.17 If the data in columns 1 and 3 are two replications, compute a pooled mean value for the process diameter. State the range for the best estimate in mean diameter at 95% confidence based on these two data sets.

4.18 For the data in column 3, if one additional measurement were made, estimate the interval of values in which the value of this measurement would fall with a 95% probability.

4.19 Consider the measurand having a known true mean of 20.0 N with variance of $4.0\,N^2$. Estimate the probability of a measurement indicating a value between 23.5 and 26.0 N; between 16.0 and 18.5 N.

4.20 A professor grades students on a normal curve. For any grade x, based on a course mean and standard deviation developed over years of testing, the following applies:

$$
\begin{aligned}
\text{A:} & \quad x > \bar{x} + 1.6\sigma \\
\text{B:} & \quad \bar{x} + 0.4\sigma < x \le \bar{x} + 1.6\sigma \\
\text{C:} & \quad \bar{x} - 0.4\sigma < x \le \bar{x} + 0.4\sigma \\
\text{D:} & \quad \bar{x} - 1.6\sigma < x \le \bar{x} - 0.4\sigma \\
\text{F:} & \quad x \le \bar{x} - 1.6\sigma
\end{aligned}
$$

How many A, C, and D grades are given per 100 students?

4.21 The production of a certain polymer fiber follows a Gaussian distribution with a true mean diameter of 20 μm and a standard deviation of 30 μm. Compute the probability of a measured value greater than 80 μm. Compute the probability of a measured value of between 50 and 80 μm.

4.22 Friction linings from 10 clutch plates were removed from service in a similar application. Measurements for wear showed the following values (μm): 204.5, 231.1, 157.5, 190.5, 261.6, 127.0, 216.6, 172.7, 243.8, and 291.0. Estimate the average wear and its variance. Based on this sample, how many clutch plates out of a large set will be expected to show wear of less than 203 μm?

4.23 Determine the mean value of the life of an electric light bulb if

$$p(x) = 0.001e^{-0.001x} \quad x \geq 0$$

and $p(x) = 0$ otherwise. Here x is the life in hours.

4.24 Compare the gain in precision in estimating x' by taking a sample of 16 measurements as opposed to a sample of only four measurements. Then compare 100 measurements to 25. Explain "diminishing returns" as it applies to using larger sample sizes to increase precision.

4.25 The variance in the strength values of 270 bricks is 6.89 (MN/m^2)^2, with a mean of 6.92 MN/m^2. What is the precision error in the mean value at 95% confidence?

4.26 A sample of 61 data points of force indicates a sample mean value of 44.20 N with a sample variance of 4.0 N^2. Estimate the probability that an additional measurement would indicate a value between 45.56 and 48.20 N.

4.27 Suppose three sets of data are collected of some variable during a similar process operating condition. The statistics are found to be

$$N_1 = 16 \quad \overline{X}_1 = 32 \quad S_1 = 3 \text{ units}$$
$$N_2 = 21 \quad \overline{X}_2 = 30 \quad S_2 = 2 \text{ units}$$
$$N_3 = 9 \quad \overline{X}_3 = 34 \quad S_3 = 6 \text{ units}$$

Determine the degrees of freedom in the pooled data. Compute an estimate of the weighted mean value of this variable and the range in which the true mean should lie with 95% confidence.

4.28 Eleven core samples of fresh concrete are taken by a county engineer from the loads of 11 concrete trucks used in pouring a structural footing. After they cure, the engineer tests to find a mean compression strength of 3027 lb/in.^{2} with a standard deviation of 53 lb/in.^{2}. State codes require a minimum strength of 3000 lb/in.^{2}. Should the footing be repoured based on a 95% confidence interval of the test data?

4.29 The following data were collected during the repeated measurement of the force load acting on a small area of a beam under "fixed" conditions:

Reading Number	Output [N]	Reading Number	Output [N]
1	923	6	916
2	932	7	927
3	908	8	931
4	932	9	926
5	919	10	923

Determine if any of these data points can be considered as outliers. If so, reject the data point. Estimate the true mean value from this data set.

4.30 An engineer measures the diameter of 20 tubes selected at random from a large shipment. The sample yields $\bar{x} = 47.5$ mm and $S_x = 8.4$ mm. The manufacturer of the tubes claims that $x' = 42.1$ mm. What can the engineer conclude about this claim?

4.31 In the manufacture of particular precision motor shafts, only shafts diameters of between 38.10 and 37.58 mm are usable. If the process mean is found to be 37.84 mm with a standard deviation of 0.13 mm, what percentage of the manufactured shafts are usable?

For Problems 4.32 through 4.35, it would be helpful either to use a regression analysis program, a spreadsheet program, or a calculator.

4.32 Determine the static sensitivity for the following data by using a least-squares regression analysis. Plot the data and fit including the 95% precision interval.

Y:	2.7	3.6	4.4	5.2	9.2
X:	0.4	1.1	1.9	3.0	5.0

4.33 Using the following data, determine a suitable least-squares fit of the data. Which order polynomial best fits this data set? Plot the data and the fit with a 95% confidence band on an appropriate graph.

Y:	1.4	4.7	17.3	82.9	171.6	1227.1
X:	0.5	1.1	2.0	2.9	5.1	10.0

4.34 The following data have the form $y = ax^b$. Plot the data and their best fit with a 95% confidence band. Estimate the static sensitivity at each value of X.

Y:	0.14	2.51	15.30	63.71
X:	0.5	2.0	5.0	10.0

4.35 A fan performance test yields the following data:

Q:	2000	6000	10000	14000	18000	22000
h:	5.56	5.87	5.73	4.95	3.52	1.08

where Q is flow rate in cubic meters per second and h is the static head in centimeters of water. Find the lowest degree of polynomial that best fits the data. Explain your choice.

4.36 A camera manufacturer purchases glass to be ground into lenses. From experience it is known that the variance in the refractive index of the glass is 1.25×10^{-4}. The company will reject any shipment of glass if the sample variance of 30 lenses measured at random exceeds 2.10×10^{-4}. What is the probability that such a batch of glass will be rejected even though it is actually within the normal variance?

4.37 The setup for grinding a type of bearing is considered under control if the bearings have a mean diameter of 5.000 mm. The normal procedure is to measure 30 bearings each for 10 min to monitor production. Production rates are 1000/min. What action do you recommend if such a sample shows a mean of 5.060 mm with a standard deviation of 0.0025 mm?

4.38 Referring to the information in Problem 4.37, suppose the mean diameter of the bearings must be held to within 0.2%. Based on the given information, how many bearings should be measured at each 10-min interval?

4.39 The manufacturer of aircraft vacuum pumps wishes to estimate the mean failure time of its product at 95% confidence. Six pumps are tested to failure with these results (in hours): 1272, 1384, 1543, 1465, 1250, and 1319. Estimate the sample mean and its 95% confidence value. How many more data points would be needed to improve the confidence value to 50 h?

4.40 Strength tests on a batch of cold-drawn steel yield the following:

Strength (MPa)	Occurrences
421–480	4
481–540	8
541–600	12
601–660	6

Estimate the statistics from this sample. Test the hypothesis that the data are a random sample that follows a normal distribution.

4.41 Referring to the strength test of Problem 4.40, how many test measurements should actually be made if the mean strength estimate should be accurate to within ±3%?

4.42 Estimate the number of specimens that should be tested to determine the mean failure strength of an alloy to within a precision interval with a range no greater than 5% of the mean, using ASTM standard procedures. A sample run of six alloy specimens suggests a mean strength of 71,327 lb/in.^{2} with a standard deviation of 8345 lb/in.^{2}. What sources contribute to this variation in "measured" strength?

4.43 Estimate the number of measurements of a time-dependent acceleration signal obtained from a vibrating vehicle that would lead to an acceptable precision interval about the mean of 0.1 g, if the standard deviation of the signal is expected to be 2 g.

4.44 The sample standard deviation of a time-dependent electrical signal, based on 60 measurements, is estimated to be 1.52 V. How many more measurements would be required to provide a precision interval in the mean of 0.28 V?

4.45 Estimate the probability with which an observed $\chi^2 = 19.0$ will be exceeded if there are 10 degrees of freedom.

4.46 A conductor is insulated by using an enameling process. It is known that the number of insulation breaks per meter of wire is 0.07. What is the probability of finding x breaks in a piece of wire 5 m long? Use the Poisson distribution.

4.47 It is known that 2% of the screws made by a certain machine are defective, with the defects occurring randomly. The screws are packaged 100 to a box. Estimate the probability that a box will contain x defective screws. (i) use the binomial distribution. (ii) use the Poisson distribution.

4.48 An optical velocity measuring instrument provides an updated signal on the passage of small particles through its focal point. Suppose the average number of particles passing in a given time interval is four. Estimate the probability of observing x particle passages in a given time. Use the Poisson distribution for $x = 1$ through 10.

Chapter 5

Uncertainty Analysis

5.1 INTRODUCTION

Suppose the competent dart thrower of Chapter 1 tossed several practice rounds of darts at a bull's-eye. This would give us a good idea of the thrower's accuracy. Then let the thrower toss another round. Without looking, can you guess where the darts hit? Measurements that include bias and random error components are much like this. We can calibrate a measurement system to get a good idea of its accuracy. However, from the calibration we can only estimate the probable error in any subsequent measurement. In other words, we need to quantify, on average, how closely the measured value agrees with the true value.

We have defined measurement as the process of assigning a value to a physical variable. The *error* in the measurement is simply the difference between the true value of the variable and the value assigned by our measurement. However, in any measurement, the true value is not known. Thus instead of the actual error, we estimate the probable error in the measurement, called the *uncertainty*. It defines an interval about the measured value within which we suspect the true value must fall. It is the process of identifying and quantifying errors that is called *uncertainty analysis*.

An uncertainty analysis provides a powerful design tool for evaluating different measurement systems and methods and for reporting measured data. It is a principal tool used during the development of a test plan. In this chapter, a systematic approach to identifying and quantifying errors that is used during the design, execution, and interpretation of tests is presented.

5.2 MEASUREMENT ERRORS

Although no general discussion of errors can be complete in listing the elements contributing to error in a particular measurement, certain generalizations for error sources can be made to help in their identification. In the discussion that follows, errors will be grouped into two very general categories: bias error and precision error. We will not consider measurement blunders that result in obviously fallacious data. Such data should be discarded.

Consider the repeated measurement of a variable under conditions that are expected to produce the same value of the measured variable. The relationship between the true value and the measured data set, containing both bias and precision errors, is illustrated in Figure 5.1. The total error in any single measurement is the sum of the bias and the precision errors in that measurement. The total error contained in

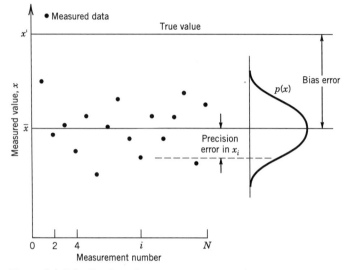

Figure 5.1 Distribution of errors upon repeated measurements.

a set of measurements obtained under seemingly fixed conditions can be described by an average bias error and a statistical estimate of the precision errors in the measurements. The bias errors will shift the sample mean away from the true mean of the measured variable by a fixed amount. Within a sample of many measurements, the precision errors bring about a normal distribution of measured values about the sample mean. Because accuracy deals specifically with the difference between the true value and the measured value, an accurate measurement has both small bias and precision errors.

The importance of measurement error is that it obscures our ability to ascertain the information that we desire: the true value of the variable measured. Because of errors, the accuracy of a measurement can never be certain. In Chapter 4, we stated that the best estimate of the true value sought in a measurement is provided by its sample mean value and the uncertainty in that value:

$$x' = \bar{x} \pm u_x \quad (P\%) \tag{4.1}$$

Uncertainty analysis is the method used to quantify the u_x term.

Certain assumptions are implicit in an uncertainty analysis:

1. The test objectives are known.
2. The measurement itself is a clearly defined process in which all known calibration corrections for bias error have already been applied.
3. Data are obtained under fixed operating conditions.
4. The engineers have some experience with the system components.

Experience allows an estimate of component bias and precision errors based on some evidence, such as personal experience through previous or simulated tests and calibrations, or someone else's experience, such as the manufacturer's performance literature, an NIST bulletin, a professional test code, or performance information discussed in the technical literature.

5.3   DESIGN-STAGE UNCERTAINTY ANALYSIS

We begin the design of a measurement system with an idea and some catalogs, and we end the project after data have been obtained and analyzed. As with any part of the design process, the uncertainty analysis will evolve as the design of the measurement system and process matures. We will study uncertainty analysis for the following measurement situations: (1) design stage, (2) advanced stage and/or single measurement, and (3) multiple measurement.

Design-stage uncertainty analysis refers to an initial analysis performed prior to the measurement. It is useful for selecting instruments, selecting measurement techniques, and obtaining an approximate estimate of the uncertainty likely to exist in the measured data. At this point, the measurement system and associated procedures are but a concept. Usually little is known about the instruments, and in many cases they are still pictures in a catalog. Major facilities may need to be built and equipment ordered with considerable lead time. Uncertainty analysis should be used to assist in the selection of equipment and procedures based on their relative performance and even cost. In the design stage, distinguishing between bias and precision errors is too difficult to be of concern. Instead, consider only sources of uncertainty in general.

The initial step in a design-stage analysis is to determine the minimum uncertainty in the measured value that would result from the measurement. However, a measurement system will usually consist of sensors and instruments, each with their respective contributions to system uncertainty—so let's first talk about individual contributions to uncertainty.

Even when all errors are otherwise zero, the value of the measurand must be affected by the ability to resolve the information provided by the instrument. We call this the *zero-order uncertainty* of the instrument, u_0. At the zero order, we assume that the variation expected in the measurand will be less than that caused by instrument resolution and that all other aspects of the measurement are perfectly controlled. Basically, u_0 is an estimate of the expected uncertainty caused by the data scatter that results when the instrument is read.

As an arbitrary rule, assign a numerical value to u_0 of one-half of the instrument resolution[1] with a probability of 95%:

$$u_0 = \pm 1/2 \text{ resolution} \quad (95\%) \tag{5.1}$$

At 95% probability, we assume that only 1 measured value in 20 (20:1 odds) would have a value exceeding the interval defined by u_0.

The second piece of information that is usually available is the manufacturer's statement concerning instrument error. In a catalog, the value stated will be some typical value for that type of instrument under ideal conditions. We can assign this stated value as the uncertainty that is due to the instrument, u_c. Sometimes the instrument errors will be stated in parts, each part caused by some contributing factor (e.g., Table 1.1). A probable estimate in u_c can be made by combining all the errors in some reasonable manner, such as in equation (1.11). This is an estimate of the bias error caused by the instrument.

A reasonable means to estimate the effects of more than one error on a particular measurement must be established. The accepted manner of combining errors is termed the root-sum-squares (RSS) method.

---

[1]In some situations, it may be possible to assign a value for u_0 that is smaller than one-half the scale resolution. Discretion should be used.

Combining Elemental Errors: RSS method

Each elemental error will combine in some manner with other elemental errors to increase the uncertainty of the measurement. We will refer to each individual error as an "element of error." Consider a measurement of x that is subject to, say, K elements of error, e_j, where $j = 1, 2, \ldots, K$. A realistic estimate of the uncertainty in the measurement, u_x, caused by these elemental errors can be computed by using the RSS method:

$$u_x = \pm\sqrt{e_1^2 + e_2^2 + \cdots + e_K^2}$$

$$= \pm\sqrt{\sum_{j=1}^{K} e_j^2} \quad (P\%)$$

(5.2)

This was the approach used to develop the instrument uncertainty in equation (1.12). It is imperative to maintain consistency in the units of each error. Ideally, each error should be estimated at the same probability level.

A general, albeit somewhat arbitrary, rule is to use the 95% probability level ($P\% = 95\%$) throughout all uncertainty calculations. Engineers follow this 95% rule for consistency. We will adhere to this rule here but point out that in rare situations this may be neither possible nor desirable.

Design-Stage Uncertainty

The RSS method of combining errors is based on the assumption that the possible variations in the values of an error encountered over repeated measurements will tend to follow a Gaussian distribution. As such, the RSS estimate of all contributing errors should provide a "probable" measure of the error in any measurement. Although the simple arithmetic sum of the elemental errors would provide a larger estimate of possible measurement error, that method assumes that all errors can occur in their worst possible way for each and all measurements. As such, the method has no basis in probability, and it would be appropriate in very few applications (such as catastrophic prediction).

The design-stage uncertainty, u_d, for the instrument can now be approximated by combining the instrument uncertainty with the zero-order uncertainty:

$$u_d = \sqrt{u_0^2 + u_c^2} \quad (P\%)$$

(5.3)

This procedure for estimating the design-stage uncertainty is outlined in Figure 5.2.

The design-stage uncertainty estimate is not meant to be a statement of the total uncertainty in a measurement. It is to be used only as a guide for selecting equipment and procedures. For example, design-stage uncertainty assumes perfect control

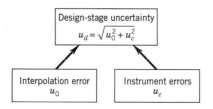

Figure 5.2 Design-stage uncertainty procedure in combining uncertainties.

of the test conditions and measurement procedure. As presented, it includes only instrument errors, it and may be interpreted as a reasonable estimate of the minimum uncertainty that can be achieved with a measurement system. *If additional information about the measurement is known, this should be used to adjust equation (5.3).*

A measurement system typically consists of a chain of sensors and instruments. The design-stage uncertainty for the measurement system is arrived at by combining each of the design-stage uncertainties for each component in the system, using the RSS method. When using the RSS method for several components, take care to use consistent units between each error or uncertainty!

EXAMPLE 5.1

Consider the force measuring instrument described by the catalog data that follows. Provide an estimate of the uncertainty attributable to this instrument and the instrument design stage uncertainty.

Resolution:	0.25 N
Range:	0–100 N
Linearity:	within 0.20 N over range
Repeatability:	within 0.30 N over range

KNOWN

Catalog specifications

ASSUMPTIONS

Values representative of instrument

FIND

u_c, u_d

SOLUTION

We will follow the procedure outlined in Figure 5.2. An estimate of the instrument uncertainty depends on each of the contributing elemental errors of linearity, e_1, and repeatability, e_2:

$$e_1 = 0.20 \text{ N} \qquad e_2 = 0.30 \text{ N}$$

Then using equation (5.2) with $K = 2$ yields

$$u_c = \pm\sqrt{(0.2)^2 + (0.3)^2}$$
$$= \pm 0.36 \text{ N} \quad (95\%)$$

The instrument resolution is given as 0.25 N, from which $u_0 = \pm 0.125$ N. From equation (5.3), the design-stage uncertainty of this instrument would be

$$u_d = \sqrt{u_0^2 + u_c^2} = \sqrt{(0.36)^2 + (0.125)^2}$$
$$= \pm 0.38 \text{ N} \quad (95\%)$$

COMMENT

The design-stage uncertainty for this instrument is simply an estimate based on the "experience" at hand. Additional information, such as calibrations or prior use of the instrument, would provide justification for modifying this estimate.

EXAMPLE 5.2

A voltmeter is to be used to measure the output from a pressure transducer that outputs an electrical signal. The nominal pressure expected will be ~ 3 psi (3 lb/in.^{2}). Estimate the design-stage uncertainty in this combination. The following information is available:

Voltmeter
    Resolution: 10 μV
    Accuracy: within 0.001% of reading
Transducer
    Range: ±5 psi
    Sensitivity: 1 V/psi
    Input power: 10 V dc ± 1%
    Output: ±5 V
    Linearity: within 2.5 mV/psi over range
    Repeatability: within 2 mV/psi over range
    Resolution: negligible

KNOWN

Instrument specifications

ASSUMPTIONS

Values representative of instrument at 95% probability

FIND

u_c for each device and u_d for the measurement system

SOLUTION

The procedure in Figure 5.2 will be used for both instruments to estimate the design stage uncertainty in each. The resulting uncertainties will then be combined by using the RSS approximation to estimate the system u_d.

The uncertainty in the voltmeter at the design stage is given by equation (5.3) as

$$(u_d)_E = \pm\sqrt{(u_0)_E^2 + (u_c)_E^2}$$

From the information available,

$$(u_0)_E = \pm 5\,\mu V \quad (95\%)$$

For a nominal pressure of 3 psi, we expect to measure an output of 3 V. Then,

$$(u_c)_E = \pm(3 \text{ V} \times 0.00001) = \pm 30\,\mu V \quad (95\% \text{ assumed})$$

so that the design-stage uncertainty in the voltmeter is

$$(u_d)_E = \pm 30.4 \, \mu V \quad (95\%)$$

The uncertainty in the pressure transducer output at the design stage is also given by equation (5.3). Assuming that we operate within the input power range specified, the instrument output uncertainty can be estimated by considering each of the instrument elemental errors of linearity, e_1, and repeatability, e_2:

$$(u_c)_P = \sqrt{e_1^2 + e_2^2} \quad (95\% \text{ assumed})$$
$$= \pm\sqrt{(2.5 \, mV/psi \times 3 \, psi)^2 + (2 \, mV/psi \times 3 \, psi)^2}$$
$$= \pm 9.61 \, mV \quad (95\%)$$

Since $(u_0) \approx 0$ V/psi, then the design-stage uncertainty in the transducer in terms of indicated voltage is $(u_d)_P = \pm 9.61$ mV (95%); but since the sensitivity is 1 V/psi, this uncertainty can be stated as $(u_d)_P = \pm 0.0096$ psi (95%).

Finally, u_d for the combined system is found by use of the RSS method, using the design-stage uncertainties of the two devices. The design-stage uncertainty in pressure as indicated by this measurement system is estimated to be

$$u_d = \sqrt{(u_d)_E^2 + (u_d)_P^2}$$
$$= \pm\sqrt{(0.030 \, mV)^2 + (9.61 \, mV)^2} \quad (95\%)$$
$$= \pm 9.61 \, mV \quad (95\%) \quad \text{or} \quad \pm 0.0096 \, psi \quad (95\%)$$

COMMENT

Note that essentially all of the uncertainty is due to the transducer. A better voltmeter will simply not improve the uncertainty in this measurement! If better pressure results are desired, a better transducer will have to be found.

5.4   ERROR SOURCES

Although design-stage uncertainty analysis is an essential first step in the design of a measurement system and will serve to guide feasibility assessment and instrument selection, a more detailed analysis is required to assign uncertainty values to an actual measured result. As a guide to looking for measurement errors, consider the measurement process as consisting of three distinct steps: calibration, data acquisition, and data reduction. Errors that enter during each of these steps may be grouped under their respective error source heading: (1) calibration errors, (2) data-acquisition errors, and (3) data-reduction errors.

Within each of these three *error source groups*, an objective is to list the types of errors encountered. Such errors are the elemental errors of the measurement. Do not become preoccupied with these groupings. They are meant for convenience in establishing the errors encountered; the important thing is that those errors encountered be listed, regardless of grouping. If a step is not included in the actual measurement procedure (e.g., no calibration is explicitly performed), that source group can be excluded from consideration in the error analysis. If you place an error in an "incorrect" group, it is okay. The final uncertainty will not be changed!

Table 5.1 Calibration Error Source Group

Element (j)	Error Source[a]
1	Primary to interlab standard
2	Interlab to transfer standard
3	Transfer to lab standard
4	Lab standard to measurement system
5	Calibration technique
Etc.	

[a] Bias and/or precision in each element.

Calibration Errors

Calibration in itself does not eliminate measurement system errors; it merely reduces these errors to more acceptable values. *Calibration errors* enter through two principal sources: (1) the bias and precision errors in the standard used in the calibration and (2) the manner in which the standard is applied to the measuring system or system component. For example, the typical laboratory standard used for calibration is approximate. Accordingly, there may be a difference between the value of the standard used and the value of the primary standard that it represents. Hence, an uncertainty in the input value on which the calibration is based will arise. In addition, there can be a difference between the value supplied by the standard and the calibration value actually sensed by the measuring system. Either of these effects will be built into the calibration data. In Table 5.1, a listing of common elemental errors contributing to this error source group is given.

Data-Acquisition Errors

All errors that arise during the actual act of measurement are referred to as data-acquisition errors. These errors include sensor and instrument errors, uncontrolled variables, such as changes or unknowns in measurement system operating conditions, power settings and environmental conditions that affect system performance, and sensor installation effects on the measurand. In addition, the measured variable temporal and spatial variations contribute errors, and these may be quantified through the application of finite statistics. A list of some elemental errors common to this source is given in Table 5.2.

Table 5.2 Data-Acquisition Error Source Group

Element (j)	Error Source[a]
1	Measurement system operating conditions
2	Sensor-transducer stage (instrument error)
3	Signal conditioning stage (instrument error)
4	Output stage (instrument error)
5	Process operating conditions
6	Sensor installation effects
7	Environmental effects
8	Spatial variation error
9	Temporal variation error
Etc.	

[a] Bias and/or precision in each element.
Note: A total input-to-output measurement system calibration will combine elements 2, 3, 4, and possibly 1 within this error source group.

Table 5.3 Data-Reduction Error
Source Group

Element (j)	Error Source[a]
1	Calibration curve fit
2	Truncation error
Etc.	

[a] Bias and/or precision in each element.

Data-Reduction Errors

The use of curve fits and correlations with their associated unknowns (Section 4.5) introduces data-reduction errors into the reported test results. The resolution of computational operations required to reduce the data into some desired result is another common contribution to this error source. A list of the elemental errors typical of this error source group is found in Table 5.3.

5.5   BIAS AND PRECISION ERRORS

In general, the classification of an error element as containing bias error, precision error, or both can be simplified by considering the methods used to quantify the error. Treat an error as a precision error if it can be statistically estimated in some manner; otherwise treat the error as a bias error.

Bias Error

A bias error remains constant during a given series of measurements under fixed operating conditions. Thus, in a series of repeated measurements, each measurement would contain the same amount of bias. Being a fixed value, the bias error cannot be directly discerned by statistical means alone. It can be difficult to estimate the value of bias or in many cases recognize the presence of a bias error. A bias error may cause either a high or a low estimate of the true value. Accordingly, an estimate of the bias error must be represented by a bias limit, noted as $\pm B$, a range within which the true value is expected to lie.

The reader may be familiar with some bias errors in measurements. The measurement of the barefoot height of a person taken while the person is wearing shoes represents a simple example of one clear bias error in a measurement. In this case this bias error, a data-acquisition error, is the height of the heels. But that is too simple! The real world is more subtle.

Consider the home bathroom scale, which may have a bias error. Does it? How might you estimate the magnitude of bias in an indicated weight? Perhaps you can calibrate the scale by using calibrated masses, account for local gravitational acceleration, and correct the output, thereby estimating the bias error of the measurement (direct calibration against a local standard). Or perhaps you can compare it to a measurement in a physician's office or one at the gym and can compare each reading (a sort of interlaboratory comparison). Or perhaps you can carefully measure your volume displacement in water and compare the results to estimate bias (concomitant methodology). Without any of these, what value would you assign? Would you even suspect bias error?

Let us think about this. An insidious aspect of bias error has been revealed. Why should one doubt a measurement indication and suspect a bias error? Figuratively speaking, there will be no shoe heels staring at you. Experience teaches us to think through each measurement carefully because bias error is always present at some magnitude. We see that it is difficult to estimate bias error without comparison, so a good design should include some means to estimate bias.

Bias can only be estimated by comparison. Various methodologies can be utilized: (1) calibration, (2) concomitant methodology, (3) interlaboratory comparisons, and (4) experience. When available, the most direct method is by calibration using a suitable standard or, alternatively, using calibration methods of inherently negligible or small bias. Another procedure is the use of concomitant methodology, which is using different methods of estimating the same thing and comparing the results. Concomitant methods that depend on different physical measurement principles are preferable, as are methods that rely on calibrations that are independent of each other. In this regard, even accepted analytical methods could be used for comparison[2] or at least to estimate the bias that is due to influential sources such as environmental conditions, instrument response errors, and loading errors. Lastly, an elaborate but good approach is through interlaboratory comparisons of similar measurements, an excellent replication method. This approach introduces different instruments, facilities, and personnel into an otherwise similar measurement procedure in an effort to compare the bias errors between measuring facilities [1]. In lieu of these methods, a value based on experience may have to be assigned.

Calibration against a standard provides the most direct method to estimate bias, but calibration cannot eliminate all bias; it can only reduce it. Consider the calibration of a temperature transducer against an NIST standard certified to be correct to within $0.01°C$. If the calibration data show that the transducer output has a bias error of $0.2°C$ relative to the standard, then we would just correct all the data obtained with this transducer by $0.2°C$. Simple enough! But the standard itself still has an intrinsic bias limit of $0.01°C$, and this uncertainty must remain in the calibrated transducer.

Precision Error

When repeated measurements are made under nominally fixed operating conditions, precision errors will manifest themselves as scatter of the measured data. Precision error is affected by the following:

Measurement System
• Repeatability and resolution

Measured Variable
• Temporal and spatial variations

Process
• Variations in operating and environmental conditions

Measurement Procedure and Technique
• Repeatability

---

[2]Smith and Wenhofer [2] provide examples for determining jet engine thrust; several complementary measurements are used with an energy balance to estimate bias error.

Because of the varying nature of precision errors, exact values cannot be given, but probable estimates can be made through statistical analyses. For example, the precision interval in any measured value of x can be estimated by $t_{v,P}S_x$, and the precision interval in the sample mean of x can be estimated by $t_{v,P}S_{\bar{x}}$.

In summary, the most direct method of both reducing and estimating bias error is through calibration. Precision errors are identified and quantified through repeated measurements and a statistical analysis of the results.

5.6   UNCERTAINTY ANALYSIS: ERROR PROPAGATION

Suppose we wished to determine how long it would take to fill a swimming pool from a garden hose. We might time the filling of a 5-gal container with a stopwatch to estimate the flow rate from the garden hose. Then with an estimate of the volume of the pool, the time to fill the pool could be estimated. Clearly, small errors in estimating the flow rate from our garden hose would translate into large differences in the time required filling the pool, because of the extremely large volume of water required! Here we are using a measured value, the flow rate, to estimate a result, the time required to fill the swimming pool.

Measured variables are commonly used with a functional relationship to determine some resultant value. For example, the estimation of the flow rate, Q, just described required measured information about the volume, $\mathcal{V}$, and the time, t, since $Q = f(\mathcal{V}, t)$. But how would uncertainties in volume and time contribute to uncertainty in the flow rate? Is the uncertainty in Q more sensitive to errors in $\mathcal{V}$ or in t? In general, how are errors propagated to a calculated result? These issues are explored below.

Propagation of Error

A general relationship $y = f(x)$ between a dependent variable y and a measured variable x is illustrated in Figure 5.3. Now suppose we measure x a number of times at some operating condition so as to establish its sample mean value and the precision of its mean value, $t_{v,P}S_{\bar{x}}$. This implies that, neglecting other precision and bias errors, the true value for x lies somewhere within the interval $\bar{x} \pm tS_{\bar{x}}$. It is reasonable to assume that the true value of y, which is determined from the measured values of x, falls within the interval defined by

$$\bar{y} \pm \delta y = f(\bar{x} \pm tS_{\bar{x}}) \tag{5.4}$$

Expanding this as a Taylor series yields

$$\bar{y} \pm \delta y = f(\bar{x}) \pm \left[\left(\frac{dy}{dx} \right)_{x=\bar{x}} tS_{\bar{x}} + \frac{1}{2} \left(\frac{d^2y}{dx^2} \right)_{x=\bar{x}} (tS_{\bar{x}})^2 + \cdots \right] \tag{5.5}$$

By inspection, the mean value for y must be $f(\bar{x})$ so that the term in brackets estimates $\pm\delta y$. A linear approximation for δy can be made, which is valid when $tS_{\bar{x}}$ is small and neglects the higher-order terms in equation (5.5), as

$$\delta y \approx \left(\frac{dy}{dx} \right)_{x=\bar{x}} tS_{\bar{x}} \tag{5.6}$$

The derivative term, $(dy/dx)_{x=\bar{x}}$, defines the slope of a line that passes through the point specified by $\bar{x}$. For small deviations from the value of $\bar{x}$, this slope predicts

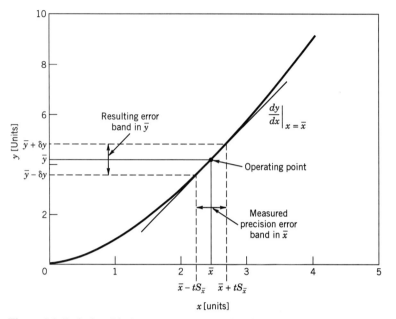

Figure 5.3 Relationship between a measured variable and a resultant calculated by using the value of that variable.

an acceptable, approximate relationship between $tS_{\bar{x}}$ and δy. It is a measure of the sensitivity of y to changes in x. Since the slope of the curve can be different for different values of x, it is important to evaluate the slope by using a representative value. The width of the interval defined by $\pm tS_{\bar{x}}$ defines a precision interval about y, that is, $\pm \delta y$, within which we should expect the true value of y to lie.

Figure 5.3 illustrates the concept that errors in a measured variable are propagated through to a resultant variable in a predictable way. In general, we apply this analysis to the errors that contribute to the uncertainty in x, u_x. The uncertainty in x will be related to the uncertainty in the resultant y by

$$u_y = \left(\frac{dy}{dx}\right)_{x=\bar{x}} u_x \tag{5.7}$$

Compare equations (5.6) and (5.7) with Figure 5.3. This idea can be extended to multivariable relationships. Consider a result, R, which is determined through some functional relationship between independent variables, $x_1, x_2, \ldots, x_L$, defined by

$$R = f_1\{x_1, x_2, \ldots, x_L\} \tag{5.8}$$

where L is the number of independent variables involved. Each variable will contain some measure of uncertainty that will affect the result. The best estimate of the true mean value, R', would be stated as

$$R' = \bar{R} \pm u_R \quad (P\%) \tag{5.9}$$

where the sample mean of R is found from

$$\bar{R} = f_1\{\bar{x}_1, \bar{x}_2, \ldots, \bar{x}_L\} \tag{5.10}$$

and the uncertainty in $\bar{R}$ is found from

$$u_R = f_2\{u_{x_1}, u_{x_2}, \ldots, u_{x_L}\} \tag{5.11}$$

In equation (5.11), each u_{x_i} $i = 1, 2, \ldots, L$ represents the uncertainty associated with the best estimate of x_1 and so forth through x_L. The value of u_R reflects the individual contributions of the individual uncertainties as they are propagated through to the result. A general rule is to assign the uncertainties at the same 95% probability level.

The most probable estimate of u_R is generally accepted as that value given by the Kline–McClintock second power relation [5]. The second power relation can be derived from the linearized approximation of the Taylor series expansion of the multivariable function defined by equation (5.8). A sensitivity index, θ_i, results from the Taylor series expansion and is given by

$$\theta_i = \left. \frac{\partial R}{\partial x_i} \right|_{x=\bar{x}} \qquad i = 1, 2, \ldots, L \tag{5.12}$$

The sensitivity index relates how changes in each x_i affect R. The use of the partial derivative in equation (5.12) is necessary when R is a function of more than one variable. This index is evaluated by using either the mean values or, lacking these estimates, the expected nominal values of the variables. From a Taylor series expansion, the contribution of the uncertainty in x to the result, R, is estimated by the term $\theta_i u_{x_i}$, which is an extension of equation (5.12). The propagation of uncertainty in the variables to the result will yield an uncertainty estimate given by

$$u_R = \pm \sqrt{\sum_{i=1}^{L} (\theta_i u_{x_i})^2} \quad (P\%) \tag{5.13}$$

Sequential Perturbation

A numerical approach can also be used to estimate the propagation of uncertainty through to a result [6]. Often referred to as *sequential perturbation*, it is generally the preferred method when direct partial differentiation is too cumbersome or intimidating, or the number of variables involved is large. It is easily programmed by using spreadsheet software to reduce data already stored in discrete form, but it can also be executed on a hand calculator.

The method is straightforward and uses a finite-difference method to approximate the derivatives:

1. Based on measurements for the independent variables under some fixed operating condition, calculate a result R_o where $R_o = f(x_1, x_2, \ldots, x_L)$. This value fixes the operating point for the numerical approximation (e.g., see Figure 5.3).

2. Next, increase the independent variables by their respective uncertainties and recalculate the result based on each of these new values. Call these values R_i^+. That is,

$$\begin{aligned} R_1^+ &= f(x_1 + u_{x_1}, x_2, \ldots, x_L), \\ R_2^+ &= f(x_1, x_2 + u_{x_2}, \ldots, x_L), \ldots \\ R_L^+ &= f(x_1, x_2, \ldots, x_L + u_{x_L}) \end{aligned} \tag{5.14}$$

3. Next, in a similar manner, decrease the independent variables by their respective uncertainties and recalculate the result based on each of these new values. Call these values R_i^-.

4. Calculate the differences δR_i^+ and δR_i^- for: $i = 1, 2, \ldots, L$

$$\begin{aligned} \delta R_i^+ &= R_i^+ - R_o \\ \delta R_i^- &= R_i^- - R_o \end{aligned} \tag{5.15}$$

5. Finally, evaluate the approximation of the uncertainty contribution from each variable:

$$\delta R_i = \frac{|\delta R_i^+| + |\delta R_i^-|}{2} \approx \theta_i u_i \tag{5.16}$$

Then, the uncertainty in the result is

$$u_R = \pm \left[\sum_{i=1}^{L} (\delta R_i)^2 \right]^{1/2} \quad (P\%) \tag{5.17}$$

Equations (5.13) and (5.17) provide two methods for estimating the propagation of uncertainty to a result. In most cases, use of either equation will yield nearly identical results, and the choice of method is left to the user. Sometimes, however, unreasonable estimates of u_R may be generated by either method. When such unreasonable propagation of uncertainties into results occurs, the cause can generally be traced to sensitivity indices that change rapidly with small changes in the independent variable, x_i, coupled with large value of the uncertainty u_{x_i}. In these situations, the engineer needs to closely examine the cause and extent of the variation in sensitivity.

In subsequent sections, we develop methods to estimate the uncertainty values from available information.

EXAMPLE 5.3

For a displacement transducer having a calibration curve $y = KE$, estimate the uncertainty in displacement y for $E = 5.00$ V, if $K = 10.10$ mm/V with $u_K = \pm 0.10$ mm/V and $u_E = \pm 0.01$ V at 95% confidence.

KNOWN

$y = KE$

$E = 5.00$ V $\qquad u_E = \pm 0.01$ V

$K = 10.10$ mm/V $\qquad u_K = \pm 0.10$ mm/V

FIND

u_y

SOLUTION

Based on equations (5.10) and (5.11), respectively,

$$\bar{y} = f(\bar{E}, \bar{K}) \quad \text{and} \quad u_y = f(u_E, u_K)$$

From equation (5.13), the uncertainty in the displacement at $y = KE$ is

$$u_y = \pm[(\theta_E u_E)^2 + (\theta_K u_K)^2]^{1/2}$$

where the sensitivity indices are evaluated from equation (5.12) as

$$\theta_E = \frac{\partial y}{\partial E} = K \qquad \theta_E = \frac{\partial y}{\partial K} = E$$

or we can write equation (5.13) as

$$u_y = \pm[(Ku_E)^2 + (Eu_K)]^{1/2}$$

The operating point occurs at $E = 5.00$ V and $y = 50.50$ mm. With $E = 5.00$ V and $K = 10.10$ mm/V and substituting for u_E and u_K, evaluate (u_y) at its operating point:

$$u_y\Big|_{y=50.5\ =\ KE} = \pm[(0.10)^2 + (0.5)^2]^{1/2} = \pm0.51 \text{ mm} \quad (95\%)$$

Alternatively, we can use sequential perturbation. The operating point for the perturbation will again be $y = R_o = 50.5$ mm. Using equations (5.14)–(5.16) gives:

i	x_i	R_i^+	R_i^-	δR_i^+	δR_i^-	δR_i
1	E	50.60	50.40	0.1	−0.1	0.1
2	K	51.00	50.00	0.5	−0.5	0.5

Then, using equation (5.17) gives

$$u_y\Big|_{y=50.5} = \pm[(0.1)^2 + (0.5)^2]^{1/2} = \pm0.51 \text{ mm} \quad (95\%)$$

The two methods give the identical result. We state the calculated displacement in the form of equation (5.9) as

$$y' = 50.50 \pm 0.51 \text{ mm} \quad (95\%)$$

5.7   ADVANCED-STAGE AND SINGLE-MEASUREMENT UNCERTAINTY ANALYSIS

In designing a measurement system a pertinent question is, how would it affect the result if this particular aspect of the technique or equipment were changed? In design-stage uncertainty analysis, only errors caused by a measurement system's resolution and estimated intrinsic errors are considered. Single-measurement uncertainty analysis permits taking that treatment further by considering procedural and test control errors that affect the measurement. We will consider a method for a thorough uncertainty analysis when a large data set is not available. Such an advanced-stage analysis, known as *single-measurement uncertainty analysis* [6],[3] is to be used as follows: (1) in the advanced design stage of a test to estimate the expected uncertainty, beyond the initial design-stage estimate, and (2) to report the results of a test program that

---

[3]It is called *replication-level analysis* in [3, 6].

involved measurements over a range of one or more parameters but with no or relatively few replications at each test condition.

In this section, the goal will be to estimate the uncertainty in some measured value, x, or in some general result, R, through an estimation of the uncertainty in each of the factors that may affect x or R. If all factors that influence a measurement were held constant, we might expect the measured value to remain constant upon repeated measurements. However, measurement errors will replicate; that is, they will affect the measured value somewhat differently with each measurement. The overall uncertainty in any measurement is affected by this replication of errors and is an estimate of the possible value of these errors. A technique that uses a step-by-step approach for identifying and estimating errors is presented. We seek the combined value of the estimates at each step.

Zero-Order Uncertainty

At zero-order uncertainty, all variables and parameters that affect the outcome of the measurement, including time, are assumed to be fixed except for the physical act of observation itself. Under such circumstances, any data scatter introduced upon repeated observations of the output value will be the result of instrument resolution alone. The value u_0 estimates the extent of variation expected in the measured value when all influencing effects are controlled. An estimate of u_0 is found by using equation (5.1).

At this level, the uncertainty value calculated would be the minimum possible as the use of zero-order analysis provides an estimate only of the effect of instrument resolution on the measurement. Obviously, a zero-order uncertainty analysis is inadequate for the reporting of test results.

Higher-Order Uncertainty

Higher-order uncertainty estimates consider the controllability of the test operating conditions. For example, at the first-order level, the effect of time as an extraneous variable in the measurement might be considered. That is, what would happen if we started the test, set the operating conditions, and sat back and watched? If a variation in the measured value is observed then time is a factor in the test, presumably as a result of some extraneous variable.

In practice, the uncertainty at this first level would be evaluated for a particular measurand by operating the test facility at some single operating condition that would be within the range of conditions to be used during the actual tests. A set of data (say, $N \geq 30$) would be obtained under some set operating condition. The first-order uncertainty of that measurand could be estimated as

$$u_1 = \pm t_{v,95} S_x \tag{5.18}$$

Only when $u_1 = u_0$ is time not a factor in the test. In itself, the first-order uncertainty is inadequate for the reporting of test results.

At each successive order, each factor identified as affecting the measured value is introduced into the analysis, thus giving a higher but more realistic estimate of the uncertainty. For example, at the second level it might be appropriate to assess the limits of the ability to duplicate the exact operating conditions and the consequences on the test outcome.

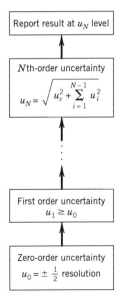

Figure 5.4 Advanced-stage and single-measurement uncertainty procedure in combining uncertainties.

Nth-Order Uncertainty

At the Nth order estimate, instrument calibration characteristics are entered into the scheme through the instrument uncertainty, u_c. A practical estimate of the Nth-order uncertainty, u_N, is given by

$$u_N = \sqrt{(u_c)^2 + \sum_{i=1}^{N-1} u_i^2} \quad (95\%) \tag{5.19}$$

Uncertainty estimates at the Nth order allow for the direct comparison between re-sults of similar tests obtained either by using different instruments or at different test facilities. The procedure for a single measurement analysis is outlined in Figure 5.4.

Ideally, the value for u_c would be based on the error estimates found during a series of calibrations for instruments representative of a pertinent type. Uncertainty estimates at this level are complicated by the fact that the u_c on hand may be based on a single model of instrument only. But so long as u_c is not the dominant term in equation (5.19), this approximation can be considered as reasonable.

Note that as a minimum design-stage analysis includes only the effects found as a result of u_0 and u_c. It is those in-between levels that allow measurement procedure and control effects to be considered in the uncertainty analysis scheme. Only when $u_N = u_d$ are the test conditions under complete control. As such, *the Nth-order uncertainty estimate provides the uncertainty value sought in advanced design stage or in single-measurement analyses*. It is the appropriate value to be used in the reporting of results.

EXAMPLE 5.4

As an exercise, obtain and examine a dial oven thermometer. How would you assess the zero- and first-order uncertainty in the measurement of the temperature of a kitchen oven by using the device?

KNOWN

Dial thermometer

ASSUMPTION

Negligible bias in the instrument

FIND

Estimate u_0 and u_1 in oven temperature

SOLUTION

The zero-order uncertainty would be that contributed by the interpolation error of the measurement system only. For example, most gauges of this type have a resolution of $10°C$. So at the zero order from equation (5.1), we would estimate

$$u_0 = \pm 5°C \quad (95\%)$$

At the first order, the uncertainty would be affected by any time variation in the measurand and variations in the operating conditions. If we placed the thermometer in the center of an oven, set the oven to some relevant temperature, allowed the oven to preheat to a steady condition, and then proceeded to record the temperature indicated by the thermometer at random intervals, we could estimate our ability to control the oven temperature. For N measurements, the first-order uncertainty of the measurement in this oven's temperature by using this technique and this instrument would be from equation (5.18)

$$u_1 = \pm t_{N-1,95} S_T \quad (95\%)$$

COMMENT

We could estimate the next order of uncertainty, u_2, by checking our ability to repeatedly set the oven to a desired mean temperature (time-averaged temperature). This could be done by changing the oven setting and then resetting the thermostat back to the original operating setting. If we were to replicate the sequence M times (reset thermostat, measure a set of data, reset thermostat, etc.), we could compute u_2 by

$$u_2 = \pm t_{M(N-1),95} \langle S_{\bar{T}} \rangle \quad (95\%)$$

The effects of instrument bias would enter at the Nth order through u_c. The uncertainty in oven temperature at some setting would be well approximated by the estimate from equation (5.19):

$$u_N = \pm \sqrt{u_1^2 + u_2^2 + u_c^2} \quad (95\%)$$

By inspection of the values u_0, u_c, and u_i, where $i = 1, 2, \ldots, N-1$, a single-measurement analysis provides the capability to pinpoint those aspects of a test that contribute most to the overall uncertainty in the measurement.

EXAMPLE 5.5

A stopwatch is to be used to estimate the time between the start of an event and the end of this event. Event duration might range from several seconds to 10 min.

Estimate the probable uncertainty in a time estimate using a hand-operated stopwatch that claims an accuracy of 1 min/month and a resolution of 0.01 s.

KNOWN

$u_0 = \pm 0.005$ s   (95%)
$u_c = \pm 60$ s/month   (95% assumed)

FIND

u_d, u_N

SOLUTION

The design-stage uncertainty will give an estimate of the suitability of an instrument for a measurement. At 60 s/month, the instrument accuracy works out to $\sim$0.01 s per 10 min of operation. This would suggest a design-stage uncertainty of

$$u_d = \left(u_0^2 + u_c^2 \right)^{1/2} = 0.01\,\text{s}\quad(95\%)$$

for an event lasting 10 min versus $\pm$0.005 s (95%) for an event lasting 10 s. Accuracy controls the longer-duration measurement, whereas resolution controls the short-duration measurement.

But do instrument resolution and accuracy control the uncertainty in this measurement? The design-stage analysis does not include the data-acquisition error involved in the act of physically turning the watch on and off. However, a first-order analysis might be run to estimate the uncertainty that enters through the procedure of using the watch. Suppose a typical trial run of 20 tries of simply turning a watch on and off suggests that the uncertainty in determining the duration of an occurrence is estimated by

$$u_1 = 0.15\,\text{s}\quad(95\%)$$

The uncertainty in measuring the duration of an event would then be better estimated by equation (5.19).

$$u_N \approx \sqrt{u_0^2 + u_1^2 + u_c^2} = 0.15\,\text{s}\quad(95\%)$$

This estimate will hold for periods of up to $\sim$2 h. Now a better decision as to whether the watch and procedure are suitable for the intended test can be made.

EXAMPLE 5.6

A flow meter can be calibrated by providing a known flow rate to the meter and measuring the meter output. One method of calibration with liquid systems is the use of a catch-and-time technique whereby a volume of liquid, after passing through the meter, is diverted to a tank for a measured period of time from which the flow rate, volume/time, is computed. There are two procedures that can be used to determine the known flow rate, Q, in, say, ft^3/min:

1. The volume of liquid, V, caught in known time, t, can be measured.

Suppose we arbitrarily set $t = 6$ s and assume that our available facilities can determine volume (at Nth order) to $\pm$0.001 ft^3. Note: The chosen time value depends on how much liquid we can accommodate in the tank.

2. The time, t, required to collect 1 ft^3 of liquid can be measured.

Based on Example 5.5, assume an Nth order uncertainty in time of ± 0.15 s. This uncertainty includes our ability to control the on–off time and watch accuracy, but it assumes that the diversion is instantaneous. In either case the same instruments are used. Determine which method is better to minimize uncertainty over a range of flow rates based on these preliminary estimates.

KNOWN

$u_N = \pm 0.001$ ft^3

$u_t = 0.15$ s

$Q = f(\forall, t) = \forall / t$

ASSUMPTIONS

Diversion instantaneous

FIND

Preferred method

SOLUTION

From the available information, the propagation of probable uncertainty to the result, Q, is estimated from equation (5.13):

$$u_Q = \sqrt{\left(\frac{\partial Q}{\partial \forall}u_\forall\right)^2 + \left(\frac{\partial Q}{\partial t}u_t\right)^2}$$

or dividing through by Q, the percent uncertainty in flow rate, u_Q/Q, is

$$\frac{u_Q}{Q} = \sqrt{\left(\frac{u_\forall}{\forall}\right)^2 + \left(\frac{u_t}{t}\right)^2}$$

Representative values of Q are needed to solve this relation. Consider a range of flow rates, say, 1, 10, and 100 ft^3/min; the results are listed in the table that follows for both methods.

Q [ft^3/min]	t [s]	$\forall$ [ft^3]	$\dfrac{u_\forall}{\forall}$	$\dfrac{u_t}{t}$	$\dfrac{u_Q}{Q}$
			method 1		
1	6	0.1	0.01	0.025	0.027
10	6	1.0	0.001	0.025	0.025
100	6	10.0	0.0001	0.025	0.025
			method 2		
1	60.0	1.0	0.001	0.0025	0.0025
10	6.0	1.0	0.001	0.025	0.025
100	0.6	1.0	0.001	0.25	0.25

In method 1, it is clear the uncertainty in time contributes the most to the relative uncertainty in Q, provided that the flow diversion is instantaneous. But in method 2,

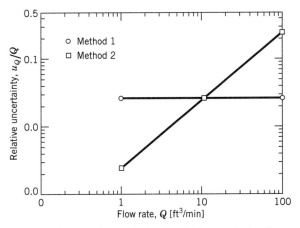

Figure 5.5 Uncertainty plot for the design analysis of Example 5.6.

uncertainty in either time or volume can contribute more to the uncertainty in Q, depending on the time sample length. The results for both methods are compared in Figure 5.5. It is apparent that for the conditions selected and preliminary uncertainty values used that method 2 would be a better procedure for flow rates up to 10 ft^3/min. Beyond 10 ft^3/min, method 1 would be better. However, the minimum uncertainty in method 1 is limited to 2.5%. The engineer may be able to reduce this uncertainty by trial-and-error selection of the operating parameters of t (method 1) and Ψ (method 2) or improvements in the time procedure.

COMMENT

These results are without consideration of some other elemental errors that are present in the experimental procedure. For example, the diversion of the flow is assumed to occur instantaneously. A first-order uncertainty estimate could be used to estimate the added uncertainty intrinsic to the time required to divert the liquid to and from the catch tank. Operator influence should be randomized in actual tests. Accordingly, the calculations given here may be used as a guide in procedure selection but should not be used as the final estimate in the overall uncertainty in any results obtained from the actual calibration.

EXAMPLE 5.7

Repeat Example 5.6, using sequential perturbation for the operating conditions $\Psi = 1$ ft^3 and $t = 6$ s.

SOLUTION

The operating point is $R_o = Q = \Psi/t = 0.1667$ ft^3/s. Then, solving equations (5.14)–(5.17) gives

i	x_i	R_i^+	R_i^-	δR_i^+	δR_i^-	δR_i
1	Ψ	0.1668	0.1665	0.000167	0.000167	0.000167
2	t	0.1626	0.1709	−0.00407	0.00423	0.00415

Applying equation (5.17) gives the uncertainty about this operating point:

$$u_Q = \pm[0.00415^2 + 0.000167^2]^{1/2} = \pm 4.15 \times 10^{-3} \, \text{ft}^3/\text{s}$$
$$u_Q/Q = \pm 0.025$$

COMMENT

These last two examples give similar results for the uncertainty aside from insignificant differences caused by roundoff in the computations.

5.8 MULTIPLE-MEASUREMENT UNCERTAINTY ANALYSIS

This section develops a method for the estimation of the uncertainty in the value assigned to a measured variable based on a set of measurements that are obtained under fixed operating conditions. The method parallels the uncertainty standards approved by professional societies and by NIST in the United States [3, 4]. The procedures assume that the errors follow a normal probability function, although the procedures are actually quite insensitive to deviations away from such behavior [5]. Sufficient repetitions and replications must be present in the measured data; otherwise some estimate of the magnitude of expected variation must be provided in the analysis.

Propagation of Elemental Errors

The procedures for a multiple-measurement uncertainty analysis consist of the following steps:

- Identify the elemental errors in each of the three source groups (calibration, data acquisition, and data reduction).
- Estimate the magnitude of bias and precision error in each of the elemental errors.
- Estimate any propagation of uncertainty through to the result.

Considerable guidance can be obtained from Tables 5.1–5.3 and from the design-stage analysis for identifying the elemental errors. In multiple-measurement analysis, it is possible to divide the estimates for elemental errors into precision and bias. Statistics are used to estimate the precision in each error.

Consider the measurement of variable x, which is subject to elemental precision errors, P_{ij}, and bias, B_{ij}, in each of the three error source groups. Let subscript i, where $i = 1, 2, 3$, refer to the error source groups (calibration error, $i = 1$; data-acquisition error, $i = 2$; data-reduction error, $i = 3$), and subscript j, where $j = 1, 2, \ldots, K$, refer to each of up to any K elements of error, e_{ij}, within each group. Then, $e_{ij} = P_{ij} + B_{ij}$, as indicated in Figure 5.6. A method to estimate the uncertainty in x based on the elemental precision and bias errors is given below and outlined in Figure 5.7.

The propagation of precision errors within a source group is given by the precision index for the group, called the *source precision index*, P_i. P_i is estimated by the RSS method of equation (5.2):

$$P_i = \sqrt{P_{i1}^2 + P_{i2}^2 + \cdots + P_{iK}^2} \qquad i = 1, 2, 3 \qquad (5.20)$$

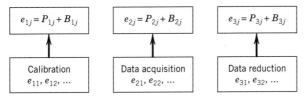

Figure 5.6 Multiple-measurement uncertainty method separates elemental errors into precision and bias errors.

The *measurement precision index*, P, represents a basic measure of the elemental errors affecting the precision in the overall measurement of variable x. It is given by the RSS of each of the precision indices for the error groups by

$$P = \sqrt{P_1^2 + P_2^2 + P_3^2} \tag{5.21}$$

The propagation of elemental bias errors, B_{ij}, is treated in a similar manner. The *source bias limit*, B_i, for the error source groups is given by

$$B_i = \sqrt{B_{i1}^2 + B_{i2}^2 + \cdots + B_{iK}^2} \qquad i = 1, 2, 3 \tag{5.22}$$

The *measurement bias limit*, B, represents a basic measure of the elemental errors that affect the overall bias in the measurement of variable x. Let B be estimated by

$$B = \sqrt{B_1^2 + B_2^2 + B_3^2} \tag{5.23}$$

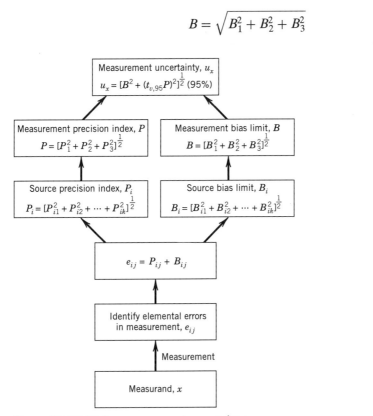

Figure 5.7 Multiple-measurement uncertainty procedure for combining uncertainties.

The measurement uncertainty in x, u_x, can be reported as a combination of the bias and precision uncertainty in x, as indicated in Figure 5.7:

$$u_x = \sqrt{(B)^2 + (t_{v,95} P)^2} \quad (95\%)$$ (5.24)

The product $t_{v,95} P$ is the estimate of the precision uncertainty in x at a 95% confidence.

Estimation of the degrees of freedom, v, in the precision index, P, requires some discussion since P is based on equations (5.20) and (5.21), each composed of elements that usually have different degrees of freedom. In this case, the degrees of freedom in the measurement precision index is estimated by using the Welch–Satterthwaite formula [3]:

$$v = \frac{\left(\sum\limits_{i=1}^{3} \sum\limits_{j=1}^{K} P_{ij}^2 \right)^2}{\sum\limits_{i=1}^{3} \sum\limits_{j=1}^{K} \left(P_{ij}^4 / v_{ij} \right)}$$ (5.25)

where i refers to the three source error groups and j to each elemental error within each source group with $v_{ij} = N_{ij} - 1$.

EXAMPLE 5.8

Ten repeated measurements of force, F, are made over time under fixed operating conditions. The data are listed as follows. Estimate the elemental error in the mean value of the force that is introduced into the measured data through the data scatter.

n	$F[N]$	n	$F[N]$
1	123.2	6	119.8
2	115.6	7	117.5
3	117.1	8	120.6
4	125.7	9	118.8
5	121.1	10	121.9

KNOWN

Measured data set
$N = 10$

ASSUMPTIONS

Error caused by data scatter (precision)

FIND

Estimate P_{ij} [equation (5.20)]

SOLUTION

The mean value of the force based on this finite data set is computed from equation (4.14) as $\bar{F} = 120.1$ N. A precision error is introduced in estimating the mean value because of data scatter. Since this error enters the measurement during data acquisition, it is listed under the data-acquisition source errors ($i = 2$). Label this precision error as P_{2j} for the jth elemental error in source group 2. Then, P_{2j} can be computed through the standard deviation of the means, equation (4.16):

$$P_{2j} = \frac{S_F}{N^{1/2}}$$
$$= \frac{3.2}{10^{1/2}} = 1.01 \, \text{N}$$

COMMENT

It is common to call this elemental error, which arises as a result of data scatter, a temporal variation error, as listed in Table 5.2.

EXAMPLE 5.9

Suppose that the force measuring device of Example 5.1 was used to generate the data set of Example 5.8. Estimate the bias in the measurement instrument.

KNOWN

$e_1 = 0.20 \, \text{N}$
$e_2 = 0.30 \, \text{N}$

ASSUMPTIONS

Manufacturer specifications reliable at 95% probability

FIND

B_{ij}

SOLUTION

Elemental errors caused by instruments are considered as data-acquisition source errors ($i = 2$). And since we have no information as to the statistics used to generate the numbers, e_1 and e_2, they must be considered as bias errors. Accordingly, the transducer bias error ($j = 2$, Table 5.2) in the measurement device, B_{22}, is estimated by

$$B_{22} = \sqrt{e_1^2 + e_2^2}$$
$$= 0.36 \, \text{N}$$

COMMENT

Lacking specific calibration data, manufacturer specifications are always considered as bias errors contributing to the data-acquisition error source. Note that the estimate for bias error caused by the instrument is equal in value to an estimate that would be arrived at by design-stage analysis.

EXAMPLE 5.10

In Examples 4.8 and 4.9, a set of measured data was used to generate a polynomial curve fit equation between variables x and y. If during a test, variable x were to be measured and y computed from x through the polynomial, estimate the data-reduction error involved. Assume that this is the only error in source group 3.

KNOWN

Data set of Example 4.8
Polynomial fit of Example 4.9
$i = 3$, $K = 1$ [equation (5.20)]

ASSUMPTIONS

Data fits curve $y = 1.04x + 0.02$

FIND

P_3

SOLUTION

The curve fit was found to be given by $y = 1.04x + 0.02$ with a standard error of the fit, $S_{yx} = 0.16$. Accordingly, the precision error in this element, given by P_{31} (see Table 5.3), is

$$P_{31} = 0.16$$

Because this is the only elemental error in this source group, $P_3 = P_{31}$.

COMMENT

Likewise, the precision error in the power law equation fit of Example 4.10 can be computed. For example, at $E = 4.3$ V, $P_{31} = S_{yx} = 0.03$ V or 1.3 ft/s.

EXAMPLE 5.11

The measurement of x is found to contain three elemental precision errors under source group 2 (data-acquisition errors). Each element is evaluated with the following conclusions:

$P_{21} = 0.60$ units, $v_{21} = 29$
$P_{22} = 0.80$ units, $v_{22} = 9$
$P_{23} = 1.10$ units, $v_{23} = 19$

Estimate the data-acquisition error source precision index.

KNOWN

P_{ij} with $i = 2$ and $j = 1, 2, \ldots, K$
$K = 3$

FIND

P_2 [using equation (5.20)]

SOLUTION

Using equation (5.20), the group 2 source precision index, P_2, is found by

$$P_2 = \sqrt{P_{21}^2 + P_{22}^2 + P_{23}^2}$$
$$= \sqrt{0.60^2 + 0.80^2 + 1.10^2}$$
$$= 1.49 \, \text{units}$$

From equation (5.25) with $i = 2$, the degrees of freedom in P_2 is determined by

$$\nu_2 = \frac{\left(\sum\limits_{j=1}^{3} P_{2i}^2 \right)^2}{\sum\limits_{j=1}^{3} \left(P_{2j}^4 / \nu_{2j} \right)} = \frac{(0.6^2 + 0.8^2 + 1.1^2)^2}{(0.6^4 / 29) + (0.8^4 / 9) + (1.1^4 / 19)}$$

or $\nu = 38.8 \approx 39$.

EXAMPLE 5.12

After an experiment to measure stress in a loaded beam, an uncertainty analysis reveals the following source errors in stress measurement whose magnitudes were computed from elemental errors by using equations (5.20) and (5.22):

$$B_1 = 1.0 \, \text{N/cm}^2 \qquad B_2 = 2.1 \, \text{N/cm}^2 \qquad B_3 = 0 \, \text{N/cm}^2$$
$$P_1 = 4.6 \, \text{N/cm}^2 \qquad P_2 = 10.3 \, \text{N/cm}^2 \qquad P_3 = 1.2 \, \text{N/cm}^2$$
$$\nu_1 = 14 \qquad\qquad \nu_2 = 37 \qquad\qquad \nu_3 = 8$$

If the mean value of the stress in the measurement is $\bar{\sigma} = 223.4 \, \text{N/cm}^2$, determine the best estimate of the stress.

KNOWN

Experimental error source indices

ASSUMPTIONS

All elemental errors have been included

FIND

P, B and u_σ [using equations (5.21), (5.23)–(5.25)]

SOLUTION

We seek values for the statement, $\sigma' = \bar{\sigma} \pm u_\sigma$ (95%), given that $\bar{\sigma} = 223.4 \, \text{N/cm}^2$. The uncertainty estimate in the measurement is obtained from the source error statements through equations (5.21) and (5.23). The measurement precision index is given by equation (5.21) as

$$P = \sqrt{P_1^2 + P_2^2 + P_3^2} = 11.3 \, \text{N/cm}^2$$

The measurement bias limit is given by equation (5.23) as

$$B = \sqrt{B_1^2 + B_2^2 + B_3^2} = 2.3 \text{ N/cm}^2$$

The degrees of freedom in P is found from equation (5.25) to be

$$\nu = \frac{\left(\sum\limits_{i=1}^{3} P_i^2\right)^2}{\sum\limits_{i=1}^{3} \left(P_t^4/\nu_i\right)} \approx 49$$

Therefore, the t estimator is $t_{49,95} \approx 2.0$. The uncertainty estimate is found using equation (5.24) to be

$$= \pm\sqrt{(2.3 \text{ N/cm}^2)^2 + (22.6 \text{ N/cm}^2)^2}$$

$$= \pm 22.7 \text{ N/cm}^2$$

The best estimate is given in the form of equation (4.1) as

$$\sigma' = 223.4 \pm 22.7 \text{ N/cm}^2 \quad (95\%)$$

Propagation of Uncertainty to a Result

Consider the result, R, which is determined through the functional relationship between the measured independent variables $x_i, x_1, x_2, \ldots, x_L$ as defined by equation (5.8). Again, L is the number of independent variables involved and each x_i has an associated bias, B, given by the measurement bias limit determined for that variable by equations (5.22) and (5.23), and a measurement precision index, P, determined by using equations (5.20) and (5.21). The best estimate of the true value, R', is given as

$$R' = \bar{R} \pm u_R \quad (P\%) \tag{5.26}$$

where

$$\bar{R} = f_1\{\bar{x}_1, \bar{x}_2, \ldots, \bar{x}_L\} \tag{5.27}$$

but u_R is now given by

$$u_R = f_2\{(B)_{x_1}, (B)_{x_2}, \ldots, (B)_{x_L}; (P)_{x_1}, (P)_{x_2}, \ldots, (P)_{x_L}\} \tag{5.28}$$

where subscripts x_1 through x_L refer to the measurement bias limits and measurement precision indices in each of these variables.

The propagation of precision through the variables to the result will yield a *result precision index*, given by

$$P_R = \pm\sqrt{\sum_{i=1}^{L} [\theta_i (P)_{x_i}]^2} \tag{5.29}$$

where θ_i is the sensitivity index defined by equation (5.12). The propagation of bias through the variables to the result will yield a *result bias limit*, given by

$$B_R = \pm\sqrt{\sum_{i=1}^{L} [\theta_i (B)_{x_i}]^2} \tag{5.30}$$

The terms $\theta_i P_{x_i}$ and $\theta_i B_{x_i}$ represent the individual contributions of the ith term to the uncertainty in R. Comparisons between the magnitudes of each individual contribution will identify the uncertainty terms to which the result is most sensitive.

The measurement precision index and the bias limit are then combined to yield an estimate of the uncertainty in the result, u_R, by

$$u_R = \sqrt{B_R^2 + (t_{v,95} P_R)^2} \quad (95\%) \tag{5.31}$$

Once again the t estimator is used to provide a reasonable weight to the interval defined by the precision index. However, the degrees of freedom in each of the variables, x_i, may not be the same. In such a case, the Welch–Satterthwaite formula is used to estimate the degrees of freedom in the result expressed as

$$\nu_R = \frac{\left\{ \sum_{i=1}^{L} [\theta_i (P)_{x_i}]^2 \right\}^2}{\sum_{i=1}^{L} \{ [\theta_i (P)_{x_i}]^4 / (\nu)_{x_i} \}} \tag{5.32}$$

EXAMPLE 5.13

The density of a gas, ρ, which is believed to follow the ideal gas equation of state, $\rho = p/RT$, is to be estimated through separate measurements of pressure, p, and temperature, T. The gas is housed within a rigid impermeable vessel. The literature accompanying the pressure measurement system states an accuracy to within 1% of the reading and that accompanying the temperature measuring system suggests 0.6°R. Twenty measurements of pressure, $N_p = 20$, and ten measurements of temperature, $N_T = 10$, are made with the following statistical outcome:

$$\bar{p} = 2253.91 \text{ psfa} \qquad S_p = 167.21 \text{ psfa}$$
$$\bar{T} = 560.4°R \qquad S_T = 3.0°R$$

where psfa refers to lb/ft^2 absolute. Determine a best estimate of the density. The gas constant is $R = 54.7$ ft lb/lb$_m$ °R.

KNOWN

$\bar{p}, S_p, \bar{T}, S_T$
$\rho = p/RT \quad R = 54.7$ ft lb/lb$_m$ °R

ASSUMPTIONS

Gas behaves as an ideal gas

FIND

$\rho' = \bar{\rho} \pm u_\rho \quad (95\%)$

SOLUTION

The measurement objective is to determine the density of an ideal gas through temperature and pressure measurements. The independent and dependent variables are related through the ideal gas law, $\rho = p/RT$. Equation (5.27) gives the mean value of density as

$$\bar{\rho} = \frac{\bar{P}}{R\bar{T}} = 0.074 \text{ lb}_{\text{m}}/\text{ft}^3$$

The next step must be to identify and estimate the errors and determine how they contribute to the uncertainty in the mean value of density. Since no calibrations are performed and the gas is considered to behave as an ideal gas in an exact manner, the measured values of pressure and temperature are subject only to elemental errors within the data-acquisition error source group. That is, in equations (5.21) and (5.23),

$$(B_1)_T = (B_3)_T = (P_1)_T = (P_3)_T = 0$$
$$(B_1)_p = (B_3)_p = (P_1)_p = (P_3)_p = 0$$

The tabulated value of the gas constant is not without error. However, estimating the possible error in a tabulated value is sometimes difficult. According to [7] the uncertainty (bias) in the evaluation of the gas constant is of the order of $\pm(0.33 \text{ J/kg K})/(\text{gas molecular weight})$ or ± 0.06 (ft lb/lb$_{\text{m}}$ °R)/(molecular weight). Since this yields a small value for a reasonable gas molecular weight, here we will assume a zero bias limit.

Consider the pressure measurement. Within the data-acquisition source group, this best estimate in the pressure is subject to two elemental errors: an instrument error, which will be based on the manufacturer's statement, and a temporal variation (data scatter) error, which will be based on the variation in the measured data obtained during presumably fixed operating conditions. The instrument error (Table 5.2) will be considered as a bias only (as discussed in Example 5.9):

$$(B_{22})_p = 22.5 \text{ psfa} \qquad (P_{22})_p = 0$$

where the first subscript 2 reflects a data-acquisition error. The precision error in the mean value of pressure caused by data scatter can be estimated directly from statistics:

$$(B_{29})_p = 0$$
$$(P_{29})_p = \frac{S_p}{20^{1/2}} = 37.4 \text{ psfa} \qquad (v)_p = 19$$

In a similar manner the data-acquisition source error in temperature is computed. Elemental error attributed to instrument error is considered as bias only:

$$(B_{22})_T = 0.6°\text{R} \quad (P_{22})_T = 0$$

The elemental error caused by temporal variation is considered as precision only and is estimated by

$$(B_{29})_T = 0 \qquad (P_{29})_T = S_T/10^{1/2} = 0.9°\text{R} \qquad (v)_T = 9$$

The data-acquisition source errors in pressure and temperature are estimated from equations (5.22) and (5.20):

$$(B_2)_p = \sqrt{(22.5)^2 + (0)^2} = 22.5 \text{ psfa}$$
$$(B_2)_T = 0.6°\text{R}$$

and

$$(P_2)_p = \sqrt{(0)^2 + (37.4)^2} = 37.4 \text{ psfa}$$

$$(P_2)_T = 0.9°\text{R}$$

with degrees of freedom determined from equation (5.25) to be

$$(\nu)_p = N_p - 1 = 19$$

$$(\nu)_T = 9$$

Since only data-acquisition errors are present in the problem, the measurement precision index and bias limit for pressure and temperature as found from equations (5.21) and (5.23) are

$$(B)_p = 22.5 \text{ psfa} \qquad (P)_p = 37.4 \text{ psfa} \qquad (\nu)_p = 19$$
$$(B)_T = 0.6°\text{R} \qquad (P)_T = 0.9°\text{R} \qquad (\nu)_T = 9$$

The propagation of bias and precision errors through to the result, the density, will be estimated by using the second power law of equations (5.29) and (5.30):

$$P = \sqrt{\left[\frac{\partial \rho}{\partial T}(P)_T\right]^2 + \left[\frac{\partial \rho}{\partial p}(P)_p\right]^2}$$

$$= \sqrt{(1.2 \times 10^{-4})^2 + (1.2 \times 10^{-3})^2}$$

$$= 0.0012 \text{ lb}_\text{m}/\text{ft}^3$$

and

$$B = \sqrt{\left[\frac{\partial \rho}{\partial T}(B)_T\right]^2 + \left[\frac{\partial \rho}{\partial p}(B)_p\right]^2}$$

$$= \sqrt{(8 \times 10^{-5})^2 + (7 \times 10^{-4})^2}$$

$$= 0.0007 \text{ lb}_\text{m}/\text{ft}^3$$

The degrees of freedom in the density is determined from equation (5.32):

$$\nu = \frac{\{[(-\bar{p}/R\bar{T}^2)(P)_T]^2 + [(1/R\bar{T})(P)_p]^2\}^2}{[(\bar{p}/R\bar{T}^2)(P_T)]^4/(\nu)_T + [(1/R\bar{T})(P)_p]^4/(\nu)_p} = 23$$

From Table 4.4, $t_{23,95} = 2.06$.

The uncertainty in the mean value of density is estimated from equation (5.31):

$$u_\rho = \sqrt{(B)^2 + (t_{23,95} P)^2}$$

$$= 0.0025 \text{ lb}_\text{m}/\text{ft}^3 \quad (95\%)$$

The best estimate of the density is given in the form of equation (5.26):

$$\rho' = 0.074 \pm 0.0025 \text{ lb}_\text{m}/\text{ft}^3 \quad (95\%)$$

This measurement of density has an uncertainty of $\sim$3.4%.

COMMENT

Examination of the magnitude of each group of terms in the equations for P and B shows that the pressure contributes more to the precision index and bias limit of density than does temperature. However, bias is essentially negligible compared with the precision of the measurements. The analysis reveals that the uncertainty in density could be reduced by improving the precision in the pressure measurement.

EXAMPLE 5.14

Consider the determination of the mean diameter of a shaft using a handheld micrometer. The shaft was manufactured on a lathe, presumably to a constant diameter. Identify possible elements of error that can contribute to the uncertainty in the estimated mean diameter.

SOLUTION

In the machining of the shaft, possible runout during shaft rotation can bring about an eccentricity in the shaft cross-sectional diameter. Further, as the shaft is machined to size along its length, possible runout along the shaft axis can bring about a difference in the machined diameter. To account for such deviations, the usual measurement procedure is to measure the diameter repeatedly at one location of the shaft, rotate the shaft, and repeatedly measure again. Then the micrometer is moved to a different location along the axis of the shaft and the procedure repeated.

It is unusual to calibrate a micrometer in a working machine shop, although an occasional bias check against accurate gauge blocks is normal procedure. Let us assume that the micrometer is used as is without calibration. Data-acquisition errors are introduced from at least several elements:

1. Since the micrometer is not calibrated, the reading during any measurement can be affected by a possible bias in the micrometer markings. This can be labeled as an instrument error, B_{22} (see Table 5.2). Experience shows that this value will be of the order of the resolution of the micrometer. Ideally, the effects of using different micrometers and operators could be tested by replication so as to provide a precision estimate instead. However, this is rarely done.

2. The precision upon repeated readings will be affected by the resolution of the readout, eccentricity of the shaft, and the exact placement of the micrometer on any cross section along the shaft. It is not possible to separate these errors, so they are grouped as variation errors, a precision error, P_{25}. This value, P_{25}, can be discerned from the statistics of the measurements made at any cross section (replication).

3. Spatial variations in the diameter along its length will introduce scatter between the statistical values for each cross section. Since this affects the spatial precision of the overall mean diameter, its effect must be accounted for. Label this error as a spatial error, P_{28}. It can be estimated by the pooled statistical variance of the mean values at each measurement location (replication).

Since a diameter is the direct outcome from any reading, there will be no data-reduction errors in such a measurement.

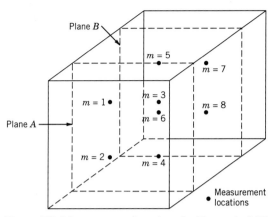

Figure 5.8 Measurement locations for Example 5.15.

COMMENT

If we were interested in the uncertainty associated with the production of these shafts, note that machinist, machine, micrometer, and possibly day of the week would be extraneous variables in need of a randomized test plan.

EXAMPLE 5.15

The mean temperature in an oven is to be estimated by using the information obtained from a temperature probe. The manufacturer offers a statement suggesting an error within 0.6°C with this probe. Determine the oven temperature.

    The measurement process is defined such that the probe is to be positioned at several strategic locations within the oven. The measurement strategy is to make four measurements within each of two equally spaced cross-sectional planes of the oven. The positions are chosen so that each measurement location corresponds to the centroid of an area of value equal for all locations so that spatial variations in temperature throughout the oven are accounted for. The measurement locations within the cubical oven are shown in Figure 5.8. In this example, 10 measurements are made at each position, with the results given in Table 5.4. Assume that a temperature readout device with a resolution of 0.1°C is used.

KNOWN

Data of Table 5.4

Table 5.4 Example 5.15: Oven Temperature Data, $N = 10$

Location	$\overline{T}_m$	S_{T_m}	Location	$\overline{T}_m$	S_{T_m}
1	342.1	1.1	5	345.2	0.9
2	344.2	0.8	6	344.8	1.2
3	343.5	1.3	7	345.6	1.2
4	343.7	1.0	8	345.9	1.1

Note: $\overline{T}_m = \frac{1}{N}\sum\limits_{n=1}^{N} T_{mn}$; $S_{T_m} = \left[\frac{1}{N-1}\sum\limits_{n=1}^{N}(T_{mn} - \overline{T}_m)^2\right]^{1/2}$

FIND

$$T' = \overline{T} \pm u_T \quad (95\%)$$

SOLUTION

The mean values at each of the locations will be averaged to yield a mean oven temperature, using pooled averaging:

$$\langle \overline{T} \rangle = \frac{1}{8} \sum_{m=1}^{8} \overline{T}_m = 344.4°C$$

Of the three error source groups, only data-acquisition errors remain, with $B_1 = B_3 = P_1 = P_3 = 0$. Elemental errors in the data-acquisition group are due to (1) the temperature probe system (instrument error), (2) spatial variation errors, and (3) temporal variation errors. Consider first the elemental error in the probe system. A statement of $0.6°C$ was provided by the manufacturer. Because no statistical information is provided, the total uncertainty caused by the probe system will be considered to be due to bias. Thus, with $j = 2$ (Table 5.2)

$$B_{22} = 0.6°C \qquad P_{22} = 0$$

Consider next the spatial error contribution to the estimate in mean temperature, $\overline{T}$. This error arises from the use of discrete measurement locations within the oven coupled with the nonuniformity in the oven temperatures. An estimate of spatial temperature distribution within the oven can be made by examining the mean temperatures at each measured location. These temperatures suggest that the oven is not uniform in temperature. The mean temperatures within the oven display a standard deviation of

$$S_T = \sqrt{\frac{\sum_{m=1}^{8} (\overline{T}_m - \langle \overline{T} \rangle)^2}{7}} = 1.26°C$$

Thus, the precision index of the oven mean temperature is

$$P_{28} = \frac{S_T}{8^{1/2}} = 0.45°C$$

with degrees of freedom $\nu = 7$. We assume no bias in this element:

$$B_{28} = 0$$

During each of the 10 measurements of temperature at each location, time variations in the probe output cause data scatter, as evidenced by the respective S_{T_m} values. Such time variations are caused by random local temperature variations as measured by the probe, probe system resolution, and oven temperature control variations during fixed operating conditions. It is rarely possible to separate these, so they are considered as one error. A precision index can be found.

Using the pooled standard deviation, equation (4.19),

$$S_T = \sqrt{\frac{\sum_{m=1}^{8} \sum_{n=1}^{10} (\overline{T}_{mn} - \langle \overline{T} \rangle)^2}{M(N-1)}} = 1.08°C$$

the precision index is estimated by

$$P_{29} = \frac{\langle S_T \rangle}{80^{1/2}} = 0.12$$

with degrees of freedom $\nu = 72$, and $B_{29} = 0$.

The data-acquisition source bias limit can now be estimated from equation (5.22); therefore,

$$B_2 = \sqrt{B_{22}^2 + B_2^2 + B_{29}^2} = 0.6°C$$

The data-acquisition source precision index is estimated from equation (5.20) to be

$$P_2 = \sqrt{P_{22}^2 + P_{28}^2 + P_{29}^2} = 0.47°C$$

with degrees of freedom from equation (5.25) of $\nu = 8$.

Because error enters through only one source, the measurement bias limit and measurement precision index are, from equations (5.23) and (5.21), simply

$$B = B_2 = 0.6°C \qquad P = P_2 = 0.47°C \qquad \text{with } \nu = 8$$

The uncertainty in the mean oven temperature is estimated from equation (5.24).

$$u_T = \sqrt{B^2 + (t_{8,95}P)^2} \quad (95\%)$$

with $t_{8,95} = 2.306$.

The best estimate of the mean oven temperature is given by equation (4.1) as

$$T' = \overline{T} \pm u_T = 344.4 \pm 1.24°C \quad (95\%)$$

5.9 SUMMARY

In this chapter the manner in which various errors can enter into a measurement have been discussed. The procedures of uncertainty analysis have been introduced as a means of estimating error propagation within a measurement and the propagation of error among independent measurands that are used to determine a result through some functional relationship. The role and use of uncertainty analysis in both the design of measurement systems, through selection of equipment and procedures, as well as in the interpretation of measured data, is discussed. The procedures will provide reasonable estimates of the uncertainty to be expected in measurements.

The reader is advised that measurement uncertainty by its very nature evades exact estimates. The important thing is that the uncertainty analysis be performed during the design and development stages of a measurement, as well as at its completion.

REFERENCES

1. International Standards Organization, ISO 5168, Measurement of fluid flow—Evaluation of uncertainty, New York, 1998.
2. Smith, R. E., and Wenhofer, S., From measurement uncertainty to measurement communications, credibility and cost control in propulsion ground test facilities, *Journal of Fluids Engineering* 107:165–172, 1985.

3. PTC 19.1, Measurement uncertainty, *ANSI/ASME Power Test Codes*, American Society of Mechanical Engineers, New York, 1998.

4. Abernathy, R. B., and Thompson, J. W., *Measurement Uncertainty Handbook*, Instrument Society of America, Research Triangle Park, NC, 1980.

5. Kline, S. J., and McClintock, F. A., Describing uncertainties in single sample experiments, *Mechanical Engineering* vol. 75, January, 1953.

6. Moffat, R. J., Uncertainty analysis in the planning of an experiment, *Journal of Fluids Engineering* 107:173–181, 1985.

7. Kestin, J., *A Course in Thermodynamics*, revised, Hemisphere, New York, 1979.

Suggested Reading

Sigurdson, S., and Kimball, D. E., Practical method for estimating number of test readings required, *Journal of Engineering Power*, July:386–392, 1976.

Taylor, J. R., *An Introduction to Error Analysis*, University Science Books, Mill Valley, CA, 1982.

Wang, T., and Simon, T. W., Development of a special purpose test surface: Introduction of a new uncertainty analysis step, AIAA paper 88-0169, New York, 1988.

Wyler, J. S., Estimating the uncertainty of spatial and time average measurements, *Journal of Engineering Power*, October:473–480, 1975.

NOMENCLATURE

e_j	jth elemental error	F	force $[m\,l\,t^{-2}]$
$f(\ )$	general functional relation	N	number of measured values
p	pressure $[m\,l^{-1}\,t^{-2}]$	R	resultant value
$t_{v,95}$	t variable at 95% probability	S_x	sample standard deviation of x
t	time $[t]$	$S_{\bar{x}}$	sample standard deviation of
u_x	uncertainty of a measured variable		the means
u_d	design-stage uncertainty	P	measurement precision index
u_c	instrument calibration uncertainty	P_{ij}	jth elemental precision error in ith
u_0	zero-order uncertainty		source group
u_1	first-order uncertainty	P_1, P_2, P_3	source precision index
u_N	Nth-order uncertainty	$(P\%)$	percent probability
x	measured variable	Q	flow rate $[l^3\,t^{-1}]$
x'	true value of x	T	temperature $[^\circ]$
$\bar{x}$	sample mean value of x	$V, \mathcal{V}$	volume $[l^{-3}]$
B	measurement bias limit	θ_i	sensitivity index
B_{ij}	jth elemental bias error in ith	ρ	gas density $[m\,l^{-3}]$
	source group	v	degrees of freedom
B_1, B_2, B_3	source bias limit	$\langle\ \rangle$	pooled statistic

PROBLEMS

5.1 In Chapter 1 the development of a test plan was discussed. Clearly discuss how a test plan should account for the presence of bias and precision errors. You may wish to include calibration, randomization, and repetition in your discussion.

5.2 How can bias error be estimated for a measurement system? How is precision error estimated?

5.3 Explain what is meant by true value, best estimate, mean value, uncertainty, and confidence interval.

5.4 Consider a common tire pressure gauge. How would you estimate the uncertainty in a measured pressure at the design stage and at the Nth order? Should the estimates differ? Explain.

5.5 A micrometer has graduations scribed at 0.001-in. (0.025-mm) intervals. Estimate the uncertainty that is due to resolution at a 95% probability.

5.6 A tachometer has an analog display dial graduated in 5 revolutions per minute (rpm) increments. The user manual states an accuracy of 1% of reading. Estimate the uncertainty in the reading at 10, 500, and 5000 rpm.

5.7 An automobile speedometer is graduated in 5-mph (8-kph) increments and has an accuracy rated to be within ±4%. Estimate the uncertainty in indicated speed at 60 mph (90 kph).

5.8 A temperature measurement system is composed of a sensor and a readout device. The readout device has an accuracy of 0.8°C with a resolution of 0.1°C. The sensor has an off-the-shelf accuracy of 0.5°C. Estimate the design-stage uncertainty in the temperature indicated by this combination of sensor and readout.

5.9 Two resistors are to be combined to form an equivalent resistance of 1000 Ω. Readily available are two common resistors rated at 500 ± 50 Ω and two common resistors rated at 2000 Ω ± 5%. What combination of resistors (series or parallel) would provide the smaller uncertainty in an equivalent 1000-Ω resistance? *parallel*

5.10 An equipment catalog boasts that a pressure transducer system comes in 3 1/2 digit (e.g. 19.99) or 4 1/2 digit (e.g., 19.999) displays. The 4 1/2 digit model costs 50% more. Both units are otherwise identical. The specifications are as follows:

Linearity error:	0.15% FSO
Hysteresis error:	0.20% FSO
Repeatability error:	0.25% FSO

For a FSO of 200 kPa, select a readout based on appropriate uncertainty calculations. Explain.

5.11 The shear modulus, G, of an alloy can be determined by measuring the angular twist, θ, resulting from a torque applied to a cylindrical rod made from the alloy. For a rod of radius R and a torque applied at a length L from a fixed end, the modulus is found by $G = 2LT/\pi R^4 \theta$. Examine the effect of the relative uncertainty of each measured variable on the shear modulus. If during test planning all of the uncertainties are set at 1%, what is the uncertainty in G?

5.12 An ideal heat engine operates in a cycle and produces work as a result of heat transfer from a thermal reservoir at an elevated temperature, T_h, and by rejecting energy to a thermal sink at T_c. The efficiency for such an ideal cycle, termed a "Carnot cycle," is

$$\eta = 1 - (T_c/T_h)$$

Determine the required uncertainty in the measurement of temperature to yield an uncertainty in efficiency of 1%. Use nominal values of temperature of $T_h = 1000$ K and $T_c = 300$ K.

5.13 Heat transfer from a rod of diameter D immersed in a fluid can be described by the Nusselt number, $\text{Nu} = hD/k$, where h is the heat-transfer coefficient and k is the thermal conductivity of the fluid. If h can be measured to within ±7%, (95%), estimate the uncertainty in Nu for the nominal value of $h = 150$ W/m^2 K. Let $D = 20 ± 0.5$ mm and $k = 0.6 ± 2$% W/m K.

5.14 Estimate the design-stage uncertainty in determining the voltage drop across a heater. The device has a nominal resistance of 30 Ω and power rating of 500 W. Available is an ohmmeter (accuracy: within 0.5%; resolution: 1 Ω) and ammeter (accuracy: within 0.1%; resolution: 100 mA). Recall $E = IR$.

5.15 Explain the critical difference(s) between a design-stage uncertainty analysis and an advanced-stage uncertainty analysis.

5.16 From an uncertainty analysis perspective, what important information does replication provide that is not found by repetition alone? How is this information included in an uncertainty analysis?

5.17 A displacement transducer has the following specifications:

Linearity:	$\pm0.25\%$ reading
Drift:	$\pm0.05\%/^\circ C$ reading
Repeatability:	$\pm0.25\%$ reading
Excitation:	10–25 V dc
Output:	0–5 V dc
Range:	0–5 cm

The transducer output is to be indicated on a voltmeter having a stated accuracy of $\pm0.1\%$ reading with a resolution of 10 μV. The system is to be used at room temperature, which can vary by $\pm10^\circ C$. Estimate an uncertainty in a nominal displacement of 2 cm at the design stage.

5.18 The displacement transducer of Problem 5.17 is used in measuring the displacement of a body impacted by a mass. Twenty measurements are made, which yields

$$\bar{x} = 172 \text{ mm} \qquad S_x = 17 \text{ mm}$$

Determine a best estimate for the mass displacement at 95% probability.

5.19 A pressure transducer outputs a voltage to a readout device that converts the signal back to pressure. The device specifications are as follows:

Resolution:	0.1 psi
Repeatability:	0.1 psi
Linearity:	within 0.1% of reading
Drift:	less than 0.1 psi/6 months (32–90°F)

The transducer has a claimed accuracy of within 0.5% of reading. For a nominal pressure of 100 psi at 70°F, estimate the uncertainty in a measured pressure.

5.20 For a thin-walled pressure vessel of diameter, D, and wall thickness, t, subjected to an internal pressure, p, the tangential stress is given by $\sigma_t = pD/2t$. During one test, 10 measurements of pressure yielded a mean of 8610 lb/ft^2 with a standard deviation of 273.1. Cylinder dimensions are to be based on a set of 10 measurements, which yielded: $\bar{D} = 6.2$ in., $S_D = 0.18$, $\bar{t} = 0.22$ in., and $S_t = 0.04$. Determine the best estimate of stress.

5.21 Suppose the pressure measurand p contains three elemental precision errors in its calibration:

$$P_{11} = 0.90 \text{ N/m}^2 \qquad P_{12} = 1.10 \text{ N/m}^2 \qquad P_{13} = 0.09 \text{ N/m}^2$$
$$v_{11} = 21 \qquad\qquad v_{12} = 10 \qquad\qquad v_{13} = 15$$

Estimate the source precision index for calibration errors.

5.22 An experiment to determine force acting at a point on a member contains the following source errors:

$$B_1 = 2 \text{ N} \qquad B_2 = 4.5 \text{ N} \qquad B_3 = 3.6 \text{ N}$$
$$P_1 = 0 \qquad\quad P_2 = 6.1 \text{ N} \qquad P_3 = 4.2 \text{ N}$$
$$\qquad\qquad\quad v_2 = 17 \qquad\quad v_3 = 19$$

If the mean value for force is estimated to be 200 N, determine the uncertainty in the mean value at 95% confidence.

5.23 The area of a flat, rectangular parcel of land is computed from the measurement of the length of two adjacent sides, X and Y. Measurements are made by using a scaled chain accurate to within 0.5% over its indicated length. The two sides are measured several times with the following results:

$$\bar{X} = 556 \text{ m} \qquad \bar{Y} = 222 \text{ m}$$
$$S_x = 5.3 \text{ m} \qquad S_y = 2.1 \text{ m}$$
$$v = 8 \qquad\qquad v = 7$$

Estimate the area of the land and state the confidence interval of that measurement at 95%.

5.24 Estimate the precision error in the measured value of stress for a data set of 23 measurements, if $\overline{\sigma} = 1061$ psi and $S_\sigma = 22$ psi.

5.25 Estimate the uncertainty at 95% confidence in the strength of a metal alloy. Six separate specimens are tested with a standard deviation of 1.23 MPa in the results. A bias limit of 1.48 MPa is estimated by the operator.

5.26 A pressure measuring system outputs a voltage that is proportional to pressure. It is calibrated against a transducer standard (certified accuracy: within ±0.5 psi) over its 0–100 psi range with the results given as follows. The voltage is measured with a voltmeter (instrument error: within ±10 μV; Resolution: 1 μV). The engineer intending to use this system estimates that installation effects can cause the indicated pressure to be off by another ±5 psi. Estimate the uncertainty at 95% confidence in using the system based on the known information.

E [mV]:	0.004	0.399	0.771	1.624	2.147	4.121
p [psi]:	0.1	10.2	19.5	40.5	51.2	99.6

5.27 The density of a metal composite is to be determined from the mass of a cylindrical ingot. The volume of the ingot is determined from diameter and length measurements. It is estimated that mass, m, can be determined to within 0.1 lb$_m$ using an available balance scale; length, L, can be determined to within 0.05 in. and diameter, D, to within 0.0005 in. Estimate the zero-order uncertainty in the determination of the density. Which measurement contributes most to the uncertainty in the density? Which measurement method should be improved first if the uncertainty in density? Which measurement method should be improved first if the uncertainty in density is unacceptable? Use the nominal values of $m = 4.5$ lb$_m$, $L = 6$ in., $D = 4$ in.

5.28 For the ingot of Problem 5.27, the diameter of the ingot is measured 10 independent times at each of three cross sections. For the results here, give a best estimate in the diameter with its uncertainty:

$$\overline{d}_1 = 3.9920 \text{ in.} \qquad \overline{d}_2 = 3.9892 \text{ in.} \qquad \overline{d}_3 = 3.9961 \text{ in.}$$
$$S_{d_1} = 0.005 \text{ in.} \qquad S_{d_2} = 0.001 \text{ in.} \qquad S_{d_3} = 0.0009 \text{ in.}$$

5.29 For the ingot of Problem 5.27, the following measurements are obtained from the ingot:

$$\overline{m} = 4.4 \text{ lb}_m \qquad \overline{L} = 5.85 \text{ in.}$$
$$S_m = 0.1 \text{ lb}_m \qquad S_L = 0.1 \text{ in.}$$
$$N = 21 \qquad N = 11$$

From the results of Problem 5.28, estimate the density and the uncertainty in this value. Compare your answer with that of 5.27. Explain any difference.

5.30 A temperature measurement system is calibrated against a standard system certified to an uncertainty of ±0.05°C at 95%. The system sensor is immersed alongside the standard within a temperature bath so that the two are separated by ~10 mm. The temperature uniformity of the bath is estimated at ~5°C/m. The temperature system sensor is connected to a readout that indicates the temperature in terms of voltage. The following are calibration results between the temperature indicated by the standard and the indicated voltage:

T [°C]	E [mV]	T [°C]	E [mV]
0.1	0.004	40.5	1.624
10.2	0.399	51.2	2.147
19.5	0.771	99.6	4.121

a. Compute the calibration curve fit.
b. Estimate the uncertainty in using the output from the temperature measurement system for temperature measurements.

5.31 The power usage of a strip heater is to be determined by measuring heater resistance and heater voltage drop simultaneously. The resistance is to be measured by using an ohmmeter having a resolution of 1 Ω and an error stated to be within 1% of its reading, and voltage is to be measured by using a voltmeter having a resolution of 1 V and a stated error of 1% of reading. It is expected that the heater will have a resistance of 100 Ω and use 100 W of power. Determine the uncertainty in power determination to be expected with this equipment at the zeroth-order level. Recompute at the design stage.

5.32 The power usage of a dc strip heater can be determined in either of two ways: (1) heater resistance and voltage drop can be measured simultaneously and power computed, or (2) heater voltage drop and current measured simultaneously and power computed. Instrument manufacturer specifications are listed:

Instrument	Resolution	Error (% reading)
Ohmmeter	1 Ω	0.5
Ammeter	0.5 A	1
Voltmeter	1 V	0.5

For loads of 10 W, 1 kW, and 10 kW, determine the best method based on an appropriate uncertainty analysis. Assume nominal values as necessary for resistance, current, and voltage.

5.33 A set of tests is to be done to study how the cure temperature of a composite material will affect its strength. Tests will be done over a range of temperatures (20–60°C). Temperature is to be measured at each of four quadrants within the cure chamber by using temperature equipment for which the accuracy can be estimated. The test specimens are expensive, so only one test at any four temperatures can be performed. Suppose before any composite is tested, the chamber thermostat is set to, say, 30°C, and 20 measurements over time are taken. This exercise is then replicated five times so that the statistics can be computed. Compare this method to one in which a composite is placed in the chamber and cured while 100 measurements of chamber temperature are made (25 at each of the four locations). Would the uncertainty estimates for the cure temperature based on the two methods include the same effects? Explain.

5.34 Time variations in a signal require that the signal be measured often in time to determine its mean value. A preliminary sample is obtained by measuring the signal 50 times. The statistics from this sample show a mean value of 2.112 V with a standard deviation of 0.387 V. If it is desired to state the precision of the mean value to within 0.100 V at 95% confidence, how many measurements of the signal will be required based on the data variations?

5.35 The diameter of a shaft is estimated in a manner consistent with Example 5.14. A handheld micrometer (resolution: 0.001 in.; accuracy: <0.001 in.), is used to make measurements about four selected cross-section locations. Ten measurements of diameter are made around each location with the results noted below. Provide a reasonable statement as to the diameter of the shaft and its tolerance (the uncertainty in that estimate).

Location:	1	2	3	4
$\bar{D}$ (in.):	4.494	4.499	4.511	4.522
S_D:	0.006	0.009	0.010	0.003

5.36 The pressure in a large vessel is to be maintained at some set pressure for a series of tests. A compressor is to supply air through a regulating valve that is set to open and close at the set pressure. A dial gauge (resolution: 1 psi; accuracy: 0.5 psi) is used to monitor pressure in the vessel. Thirty trial replications of pressurizing the vessel to a set pressure of 50 psi are attempted to estimate pressure controllability and a standard deviation in set pressure of 2 psi is found. Estimate the uncertainty to be expected in the vessel set pressure.

5.37 The linear displacement of a vehicle caused by an applied impact force is measured with a transducer. Transducer specifications are as follows:

Input range:	0–5 m
Output range:	0–5 V
Linearity:	±0.25% reading
Drift:	±0.05%/°C FSO
Repeatability:	±0.25% reading
Sensitivity:	±0.10% FSO

The transducer output is indicated on a voltmeter (Accuracy: within ±0.1% reading; resolution: 10 μV). The expected nominal displacement of the vehicle is to be 4 m for an impact force of 2000 ± 100 N (95%). The force-displacement relation can be assumed linear. Four replications consisting of 10 measurements are made over the course of 1 day. The results are given with the ambient temperature for each run.

Test Run	N	Mean Value [m]	S_x [m]	$T_{ambient}$ [°C]
1	10	4.3	0.22	21
2	10	3.8	0.27	24
3	10	4.2	0.31	27
4	10	4.0	0.15	22

Report the mean displacement of the vehicle. Include a complete uncertainty estimate. Compare your uncertainty estimate to that from a simple design-stage analysis. Discuss why these differ.

5.38 The cooling of a thermometer (e.g., Examples 3.3 and 3.4) can be modeled as a first-order system with $\Gamma = e^{-t/\tau}$. If Γ can be measured to within 2% and time within 1%, explain the uncertainty in τ over the range $0 \le \Gamma \le 1$.

5.39 A J-type thermocouple monitors the temperature of air flowing through a duct. Its signal is measured by a thermostat. The air temperature is maintained constant by an electric heater whose power is controlled by the thermostat. To test the control system, 20 measurements of temperature were taken over a reasonable time period during steady operation. This was repeated three times with these as:

Run	N	$\overline{T}$ [°C]	S_T [°C]
1	20	181.0	3.01
2	20	183.1	2.84
3	20	182.1	3.08

The thermocouple itself has an error within 1°C (95%). It has a 90% rise time of 20 ms. Thermocouple insertion errors are estimated to be <1.2°C (95%). What information is found by performing replications? Identify the elemental errors that affect the system's control of the air temperature. What is the uncertainty in the set temperature?

5.40 Based on everyday experience, estimate the bias error in the following measuring instruments: bathroom scale; plastic ruler scale; micrometer; kitchen window bulb thermometer; automobile speedometer.

5.41 A tank is pressurized with air at 25 ± 2°C (95%). Determine the uncertainty in the tank's air density if tank pressure is known to within ±1% (95%). Assume ideal gas behavior.

5.42 The density of air must be known to within ±0.5%. If the air temperature can be determined to within ±1°C (95%), what uncertainty can be tolerated in the pressure measurement if air behaves as an ideal gas?

5.43 In pneumatic conveying, solid particles such as flour or coal are carried through a duct by a moving air stream. Solids density at any duct location can be measured by passing a laser beam of known intensity I_o through the duct and measuring the light intensity transmitted to the other side, I. A transmission factor is found by

$$T = I/I_o = e_{.}^{-KEW} \qquad 0 \leq T \leq 1$$

Here W is the width of the duct, K is the solids density, and E is a factor taken as 2.0 $\pm$0.4 (95%) for spheroid particles. Determine how u_K/K is related to the relative uncertainies of the other variables. If the transmission factor and duct width can be measured to within $\pm$1%, can solids density be measured to within 5%? 10%? Discuss your answer, remembering that T varies from 0 to 1.

5.44 A step test is run to determine the time constant of a first order instrument (see Chapter 3). If the error fraction, $\Gamma(t)$, can be estimated to within $\pm$2% (95%) and time, t, can be estimated in seconds to within $\pm$0.5% (95%), plot U_τ/τ versus $\Gamma(t)$ over its range, $0 \leq \Gamma(t) \leq 1$.

As described in Section 5.6, sequential perturbation employs finite-difference estimates of the sensitivity indices, θ_i, to propagate uncertainties to a result. The following problems illustrate the use of experimental data in determining the sensitivity indices, θ_i, through numerical approximation.

5.45 Golf balls are often tested by using a mechanical player called an "Iron Byron" because the robotic golfer's swing was patterned after Byron Nelson, a famous and successful golf professional. It is proposed to use initial velocity and launch angle from such testing to predict carry distance. The following data represent the effect of initial speed and launch angle on distance a particular golf ball flies when struck with a driver.

Initial Velocity (mph)	Launch Angle (deg)	Carry Distance (yds)
165.5	8	254.6
167.8	8	258.0
170	8	261.4
172.2	8	264.8
165.5	10	258.2
167.8	10	261.6
170	10	264.7
172.2	10	267.9
165.5	12	260.6
167.8	12	263.7
170	12	266.8
172.2	12	269.8

If the initial velocity can be measured to within 1 mph, and the launch angle to within 0.1°, estimate the resulting uncertainty in the carry distance as a function of initial velocity and launch angle. Over this range of initial velocity and launch angle, can a single value of uncertainty be assigned?

5.46 A particular flow meter allows the volumetric flow rate, Q, to be inferred from a measured pressure drop, ΔP. Theory predicts $Q \propto \sqrt{\Delta P}$, and a calibration yields the following data:

Q (m³/min)	ΔP (Pa)
10	1000
20	4271
30	8900
40	16023

What uncertainty in the measurement of ΔP is required to yield an uncertainty of 0.25% in Q over the range of flow rates from 10 to 40 m³/min?

5.47 A geometric stress concentration factor, K_t, is used to relate the actual maximum stress to a

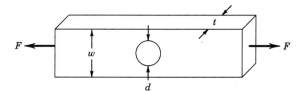

Figure 5.9 Structural member for Problem 5.47.

well-defined nominal stress. The maximum stress is given by

$$\sigma_{\text{max}} = K_t \sigma_o$$

where σ_{max} represents the maximum stress and σ_o represents the nominal stress. The nominal stress is most often defined at the minimum cross section.

Consider the axial loading shown in Figure 5.9, where the structural member is in axial tension and experiences a stress concentration as a result of a transverse hole. In this geometry, $\sigma_o = F/A$, where $A = (w - d)t$. If $d = 0.5w$, then $K_t = 2.2$. Suppose $F = 10,000 \pm 500$ N, $w = 1.5 \pm 0.02$ cm, $t = 0.5 \pm 0.02$ cm, and the uncertainty in the value of d is 3%. Neglecting the uncertainty in the stress concentration factor, determine the uncertainty in the maximum stress experienced by this part.

Chapter 6

Analog Electrical Devices and Measurements

6.1 INTRODUCTION

The output from any stage of a measurement system is often an electrical signal. This signal typically originates from the measurement of a physical variable using some fundamental electromagnetic or electrical phenomenon and then propagates from stage to stage of the measurement system. Data acquisition, data transmission, and data processing are mostly accomplished by using electronic devices.

This chapter provides an introduction to basic electrical measurement devices for analog signals. These devices have historically served as the basis for a variety of measurement systems, but more importantly demonstrate measurement principles for electrical analog signals.

The integration of discrete or digital devices with analog electrical devices is the most common means of recording and transmitting data. As such, the issues arising in the transmission and processing of analog electrical signals, including useful devices, are discussed in this chapter.

6.2 ANALOG DEVICES: CURRENT MEASUREMENTS

This section deals specifically with the analog measurement of time-continuous static (dc) or dynamic (ac) current signals. Although a number of different methods are available for the measurement of current, the most common methods use a galvanometer with associated circuitry. In fact, most analog meters that may be calibrated to indicate some other variable actually sense and respond to current flow through a galvanometer.

Direct Current

The most basic analog measurements of electrical current are accomplished as a result of the force exerted on a current-carrying conductor in a magnetic field. If a positive electric charge in motion experiences a force that tends to deflect it from a straight line path, a magnetic field must be present. Because an electric current is a series of moving charges, a magnetic field will exert a force on a current-carrying conductor. The force exerted by a magnetic field can be utilized to measure the flow of current in a conductor and to move a pointer on a display. Consider a straight

192

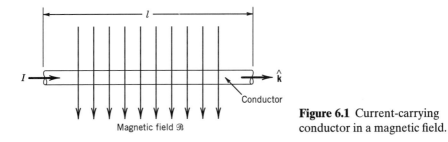

Figure 6.1 Current-carrying conductor in a magnetic field.

length of a conductor through which a current, I, flows, as shown in Figure 6.1. The force on this conductor that is due to a magnetic field strength $\mathcal{B}$ is

$$F = Il\mathcal{B} \tag{6.1}$$

This relation is valid only when the direction of the current flow and the magnetic field are at right angles. In the general case

$$\mathbf{F} = Il\hat{k} \times \mathcal{B} \tag{6.2}$$

where $\hat{k}$ is a unit vector along the direction of the current flow, and the force $\mathbf{F}$ and the magnetic field $\mathcal{B}$ are also vector quantities. Equation (6.2) provides the magnitude of the developed force $\mathbf{F}$, and, by the right-hand rule, also the direction in which the force on the conductor acts.

Similarly, a current loop in a magnetic field will experience a torque if the loop is not aligned with the magnetic field, as illustrated in Figure 6.2. The torque[1] on a loop composed of N turns is given by

$$T_\mu = NIA\mathcal{B} \sin\alpha \tag{6.3}$$

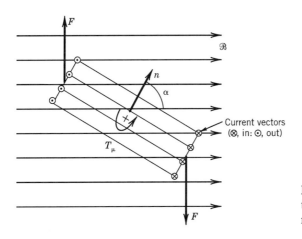

Figure 6.2 Forces and resulting torque on a current loop in a magnetic field.

---

[1] In vector form, the torque on a current loop can be written

$$\mathbf{T}\mu = \mathbf{\mu} \times \mathcal{B}$$

where $\mathbf{\mu}$ is the magnetic dipole moment.

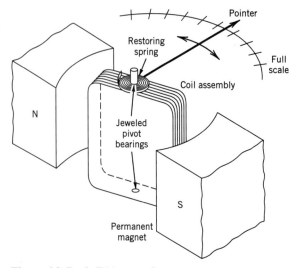

Figure 6.3 Basic D'Arsonval meter movement.

where

A = cross-sectional area defined by the perimeter of the current loop

$\mathscr{B}$ = magnetic field strength

I = current

α = angle between the normal to the cross-sectional area of the current loop and the magnetic field

A galvanometer is a device that is sensitive to current flow through the torque exerted on a current loop in a magnetic field; the basic design of a galvanometer is shown in Figure 6.3. In this arrangement, the uniform radial magnetic field and torsional spring result in a steady angular deflection of the coil, which corresponds to the existing current in the coil. The coil and fixed permanent magnet are called a D'Arsonval movement. In this arrangement, the normal direction to the current loop is such that $\alpha = 90°$. For increased sensitivity, the pointer can be replaced by a mirror and light beam arrangement. The highest sensitivity of commercial galvanometers ranges from approximately 10^{-5} μA/div for light beam galvanometers, to approximately 0.1 μA/div for a galvanometer with a pointer.

The sensitivity of the D'Arsonval movement to the flow of direct current makes it a suitable method for transforming an electrical signal into a visually discernible signal, the movement of a pointer on a dial. A typical circuit for an analog current measuring device, an ammeter, is shown in Figure 6.4. The range of current that can be measured is determined by selection of the combination of shunt resistor and the internal resistance of the galvanometer. The shunt resistor provides a bypass for current flow, reducing the current that flows through the galvanometer.

Errors inherent to a mechanical D'Arsonval movement include hysteresis and repeatability errors caused by mechanical friction in the pointer-bearing movement, and linearity errors in the spring that provides the restoring force for equilibrium. Also, in developing a torque, the D'Arsonval movement must extract energy from the current flowing through it. This draining of energy from the signal being measured changes the measured signal. Such an effect is a loading error. A quantitative analysis of loading errors will be provided later.

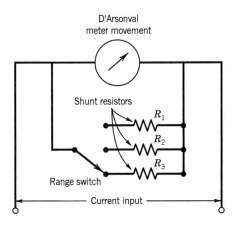

Figure 6.4 Simple multirange ammeter.

Alternating Current

The measurement of alternating current is accomplished in several ways in analog instruments. Common ac analog meter arrangements employ electromagnets whose magnetic field strengths increase with increasing current flow. Either an iron plunger similar to the design of a solenoid switch, or an electrodynamometer may be used to transform the electromagnetic force into movement of a pointer.

An electrodynamometer is basically a D'Arsonval movement modified for use with ac current by replacing the permanent magnet with an electromagnet in series with the current coil. These ac meters have upper limits on the frequency of the alternating current that they can effectively measure; many basic instruments are intended for use with standard line frequency.

EXAMPLE 6.1

A galvanometer consists of N turns of a conductor wound about a core of length l and radius r that is perpendicular to a magnetic field of uniform flux density $\mathscr{B}$. A dc current passes through the conductor as a result of an applied potential, $E_i(t)$. The output of the device is the rotation of the core and pointer, θ, as shown in Figure 6.5. Develop a model relating pointer rotation and current.

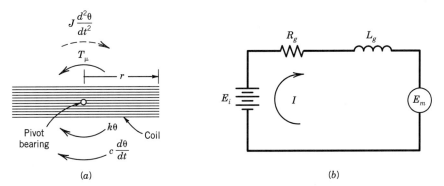

Figure 6.5 Circuit and free-body diagram for Example 6.1.

FIND

A dynamic model relating θ and I.

SOLUTION

The galvanometer is a rotational system consisting of a torsional spring, a pointer, and core that are free to deflect; the rotation is subject to frictional damping. The mechanical portion of the device consists of a coil having moment of inertia, J, bearings that provide damping from friction, with a damping coefficient, c, and a torsional spring of stiffness, k. The electrical portion of the device consists of a coil with a total inductance, L_g, and a resistance, R_g.

Application of Newton's second law can be accomplished by examining the mechanical free-body diagram shown in Figure 6.5(a). As a current passes through the coil, a force is developed that produces a torque on the coil:

$$T_\mu = (2N\mathcal{B}lr)I \tag{6.4}$$

This torque, which tends to rotate the coil, is opposed by an electromotive force, E_m, caused by the Hall effect

$$E_m = \left(\mathbf{r}\frac{d\theta}{dt} \times \mathcal{B}\right)l\hat{k} = (2N\mathcal{B}r)\frac{d\theta}{dt} \tag{6.5}$$

which produces a current in the direction opposite to that produced by E_i. The electrical free body is shown in Figure 6.5(b). Application of Newton's law to the free body shown in Figure 6.5(a) yields

$$J\frac{d^2\theta}{dt^2} + c\frac{d\theta}{dt} + k\theta = T_\mu \tag{6.6}$$

Kirchhoff's law provides from Figure 6.5(b)

$$L_g\frac{dI}{dt} + R_g I = E_i - E_m \tag{6.7}$$

Equations (6.4) and (6.5) define the coupling relations between mechanical equation (6.6) and electrical equation (6.7). We see that the current that is due to potential, E_i, brings about developed torque, T_μ, that moves the galvanometer pointer, and that this motion is opposed by the mechanical restoring force of the spring and by the development of an opposing electrical potential, E_m. The system damping allows the pointer to settle to an equilibrium position.

COMMENT

The system output, the pointer movement, is governed by the second-order system response described in Chapter 3. However, the torque input to the mechanical system is a function of the electrical response of the system. In this measurement system, the input signal, the applied voltage, is transformed into a mechanical torque to provide the output deflection.

6.3 ANALOG DEVICES: VOLTAGE MEASUREMENTS

The output signal from many transducers and the signal between various stages of a measurement system are often time-dependent electrical voltages. The magnitude of these signals may range from the microvolt level for many transducers to a level

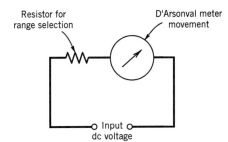

Figure 6.6 A dc voltmeter circuit.

of up to several volts at the output stage of a measurement system. Power systems may deal with voltage levels in the kilovolt or larger range. The frequency content of dynamic voltage signals is often of interest as well. Therefore, a wide variety of measurement systems have been developed for voltage measurement of static and dynamic signals. In this section, several convenient and common methods to indicate voltage in measurement systems are discussed.

Analog Meters

The measurement of dc voltage by using analog instruments may be accomplished through the circuit shown in Figure 6.6, where a D'Arsonval movement is employed in series with a resistor. The D'Arsonval movement is fundamentally sensitive to current flow, but in conjunction with an appropriate known fixed resistance can be calibrated in terms of voltage. This basic circuit is employed in the construction of analog voltage dials, volt ohmmeters (VOM), and the vacuum tube voltmeter (VTVM), which for many years served as common measurement devices for current, voltage, and resistance.

The measurement of ac voltage can be accomplished through rectification of the ac signal or through use of an electromagnet, either in an electrodynamometer, or with a movable iron vane. These instruments are basically sensitive to the rms value of a simple periodic ac current and can be calibrated in terms of voltage; shunt resistors can be used to establish the appropriate scale. The circuit shown in Figure 6.6 can also be used to measure an ac voltage if the input voltage is rectified prior to input to this circuit. The waveform output of the rectifier must be considered if a steady meter reading is to be obtained. An ac meter indicates a true rms value for a simple periodic signal only, but a *true rms* ac voltmeter performs the signal integration [i.e. equation (2.4)] required to accurately determine the rms value in a signal-conditioning stage and indicates true signal rms regardless of waveform.

Oscilloscope

The cathode ray oscilloscope provides a practical means of measuring dynamic waveforms at high frequencies for electrical signals having a time-varying voltage [1]. The oscilloscope provides extremely versatile voltage measurements, with accuracies of 3% over the frequency band extending from dc to several megahertz (sampling oscilloscopes may extend the frequency range to the gigahertz range; [2]). A schematic and a photograph of an oscilloscope are shown in Figure 6.7. A beam of electrons is emitted by the cathode ray tube. The output of the oscilloscope is a signal trace on the

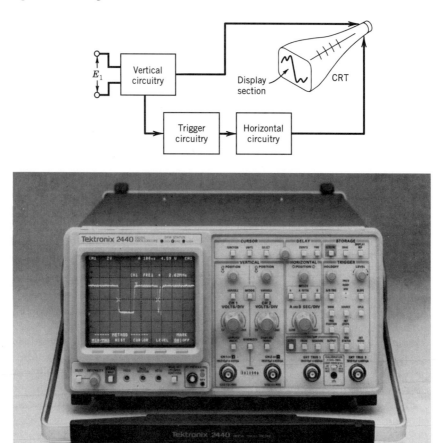

Figure 6.7 Oscilloscope. (Photograph courtesy of Tektronix, Inc.)

screen of the instrument, created by the impact of the electrons on a phosphorescent coating on the screen. Because an electron beam is composed of charged particles, it can be guided by an electrical field. In the case of the oscilloscope, pairs of plates are oriented horizontally and vertically, to control the location of the impact of the electron beam on the screen. The beam sweeps horizontally across the screen at a known speed or frequency. Input voltages to the oscilloscope result in vertical deflections of the beam and produce a trace of the voltage variations in time on the screen. A typical trace from an oscilloscope is shown in Figure 6.8. The horizontal sweep frequency can be varied over a wide range, and the operation of the oscilloscope is such that high-frequency waveforms can be resolved.

The oscilloscope provides an important test instrument for examining input voltage waveforms to measurement systems. The visual image provides a direct means to detect the superposition of noise and interference on a measured signal. This is important since nonvisual current and voltage metering devices cannot delineate noise and interference from a measured signal. By scanning over the horizontal sweep range of the oscilloscope, the contributing frequencies of such extraneous information can be detected, their sources diagnosed, and appropriate action taken to reduce their effects.

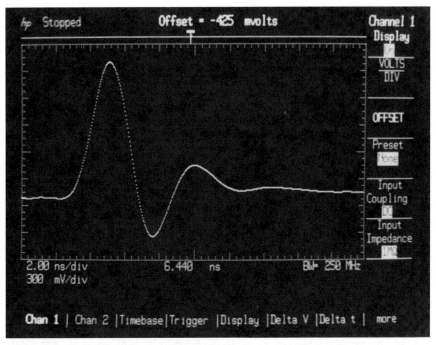

Figure 6.8 Oscilloscope output. (Photograph courtesy of Hewlett-Packard Company.)

Potentiometer

For measurements of dc voltage, in the microvolt to millivolt range, the potentiometer is a useful measuring device. Equivalent to a balance scale, a potentiometer balances an unknown input voltage against a known internal voltage. At balanced conditions there will be zero current flow in the circuit, and this condition provides for the direct determination of the unknown voltage. Thus, it is classified as a null balance instrument. A potentiometer can also serve as a laboratory voltage standard for calibration purposes. Before a practical potentiometer circuit is described, an important component in that circuit, the voltage divider, is introduced.

Voltage Divider Circuit

The circuit shown in Figure 6.9 is a simple voltage dividing circuit. The voltage dividing circuit is constructed by using a resistor made of a length of wire wound around an appropriate insulating support. The point labeled A in Figure 6.9 represents a sliding contact, which makes an electrical connection with the resistor R at any point along its length. The resistance between point A and point B in the circuit is a linear function of the distance from A to B, for an ideal resistor. Thus, the output voltage sensed by the voltmeter is given by

$$E_o = \frac{L_x}{L_T} E_i = \frac{R_x}{R_T} E_i \qquad (6.8)$$

if the internal resistance of the meter is very large relative to R_T so as to ignore the current draw of the meter.

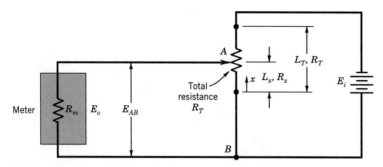

Figure 6.9 Voltage divider circuit.

Potentiometer Circuit

A simple potentiometer[2] circuit that is derived from the voltage divider circuit is shown in Figure 6.10. In this circuit a galvanometer is used to detect current flow; any current flow through the galvanometer, G, would be a result of an imbalance in the measured voltage, E_m, and the voltage imposed across points A to B, E_{AB}. If E_m is not equal to E_{AB}, a current will flow through the galvanometer, G. From the voltage divider circuit it is clear that E_{AB} can be varied by moving the sliding contact A. A null balance, corresponding to zero current flow through G, will occur only when $E_m = E_{AB}$. The measurement of E_m is accomplished by moving A until the deflection of the galvanometer reduces to zero. With a known and constant supply voltage E_i, the position of A can be calibrated to indicate E_m directly as suggested by equation (6.8).

A circuit for a practical potentiometer must include a means for determining or standardizing the supply voltage. This is accomplished by the arrangement shown in Figure 6.11 in which the output of a standard voltage cell is compared to the output voltage of the working voltage supply, usually a commercially available battery. As the battery voltage changes over time, comparison with the known standard cell voltage, and adjustment of the standardization resistance, R_c, allows for continued maintenance of E_i. To minimize current drain from the standard cell and to protect it from damaging polarization, the standard cell can be only momentarily connected in the circuit. The switches in the potentiometer circuit are normally of a spring-loaded type, which are manually closed, and open upon release. When such a switch is closed momentarily, the deflection of the galvanometer can be observed, and appropriate adjustments made to the circuit. A similar procedure should be used during measurement to protect the battery.

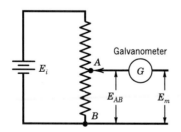

Figure 6.10 Basic potentiometer circuit.

[2]The term "potentiometer" has several common uses: a sliding contact precision variable resistor, the circuit in Figure 6.9, and the voltage measuring instrument described.

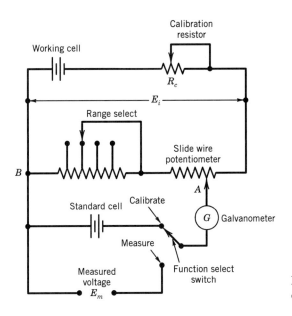

Figure 6.11 Practical potentiometer circuit, including standard cell.

Potentiometers have capabilities that make them a good choice for the accurate measurement of low voltages, with accuracies to the microvolt range. The precision of a potentiometer depends strongly on the precision of the galvanometer. However, bias errors of $<10\ \mu V$ and precision errors $<2\ \mu V$ are common in moderately priced units. Laboratory-grade units reduce these errors by 1 order of magnitude.

6.4   ANALOG DEVICES: RESISTANCE MEASUREMENTS

The measurement of electrical resistance has many important applications ranging from the simple determination of continuity in an electrical circuit to the precise measurement of changes in resistance of the order of $10^{-6}\ \Omega$ and absolute levels ranging from 10^{-5} to $10^{15}\ \Omega$. As a result of the tremendous range of resistance values that have practical application, numerous measurement techniques that are appropriate for specific ranges of resistance or resistance changes have been developed. Some of these measurement systems and circuits provide necessary protection for delicate instruments, allow compensation for changes in ambient conditions, or accommodate changes in transducer reference points. The working principle in many transducers is a change in resistance relative to a change in the measured variable. We will discuss the measurement of resistance using basic techniques of voltage and current measurement in conjunction with Ohm's law.

Ohmmeter Circuits

The determination of resistance can be accomplished by imposing a voltage across an unknown resistance and measuring the resulting current flow, using a galvanometer. Clearly, from Ohm's law the value of resistance can be determined in a circuit such as in Figure 6.12, which forms the basis of an analog ohmmeter. Such analog ohmmeters would employ circuits similar to those shown in Figure 6.13, which use shunt resistors and a D'Arsonval mechanism for accurately measuring a wide range of resistance while limiting the flow of current through the ammeter.

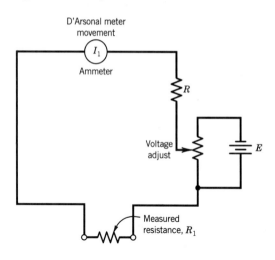

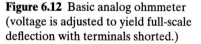

Figure 6.12 Basic analog ohmmeter (voltage is adjusted to yield full-scale deflection with terminals shorted.)

The lower limit on resistance, R_1, that can be measured is determined by the current flow through R_1, that is, I_1. Practical limitations on the maximum current flow through a resistance are imposed by the ability of the particular resistance to dissipate the power generated by the flow of current, (I^2R heating). For example, the ability of a thin metallic conductor to dissipate heat is the principle on which fuses are based. At too large a value of I_1, they melt. The same is true for many delicate, small resistance sensors!

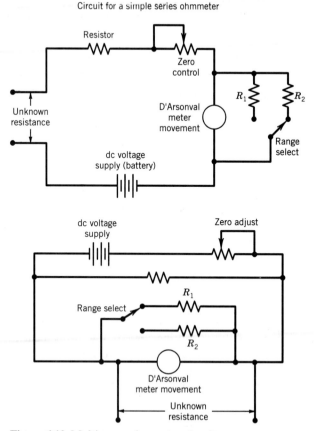

Figure 6.13 Multirange ohmmeter circuits.

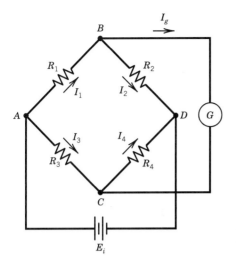

Figure 6.14 Basic Wheatstone bridge circuit (G, galvanometer).

Bridge Circuits

A variety of bridge circuits have been devised for the measurement of capacitance, inductance, and most often for the measurement of resistance. A purely resistive bridge, called a *Wheatstone bridge*, provides a means for accurately measuring resistance and for detecting small changes in resistance. Figure 6.14 shows the basic arrangement for a bridge circuit, where R_1 may be a transducer that experiences a change in resistance with a change in the measured variable. A dc voltage is input across nodes A to D, and the bridge forms a parallel circuit arrangement between these two nodes. The currents flowing through the resistors R_1 to R_4 are I_1 to I_4, respectively. Under the condition that the current flow through the galvanometer, I_g, is zero, the bridge is said to exist in a balanced condition. A specific relationship exists between the resistances that form the bridge at balanced conditions. To find this relationship, a circuit analysis is performed with $I_g = 0$. Under this balanced condition, there is no voltage drop from B to C and

$$I_1 R_1 - I_3 R_3 = 0$$
$$I_2 R_2 - I_4 R_4 = 0$$

(6.9)

Under balanced conditions, the current through the galvanometer is zero, and the currents through the arms of the bridge are equal:

$$I_1 = I_2 \quad \text{and} \quad I_3 = I_4$$

(6.10)

Solving equations (6.9) simultaneously, with the condition stated in equation (6.10), yields the relationship among the resistances necessary for a balanced bridge:

$$\frac{R_2}{R_1} = \frac{R_4}{R_3}$$

(6.11)

Measurement of resistance using this bridge circuit can be performed in two basic ways. If the resistor R_1 varies with changes in the measured physical variable, one of the other arms of the bridge can be adjusted to null the circuit. Another method for measurement using the Wheatstone bridge is to replace the galvanometer with a voltage measuring device and to sense the unbalance in the bridge as an indication of the change in resistance. Both of these methods will be analyzed further.

Null Method

Consider the circuit shown in Figure 6.14, where R_2 is a variable resistance. If the resistance R_1 changes as a result of a change in the measured variable, the resistance R_2 can be adjusted to compensate so that the bridge is once again balanced. In this null method of operation, the resistance R_2 must be a calibrated variable resistor, such that adjustments to R_2 directly indicate the value of R_1. The balancing operation may be accomplished manually or controlled automatically. An advantage of the null method is that the input voltage need not be known, and changes in the input voltage do not affect the accuracy of the measurement. In addition, the galvanometer need only detect if there is a flow of current, not measure its value.

However, even null methods are limited. Earlier in this section, we assumed that the galvanometer current was exactly zero when the bridge was balanced. In fact, because the resolution of the galvanometer is limited, the current cannot be set exactly to zero. Consider a bridge that has been balanced within the sensitivity of the galvanometer, such that I_g is smaller than the smallest current detectable. This current flow will create an uncertainty in the measured resistance, R_1, caused by the galvanometer resolution error, u_R. A basic analysis of the circuit with a current flow through the galvanometer yields

$$\frac{u_R}{R_1} = \frac{I_g(R_1 + R_g)}{E_i} \tag{6.12}$$

where R_g is the internal resistance of the galvanometer. This equation can serve to guide the choice of a galvanometer and a battery voltage for a particular application. Clearly, the error is reduced by increased input voltages. However, the input voltage is limited by the power dissipating capability of the resistance device, R_1. The power that must be dissipated by this resistance is $I_1^2 R_1$.

Deflection Method

In an unbalanced condition, the magnitude of the current or voltage drop for the meter or galvanometer portion of a bridge circuit is a direct indication of the change in resistance of one or more of the arms of the bridge. Consider first the case in which the voltage drop from node B to node C in the basic bridge is measured by a meter having an infinite internal impedance, so that there is no current flow through the meter, as shown in Figure 6.15. Knowing the conditions for a balanced bridge given in equation (6.10), the voltage drop from B to C can be determined, since the current I_1 must equal the current I_2, as

$$E_o = I_1 R_1 - I_3 R_3 \tag{6.13}$$

Under these conditions, substituting equations (6.9)–(6.11) into equation (6.13) yields

$$E_o = E_i \left(\frac{R_1}{R_1 + R_2} - \frac{R_3}{R_3 + R_4} \right) \tag{6.14}$$

For many transducers, the bridge may be balanced at a reference condition. The transducer resistance changes, as a result of changes in the measured variable, would then cause a deflection in the bridge voltage away from the balanced condition. Assume that from an initially balanced condition, where $E_o = 0$, a change in R_1

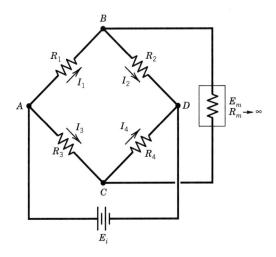

Figure 6.15 Voltage-sensitive Wheatstone bridge.

occurs to some new value, $R_1' = R_1 + \delta R$. The output from the bridge becomes

$$E_o + \delta E_o = E_i\left(\frac{R_1'}{R_1' + R_2} - \frac{R_3}{R_3 + R_4}\right) = E_i\frac{R_1' R_4 - R_3 R_2}{(R_1' + R_2)(R_3 + R_4)} \tag{6.15}$$

In many designs the bridge resistances are initially equal. Setting $R_1 = R_2 = R_3 = R_4 = R$ allows equation (6.15) to be reduced to

$$\frac{\delta E_o}{E_i} = \frac{\delta R/R}{4 + 2(\delta R/R)} \tag{6.16}$$

In contrast with the null method of operation of a Wheatstone bridge, the deflection bridge requires a meter capable of accurately indicating the output voltage, as well as a stable and known input voltage; but a big advantage of the deflection bridge is associated with its frequency response. The bridge output should follow any resistance changes over any frequency input, up to the frequency limit of the detection device! Such a bridge circuit may be connected to an oscilloscope to allow the observation and recording of signals having very high frequency content. So a deflection mode is often used to measure time-varying signals.

If the high-impedance voltage measuring device in Figure 6.15 is replaced with a relatively low-impedance current measuring device and the bridge is operated in an unbalanced condition, a current-sensitive bridge circuit results. Consider Kirchhoff's laws applied to the Wheatstone bridge circuit for a galvanometer resistance R_g. The input voltage is equal to the voltage drop in each arm of the bridge,

$$E_i = I_1 R_1 + I_2 R_2 \tag{6.17}$$

but

$$I_2 = I_1 - I_g$$

which implies

$$E_i = I_1(R_1 + R_2) - I_g R_2 \tag{6.18}$$

If we consider the voltage drops in the path through R_1, R_g, and R_3, the total voltage drop must be zero:

$$I_1 R_1 + I_g R_g - I_3 R_3 = 0 \tag{6.19}$$

For the circuit formed by R_g, R_4, and R_2,

$$I_g R_g + I_4 R_4 - I_2 R_2 = 0 \qquad (6.20)$$

or with $I_2 = I_1 - I_g$ and $I_4 = I_3 + I_g$,

$$I_g R_g + (I_3 + I_g)R_4 - (I_1 - I_g)R_2 = 0 \qquad (6.21)$$

Equations (6.18)–(6.21) form a set of three simultaneous equations in the three unknowns I_1, I_g, and I_3. Solving these three equations for I_g yields

$$I_g = \frac{E_i(R_3 R_2 - R_1 R_4)}{R_3(R_1 + R_2)(R_g + R_2 + R_4) + R_1 R_2 R_4 - R_3 R_2^2 + R_g R_4(R_1 + R_2)} \qquad (6.22)$$

Then, the change in resistance of R_1 can be found in terms of the bridge deflection voltage, E_o, by

$$\frac{\delta R}{R_1} = \frac{(R_3/R_1)[E_o/E_i + R_2/(R_2 + R_4)]}{1 - E_o/E_i - R_2/(R_2 + R_4)} - 1 \qquad (6.23)$$

Consider the case when all of the resistances in the bridge are initially equal to R, and subsequently R_1 changes by an amount δR. The current through the meter is given by

$$I_g = E_i \frac{\delta R/R}{4(R + R_g)} \qquad (6.24)$$

and the output voltage is given by $E_o = I_g R_g$:

$$E_o = E_i \frac{\delta R/R}{4(1 + R/R_g)} \qquad (6.25)$$

The bridge impedance can affect the output from a constant voltage source having an internal resistance R_s. The effective bridge resistance, based on a Thévenin equivalent circuit analysis, is given by

$$R_B = \frac{R_1 R_3}{R_1 + R_3} + \frac{R_2 R_4}{R_2 + R_4} \qquad (6.26)$$

such that for a power supply of voltage E_s

$$E_i = \frac{E_s R_B}{R_s + R_B} \qquad (6.27)$$

In a similar manner, the bridge impedance can affect the voltage indicated by the voltage measuring device. For a voltage measuring device of internal impedance R_g, the actual bridge deflection voltage, relative to the indicated voltage, E_m, is

$$E_o = \frac{E_m}{R_g} \left(\frac{R_1 R_2}{R_1 + R_2} + \frac{R_3 R_4}{R_3 + R_4} + R_g \right) \qquad (6.28)$$

The difference between the measured voltage, E_m, and the actual voltage, E_o, is called a loading error, in this case caused by the bridge impedance load. Such errors are discussed next.

EXAMPLE 6.2

A certain temperature sensor experiences a change in electrical resistance with temperature according to the equation

$$R = R_0[1 + \alpha(T - T_0)] \qquad (6.29)$$

where

R = sensor resistance $[\Omega]$
R_0 = sensor resistance at the reference temperature, T_0 $[\Omega]$
T = temperature $[^\circ C]$
T_0 = reference temperature, $0^\circ C$
α = constant: $0.00395^\circ C^{-1}$

This temperature sensor is connected in a Wheatstone bridge like the one shown in Figure 6.14, where the sensor occupies the R_1 location, and R_2 is a calibrated variable resistance. The bridge is operated using the null method. The fixed resistances R_3 and R_4 are each equal to 500 Ω. If the temperature sensor has a resistance of 100 Ω at $0^\circ C$, determine the value of R_2, that would balance the bridge at $0^\circ C$.

KNOWN

$R_1 = 100\ \Omega$
$R_3 = R_4 = 500\ \Omega$

FIND

R_2 for null balance condition

SOLUTION

From equation (6.11), a balanced condition for this bridge would be achieved when $R_2 = R_1$. Thus, a value of 100 Ω for R_2 would create a balanced condition. Notice that to be a useful circuit, R_2 must be adjustable and provide an indication of its resistance value at any setting.

EXAMPLE 6.3

Consider a deflection bridge, which initially has all arms of the bridge equal to 100 Ω, with the temperature sensor described in Example 6.2 again as R_1. The input or supply voltage to the bridge is 10 V. If the temperature of R_1 is changed such that the bridge output is 0.569 V, what is the temperature of the sensor? How much current flows through the sensor, and how much power must it dissipate?

KNOWN

$E_i = 10$ V

Initial state:
$R_1 = R_2 = R_3 = R_4 = 100\ \Omega$
$E_o = 0$ V

Deflection state:
$E_o = 0.569$ V

ASSUMPTION

Voltmeter has infinite input impedance but source has negligible impedance ($E_s = E_i$).

FIND

T_1, I_1, P_1

SOLUTION

The change in the sensor resistance can be found from equation (6.16) as

$$\frac{\delta E_o}{E_i} = \frac{\delta R/R}{4 + 2(\delta R/R)} \Rightarrow \frac{0.569}{10} = \frac{\delta R}{400 + 2\delta R}$$

$$\delta R = 25.67\ \Omega$$

This results in a sensor resistance of $R_1 = 125.7\ \Omega$, and implies a sensor temperature of $T_1 = 65°C$.

To determine the current flow through the sensor, consider first the balanced case in which all the resistances are equal to $100\ \Omega$. The equivalent bridge resistance, R_B, is simply $100\ \Omega$, and the total current flow from the supply, $E_i/R_B = 100$ mA. Thus, at the initially balanced condition, through each arm of the bridge and through the sensor the current flow is 50 mA. If the sensor resistance changes to $125.67\ \Omega$, the current will be reduced. If the output voltage is measured by a high-impedance device, such that the current flow to the meter is negligible, the current flow through the sensor is given by

$$I_1 = E_i \frac{1}{R_1 + R_2} \qquad (6.30)$$

This current flow is then 44.3 mA.

The power, $P_1 = I_1^2 R_1$, that must be dissipated from the sensor is 0.25 W, which, depending upon the surface area and local heat transfer conditions, may cause a change in the indicated temperature of the sensor.

COMMENT

The current flow through the sensor results in a sensor temperature higher than would occur with zero current flow, as a result of $I^2 R$ heating. This creates a loading error in the indicated temperature. There is a trade-off between the increased sensitivity, dE_o/dR_1, and the correspondingly increased current associated with an increased E_1. The input voltage must be chosen appropriately for a given application.

6.5   LOADING ERRORS AND IMPEDANCE MATCHING

In an ideal sense, an instrument or measurement system should not in itself affect the variable being measured. Any such effect will alter the variable and be considered as a "loading" that the measurement system exerts on the measured variable. The resulting difference between the unaltered value of the measurand and the indicated value from the measurement is a *loading error*. Loading effects can be of any form: mechanical, electrical, or optical. When the insertion of a sensor into a process somehow changes the physical variable being measured, a *process loading error* occurs. A signal loading can also occur along the signal path of a measurement system. If the output from one system stage is in any way affected by the subsequent stage, then the signal is affected by *interstage loading error*. A goal in measurement system design should be to minimize loading errors.

Consider the measurement of the temperature of a volume of a high-temperature liquid using a mercury-in-glass thermometer. Some finite quantity of energy must flow from the liquid to the thermometer to achieve thermal equilibrium between the thermometer and the liquid. As a result of this energy flow, the liquid temperature is reduced, and the measured value will not correspond to the initial liquid temperature. Because the goal of the measurement was to measure the temperature of the liquid in its initial state, this act of measurement has introduced a loading error.

Or consider the current flow that drives the galvanometer in the Wheatstone bridge of Figure 6.14. Under deflection conditions, some energy must be removed from the circuit to deflect the pointer. This reduces the current in the circuit, bringing about a loading error in the measured resistance. Under null balance conditions, there is no pointer deflection to within the resolution of the meter and a negligible amount of current is removed from the circuit. There is then negligible loading error.

In general, *null balance techniques will minimize the magnitude of loading errors*, usually to negligible levels. However, deflection methods can derive considerable energy from the process being measured and therefore need careful consideration to keep loading errors to a minimum.

Loading Errors for A Voltage Dividing Circuit

Consider the voltage dividing circuit shown in Figure 6.9, for the case in which R_m is finite. Under these conditions, the circuit can be represented by the equivalent circuit shown in Figure 6.16. As the sliding contact at point A moves, it divides the full-scale deflection resistance, R_T, into R_1 and R_2, such that $R_1 + R_2 = R_T$. The resistances R_m and R_1 form a parallel loop, yielding an equivalent resistance R_L, given by

$$R_L = \frac{R_1 R_m}{R_1 + R_m} \tag{6.31}$$

The equivalent resistance for the entire circuit, as seen from the voltage source, is R_{eq}:

$$R_{eq} = R_2 + R_L = R_2 + \frac{R_1 R_m}{R_1 + R_m} \tag{6.32}$$

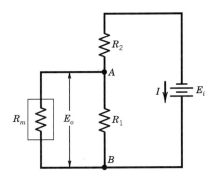

Figure 6.16 Instruments in parallel to signal path form an equivalent voltage dividing circuit.

The current flow from the voltage source is then

$$I = \frac{E_i}{R_{eq}} = \frac{E_i}{R_2 + R_1 R_m/(R_1 + R_m)}$$ (6.33)

and the output voltage is given by

$$E_o = E_i - \dot{I}R_2$$ (6.34)

This result can be expressed as

$$\frac{E_o}{E_i} = \frac{1}{1 + (R_2/R_1)(R_1/R_m + 1)}$$ (6.35)

The limit of this expression as R_m tends to infinity is

$$\frac{E_o}{E_i} = \frac{R_1}{R_1 + R_2}$$ (6.36)

which is equation (6.8). Also, as R_2 approaches zero, the output voltage approaches the supply voltage, as expected. With E_o/E_i found from equation (6.8) expressed as $(E_o/E_i)'$, the loading error, e_I, may be given by

$$e_I = E_i \left[\left(\frac{E_o}{E_i} \right)' - \frac{E_o}{E_i} \right]$$
$$= E_i \frac{R_1 - R_T + (R_T - R_1)[(R_1/R_m) + 1]}{R_T + [(R_T^2/R_1) - R_T][(R_1/R_m) + 1]}$$ (6.37)

The loading error goes to zero as $R_m \to \infty$

Interstage Loading Errors

Consider the common situation of Figure 6.17, in which the output voltage signal from one measurement system device provides the input to the following device. The open circuit potential, E_1, is present at the output terminal of device 1 with output impedance, Z_1. However, the output signal from device 1 provides the input to a second device, which, at its input terminals, has an input impedance, Z_m. As shown, the Thévenin equivalent circuit of device 1 consists of a voltage generator, of open circuit voltage E_1, with internal series impedance, Z_{th}. The potential sensed by device 2 will be

$$E_m = IZ_{th} = E_1 \frac{1}{1 + Z_1/Z_m}$$ (6.38)

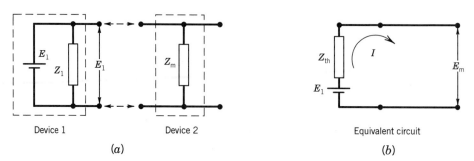

Device 1   Device 2   Equivalent circuit

(a)   (b)

Figure 6.17 Equivalent circuit formed by interstage (parallel) connections.

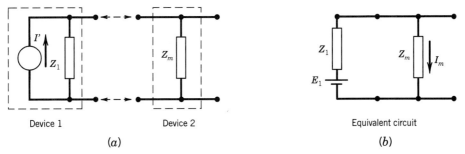

Device 1   Device 2   Equivalent circuit

(a)   (b)

Figure 6.18 Instruments in series with signal path.

The original potential has been changed due to the interstage connection which has caused a loading error, $e_I = E_1 - E_m$,

$$e_I = E_1\left(1 - \frac{1}{1 + Z_1/Z_m}\right) \tag{6.39}$$

For a maximum in the voltage potential between stages, it is required that $Z_m \gg Z_1$ such that $e_I \to 0$. As a practical matter, this can be difficult to achieve at reasonable cost and some design compromise is inevitable.

When the signal is current driven, the maximum current transfer between devices 1 and 2 is desirable. Consider the circuit shown in Figure 6.18 in which device 2 is current sensitive, such as with a current-sensing galvanometer. The current through the loop indicated by device 2 is given by

$$I_m = \frac{E_1}{Z_1 + Z_m} \tag{6.40}$$

However, if the measurement device is removed from the circuit, the current is given by the short-circuit value

$$I' = \frac{E_1}{Z_1} \tag{6.41}$$

The loading error, $e_I = I' - I_m$, is

$$e_I = E_1\frac{Z_m}{Z_1^2 + Z_1 Z_m} \tag{6.42}$$

From equation (6.42) or by inspection of equations (6.40) and (6.41), it is clear that for maximum current transfer in which $I_m \to I'$, it is required that $Z_m \ll Z_1$, such that $e_I \to 0$.

EXAMPLE 6.4

Consider the Wheatstone bridge shown in Figure 6.19. Find the open circuit output voltage (when $R_m \to \infty$) if the four resistances change by

$$\delta R_1 = +40\,\Omega \qquad \delta R_2 = -40\,\Omega \qquad \delta R_3 = +40\,\Omega \qquad \delta R_4 = -40\,\Omega$$

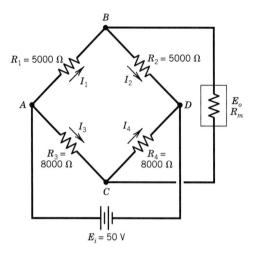

Figure 6.19 Bridge circuit for Example 6.4.

KNOWN

Measurement device resistance given by R_m

ASSUMPTIONS

$R_m \to \infty$
Negligible source impedance

FIND

E_o

SOLUTION

From equation (6.14)

$$E_o = E_i \left(\frac{R_1}{R_1 + R_2} - \frac{R_3}{R_3 + R_4} \right)$$

$$= E_i \left(\frac{5040\,\Omega}{5040 + 4960\,\Omega} - \frac{8040\,\Omega}{8040 + 7960\,\Omega} \right)$$

$$= 50\,\text{V}\,(0.5040 - 0.5025) = +0.075\,\text{V}$$

EXAMPLE 6.5

Consider the case in which the output of the bridge in Example 6.4 is measured by a meter with internal impedance $R_m = 20,000\,\Omega$. What is the output voltage with this particular meter in the circuit? Estimate the loading error.

KNOWN

$R_m = 20,000\,\Omega$

FIND

E_o, e_I

SOLUTION

Because the output voltage, E_o, is equal to $I_m R_m$, equation (6.22) can be used:

$$I_m = \frac{E_i(R_3 R_2 - R_1 R_4)}{R_3(R_1 + R_2)(R_m + R_2 + R_4) + R_1 R_2 R_4 + R_m R_4(R_1 + R_2) - R_3 R_2^2}$$

For $R_m = 20{,}000\ \Omega$, this gives $E_o = -0.0566$ V.

The loading error, e_I, will be the difference between the output voltage assuming infinite impedance in the meter (Example 6.4) and the output voltage for the finite impedance value. This gives $e_I = 18.4$ mV.

The bridge output may also be expressed as a ratio of the output voltage with $R_m \to \infty$, E_o', and the output voltage with a finite meter resistance, E_o,

$$\frac{E_o}{E_o'} = \frac{1}{1 + R_e/R_m}$$

where

$$R_e = \frac{R_1 R_2}{R_1 + R_2} + \frac{R_3 R_4}{R_3 + R_4}$$

which yields for the present example

$$\frac{E_o}{E_o'} = \frac{1}{1 + 6500/20{,}000} = 0.755$$

The percent loading error, $100 \times [1 - (E_o/E_o')]$, is 24.5%.

6.6 ANALOG SIGNAL CONDITIONING: AMPLIFIERS

An amplifier is a device that scales the magnitude of an analog input signal according to the relation

$$E_o(t) = h\{E_i(t)\} \tag{6.43}$$

where $h\{E_i(t)\}$ defines a mathematical function. The simplest amplifier is the *linear scaling amplifier* in which

$$h\{E_i(t)\} = GE_i(t) \tag{6.44}$$

where the gain G is a constant that may be any positive or negative value. Many other types of operation are possible, including the "base x" *logarithmic amplifier* in which

$$h\{E_i(t)\} = G\log_x[E_i(t)] \tag{6.45}$$

Amplifiers have a finite frequency response and limited input voltage range.

The most widely used type of amplifier is the operational amplifier. This device is characterized by (1) a very high input impedance ($Z_i > 10^7\ \Omega$), (2) a low output impedance ($Z_o < 100\ \Omega$), and (3) a high internal gain ($A \approx 10^5$–10^6). As shown in the general diagram of Figure 6.20(a), an operational amplifier has two input ports, a noninverting and an inverting input, and one output port. The signal at the output port will be in phase with a signal passed through the noninverting input port but will be 180° out of phase with a signal passed through the inverting input port. The amplifier requires dual polarity dc excitation power ranging from ±5 V to ±15 V. An internal schematic diagram of a common operational amplifier circuit, type 741, is shown in Figure 6.20(b). As shown, each input port is attached to the base of an npn transistor.

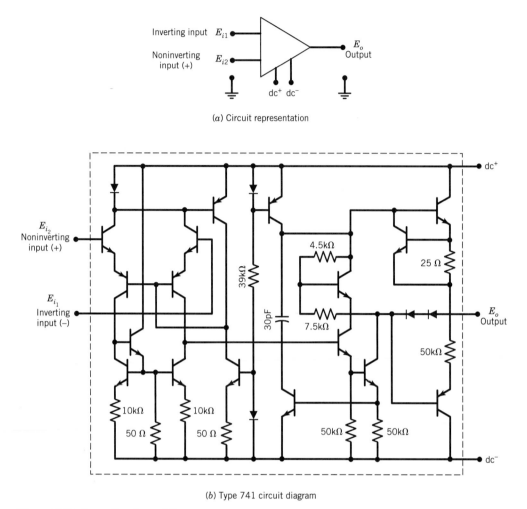

(a) Circuit representation

(b) Type 741 circuit diagram

Figure 6.20 Operational amplifier.

The high internal open-loop gain, A, of an operational amplifier is given as

$$E_o(t) = A[E_{i_2}(t) - E_{i_1}(t)]$$

The magnitude of A, flat at low frequencies, falls off rapidly at high frequencies, but this intrinsic gain curve is overcome by using external input and feedback resistors which will actually set the circuit gain, G, and circuit response.

Possible amplifier configurations using an operational amplifier are shown in Figure 6.21. Because the amplifier has a very high internal gain and negligible current draw, resistors R_1 and R_2 are used to form a feedback loop and control the overall amplifier circuit gain, called the closed loop gain, G. The noninverting linear scaling amplifier circuit of Figure 6.21(a) has a closed-loop gain of

$$G = \frac{E_o(t)}{E_i(t)} = \frac{R_1 + R_2}{R_2} \tag{6.46}$$

The inverting linear scaling amplifier circuit of Figure 6.21(b) provides a gain of

$$G = \frac{E_o(t)}{E_i(t)} = \frac{R_2}{R_1} \tag{6.47}$$

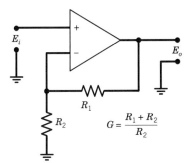

(a) Noninverting amplifier

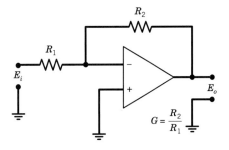

(b) Inverting amplifier

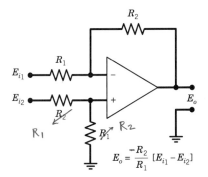

(c) Differential amplifier

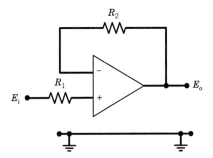

(d) Voltage follower

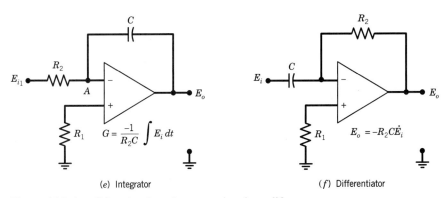

(e) Integrator

(f) Differentiator

Figure 6.21 Amplifier circuits using operational amplifiers.

By utilizing both inputs, the arrangement forms a differential amplifier, Figure 6.21(c), in which

$$E_o(t) = [E_{i_1}(t) - E_{i_2}(t)](R_2/R_1) \tag{6.48}$$

The differential amplifier circuit is effective as a voltage comparator for many instrument applications (see Section 6.7).

A voltage follower circuit is commonly used to isolate an impedance load from other stages of a measurement system, such as might be needed with a high-output impedance transducer. A schematic diagram of a voltage follower circuit is shown in Figure 6.21(d). Note that the feedback signal is sent directly to the inverting port. For such a circuit, then,

$$E_o(t) = A[E_i(t) - E_o(t)]$$

or in terms of the circuit gain, $G = E_o(t)/E_i(t)$,

$$G = \frac{A}{1+A} \approx 1 \tag{6.49}$$

Using Kirchhoff's law about the noninverting input loop yields

$$I_i(t)R_1 + E_o(t) = E_i(t)$$

Then the circuit input resistance, R_i, is

$$R_i = \frac{E_i(t)}{I_i(t)} = \frac{E_i(t)R_1}{E_i(t) - E_o(t)}$$
$$= (1+A)R_1 \tag{6.50}$$

Likewise the output resistance is found to be

$$R_o = \frac{R_2}{1+A} \tag{6.51}$$

Because A is large, equations (6.49)–(6.51) show that the input impedance, developed as a resistance, of the voltage follower can be large, its output impedance can be small, and the circuit will have near unity gain. Acceptable values for R_1 and R_2 range from 10 kΩ to 100 kΩ to maintain stable operation.

Input signal integration and differentiation can be performed with the operational amplifier circuits. For the integration circuit in Figure 6.21(e), the currents through R_2 and C are given by

$$I_{R_2}(t) = \frac{E_i(t) - E_A(t)}{R_2}$$

$$I_c(t) = C\frac{d}{dt}[E_o(t) - E_A(t)]$$

Summing currents at node A yields

$$E_o(t) = -\frac{1}{R_2 C} \int E_i(t)\, dt \tag{6.52}$$

Following a similar analysis, the differentiator circuit shown in Figure 6.21(f), will perform the operation

$$E_o(t) = -R_2 C \dot{E}_i(t) \tag{6.53}$$

6.7 ANALOG SIGNAL CONDITIONING: SPECIAL PURPOSE CIRCUITS

Analog Voltage Comparator

A voltage comparator provides an output proportional to the difference between two input voltages. As shown in Figure 6.22(a), a basic comparator consists of an operational amplifier operating in a high-gain differential mode. In this case, any difference between inputs E_{i_1} and E_{i_2} is amplified by the gain G. But the output from the operational amplifier will saturate when $E_{i_1} - E_{i_2}$ exceeds threshold voltage E_T. The value for E_T is fixed by the amplifier bias voltage, E_{bias}. So the comparator output

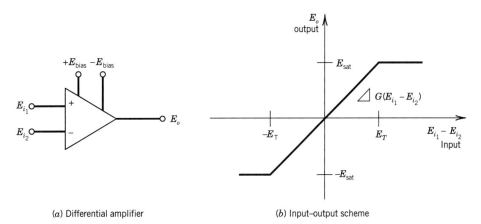

(a) Differential amplifier   (b) Input–output scheme

Figure 6.22 Analog voltage comparator.

is given by

$$E_0 = G(E_{i_1} - E_{i_2}) \quad \text{for } |E_{i_1} - E_{i_2}| \leq E_T$$
$$= +E_{\text{bias}} \qquad\qquad \text{for } E_{i_1} - E_{i_2} > E_T \qquad\qquad (6.54)$$
$$= -E_{\text{bias}} \qquad\qquad \text{for } E_{i_1} - E_{i_2} < -E_T$$

in which $E_T = E_{\text{bias}}/G$. This input–output relation is shown in Figure 6.22(b).

Often E_{i_2} will be a known reference voltage. This allows the comparator output to be used for control circuits to decide if E_{i_1} is less than or greater than E_{i_2}. One frequent use of the comparator is in an analog-to-digital converter (Chapter 7).

The output levels of a comparator are not immediately TTL (transistor-transistor logic) compatible for digital system use. However, a Zener diode circuit of 3–4V placed on the output will produce a TTL-compatible output signal.

Sample and Hold Circuit

The sample and hold circuit (SHC) is used to take a narrow-band measurement of a time-changing signal and to hold that measured value until reset. It is widely used in data-acquisition systems using A/D converters. The circuit tracks the signal until it is triggered to sample the signal and hold it. This is illustrated in Figure 6.23(a), in which the track and hold logic provides the appropriate trigger.

The basic circuit for sample and hold is shown in Figure 6.23(b). The switch is a fast analog device. When the switch is closed the "hold" capacitor, C, is charged through the source resistor R_s. When the capacitor is charged, the switch is opened. The amplifier presents a very high input impedance and very low current that, together with a very low leakage capacitor, allows for a long hold time. The typical SHC is noninverting with a unit gain ($G = 1$).

Charge Amplifier

A *charge amplifier* is used to convert a high-impedance charge, q, into an output voltage, E_o. The circuit consists of a high gain, inverting voltage operational amplifier such as shown in Figure 6.24. These circuits are commonly used with transducers that

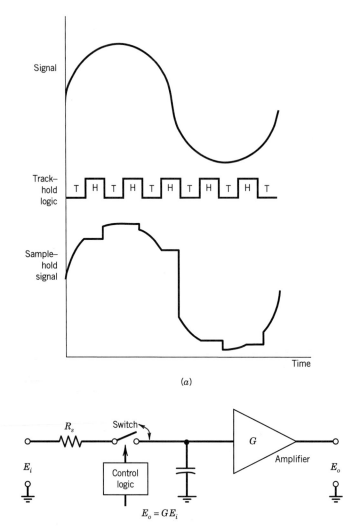

(a)

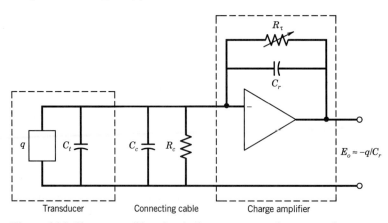

(b)

Figure 6.23 Sample and hold technique: (a) original signal and sample and hold signal, (b) circuit.

Figure 6.24 Charge amplifier circuit shown connected to a transducer.

utilize piezoelectric crystals. A piezoelectric crystal will develop a time-dependent surface charge under varying mechanical load.

The circuit output voltage is determined by

$$E_o = -q/[C_r + (C_T/A)] \tag{6.55}$$

with $C_T = C_t + C_c + C_r$ where C_t, C_c, and C_r represent the transducer, cable, and feedback capacitances, respectively, R_c is the cable and transducer resistances, and A is the amplifier open-loop gain. Because the open-loop gain of operational amplifiers is very large, equation (6.55) can be simplified to

$$E_o \approx -q/C_r \tag{6.56}$$

A variable resistance, R_τ, and fixed feedback capacitor are used to average out low-frequency fluctuations in the signal.

Current Loop: 4–20 mA

A problem with voltage signals below ~100 mV is that they are quite vulnerable to noise along the transmission lines. One means of transmitting low-level voltage signals over long distances is by signal boosting. Although this can be accomplished by using the amplification methods described, a common alternative method is a 4–20 mA current loop (read as 4 to 20). In this approach, the low-level voltage is converted into a standard current loop signal of between 4 and 20 mA, the lower constant current value for the minimum voltage and the higher value for the maximum voltage in the range. The 4–20 mA current loop can be transmitted over several hundred meters. Many transducer manufacturers offer a 4–20 mA output as an option for their devices.

A receiver converts the current back to a low-level voltage. This can be as simple as a single resistor in parallel with the loop. For example, a 4–20 mA signal can be converted back to a 100–500 mV signal by placing a 25-Ω resistor across the loop.

Multivibrator and Flip-Flop Circuits

The *multivibrator* is a switching circuit that toggles on and off continuously in response to an applied input voltage. The heart of this circuit, as shown in Figure 6.25, is the two transistors T_1 and T_2, which conduct alternately. This generates a square wave signal at the output as shown in Figure 6.26(a). The working mechanism is straightforward. The two transistors change state as the currents through capacitors C_1 and C_2 increase and decrease as a result of the applied input. For example, when T_2 turns on and its collector switches from from V_c to V_2, the base of T_1 is driven negative, which turns T_1 off. While C_2 discharges, C_1 charges. As the voltage across C_1 increases, the current across it decreases, but as long as the current through C_1 is large enough, T_2 remains on. It eventually falls to a value that turns T_2 off. This causes C_2 to charge, which turns T_1 on. The circuit continues to toggle in this manner, and the output signal toggles between a low and a high voltage (typically, between 0 V and 5 V nominal) with time.

A useful variation of the multivibrator circuit is the *monostable*. In this arrangement T_2 stays on until a positive external pulse or change in voltage is applied to the input. At that moment, T_2 will turn off for a brief period of time, providing a

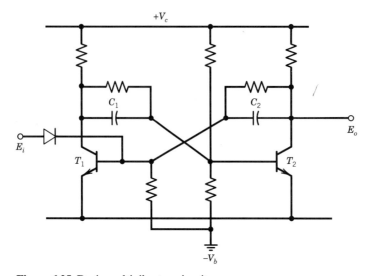

Figure 6.25 Basic multivibrator circuit.

jump in the output voltage. The monostable circuit will cycle just once while awaiting another pulse, as shown in Figure 6.26(*b*). Hence, it is often referred to as a *one-shot*. The monostable is an effective trigger.

Another variation of this circuit is the *flip-flop* or *bistable multivibrator* in Figure 6.27. This circuit is an effective electronic switch. Its operation is analogous to that of a light switch; it is either in an on or an off state. In practice, transistors T_1 and T_2 will change state every time a pulse is applied. If T_1 is on with T_2 off, a pulse will turn T_2 off, turning T_1 on, producing the output shown in Figure 6.26(*c*). So the flip-flop output level changes from low to high voltage or high to low voltage on command. The flip-flop is also the basic circuit of computer memory chips, as it is capable of holding a single bit of information (high or low state) at any instant.

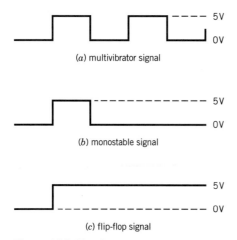

(*a*) multivibrator signal

(*b*) monostable signal

(*c*) flip-flop signal

Figure 6.26 Circuit output response to an applied input signal.

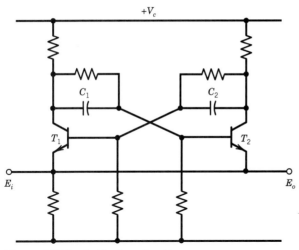

Figure 6.27 Basic flip-flop circuit.

6.8 ANALOG SIGNAL CONDITIONING: FILTERS

A *filter* is used to remove undesirable frequency information from a dynamic signal. Filters can be broadly classified as being low pass, high pass, bandpass, and notch. The ideal gain characteristics of filters can be described by the magnitude ratio plots shown in Figure 6.28. It is shown that a *low-pass* filter permits frequencies below a prescribed cutoff frequency to pass while blocking the passage of frequency information above

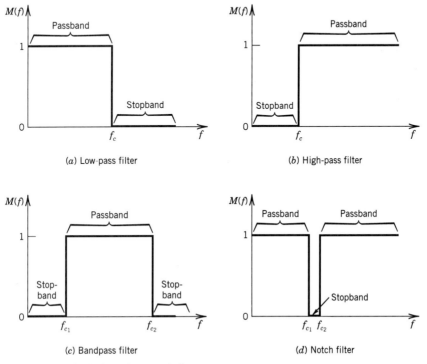

Figure 6.28 Ideal filter characteristics.

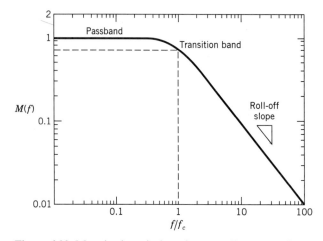

Figure 6.29 Magnitude ratio for a low-pass Butterworth filter.

the cutoff frequency, f_c. Similarly, a *high-pass* filter permits only frequencies above the cutoff frequency to pass. A *bandpass* filter combines features of both the low- and high-pass filters. It is described by a low cutoff frequency, f_{c_1}, and a high cutoff frequency, f_{c_2}, to define a band of frequencies that are permitted to pass through the filter. A *notch* filter permits the passage of all frequencies except those within a narrow frequency band. An intensive treatment of filters for analog and digital signals can be found in many specialized texts (e.g., [3–7]).

Filters perform a well-defined mathematical operation, as specified by their transfer function, on the input signal. Passive analog filter circuits consist of combinations of resistors, capacitors, and inductors. Active filters incorporate operational amplifiers into the circuit.

The sharp cutoff of the ideal filter cannot be realized in a practical filter. As an example, a plot of the magnitude ratio for a real low-pass filter is shown in Figure 6.29. All filter response curves will contain a transition band over which the magnitude ratio decreases relative to the frequency. This rate of transition is known as the filter *roll-off*, usually specified in decibels per decade. In addition, the filter will introduce a phase shift between its input and output signal. Depending on the transfer function used, filters can be designed to achieve certain desirable response features. For example, a relatively flat magnitude ratio over its passband with a moderately steep initial roll-off is a characteristic of a *Butterworth* filter response. A characteristic of a *Bessel* filter response is a linear phase shift over its passband, but with a relatively nonflat magnitude ratio and a gradual initial roll-off.

Butterworth Filter Design

A simple passive low-pass Butterworth filter can be constructed by using the resistor and capacitor (RC) circuit of Figure 6.30. Application of Kirchhoff's law about the input loop gives the model relating the input voltage, E_i, to the output voltage, E_o:

$$RC\dot{E}_o + E_o = E_i \tag{6.57}$$

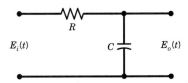

Figure 6.30 Low-pass RC Butterworth filter circuit.

This real filter is a first-order system. A first-order or one-stage filter contains only one element whose reactance varies with frequency, in this case the capacitor. The magnitude ratio for this filter when subjected to a sinusoidal waveform input is given by equation 3.10 with $\tau = RC$ and $\omega = 2\pi f$ and is described by Figure 6.29. The roll-off slope is 20 dB/decade. The phase shift is given in equation 3.9.

A filter is designed around its cutoff frequency, f_c, which is usually defined as the frequency at which the signal power is reduced to one-half. This is equivalent to the magnitude ratio being reduced to 0.707. In terms of the decibel (dB)

$$dB = 20 \log M(f) \tag{3.11}$$

f_c occurs at -3 dB. For the filter of Figure 6.30, this requires that $\tau = RC = 1/2\pi f_c$. The roll-off slope of a filter can be improved by placing several filters in series or stages, known as cascading filters. This is done by adding additional reactive elements, such as inductors and capacitors, to the circuit, as shown in Figure 6.31. A kth order or k-stage filter contains k reactive elements in k stages. The magnitude ratio and phase shift for a k-stage Butterworth filter is given by

$$M(f) = \frac{1}{[1 + (f/f_c)^{2k}]^{1/2}} \tag{6.58a}$$

$$\Phi(f) = \sum_{i=1}^{k} \phi_i(f) \tag{6.58b}$$

The attenuation at any frequency can be estimated from the dynamic error, $\delta(f)$, or directly in decibels by

$$\text{Attenuation (dB)} = 10 \log [1 + (f/f_c)^{2k}] \tag{6.59}$$

The general magnitude response curve for a k stage Butterworth filter is shown in Figure 6.32. The roll-off slope is $20 \times k$ dB/decade. Equation (6.60) is useful for specifying the required order of a filter based on magnitude response needs.

Table 6.1 lists the required values for elements C_i and L_i for a two-through five-stage Butterworth filter of the form of Figure 6.31 [6, 7]. The values in Table 6.1 assume that $R_s = R_L = 1\,\Omega$ and $\omega_c = 2\pi f_c = 1$ rad/s. For other values, L_i and C_i are scaled by

$$L = L_i R_s / 2\pi f_c \tag{6.60a}$$

$$C = C_i / (R_s 2\pi f_c) \tag{6.60b}$$

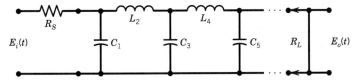

Figure 6.31 Multistage or cascading low-pass LC filters.

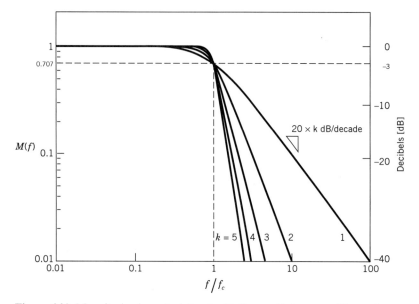

Figure 6.32 Magnitude characteristics for Butterworth low-pass filters of various stages.

Butterworth filters are quite common and can be found in consumer products, such as in audio equipment.

A Butterworth high pass-filter can be made by switching the places of the inductors and capacitors of the low-pass filter as shown in of Figure 6.33. The component values of Table 6.1 and the scaling equations (6.60a) and (6.60b) apply. While the phase shift properties remain as in equation (6.58b), the magnitude ratio for this filter is

$$M(f) = \frac{1}{\left(1 + \left(\frac{f_c}{f}\right)^{2k}\right)^{1/2}} \qquad (6.61)$$

$$\frac{f}{f_c}$$

EXAMPLE 6.6

Design a one-stage Butterworth RC low-pass filter with a cutoff frequency of -3 dB set at 100 Hz. Calculate the attenuation at 60 and 200 Hz.

KNOWN

$f_c = 100 \text{ Hz} \qquad k = 1$
$M(100 \text{ Hz}) = 0.707$

Table 6.1 Values for Low-Pass LC Butterworth Filter[1]

k	C_1	L_2	C_3	L_4	C_5
2	1.414	1.414			
3	1.000	2.000	1.000		
4	0.765	1.848	1.848	0.765	
5	0.618	1.618	2.000	1.618	0.618

[1]Values for C_i in farads and L_i in henrys are referenced to $R_s = R_L = 1 \, \Omega$ and $\omega_c = 1$ rad/s. See discussion for proper scaling.

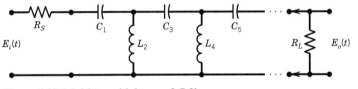

Figure 6.33 Multistage high-pass LC filters.

FIND

R, C, and δ

SOLUTION

A single-stage Butterworth RC filter circuit is shown in Figure 6.30. With the use of $f = \omega/2\pi$, the magnitude ratio for this circuit is given by

$$M(f) = \frac{1}{[1 + (2\pi f \tau)^2]^{1/2}}$$

Setting $M(f) = 0.707 = -3$ dB with $f = f_c = 100$ Hz yields, from (6.58a),

$$\tau = 1/2\pi f_c = RC = 0.0016 \, \text{s}$$

One possible combination might be $R = 800 \, \Omega$ and $C = 2 \, \mu\text{F}$.

The attenuation is found directly from the dynamic error, $\delta(f) = M(f) - 1$ or in decibels from equation (6.59).

The attenuation at 60 Hz would be

$$\delta(60) = M(60 \, \text{Hz}) - 1 = 0.86 - 1 = -0.14$$

So the amplitude of the input signal at 60 Hz will be down 14% or -1.33 dB after passing through this filter. Ideally, there would be no attenuation for frequencies below the cutoff frequency and complete attenuation above it. But practically this is not possible.

At 200 Hz,

$$\delta(200) = -0.55$$

or -46 dB. Over 55% of the signal amplitude at 200 Hz has been eliminated.

COMMENT

A two stage Butterworth filter would decrease the undesirable signal attenuation at 60 Hz to $\delta(60) = -0.06$ or 6% while providing an even better signal attenuation at 200 Hz to $\delta(200) = -0.76$ or 76%. In general, higher order Butterworth filters give better response in the passband and sharper attenuation in the stopband, but with increased phase shift.

Bessel Filter Design

A Bessel filter sacrifices a flat gain over its passband in exchange for a linear phase shift. A k-stage low-pass Bessel filter has the transfer function

$$G(s) = \frac{a_0}{a_0 + a_1 s + \cdots + a_k s^k} \tag{6.62}$$

Table 6.2 Values for Low-Pass LC Bessel Filters[1]

k	C_1	L_2	C_3	L_4	C_5
2	1.577	0.423			
3	1.255	0.553	0.192		
4	1.060	0.512	0.318	0.110	
5	0.930	0.458	0.331	0.209	0.072

[1]Values for C_i in farads and L_i in henrys are referenced to $R_s = R_L = 1\,\Omega$ and $\omega_c = 1$ rad/s. See discussion for proper scaling.

For design purposes, this can be rewritten as

$$G(s) = \frac{a_0}{D_k(s)} \tag{6.63}$$

where

$$D_k(s) = (2k - 1)D_{k-1}(s) + s^2 D_{k-2}(s)$$

and

$$D_0(s) = 1 \quad \text{and} \quad D_1(s) = s + 1$$

A k-stage LC low-pass filter is shown in Figure 6.31. Table 6.2 lists the required values for elements C_i and L_i for a two- through five-stage Bessel filter of the form of Figure 6.31 [6, 7]. The values in Table 6.2 assume that $R_s = R_L = 1\,\Omega$ and $\omega_c = 2\pi f_c = 1$ rad/s. For other values, L_i and C_i are found from equations (6.60a) and (6.60b).

Active Filters

As an alternative to using only passive components, an active analog filter uses the high-frequency gain characteristics of the operational amplifier to form an effective active analog filter. Such an active analog filter is shown in Figure 6.34(a) with the type 741 operational amplifier. This is a first order, single stage, low-pass Butterworth filter. It has a low-pass cutoff frequency given by

$$f_c = \frac{1}{2\pi R_2 C_2} \tag{6.64}$$

Because the operational amplifier is used, the filter static sensitivity is given by R_2/R_1. The filter retains the Butterworth characteristics described by equation (6.58).

A first-order, single-stage, high-pass Butterworth active filter is also shown in Figure 6.34, again with the 741 operational amplifier. The filter has a high-pass cut-off frequency of

$$f_c = \frac{1}{2\pi R_1 C_1} \tag{6.65}$$

with static sensitivity of R_2/R_1. The magnitude ratio is given by

$$KM(f) = (R_2/R_1)\frac{(f/f_c)}{[1 + (f/f_c)^2]^{1/2}} \tag{6.66}$$

6.9 GROUNDS, SHIELDING, AND CONNECTING WIRES

The type of connecting wires used between electrical devices can have a significant impact on the noise level of the signal. Low-level signals of <100 mV are particularly

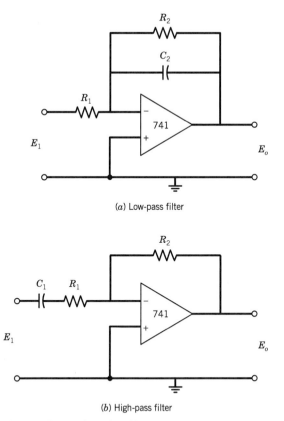

(a) Low-pass filter

(b) High-pass filter

Figure 6.34 Basic active filters.

susceptible to errors induced by noise. Three simple rules will help to keep noise levels low: (1) keep the connecting wires as short as possible; (2) keep signal wires away from noise sources; and (3) use a wire shield and proper ground.

Ground and Ground Loops

The voltage at the end of a wire that is connected to a rod driven far into the ground would likely be at the same voltage level as the earth—a reference datum called zero or *earth ground*. A *ground* is simply a return path to earth. Now suppose that wire is connected from the rod through an electrical box and then through various building routings to the ground plug of an outlet. Would the ground potential at the outlet still be at zero? The answer is maybe but probably not. The network of wires that form the return path to earth would likely act as antennae and pick up some voltage potential relative to earth ground. Any instrument grounded at the outlet would be referenced back to this voltage potential, not earth ground. Thus, an electrical ground is not an absolute value. Ground values vary between ground points because the ground returns pass through different equipment or building wiring on their return path to earth. Thus if a signal is grounded at two points, say at a power source ground and then at an earth ground, the grounds could be at different voltage levels. The difference between the two ground point voltages is called the *common-mode voltage*. This can lead to problems, as discussed next.

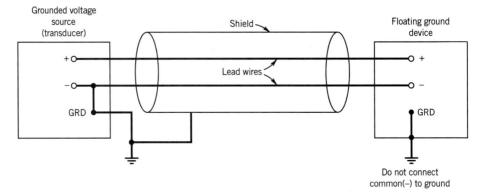

Figure 6.35 Signal grounding arrangements: grounded source signal with shield; floating signal at measuring device.

Ground loops are brought on by connecting a signal circuit to two or more grounds that are at different potentials. A ground wire of finite resistance will usually carry some current and will develop a potential. Thus two separate and different grounds, even in close proximity, can be at different potential levels. When these ground points are connected into a circuit, the potential difference itself drives a current. The result is an electrical interference induced on the signal. This is a ground loop. A ground loop can manifest itself in various forms, such as a sinusoidal signal or simply a voltage bias.

Figure 6.35 shows a proper connection between a grounded voltage source, such as a transducer, and a measuring device. Note that the common (−) line at the measuring device is not grounded. As such, it is referred to as being isolated or as a floating ground device. Incidentally, many devices are grounded through their ac power lines by means of a third prong. This can set up a ground loop relative to other circuit grounds. To create a floating ground, one must break this ground connection by using a three to two prong adapter, but be sure to take proper precautions to guard against electrical shock and to ensure that the system has only one ground point.

Shields

Long wires act as antennae and will pick up stray signals from nearby electrical fields. The most common problem is ac line noise. Electrical shields are effective against such noise. A *shield* is a piece of metal foil or wire braid wrapped around the signal wires and connected to ground. The shield intercepts external electrical fields, returning them to ground. A shield ground loop is prevented by grounding the shield at only one point, usually the signal ground at the transducer. Figure 6.35 shows such a shield-to-ground arrangement for a grounded voltage source connected to a floating ground measuring device.

A common source of electrical fields is from an ac power supply transformer. A capacitance coupling between the 60- or 50-Hz power supply wires and the signal wires is set up. For example, a 1-pF capacitance will superimpose a 40-mV interference on a 1-mV signal.

A different source of noise is from a magnetic field. When a lead wire moves within a magnetic field, a voltage is induced in the signal path. The same effect occurs with a lead wire in proximity to an operating motor. The best prevention is to separate the signal lead wires from such sources. Twisting the lead wires together also tends to cancel any induced voltage, as the currents through the two wires are in opposite directions. A final recourse is the use of a magnetic shield made from a material having a high ferromagnetic permeability.

Connecting Wires

There are several types of wires available. Single cable refers to a single length of common wire or wire strand. The conductor is coated for electrical insulation. Single cable is efficient only for connections of several centimeters and involving only a few wires. The wire is readily available and cheap, but it should not be used for low-level (millivolt level) signal connections. Flat cable is similar but consists of multiple conductors arranged in parallel strips, usually in groups of 4, 9, or 25. Flat cable is commonly used for short connections between adjacent electrical boards, but in such applications the signals are of the order of 1 V or more. Neither of these two types of wires offers any shielding.

Twisted pairs of wires consist of two electrically insulated conductors twisted about each other along their lengths. Twisted pairs are widely used to interconnect transducers and equipment. The intertwining of the wires offers some immunity to noise. Cables containing several twisted pairs are available. Shielded twisted pairs wrap the twisted pairs within a metallic foil shield. Shielded twisted pairs are one of the best choices for most applications.

Coaxial cable consists of an electrically insulated inner single conductor surrounded within an outer conductor made of stranded wire. The cable also contains a shield. In general, current flows in one direction along the inner wire and the opposite direction along the outer wire. Any electromagnetic fields generated will cancel. Coaxial cable is the choice for high-frequency signals. Signals can be sent over very long distances with little loss. A variation of coaxial cable is triaxial cable, which contains two inner conductors. It is used in applications as with twisted pairs but offers superior noise immunity.

Optical cable is widely used to transmit low-level signals over long distances. This cable may contain one or more fiber optic ribbons within a polystyrene shell. A transmitter converts the low-level voltage signal into infrared light. The light is transmitted through the cable to a receiver, which converts it back to a low-level voltage signal. The cable is virtually noise free from magnetic fields and harsh environments.

6.10   SUMMARY

This chapter has focused on classic but basic analog electrical measurement devices, including those used for signal conditioning. Devices that are sensitive to current, resistance, and voltage were presented. Signal conditioning devices that can modify or condition a signal, including amplifiers, current loops, and filters, were also presented. Because all of these devices are widely used, often in combination and often masked within more complicated circuits, the reader should strive to become familiar with the workings of each.

Loading errors, which are caused by the act of measurement or the presence of a sensor, can only be minimized by the proper choice of sensor and technique. Interstage loading errors, which occur between the connecting stages of a measurement system, can be minimized by proper impedance matching. A discussion of such impedance matching has been presented, including the use of the versatile voltage follower circuit for this purpose.

REFERENCES

1. Prentiss, S. R., *Oscilloscopes*, Reston (a Prentice-Hall company), Reston, VA, 1981.
2. Bleuler, E., and Haxby, R. O., *Methods of Experimental Physics*, 2d ed., Vol. 2, Academic, New York, 1975.
3. Money, S. A., *Microprocessors in Instrumentation and Control*, McGraw-Hill, New York, 1985.
4. Stanley, W. D., Dougherty, G. R., and Dougherty, R., *Digital Signal Processing*, 2d ed., Reston (a Prentice-Hall company), Reston, VA, 1984.
5. DeFatta, D. J., Lucas, J., and Hodgkiss, W., *Digital Signal Processing*, Wiley, New York, 1988.
6. Lam, H. Y., *Analog and Digital Filters: Design and Realization*, Prentice-Hall, Englewood Cliffs, 1979.
7. Niewizdomski, S., Filter Handbook–A Practical Design Guide, CRC Press, Boca Raton, 1989.

Suggested Reading

Ahmed, H., and Spreadbury, P. J., *Analogue and Digital Electronics for Engineers*, 2d ed., Cambridge University Press, Cambridge, UK, 1984.

Evans, Alvis J., Mullen, J. D., and Smith, D. H., *Basic Electronics Technology*, Texas Instruments Inc., Dallas, TX, 1985.

Wobschall, D., *Circuit Design for Electronic Instrumentation: Analog and Digital Devices from Sensor to Display*, McGraw-Hill, New York, 1979.

NOMENCLATURE

c	damping coefficient	$G(s)$	transfer function
e	error	I	electric current $[C\,t^{-1}]$
f	frequency $[t^{-1}]$	I_e	effective current $[C\,t^{-1}]$
f_c	filter cutoff frequency $[t^{-1}]$	I_g	galvanometer current $[C\,t^{-1}]$
f_m	maximum analog signal frequency $[t^{-1}]$	K	static sensitivity
$h\{E_i(t)\}$	function	L	inductance [H]
k	cascaded filter stage number	$M(f)$	magnitude ratio at frequency, f
$\hat{k}$	unit vector aligned with current flow	N	number of turns in a current-carrying loop
l	length $[l]$	R	resistance $[\Omega]$
$\mathbf{n}$	unit vector normal to current loop	R_B	effective bridge resistance $[\Omega]$
q	charge [C]	R_g	galvanometer resistance $[\Omega]$
$\mathbf{r}$	radius, vector $[l]$	R_{eq}	equivalent resistance $[\Omega]$
t	time $[t]$	R_m	meter resistance $[\Omega]$
u_R	uncertainty in resistance R	δR	change in resistance $[\Omega]$
A	cross-sectional area of a current-carrying loop; operational amplifier open-loop gain	$\mathbf{T}_\mu$	torque on a current-carrying loop in a magnetic field $[m\,l^2\,t^{-2}]$
E	voltage [V]	Z_1	output impedance $[\Omega]$
E_1	open circuit potential [V]	Z_m	measuring device impedance $[\Omega]$
E_i	input voltage [V]	Z_{th}	Thévenin equivalent impedance $[\Omega]$
E_m	indicated output voltage [V]	α	angle
E_o	output voltage [V]	τ	time constant $[t]$
$\mathbf{F}$	force $[m\,l\,t^{-2}]$	$\Phi(f)$	phase shift at frequency, f
G	amplifier closed-loop gain	θ	angle

PROBLEMS

6.1 Determine the maximum torque on a current loop having 20 turns, a cross-sectional area of 1 in.^{2}, and experiencing a current of 20 mA. The magnetic field strength is 0.4 Wb/m^2.

6.2 A 5000-Ω voltage-dividing circuit is used in a particular application using the arrangement of Figure 6.16. What is the minimum allowable meter internal resistance such that the loading error in measuring the open circuit potential, E_o, will be less than 12% of the full-scale value? Note: a family of solutions will result.

6.3 Determine the loading error as a percentage of the output for the voltage dividing circuit of Figure 6.16, if $R_T = R_1 + R_2$ and $R_1 = kR_T$. The parameters of the circuit are

$$R_T = 500\ \Omega \qquad E_i = 10\ \text{V} \qquad R_m = 10{,}000\ \Omega \qquad k = 0.5$$

What would the loading error be if expressed as a percentage of the full-scale output? Show that the two answers for the loading error are equal when expressed in volts.

6.4 Consider the Wheatstone bridge shown in Figure 6.14. Suppose

$$R_3 = R_4 = 200\ \Omega$$
$$R_2 = \text{variable calibrated resistor}$$
$$R_1 = \text{transducer resistance} = 40x + 100$$

a. When $x = 0$, what is the value of R_2 required to balance the bridge?
b. If the bridge is operated in a balanced condition in order to measure x, determine the relationship between R_2 and x.

6.5 For the Wheatstone bridge shown in Figure 6.14, R_1 is a sensor whose resistance is related to a measured variable x by the equation $R_1 = 20x^2$. If $R_3 = R_4 = 100\ \Omega$ and the bridge is balanced when $R_2 = 46\ \Omega$, determine x.

6.6 A force sensor has as its output a change in resistance. The sensor forms one leg (R_1) of a basic Wheatstone bridge. The sensor resistance with no force load is 500 Ω, and its static sensitivity is 0.5 Ω/N. Each arm of the bridge is initially 500 Ω.

a. Determine the bridge output for applied loads of 100, 200, and 350 N. The bridge is operated as a deflection bridge, with an input voltage of 10 V.
b. Determine the current flow through the sensor.
c. Repeat parts (a) and (b) with $R_m = 10\ \text{k}\Omega$ and $R_s = 600\ \Omega$.

6.7 A reactance bridge arrangement replaces the resistor in a Wheatstone bridge with a capacitor or inductor. Such a reactance bridge is then excited by an ac voltage. Consider the bridge arrangement shown in Figure 6.36. Show that the balance equations for this bridge are given by

$$C_2 = C_1 \frac{R_1}{R_2}$$

6.8 Circuits containing inductance- and capacitance-type elements exhibit varying impedance depending upon the frequency of the input voltage. Consider the bridge circuit of Figure 6.37. For a capacitor and an inductor connected in a series arrangement, the impedance is a function of frequency, such that a minimum impedance occurs at the resonance frequency

$$f = \frac{1}{2\pi(LC)^{1/2}}$$

where f is the frequency [Hz], L is the inductance [H], and C is the capacitance [F]. Design a bridge circuit that could be used to calibrate a frequency source at 500 Hz.

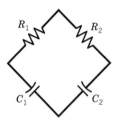

Figure 6.36 Bridge circuit for Problem 6.7.

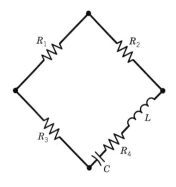

Figure 6.37 Bridge circuit for Problem 6.8.

6.9 A Wheatstone bridge initially has resistances equal to $R_1 = 200\ \Omega$, $R_2 = 400\ \Omega$, $R_3 = 500\ \Omega$, and $R_4 = 600\ \Omega$. For an input voltage of 5 V, determine the output voltage at this condition. If R_1 changes to 250 Ω, what is the bridge output?

6.10 Construct a plot of the voltage output of a Wheatstone bridge having all resistances initially equal to 500 Ω, with a voltage input of 10 V for the following cases:

a. R_1 changes over the range 500–1000 Ω.
b. R_1 and R_2 change equally, but in opposite directions, over the range 500–600 Ω.
c. R_1 and R_3 change equally over the range 500–600 Ω.

Discuss the possible implications of these plots for using bridge circuits for measurements with single and multiple transducers connected as arms of the bridge.

6.11 Consider the simple potentiometer circuit shown in Figures 6.9 and 6.10. Perform a design-stage uncertainty analysis to determine the minimum uncertainty in measuring a voltage. The following information is available concerning the circuit (assume 95% confidence):

$$E_i = 10 \pm 0.1\,\text{V} \qquad R_T = 100 \pm 1\,\Omega \qquad R_g = 100\,\Omega \qquad R_x = \text{reading} \pm 2\%$$

where R_g is the internal resistance of the galvanometer. The null condition of the galvanometer may be assumed to have negligible error. The uncertainty associated with R_x is associated with the reading obtained from the location of the sliding contact. Estimate the uncertainty in the measured value of voltage at nominal values of 2 and 8 V.

Problems 6.12 through 6.14 relate to the comparison of two sinusoidal input signals using a dual trace oscilloscope. A schematic diagram of the measurement system is shown in Figure 6.38, where one of the signals is assumed to be a reference standard signal, having a known frequency and amplitude. The oscilloscope trace for these inputs is called a Lissajous diagram.

6.12 Develop the characteristic shape of the Lissajous diagrams for two sinusoidal inputs having the same amplitude, but with the following phase relationship: (a) in phase, (b) ±90° out of phase, and (c) 180° out of phase.

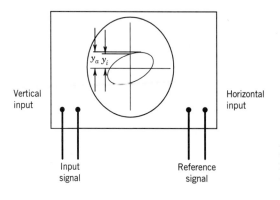

Figure 6.38 Dual-trace oscilloscope for measuring signal characteristics by using Lissajous diagrams.

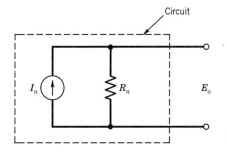

Figure 6.39 Circuit for Problem 6.19.

6.13 Show that the phase angle for two sinusoidal signals can be determined from the relationship

$$\sin \phi = y_i / y_a$$

where the values of y_i and y_a are illustrated in Figure 6.38 and represent the vertical distance to the y intercept and the maximum y value, respectively.

6.14 Draw a schematic diagram of a measurement system to determine the phase lag resulting from an electronic circuit. Your schematic diagram should include a reference signal, the electronic circuit, and the dual-trace oscilloscope. Discuss the expected result and how a quantitative estimate of the phase lag can be determined.

6.15 Construct Lissajous diagrams for sinusoidal inputs to a dual-trace oscilloscope having the following horizontal signal to vertical signal frequency ratios: (a) 1:1, (b) 1:2, (c) 2:1, (d) 1:3, (e) 2:3, and (f) 5:2. These plots can be easily developed by using spreadsheet software and plotting a sufficient number of cycles of each of the input signals.

6.16 A single-stage RC filter with $f_c = 100$ Hz is used to filter an analog signal. Determine the attenuation of the filtered analog signal at 10, 50, 75, and 200 Hz.

6.17 A three-stage LC Bessel filter with $f_c = 100$ Hz is used to filter an analog signal. Determine the attenuation of the filtered analog signal at 10, 50, 75, and 200 Hz.

6.18 Design a cascading LC low-pass filter that has a magnitude ratio flat to within 3 dB from 0 to 5 kHz but with an attenuation of at least 30 dB for all frequencies at and above 10 kHz.

6.19 Consider the circuit of Figure 6.39, which consists of an ideal current source I_n and a parallel resistor R_n. Find its Thévenin equivalents. E_{th} and R_{th}.

6.20 An electrical displacement transducer has an output impedance of 500 Ω. Its voltage is input to a voltage measuring device that has an input impedance of 100 kΩ. Estimate the ratio of true voltage to the voltage measured by the measuring device.

6.21 Consider a transducer connected to a voltage measuring device. Plot the ratio E_m/E_1 as the ratio of output impedance of the transducer to input impedance of the measuring device varies from 1:1 to 1:1,000,000 (look at each order of magnitude). Comment on the effect of input impedance on measuring a voltage.

6.22 The pH meter of Figure 6.40 consists of a transducer made of glass containing paired electrodes. This transducer can produce voltage potentials up to 1 V with an internal output impedance of up to 10^9 Ω. If the signal is to be conditioned, estimate the minimum input impedance of the op-amp shown required to keep loading error below 0.1%.

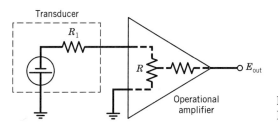

Figure 6.40 pH transducer circuit for Problem 6.22.

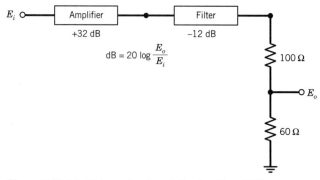

Figure 6.41 Multistage signal path for Problem 6.23.

6.23 Consider the circuit of Figure 6.41, which consists of an amplifier providing 32-dB gain, followed by a filter, which causes an attenuation of 12 dB at the frequency of interest, followed by a voltage divider. Find E_o/E_i across the 60-Ω resistor.

6.24 The temperature sensor and bridge circuit described in Example 6.2 is used to measure a temperature; the sensor output indicates a temperature of 30°C. It is desired to measure this temperature within ±1°C. Assume that the errors in temperature measurement result from two sources:

1. The galvanometer has limited sensitivity, allowing some current flow when indicating a balanced bridge condition. Take $R_g = 5\,\Omega$.
2. The ratio R_3/R_4 and the value of R_2 must be known. Uncertainties in these values contribute to the uncertainty in the value of R_1.

Specify a combination of galvanometer current at balanced conditions, and uncertainty in the bridge resistances to yield the required uncertainty in measured temperature.

6.25 The input to a subwoofer loudspeaker is to pass through a passive low-pass Butterworth filter having a cut-off frequency of 100 Hz. Specify the filter to meet this application: At 50 Hz, the signal magnitude ratio should be at least 0.95 and at 200 Hz, the magnitude ratio should be no more than 0.01. The source and load resistances are 10 Ω. You need to determine the number of stages required and the values of the components.

6.26 The input to a high frequency loudspeaker is to pass through a passive high-pass Butterworth filter having a cut-off frequency of 5000 Hz. Specify the filter to meet this application: At 8000 Hz, the signal attenuation should be no more than −0.45 dB and at 2500 Hz, the attenuation should be at least −20 dB. The source and load resistances are 10 Ω. You need to determine the number of stages required and the values of the components.

Chapter 7

Sampling, Digital Devices, and Data Acquisition

7.1 INTRODUCTION

Integrating analog electrical transducers with digital data-acquisitions systems is cost effective and commonplace on the factory floor, the testing lab, and even in our homes. There are many advantages to this hybrid arrangement, including the efficient handling and rapid processing of large amounts of data and varying degrees of artificial intelligence by using digital microprocessor systems.

There are fundamental differences between analog and digital systems, which impose some limitations and liabilities of which the engineer needs be aware. As pointed out in Chapter 2, the most important difference is that analog signals are continuous in both amplitude and time, whereas digital signals are discrete (noncontinuous) in both amplitude and time.

This chapter begins with an introduction to the fundamentals of sampling, the process by which continuous signals are made discrete. Criteria are presented that circumvent the loss or misinterpretation of signal information while undergoing the sampling process. We show how a discrete series of data can actually contain all of the information available in a continuous signal, or at least provide a very good approximation.

The discussion moves on to the devices most often involved in analog and digital systems. Analog devices interface with digital devices through an analog-to-digital converter. The reverse process of a digital device interfacing with an analog device occurs through a digital-to-analog converter. A digital device interfaces with another digital device through a digital input–output (I/O) port. These interfaces are the major components of computer-based data-acquisition systems. Necessary components and the basic layout of these systems are introduced and standard methods for communication between digital devices are presented.

7.2 SAMPLING CONCEPTS

Consider an analog signal and its discrete time series representation in Figure 7.1. There would appear to be quite a difference in the information contained in the analog and discrete representations. However, the important analog signal information concerning amplitude and frequency can be well represented by such a discrete series. Just how well it will be represented will depend on (1) the frequency content of

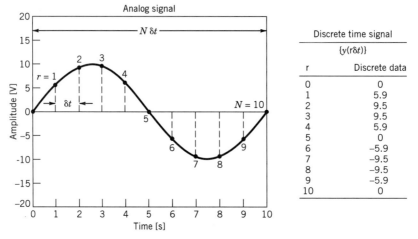

Figure 7.1 Analog and discrete representations of a time-varying signal.

the measured analog signal; (2) the size of the time increment between each discrete number; and (3) the total sample period of the measurement.

Chapter 2 discussed how a continuous dynamic signal could be represented by a Fourier series. The discrete Fourier transform (DFT) was also introduced as a method for the reconstruction of a dynamic signal from a discrete set of data. The DFT, through equation (2.40), conveys all the information needed to reconstruct the Fourier series of a continuous dynamic signal from a representative discrete time series. Hence, Fourier analysis provides certain guidelines for sampling continuous data. In this section, the concept of converting a continuous analog signal into an equivalent discrete time series is examined. Extended discussions on the subject of signal analysis can be found in many texts (e.g., [1–3]).

Sample Rate

Just how frequently must a time-dependent signal be measured to determine its frequency content? Consider Figure 7.2(a) in which the magnitude variation of a 10-Hz sine wave is plotted versus time over time period t_f. Suppose this sine wave is measured repeatedly at successive sample time increments δt. This corresponds to measuring the signal with a sample frequency or *sample rate* (in hertz) of

$$f_s = 1/\delta t \tag{7.1}$$

For this discussion, we will assume that the signal measurement occurs at a constant sample rate. At each measurement, the sine wave is converted into a number. For comparison, in Figures 7.2(b)–7.2(d) the resulting series versus time plots are given when the signal is measured by using sample time increments (or sample rates) of (b) 0.010 s ($f_s = 100$ Hz), (c) 0.037 s ($f_s = 27$ Hz), and (d) 0.083 s ($f_s = 12$ Hz). It is apparent that the sample rate has a significant effect on our perception and reconstruction of the continuous analog signal in the time domain. In Figures 7.2(b) and 7.2(c), we can still discern the 10-Hz frequency content of the original signal. But as seen in Figure 7.2(d), an interesting phenomenon occurs if the sample rate becomes too slow: The frequency of the original sine wave appears to be a lower frequency.

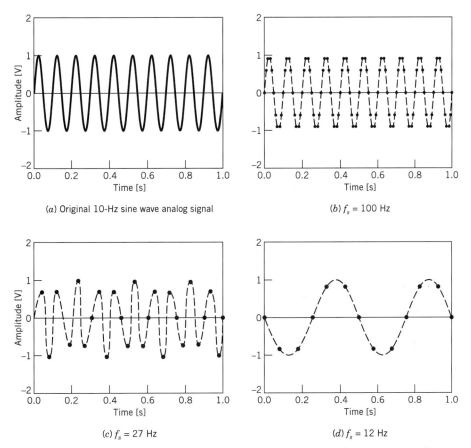

Figure 7.2 The effect of sample rate on signal frequency and amplitude interpretation.

It can be concluded that the sample time increment or the corresponding sample rate plays a significant role in signal frequency representation. The *sampling theorem* states that to reconstruct the frequency content of a measured signal accurately, *the sample rate must be more than twice the highest frequency contained in the measured signal.* Denoting the maximum frequency in the analog signal as f_m, the sampling theorem requires

$$f_s > 2 f_m \tag{7.2}$$

or in terms of sample time increment,

$$\delta t < \frac{1}{2 f_m} \tag{7.3}$$

When signal frequency content is important, equations (7.2) and (7.3) provide a criterion for the minimum sample rate or maximum sample time increment, respectively, to be used in converting data from a continuous to a discrete form. The frequencies that will be defined by the DFT of the resulting discrete series will provide an accurate representation of the original signal frequencies regardless of the sample rate used, provided that the requirements of the sampling theorem are satisfied.

Alias Frequencies

When a signal is sampled at a sample rate that is less than $2 f_m$, the higher-frequency content of the analog signal will take on the false identity of a lower frequency in the resulting discrete series. This is seen to occur in Figure 7.2(d), where, because $f_s < 2 f_m$, the 10-Hz analog signal is observed to take on the false identity of a 2-Hz signal. A misinterpretation of the frequency content of the original signal results! Such a false frequency is called an *alias frequency*.

The alias phenomenon is an inherent consequence of a discrete sampling process. To illustrate this, consider a simple periodic signal that can be described by the one-term Fourier series:

$$y(t) = C \sin[2\pi f t + \phi(f)]$$

Suppose $y(t)$ is measured with a sample time increment of δt, so that its discrete time signal is given by

$$y(r\,\delta t) = C \sin[2\pi f r\,\delta t + \phi(f)] \qquad r = 1, 2, \ldots, N \qquad (7.4)$$

Now using the identity, $\sin x = \sin(x + 2\pi q)$, where q is any integer, we rewrite $y(r\,\delta t)$ as

$$C \sin[2\pi f r\,\delta t + \phi(f)] = C \sin[2\pi f r\,\delta t + 2\pi q + \phi(f)]$$

$$= C \sin\left[2\pi\left(f + \frac{m}{\delta t}\right) r\,\delta t + \phi(f)\right] \qquad (7.5)$$

where $m = 0, 1, 2, \ldots$ (and hence, $mr = q$ is an integer). This implies that for any value of δt, the frequencies of f and $f + m/\delta t$ must be indistinguishable. Hence, all frequencies given by $f + m/\delta t$ are the alias frequencies of f. However, by adherence to the sampling theorem criterion, all $m \geq 1$ will be eliminated from the sampled signal and, thus, this ambiguity between frequencies is avoided.

This same discussion must apply to complex periodic, aperiodic, and nondeterministic waveforms. This is immediately seen by considering the general Fourier series used to represent such signals. Such a discrete series,

$$y(r\,\delta t) = \sum_{n=1}^{\infty} C_n \sin[2\pi f r\,\delta t + \phi_n(f)] \qquad (7.6)$$

for $r = 1, 2, \ldots, N$, can be rewritten as

$$y(r\,\delta t) = \sum_{n=1}^{\infty} C_n \sin\left[2\pi\left(nf + \frac{nm}{\delta t}\right) r\,\delta t + \phi_n(f)\right] \qquad (7.7)$$

which displays the same aliasing phenomenon shown in equation (7.5).

In general, the Nyquist frequency defined by

$$f_N = \frac{f_s}{2} = \frac{1}{2\delta t} \qquad (7.8)$$

represents a folding point for the aliasing phenomenon. All actual frequency content in the analog signal that is at frequencies above f_N will appear as alias frequencies of less than f_N; that is, such frequencies will be folded back and superimposed on the signal as lower frequencies.

An alias frequency, f_a, can be computed from the folding diagram of Figure 7.3 in which the original frequency axis is folded back over itself at the folding point

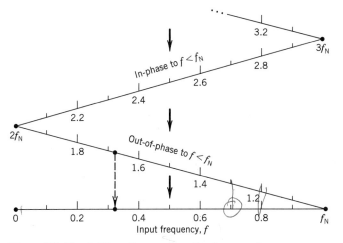

Figure 7.3 The folding diagram for alias frequencies.

of f_N and its harmonics, mf_N, where $m = 1, 2, \ldots$. For example, as noted by solid arrows in Figure 7.3, the frequencies of $f = 0.5\,f_N, 1.5\,f_N, 2.5\,f_N, \ldots$ will all appear as $0.5\,f_N$ in the discrete series $y(r\,\delta t)$. Use of the folding diagram is illustrated further in Example 7.1.

How does one avoid the alias phenomenon when sampling a signal of unknown frequency content? If the frequency range of interest is limited to a certain maximum frequency, the signal should be filtered at and above this frequency *prior* to the sampling process. Otherwise, the maximum frequency is to be limited by the maximum sample rate possible and the sampling theorem. In all cases, once a sample rate is chosen *the signal should be filtered at and above f_N to avoid the problem of alias frequencies*.

EXAMPLE 7.1

A 10-Hz sine wave is sampled at 12 Hz. Compute the maximum frequency that can be represented in the resulting discrete signal. Compute the alias frequency.

KNOWN

$f = f_m = 10$ Hz
$f_s = 12$ Hz

ASSUMPTION

Constant sample rate

FIND

f_N and f_a

SOLUTION

The Nyquist frequency, f_N, sets the maximum frequency that can be represented in a resulting data set. Using equation (7.8) with $f_s = 12$ Hz gives the Nyquist frequency,

$f_N = 6$ Hz. This is the maximum frequency that can be sampled correctly, so the measured frequency will be in error.

All frequency content in the analog signal above f_N will appear as an alias frequency, f_a, between 0 and f_N. Because $f/f_N \approx 1.67$, f will be folded back and appear in the sampled signal as a frequency between 0 and f_N. Reading Figure 7.3, we see that $f = 1.67 f_N$ is located directly above $0.33 f_N$. Thus f will be folded back to appear as $f_a = 0.33 f_N$ or 2 Hz. This folding is noted by the dotted line in Figure 7.3. Also, the 2 Hz signal will be $180°$ out of phase with the original signal as noted in Figure 7.3 and seen in Figure 7.2a and d. As a consequence, the 10-Hz sine wave sampled at 12 Hz will be completely indistinguishable from a 2-Hz sine wave sampled at 12 Hz in the discrete time series.

COMMENT

On the companion software disk, program file *Sampling* can be used to study the effects of sample rate and sample period on the resulting signal.

EXAMPLE 7.2

A complex periodic signal has the form

$$y(t) = A_1 \sin 2\pi(25)t + A_2 \sin 2\pi(75)t + A_3 \sin 2\pi(125)t$$

If the signal is sampled at 100 Hz, determine the frequency content of the resulting discrete series.

KNOWN

$f_s = 100$ Hz $f_2 = 75$ Hz
$f_1 = 25$ Hz $f_3 = 125$ Hz

ASSUMPTION

Constant sample rate

FIND

The alias frequencies f_{a_1}, f_{a_2}, and f_{a_3}, in $\{y(r \, \delta t)\}$

SOLUTION

From equation (7.8) for a sample rate of 100 Hz, the Nyquist frequency is 50 Hz. All frequency content in the analog signal above f_N will take on an alias frequency between 0 and f_N in the sampled data series. With $f_1 = 0.5 f_N$, $f_2 = 1.5 f_N$, and $f_3 = 2.5 f_N$, we can use Figure 7.3 to determine the respective alias frequencies.

i	f_i	f_{a_i}
1	25 Hz	25 Hz
2	75 Hz	25 Hz
3	125 Hz	25 Hz

So in the resulting series, $f_{a_1} = f_{a_2} = f_{a_3} = 25$ Hz. As a result of aliasing, the

75- and 125-Hz components would be completely indistinguishable from the 25-Hz component. The discrete series would be described by

$$y(n\,\delta t) = (B_1 + B_2 + B_3) \sin 2\pi(25)r\,\delta t \quad r = 0, 1, 2, \ldots$$

where $B_1 = A_1$, $B_2 = -A_2$, and $B_3 = A_3$. Clearly, this misinterprets the original analog signal.

COMMENT

Example 7.2 illustrates the potential for misrepresentation of a signal through improper sampling. For signals for which the frequency content is not known prior to their measurement, sample frequencies should be increased according to the following scheme:

- Sample the input signal at increasing sample rates by using a fixed total sample time, and examine the time plots for each signal. Look for changes in the shape of the waveform (e.g., see Figure 7.2).
- Compute the amplitude spectrum for each signal and compare the resulting frequency content.

Always use antialias analog filters set at the Nyquist frequency.

Amplitude Ambiguity

For simple and complex periodic waveforms, the discrete Fourier transform of the sampled discrete time signal will remain unaltered by a change in the total sample period, $N\,\delta t$, provided that (1) the total sample period remains an integer multiple of the fundamental period, T_1, of the measured continuous waveform, that is, $mT_1 = N\,\delta t$, where m is an integer; and (2) the sample time increment meets the sampling theorem criterion. As a consequence, the amplitudes associated with each frequency in the periodic signal, the spectral amplitudes, will be accurately represented by the DFT. This suggests that an original periodic waveform can be completely reconstructed from a discrete time series regardless of the sample time increment used. The total sample period will define the frequency resolution of the DFT:

$$\delta f = \frac{1}{N\,\delta t} = \frac{f_s}{N} \tag{7.9}$$

The frequency resolution plays a crucial role in the reconstruction of the signal amplitudes, as noted below.

An important difficulty appears when $N\,\delta t$ is not coincident with an integer multiple of the fundamental period of $y(t)$: The resulting DFT cannot exactly represent the spectral amplitudes of the sampled continuous waveform. The problem is brought on by the truncation of a complete cycle of the signal (see Figure 7.4), and from spectral resolution, because the associated fundamental frequency and its harmonics will not be coincident with a center frequency of the DFT. However, this error in the spectral amplitudes will decrease either as the value of $N\,\delta t$ more closely approximates an exact integer multiple of T_1 or as f_s becomes large relative to f_m.

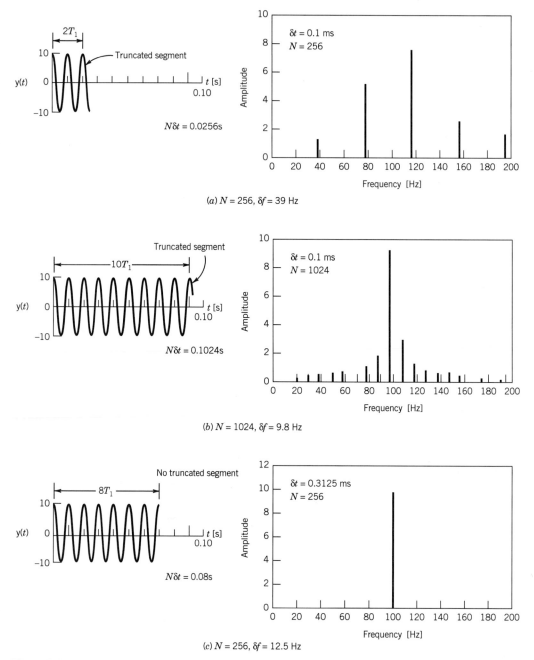

Figure 7.4 Amplitude spectra for $y(t)$.

This situation is illustrated in Figure 7.4, which compares the amplitude spectrum resulting from sampling the signal, $y(t) = 10\cos 628t$, over different sample periods. Two sample periods of 0.0256 and 0.1024 s were used with $\delta t = 0.1$ ms, and a third sample period of 0.08 s with $\delta t = 0.3125$ ms.[1] These sample periods provide for a DFT frequency resolution, δf, of about 39, 9.8, and 12.5 Hz, respectively.

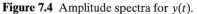

[1]Many DFT (including FFT) algorithms require that $N = 2^M$, where M is an integer. This affects the selection of δt.

The two spectra shown in Figures 7.4(a) and 7.4(b) display a spike near the correct 100 Hz with surrounding noise spikes, known as leakage, at adjacent frequencies. Note that the original signal $y(t)$ cannot be reconstructed exactly from either of these spectra. The spectrum in Figure 7.4(c) has been constructed by using a longer sample time increment (lower sample rate) than that of the spectra of Figures 7.4(a) and 7.4(b), and fewer data points than the spectrum of Figure 7.4(b). Despite this, the spectrum of Figure 7.4(c) provides an exact representation of $y(t)$. The N and δt combination used has reduced leakage to zero, which maximizes the amplitude at 100 Hz. Its sample period corresponds to exactly eight periods of $y(t)$, and the frequency of 100 Hz corresponds to the center frequency of the eighth frequency interval of the DFT. As seen in Figure 7.4, the loss of accuracy of a DFT representation occurs in the form of amplitude "leakage" to adjacent frequencies. To the DFT, the truncated segment of the sampled signal appears as an aperiodic signal. The DFT returns the correct spectral amplitudes for both the periodic and aperiodic signal portions. But as a result, the spectral amplitudes for any truncated segment will be superimposed onto portions of the spectrum adjacent to the correct frequency. Note that the effects of leakage are also apparent in the time signal of Figure 7.2(c). By varying the sample period or its equivalent, the DFT resolution, one can minimize leakage and control the accuracy of the spectral amplitudes.

If $y(t)$ is an aperiodic or nondeterministic waveform, there may not be a fundamental period. In such situations, one controls the accuracy of the spectral amplitudes by varying the DFT resolution, δf, to minimize leakage. An associated method uses a cut-and-try approach whereby δf is varied and the resulting signal statistics estimated and compared. The statistics will approach an acceptable level of precision when the DFT resolution is reduced to an adequate level.

In summary, *the reconstruction of a measured waveform from a discrete signal is controlled by the sampling rate and the DFT resolution.* By adherence to the sampling theorem, one controls the frequency content of both the measured signal and the resulting spectrum. By variation of δf, one can control the accuracy of the spectral amplitude representation.

 On the companion software disk, program file *Sampling* is useful to explore the above concept of spectral leakage. Program file *DataSpect* can be used to explore sampling concepts on user generated data series.

Selecting Sample Rate and Data Number

For an exact discrete representation in both frequency and amplitude of any periodic, analog waveform, both the number of data points and the sample rate should be chosen based on the preceding discussion by using the criteria of equations (7.3) and (7.9). Equation (7.3) sets the maximum value for δt, or minimum sample rate f_s, and equation (7.9) sets the total sampling time, $N\delta t$, from which the data number, N is estimated.

For most real signals, exact discrete representations of the input analog signal frequency and amplitude content are not possible or practical. Setting the sample rate, f_s, at five times the maximum signal frequency together with large values of $N\delta t$ is recommended to minimize spectral leakage and provide a good approximation of the original signal. An antialias filter should be used to ensure that no frequency above a desired maximum frequency is encountered. Still, the maximum sample rate available will be limited by the data-acquisition system and the maximum data number by the memory size available. A good reconstruction of the original time-based signal is possible through

$$y^*(t) = \frac{1}{\pi} \sum_{r=1}^{r=N} y(r\,\delta t) \frac{\sin[\pi(t/\delta t) - r]}{t/\delta t - r} \qquad (7.10)$$

where $y^*(t)$ is the signal reconstructed from the data sequence $\{y(r\,\delta t)\}$. Although the equation assumes a semi-infinite dataset, it is effective for finite data sets.

7.3 DIGITAL DEVICES: BITS AND WORDS

Digital signals are discrete in both time and amplitude. Nearly all digital measuring systems will use some variation of a binary numbering system to represent and to transmit the signal information in digital form. A binary system is a number system using base 2.

Binary systems use the binary digit or *bit* as the smallest unit of information. A bit will consist of only a single digit, either a 1 or a 0. Bits are like electrical switches and are used to convey both logical and numerical information. From a logic standpoint, the 1 or 0 are represented by the on or off switch settings. By appropriate action a bit can be reset to either on or off positions, thereby possessing a value of 1 or 0. By combining bits it is possible to define integer numbers greater than a 1 or 0. Such a collection of bits is used to express numerical information in the entity called a *word*. Typically, word length may range from 4 bits for a small dedicated digital device up to 32 bits or more for a large microcomputer. A word formed by 8 bits is defined as a *byte*. A particular physical location used to store a word is known as a *register*.

The number of bits that define a word determines the size of the maximum number that the word can represent. For example, since any bit can represent either a 1 or 0, then a combination of two bits can represent 2^2 or 4 possible combinations of bit arrangements: 0 0, 0 1, 1 0, or 1 1. We can alternatively reset this 2-bit word to produce these four different arrangements, which represent the decimal, that is, base 10, integer numbers 0, 1, 2, or 3, respectively. Thus, the M bits in an M-bit word can be combined to represent any of 2^M integers. An 8-bit word can represent the numbers 0 through 255; a 16-bit word can represent 0 through 65,535.

The numerical value for a word is computed by moving by bit from right to left. From the right side, each successive bit will increase the value of the word by a unit, a two, a four, an eight, and so forth through 2^{M-1}, provided that the bit is in its on (value of 1) position; otherwise, the particular bit increases the value of the word by zero. A weighting scheme of an M-bit word is given below:

Bit $M - 1$	...	Bit 3	Bit 2	Bit 1	Bit 0
2^{M-1}	...	2^3	2^2	2^1	2^0
2^{M-1}	...	8	4	2	1

With this scheme, bit $M - 1$ is known as the most significant bit (MSB) because its contribution to the numerical level of the word is the largest relative to the other bits. Bit 0 is known as the least significant bit (LSB). This type of scheme is known as the *straight binary code*.

Several binary codes are in common use. The straight binary code is the most straightforward. Consider a 4-bit word. In 4-bit straight binary, 0 0 0 0 is equivalent to analog zero and 1 1 1 1 is equivalent to 15. Because all numbers are of like sign, straight binary is considered to be a unipolar code. Bipolar codes allow for sign

Table 7.1 Binary Codes (Example: 4-Bit Words)

Bits	Straight	Offset	Twos Complement	Ones Complement	AVPS
0 0 0 0	0	−7	+0	+0	−0
0 0 0 1	1	−6	+1	+1	−1
0 0 1 0	2	−5	+2	+2	−2
0 0 1 1	3	−4	+3	+3	−3
0 1 0 0	4	−3	+4	+4	−4
0 1 0 1	5	−2	+5	+5	−5
0 1 1 0	6	−1	+6	+6	−6
0 1 1 1	7	−0	+7	+7	−7
1 0 0 0	8	+0	−8	−7	+0
1 0 0 1	9	+1	−7	−6	+1
1 0 1 0	10	+2	−6	−5	+2
1 0 1 1	11	+3	−5	−4	+3
1 1 0 0	12	+4	−4	−3	+4
1 1 0 1	13	+5	−3	−2	+5
1 1 1 0	14	+6	−2	−1	+6
1 1 1 1	15	+7	−1	−0	+7

changes. *Offset binary* is a bipolar code with $0\,0\,0\,0$ equal to $−7$ and $1\,1\,1\,1$ equal to $+7$. The comparison between these two codes and several others is shown in Table 7.1. In offset binary there are two zero levels, true zero being centered between them. Note how the MSB is used as a sign bit in a bipolar code.

The *ones-* and *twos-complement binary* codes shown in Table 7.1 are both bipolar. They differ in the positioning of the zero, with the twos-complement code containing one more negative level ($−8$) than positive ($+7$). Twos-complement code is widely used in digital computers. Also popular is the bipolar *absolute-value-plus-sign* code (AVPS), which uses, the MSB as the sign bit but uses the lower bits in a straight binary representation; for example, $1\,1\,0\,0$ and $0\,1\,0\,0$ represent $−4$ and $+4$, respectively.

A code often used for digital decimal readouts and for communication between digital instruments is the *binary coded decimal*, or BCD. In this code, each individual digit of a decimal number is represented separately by its equivalent value coded in straight binary. For example, the three digits of the base 10 number 532_{10} are represented by the BCD number $0\,1\,0\,1\ \ 0\,0\,1\,1\ \ 0\,0\,1\,0$ (i.e., binary 5, binary 3, binary 2). In this code each binary number need span but a single decade, that is, have values from 0 to 9. Hence, in digital readouts, when the limits of the lowest decade are exceeded, that is, 9 goes to 10, the code just carries into the next higher decade.

EXAMPLE 7.3

The contents of a 4-bit register is the binary word $0\,1\,0\,1$. Convert this to its decimal equivalent, assuming a straight binary code.

KNOWN

4-bit register

FIND

Convert to base 10

ASSUMPTION

Straight binary code

SOLUTION

The content of the 4-bit register is the binary word 0 1 0 1. This will represent

$$0 \times 2^3 + 1 \times 2^2 + 0 \times 2^1 + 1 \times 2^0 = 5$$

The equivalent of 0 1 0 1 in decimal value is 5.

7.4   TRANSMITTING DIGITAL NUMBERS: HIGH AND LOW SIGNALS

Electrical devices indicate binary code by using differing voltage levels. Because a bit can have a value of a 1 or 0, the presence of a particular voltage (call it HIGH) at a junction could be used to represent a 1-bit value, whereas a different voltage (LOW) could represent a 0. Most simply, this voltage can be affected by use of an open or closed switch using a flip-flop (Chapter 6), such as depicted in Figure 7.5(*a*).

For example, a common method used in computers is a +5 V form of TTL, which uses a nominal +5 V HIGH/0 V LOW scheme for representing the value of a bit. To avoid any ambiguity that is due to voltage fluctuations and to provide for higher definition, a specific range of voltage levels rather than exact values are accepted as representing a HIGH signal and a LOW signal. In TTL, the presence of a voltage level between +2 and +5.5 V might be taken as HIGH and therefore a bit value of 1, whereas a voltage between −0.6 and +0.8 V is taken as LOW or a 0 bit value. But this scheme is not unique. Many modem-to-computer communications rely on a scheme that accepts a positive voltage between +3 and +25 V to represent a HIGH value and a negative one between −3 and −25 V voltage range for a LOW value.

Binary numbers can be formed through a combination of HIGH and LOW voltages. Several switches can be grouped in parallel to form a register. Such an *M*-bit register forms a number based on the value of the voltages at its *M* output lines. This is illustrated in Figure 7.5(*b*) for a 4-bit register forming the number 1 0 1 0. So the bit values can be changed simply by changing the opened or closed state of the switches that connect to the output lines. This defines a *parallel* form in which all of the bits needed to form a word are available simultaneously. Another form is *serial*, in which the bits are separated in a time sequence of HIGH/LOW pulses, each pulse lasting for only one predefined time duration, δt. This is illustrated in the pulse sequence of Figure 7.5(*c*), which also forms the number 1 0 1 0.

Communication between components internal to a digital device is usually in a bit parallel, but word serial form. That is, groups of bits are transmitted one group at a time. Communication between digital devices may take this same form or be totally serial. Communication is discussed in detail later in this chapter.

7.5   VOLTAGE MEASUREMENTS

Digital measurement devices will consist of several components that interface the digital device with the analog world. In particular, the digital-to-analog converter

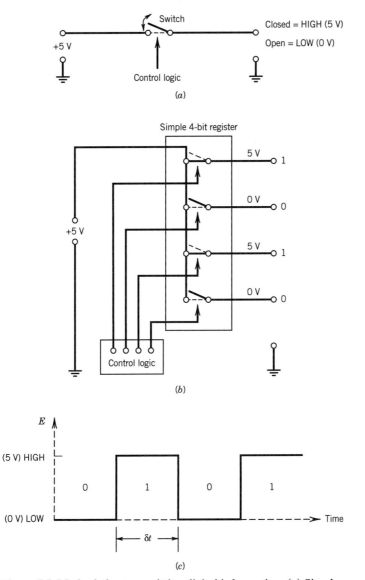

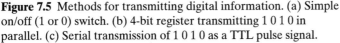

Figure 7.5 Methods for transmitting digital information. (a) Simple on/off (1 or 0) switch. (b) 4-bit register transmitting 1 0 1 0 in parallel. (c) Serial transmission of 1 0 1 0 as a TTL pulse signal.

and the analog-to-digital converter are discussed below. These form the major components of a digital voltmeter and a data-acquisition system.

Digital-to-Analog Converter

A digital-to-analog (D/A) converter is an M-bit digital device that converts a digital binary word into an analog voltage (e.g., [4–7]). One possible scheme uses an M-bit register with a weighted resistor network and operational amplifier, as depicted in Figure 7.6. The network consists of M binary weighted resistors having a common summing point. The resistor associated with the register MSB will have a value of R.

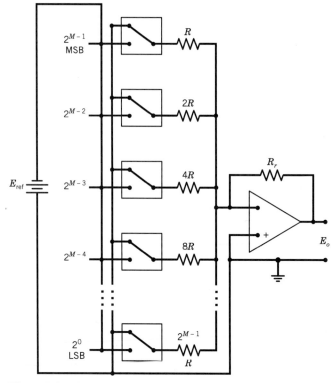

Figure 7.6 D/A converter.

At each successive bit the resistor value is doubled, so that the resistor associated with the LSB will have a value of $(2^{M-1}R)$. The network output is a current, I, given by

$$I = E_{\text{ref}} \sum_{m=1}^{M} \frac{c_m}{2^{m-1} R} \tag{7.11}$$

The values for c_m are either 0 or 1 depending on the associated mth bit value of the register that controls the switch setting. The output voltage from the amplifier is

$$E_o = I R_r \tag{7.12}$$

Note that this is equivalent to an operation in which an M-bit D/A converter would compare the magnitude of the actual input binary number, X, contained within its register to the largest possible number 2^M. This ratio determines E_o from

$$E_o = \frac{X}{2^M} \tag{7.13}$$

The D/A converter will have both digital and analog specifications, the latter expressed in terms of its full-scale analog voltage range output (E_{FSR}). Typical values for E_{FSR} are 0 to 10 V and ±5 V and for M are 8, 12, 16, and 18 bits.

Analog-to-Digital Converter

An analog-to-digital converter converts an analog voltage value into a binary number through a process called *quantization*. The conversion is discrete, taking place one number at a time.

The A/D converter is a hybrid device having both an analog side and a digital side. The analog side is specified in terms of a full-scale voltage range, E_{FSR}. The E_{FSR} defines the voltage range over which the device will operate. The digital side is specified in terms of the number of bits of its register. An M-bit A/D converter will output an M-bit binary number. It can represent 2^M different binary numbers. For example, a typical 8-bit A/D converter with an $E_{\mathrm{FSR}} = 10$ V would be able to represent analog voltages in the unipolar range between 0 and 10 V or the bipolar range between ± 5 V with $2^8 = 256$ different binary values. Principal considerations in selecting a type of A/D converter will include resolution, voltage range, and conversion speed.

Primary sources of error intrinsic to any A/D converter are the (a) resolution and associated quantization error, (b) saturation error, and (c) conversion error. Each will be discussed as follows.

Resolution

The resolution of an A/D converter is defined in terms of the smallest voltage increment that will cause a bit change. Resolution is specified in volts and determined by

$$Q = E_{\mathrm{FSR}}/2^M \tag{7.14}$$

Quantization Error

The analog input to digital output relationship for a 0–4 V, 2-bit A/D converter is represented in Figure 7.7. From equation (7.14), the resolution of this device is $Q = 1$ V. This limited resolution brings about the possibility of an error between the actual input analog value and the binary value assigned by the A/D converter. This error is referred to as the *quantization error*. For such a converter, an input voltage of 0.0 V would result in the same binary output of 00 as would an input of 0.9 V.

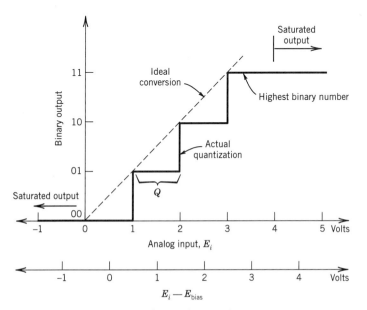

Figure 7.7 Binary quantization and saturation.

Table 7.2 Conversion Resolution

Bits M	Q^a [V]	SNR [dB]
2	2.5	12
4	0.625	24
8	0.039	48
12	0.0024	72
16	$0.15 \, (10^{-3})$	96
18	$0.0381 \, (10^{-3})$	108

^{a} Assumes $E_{\text{FSR}} = 10$ V.

Yet an input of 1.1 V results in an output of 0 1. Clearly, the output error is directly related to the resolution. This error behaves as noise imposed on the digital signal.

The resolution of the A/D converter represents the value of its LSB. With $Q = 1$ V in our example of Figure 7.7, 1 LSB is equivalent to 1 V. In this encoding scheme, e_Q is bounded between 0 and 1 LSB above E_i, so we estimate $e_Q = Q$. This scheme is widely used in digital readout devices.

However, a second common scheme makes the quantization error symmetric about the input voltage. For this, the analog voltage is shifted internally within the A/D converter by a bias voltage, E_{bias}, of an amount equivalent to 1/2 LSB. This shift is transparent to the user. The effect of such a shift is shown on the second lower axis in Figure 7.7. This now makes e_Q bounded by $\pm 1/2$ LSB about E_i, that is, $e_Q = \pm 1/2Q$. Regardless of the scheme used, the span of the quantization error remains 1 LSB and its effect is significant at small voltages.

A/D converter resolution is sometimes specified in terms of *signal-to-noise ratio* (SNR). The SNR relates the power of the signal, given by Ohm's law as E^2/R, to the power that can be resolved by quantization, given by $E^2/R2^M$. The SNR is just the ratio of these values. Defined in terms of the decibel this gives

$$\text{SNR [dB]} = 20 \log 2^M \tag{7.15}$$

The effect of bit number on resolution and SNR is given in Table 7.2.

Saturation Error

An A/D converter is limited by both the minimum and maximum analog voltage that it can convert. If either limit is exceeded, the A/D converter output will not be able to represent the voltage in a correct digital form. Instead, the output from the A/D converter becomes saturated and will no longer change with an increase in input level. As noted in Figure 7.7, an input to the 0–4 V, 2-bit A/D converter above 4 V results in an output of 11 and below 0 V of 0 0. A *saturation error* is defined by the difference between these input signal limits and the equivalent digital value assigned by the A/D converter. If an analog signal level is unbounded, then the saturation error associated with it will be unbounded. Saturation error can be avoided by quantizing signals that remain within the limits of the A/D converter. This may require analog signal conditioning, particularly for highly dynamic signals.

Conversion Error

The ability of any measurement device to output a magnitude representative of its input value determines the device precision and bias. An A/D converter is not immune to elemental errors arising during the conversion process that can lead to a misrepresentation of the input value. As with any device, the A/D converter *conversion errors* can be delineated into hysteresis, linearity, sensitivity, zero, and repeatability errors. The extent of such errors depends on the particular method of A/D conversion. Factors that contribute to conversion error include A/D converter settling time, signal noise during the analog sampling, temperature effects, and excitation power fluctuations [4–6].

EXAMPLE 7.4

Compute the relative effect of quantization error (e_Q/E_i) in the quantization of a 100-mV and a 1-V analog signal by using an 8-bit and a 12-bit A/D converter, both having a full-scale range of 0–10 V.

KNOWN

$E_i = 100$ mV and 1 V
$M = 8$ and 12
$E_{FSR} = 0$–10 V

FIND

e_Q/E_i, where e_Q is the quantization error.

SOLUTION

The resolutions for the 8-bit and 12-bit converters can be estimated from equation (7.14) as

$$Q_8 = \frac{E_{FSR}}{2^8} = \frac{10}{256} = 39 \text{ mV}$$

$$Q_{12} = \frac{E_{FSR}}{2^{12}} = \frac{10}{4096} = 2.4 \text{ mV}$$

Assume the A/D converter is designed so that the absolute quantization error is given by $\pm\frac{1}{2}Q$. The relative effect of the quantization error can be computed by e_Q/E_i. The results are tabulated as follows.

E_i	M	e_Q	$100 \times e_Q/E_i$
100 mV	8	±19.5 mV	19.5%
100 mV	12	±1.2 mV	1.2%
1 V	8	±19.5 mV	1.95%
1 V	12	±1.2 mV	0.12%

From a relative point of view, the error in converting a voltage is more significant at lower voltage levels.

EXAMPLE 7.5

The A/D converter with the following specifications listed is to be used in an environment in which the A/D converter temperature may change by ±10°C. Estimate the contributions of conversion and quantization errors to the uncertainty in the digital representation of an analog voltage by the converter.

	A/D Converter
E_{FSR}	0–10 V
M	12 bits
Linearity	±3 bits
Temperature drift	1 bit/ 5°C

KNOWN

See previous table

FIND

$(u_c)_E$ measured

SOLUTION

We can estimate the instrument uncertainty, u_c, as a combination of uncertainty caused by quantization errors, e_Q, and by conversion errors, e_C:

$$(u_c)_E = \sqrt{e_Q^2 + e_C^2}$$

The resolution of a 12-bit A/D converter with full-scale range of 0–10 V is given by

$$Q = \frac{E_{FSR}}{2^{12}} = \frac{10}{4096} = 2.4 \text{ mV}$$

The quantization error is found to be

$$e_Q = \pm\tfrac{1}{2}Q = \pm1.2 \text{ mV}$$

Now the conversion error is affected by two elements:

$$\text{linearity error} = \pm3 \text{ bits} \times 2.4 \text{ mV}$$
$$= \pm7.2 \text{ mV}$$
$$\text{temperature error} = e_2 = \frac{1 \text{ bit}}{5°C} \times 10°C \times 2.4 \text{ mV}$$
$$= \pm4.8 \text{ mV}$$

An estimate of the conversion error is found by using the RSS method:

$$e_C = \pm\sqrt{e_1^2 + e_2^2}$$
$$= \pm\sqrt{(7.2 \text{ mV})^2 + (4.8 \text{ mV})^2} = \pm8.6 \text{ mV}$$

The combined uncertainty in the digital representation of the analog value caused by these two errors is then

$$(u_c)_E = \pm\sqrt{(1.2 \text{ mV})^2 + (8.6 \text{ mV})^2}$$
$$= \pm8.7 \text{ mV} \quad (95\% \text{ assumed})$$

Note that the conversion errors dominate the uncertainty.

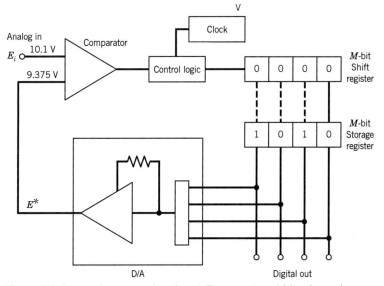

Figure 7.8 Successive approximation A/D converter; 4-bit scheme is shown with register = 1 0 1 0 with E_i = 10.1 V.

Successive Approximation Converters

There are a number of different methods to perform the A/D conversion [4, 6], but several common methods for converting voltage signals into binary words are now discussed.

The most common type of A/D converter uses the *successive approximation* technique. This technique uses a trial-and-error approach for estimating the input voltage to be converted. Basically, the successive approximation A/D converter guesses successive binary values as it narrows in on the appropriate binary representation for the input voltage. As depicted in Figure 7.8, this A/D converter uses an M-bit register to generate a trial binary number, a D/A converter to convert the register contents into an analog voltage, and a voltage comparator to compare the input voltage with the internally generated voltage. The conversion sequence is as follows.

1. The MSB is set to 1. All other bits are set to 0. This produces an equivalent value of E^* at the D/A converter output equivalent to the register setting. If $E^* > E_i$, the comparator goes LOW, causing the MSB to be reset to 0. If $E^* < E_i$, the MSB is kept HIGH at 1.
2. The second MSB is set to 1. Again, if $E^* > E_i$, it is reset to 0; otherwise its value remains 1.
3. The process continues through to the LSB. The final register value gives the quantization of E_i.

The process requires one clock tick per bit.

An example of this sequence is shown in Figure 7.9 for an input voltage of E_i = 10.1 V and using the 0–15 V, 4-bit successive approximation A/D converter of Figure 7.8. For this case, the converter has a resolution of 0.9375 V. The final register count of 1 0 1 0 or its equivalent of 9.375 V is the output from the A/D converter. Its value differs from the input voltage of 10.1 V as a result of the quantization error. It should be clear that this error can be reduced by increasing the bit size of the register.

	Register	E^*	E_i	Comparator
Initial status	0000	0	10.1	
MSB set to 1	1000	7.5	10.1	High
Leave at 1	1000	7.5		
Second MSB set to 1	1100	11.25		Low
Reset to 0	1000	7.5	10.1	
Third MSB set to 1	1010	9.375		High
Leave at 1	1010	9.375		
LSB set to 1	1011	10.3125	10.1	Low
Reset to 0	1010	9.375		

Figure 7.9 Example of successive approximation conversion sequence for a 4-bit converter.

The successive approximation converter is typically used when conversion speed at a reasonable cost is important. The number of steps required to perform a conversion equals the number of bits in the A/D converter register. With one clock tick per step, a 12-bit A/D converter operating with a typical 1-MHz clock would require a maximum time of 12 μs per conversion, but this also reveals the trade-off between increasing the number of bits to lower quantization error and the resulting increase in conversion time.

Sources of conversion error originate in the accuracies of the D/A converter and the comparator. Noise is the principle weakness of this type of converter, particularly at the decision points for the higher-order bits. The successive approximation process requires that the voltage remain constant during the conversion process. Because of this, a sample and hold circuit (SHC), discussed in Chapter 6, is used ahead of the converter input to measure and to hold the input voltage value constant throughout the duration of the conversion. The SHC also minimizes noise during the conversion.

Ramp Converters

High accuracy, low level (<1 mV) measurements usually rely on ramp converters. Ramp converters use the voltage level of a linear reference ramp signal to discern the voltage level of the analog input signal and convert it to its binary equivalent. Principal components, as shown in Figure 7.10, consist of an analog comparator, ramp function generator, and counter and M-bit register. The reference signal, initially at zero, is increased at set time steps, within which the ramp level is compared with the input voltage level. This comparison is continued until the two are equal.

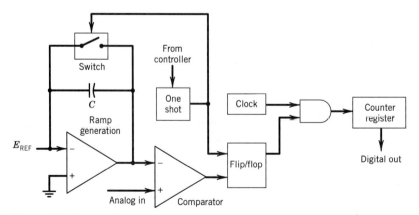

Figure 7.10 Ramp A/D converter.

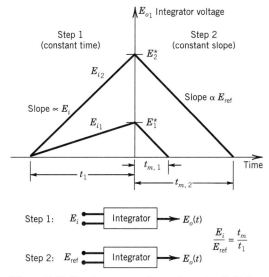

Figure 7.11 Dual-ramp analog voltage to digital conversion.

The usual method for generating the ramp signal is to use a capacitor of known fixed capacitance, C, which is charged from a constant current source of amperage, I. Because the charge is related to current by

$$q = It \tag{7.16}$$

the reference ramp voltage will be linearly related to the elapsed time by

$$E_{\text{ref}} = \frac{q}{C} = \text{constant} \times t \tag{7.17}$$

Time is integrated by a counter that increases the register value by 1 bit at each time step. Time step size depends on the value of 2^M. When the input voltage and ramp voltage magnitudes cross during a time step, the comparator output will go to zero, which flips a flip-flop, halting the process. The register count value will then indicate the digital binary equivalent of the input voltage.

The accuracy of this single-ramp operation is limited by the accuracy of the timing clock and the known values and constancy of the capacitor and the charging current. Increased accuracy can be achieved by using a dual-ramp converter, sometimes referred to as a dual-slope integrator, in which the measurement is accomplished in two steps, as illustrated in Figure 7.11. In the first step, the input voltage, E_i, is applied to an integrator for a fixed time period, t_1. The integrator output voltage increases in time with a slope proportional to the input voltage. At the end of the time interval, the output of the integrator has reached the level E^*. The second step in the process involves the application of a fixed reference voltage, E_{ref}, to the integrator. This step occurs immediately at the end of the fixed time interval in the first step, with the integrator voltage at exactly the same level established by the input. The reference voltage has the opposite polarity to the input voltage, and thus reduces the output of the integrator at a known rate. The time required for the output to return to zero is a direct measure of the input voltage. The time intervals are measured with a digital counter, which accumulates pulses from a stable oscillator. The input voltage is given by the relationship

$$E_i / E_{\text{ref}} = t_m/t_1 \tag{7.18}$$

where t_m is the time required for step two. Dual-ramp converter accuracy depends on the stability of the counter during conversion and requires a very accurate reference voltage.

Ramp converters are inexpensive but relatively slow devices. However, their integration process tends to average out any noise in the input signal. This feature is particularly attractive for measuring low-level signals. The maximum conversion time using a ramp converter is estimated as

$$\text{maximum conversion time} = 2^M/\text{clock speed} \tag{7.19}$$

For a dual-ramp converter, this time is doubled. If one assumes a normal distribution of input values are applied over the full range, the average conversion time can be taken to be one-half of the maximum value. For a 12-bit register and its 1-MHz clock required to count to full 4096 (2^{12}) pulses, a dual-ramp converter would require ~8 ms for a conversion.

Parallel Converters

The fastest type of A/D converter is the *parallel* or *flash* converter depicted in Figure 7.12. These converters are common to high-end stand-alone digital oscilloscopes and spectral analyzers. An M-bit parallel converter uses $2^M - 1$ separate voltage comparators to compare a reference voltage to the applied input voltage. As

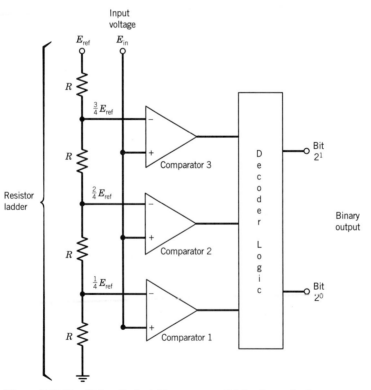

Figure 7.12 Parallel or flash A/D converter; 2-bit scheme is shown.

Table 7.3 Logic Scheme of a 2-Bit Parallel A/D Converter

Comparator States			Binary Output
HIGH	HIGH	HIGH	1 1
LOW	HIGH	HIGH	1 0
LOW	LOW	HIGH	0 1
LOW	LOW	LOW	0 0

indicated in Figure 7.12, the reference voltage applied to each successive comparator is increased by the equivalent of 1 LSB by using a voltage dividing resistor ladder. If the input voltage is less than its reference voltage, a comparator will go LOW; otherwise it will go HIGH.

Consider the 2-bit converter shown in Figure 7.12. If $E_{in} \geq (1/2)E_{ref}$ but $E_{in} < (3/4)E_{ref}$, then comparators 1 and 2 will be HIGH but comparator 3 will be LOW. In this manner, there can only be $2^M = 2^2$ different HIGH/LOW combinations from $2^2 - 1$ comparators, as noted in Table 7.3. The table shows how these combinations correspond to the 2^2 possible values of a 2-bit binary output. Logic circuits transfer this information to a register.

Because all the comparators act simultaneously, the conversion occurs in a single clock count. So the attraction of this converter is its speed, with typical sampling rates of 100 MHz or more. Its disadvantage lies in the cost associated with its $2^M - 1$ comparators and the associated logic circuitry. For example, an 8-bit converter will require 255 comparators.

Digital Voltmeters

The digital voltmeter must convert an analog input signal to a digital output. One can accomplish this conversion through the use of several basic techniques [6]. The most common method used in digital meters is with dual-ramp converters. The limits of performance for digital voltmeters can be significantly higher than for analog meters. They also have a high-input impedance reducing loading errors at low levels. The resolution (1 LSB) of these meters may be significantly better than their accuracy, so care must be taken in estimating uncertainty in their output. These devices are able to perform integration in both dc and true rms ac architecture.

EXAMPLE 7.6

A 0–10 V, three-digit digital voltmeter is built around a 10-bit, single-ramp A/D converter. A 100-kHz clock is used. An input voltage of 6.372 V is applied. What should be the digital voltmeter output value? How long will the conversion process take?

KNOWN

$E_{FSR} = 10$ V
$M = 10$
Clock speed $= 100$ kHz
$E_i = 6.372$ V

FIND

Digital output display value
Conversion time

SOLUTION

The resolution of the A/D converter is

$$Q = E_{FSR}/2^M = 10 \text{ V}/1024 = 9.77 \text{ mV}$$

Because each ramp step increases the counter value by 1 bit, the ramp converter will have a slope

$$\text{slope} = Q \times 1 \text{ bit/step} = 9.77 \text{ mV/step}$$

The number of steps required for conversion is

$$\text{steps required} = \frac{E_i}{\text{slope}} = \frac{6.372 \text{ V}}{0.00977 \text{ V/step}}$$
$$= 652.29 = 653 \text{ steps}$$

so the indicated output voltage is

$$E_o = \text{slope} \times \text{number of steps}$$
$$= 9.77 \text{ mV/step} \times 653 \text{ steps} = 6.3769 \text{ V} \approx 6.38 \text{ V}$$

A three-digit display will round E_o to 6.38 V. The difference between the input and output values is attributed to quantization error. With a clock speed of 100 kHz, each ramp step requires 10 μs/step, and

$$\text{conversion time} = 653 \text{ steps} \times 10 \text{ μs/step} = 6530 \text{ μs}.$$

COMMENT

A similar dual-ramp converter would require twice as many steps and twice the conversion time.

7.6 DATA-ACQUISITION SYSTEMS

A *data-acquisition system* is the portion of a measurement system that quantifies and stores data. There are many ways to do this. An engineer who reads a transducer dial, associates a number with the dial position, and records the information in a log book performs all of the tasks germane to a data-acquisition system. In this section, we focus on microprocessor-based data-acquisition systems, which are used to perform data quantification and storage.

Figure 7.13 shows how a data-acquisition system (DAS) might fit into the general measurement scheme between the actual measurement and the subsequent data reduction. A typical signal flow scheme is shown in Figure 7.14 for multiple input signals to a single microprocessor-based/controller DAS.

Dedicated microprocessor systems can continuously perform their programming instructions to measure, store, interpret, and provide process control without any intervention. Such microprocessors have I/O ports to interface with other devices to measure and to output instructions. Programming allows for operations such as, which

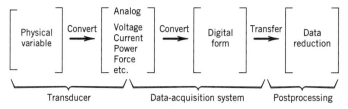

Figure 7.13 Typical analog signal and measurement scheme.

sensors to measure, and when and how often, and for data reduction. Programming can allow for decision-making and feedback to control process variables.

Personal computer-based data-acquisition systems are hybrid systems combining a data-acquisition package with both the microprocessor and human interface capability of a personal computer (PC). The interface between external instruments and the PC is done by using I/O plug-in boards, which mate either to an expansion slot in the computer or through an external board connected to the computer's bus. This provides direct access to the computer's bus, the main path used for all computer operations. Because the interior of a computer is electrically noisy, the external board is the preferred choice for low-level measurements. Cables connect between remote measuring devices and the plug-in board I/O ports. Software programs called "drivers" are required to drive the interface communication and are readily available.

Another approach is using a dedicated microprocessor. It is well suited to handling repetitive tasks involving measurement, recording, and control and portable. Direct serial or parallel communication with a host computer is possible. In contrast, availability and flexibility are the main attractions of a PC-based system. Not only can data be recorded, but the computer can be used to interface with control equipment, to make programmed decisions to control the measured process, to send data over a network line, to reduce the data into results, and to write the final report.

7.7   DATA-ACQUISITION SYSTEM COMPONENTS

Signal Conditioning: Filters and Amplification

Analog signals will usually require some type of signal conditioning for the proper interface with a digital system. Filters and amplifiers are the most common components used.

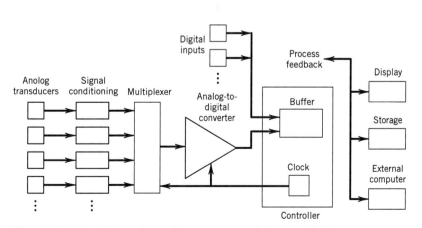

Figure 7.14 Signal flow scheme for an automated data-acquisition system.

Filters

Analog filters are used to control the frequency content of the signal being sampled. When signal frequency content is important, anti-alias analog filters are required to remove signal information above the Nyquist frequency before sampling. Not all data-acquisition boards contain on-board analog filters, so these necessary components are often overlooked.

Digital filters, which are available in most software packages, are effective for signal analysis and removing unwanted components from the sampled signal. However, they cannot be used to prevent aliasing or to remove its effects. A typical digital filtering scheme involves taking the Fourier transform of the sampled signal, multiplying the signal amplitude in the frequency domain by the desired frequency response (i.e., type of filter and desired filter settings), and transforming the signal back into the time domain by using the inverse Fourier transform [1, 2].

A simple digital filter is the *moving average* or *smoothing filter*, which is used for removing broadband noise. Essentially, this filter replaces a current data point value with an average based on a series of successive data point values. A center-weighted moving averaging scheme takes the form

$$y_i^* = (y_{i-n} + \cdots + y_{i-1} + y_i + y_{i+1} + \cdots + y_{i+n})/(2n + 1) \qquad (7.20)$$

where y_i^* is the averaged value that is calculated and used in place of y_i, and $(2n + 1)$ represents the number of successive values used for the averaging. For example, if $y_4 = 3$, $y_5 = 4$, and $y_6 = 2$, then a three-term average of y_5 produces $y_5^* = 3$.

In a similar manner, a forward-moving averaging smoothing scheme takes the form

$$y_i^* = (y_i + y_{i+1} + \cdots + y_{i+n})/(n + 1) \qquad (7.21)$$

and a backward-moving averaging smoothing scheme takes the form

$$y_i^* = (y_{i-n} + \cdots + y_{i-1} + y_i)/(n + 1) \qquad (7.22)$$

Light filtering might use three terms whereas heavy filtering might use 10 or more terms. This filtering scheme is easily accomplished in the data-analysis software or within a spreadsheet program.

Amplifiers

All automated data-acquisition systems are input range limited, that is, there is a minimum value of signal that they can resolve, and a maximum value that is below saturation. This is usually mandated by the A/D converter. Therefore, many transducer signals will need amplification or attenuation prior to conversion. Most data-acquisition systems contain on-board instrumentation amplifiers. For cost reasons, they may use a variation to the general layout of Figure 7.14, placing a single amplifier following the system multiplexer. This reduces the required number of amplifiers to just one, but the general purpose instrument amplifier used tends to be gain and bandwidth limited. Gain is varied manually by a resistor jumper or by logic switches set by software. The latter method allows the gain to vary by channel.

Although instrument amplifiers offer good output impedance characteristics, voltages larger than the input range for an A/D converter can also be attenuated using a voltage divider. The output voltage from the divider circuit of Figure 7.15 is

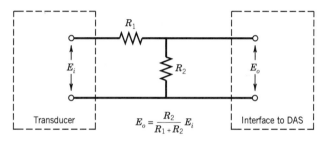

Figure 7.15 Voltage divider circuit for signal amplitude attenuation.

determined by

$$E_o = E_i \frac{R_2}{(R_1 + R_2)} \tag{7.23}$$

For example, a 0–50 V signal can be measured by a 0–10 V A/D converter using $R_1 = 40\,k\Omega$ and $R_2 = 10\,k\Omega$.

When only the dynamic content of time-dependent signals is important, amplification may require some strategy. For example, suppose the mean value of a voltage signal is large but the dynamic content is small, such as with the 5 Hz signal

$$y(t) = 2 + 0.05 \sin 10\pi t \ [V] \tag{7.24}$$

Setting the amplifier gain at $G = 5$ or above to improve the resolution of the dynamic content would saturate a 0–10 V A/D converter. In such situations, it is necessary to remove the mean component from the signal prior to amplification usually (1) by adding an equal, but of opposite sign, mean voltage to the signal, such as -2 V here, or (2) by passing the signal through a high-pass filter having a very low cutoff frequency, such as well below 5 Hz here.

Shunt Circuits

Many common transducers produce a current as their output; however, an A/D converter requires a voltage signal at its input. It is straightforward to convert current signals into voltage signals by using a shunt resistor. The circuit in Figure 7.16 provides a voltage

$$E_o = IR_{\text{shunt}} \tag{7.25}$$

for signal current I. For a standard 4–20 mA current loop, a $500\,\Omega$ shunt would provide a 2–10 V signal to a 10-V analog input.

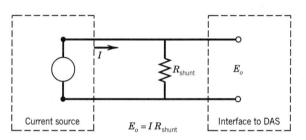

Figure 7.16 Simple shunt resistor circuit.

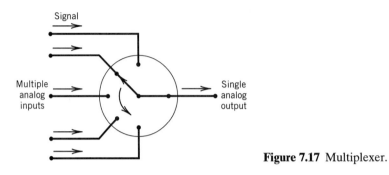

Figure 7.17 Multiplexer.

Analog Multiplexers

A multiplexer is simply a multiple-port switch that permits several analog input lines to be connected through to a common output line. This is best illustrated by the mechanical multiplexer depicted in Figure 7.17. Each analog signal enters through an individual switch port that is in-turn sequentially connected to the output line at a switching rate determined by the conversion timing control logic. An integrated circuit multiplexer behaves in a similar manner except that it uses parallel flip-flops to connect several analog inputs to a single output.

A/D Converters

High-speed data-acquisition boards use successive approximation converters with maximum conversion rates typically in the 1-kHz to 1-MHz range or parallel converters for rates up to and over 100 MHz. Low-level voltage measurements require the high noise rejection of the dual-ramp converter. Here the trade-off is in speed, with maximum conversion rates of 1–100 Hz more common.

D/A Converters

A D/A converter permits a DAS to convert digital numbers into voltages, which might be used for process control, such as to change a process variable or to drive a sensor positioning motor. The D/A signal is initiated by computer software.

Digital Input–Output

Digital signals represent a discrete state, either high or low. Digital input–output lines may be used to communicate between instruments (see Section 7.9), control relays (to power equipment on or off), or indicate the state or status of a device. A common way to transmit digital information is the use of a 5-V TTL signal (e.g., see Figure 7.5). If a signal larger than the TTL is needed, an amplifier can be used.

Digital I/O signals may be as a single state (HIGH or LOW) or as a series of pulses of HIGH/LOW states. The single-state signal might be used to operate a switch or relay. A series of pulses transmits a data series. Gate pulse counting entails counting pulses that occur over a specified period of time. This enables frequency determination (number of pulses/unit time) and counting/timing applications. Pulse stepping, sending a predetermined number of pulses in a series, is used to drive stepper motors and servos.

Central Processing Unit: Microprocessor

A common laboratory or industrial approach to data acquisition is built around the use of a personal computer. The microprocessor system of a PC consists of memory, control circuits, I/O peripherals, and a clock built around a *central processing unit* (CPU). The CPU controls the computer and sends and receives information via a bus. The CPU performs arithmetic operations, comparisons, logical operations, and data management by combining the contents of registers. Each register contains bits in blocks of eight, called *bytes*, or in multiple blocks forming words. The CPU moves bits between registers in such blocks. For example, a 32-bit CPU can access blocks of 32 bits representing 2^{32} different integer numbers.

The processing speed of the CPU is a function of both its clock speed and its bus size. The clock provides timing for each operation. A CPU can perform only so many operations within a clock tick. The actual number of operations depends on the particular microprocessor. The bus is an electrical pathway along which data and commands are sent to different areas of the computer and to any peripheral devices connected to the bus. Bus size affects speed because a larger bit-sized bus allows more data to be transferred per operation. It also reflects computer power, as a larger bus allows the computer to address more memory. A microprocessor may have more than one bus to further enhance its speed. Thus, different microprocessors using similar clocks may differ in the time required for a given task.

Special purpose, direct application microprocessors can operate independently from a PC. These are found in many devices, including stand-alone data-acquisition systems, and are quite flexible in their use. To illustrate this class of device, Figure 7.18 shows the BASIC Stamp2[2] that is useful for many engineering applications involving sensors and devices [7]. This small 16-bit, 4-MHz battery-operated unit offers fully programmable I/O ports that are easily interfaced with 5-V TTL-level devices and, with a control card, are interfaced with non-TTL devices. This permits control of devices ranging from data registers, A/D converters, and alarms to sensors, servos, relays, and serial lines. Communication with the microprocessor is also available through a synchronous three-wire interface that can couple with a serial communication port, such as on a PC. It can be programmed by using the serial port to run interpreter-based software algorithms, which are stored in a nonvolatile, reprogrammable memory called EEPROM (see the Memory section that follows). The utility and portability of such microprocessors is remarkable, with applications ranging from multipurpose robots to remote sensing stations.

Memory

Memory provides a means to store and retrieve information. Memory consists of registers. Registers are locations for the storage of numbers, the basic form of information storage on a computer. Computers contain three types of memory: *Random access memory* (RAM), *read-only memory* (ROM), and *external memory*. Memory is measured in quantities of kilobytes (KB), megabytes (MB), or gigabytes (GB).

---

[2]BASIC Stamp is a registered trademark of Parallax, Inc.

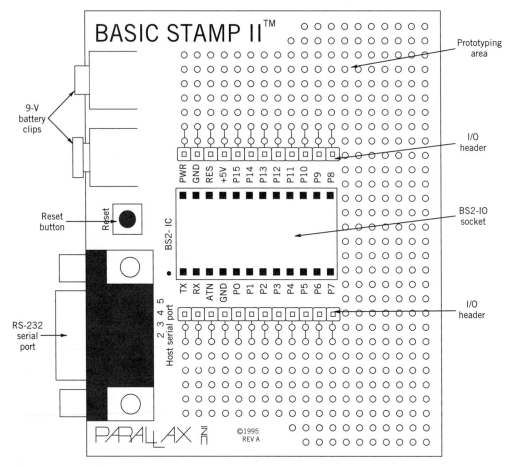

Figure 7.18 BASIC Stamp typical of special purpose, reprogrammable, multiport microprocessors.

RAM and ROM are both contained on computer chips. Access times to store or retrieve information are rapid, of the order of 10 ns. RAM is memory used to store programs and data during computer operation. RAM is reusable memory. It can be overwritten and normally is volatile; that is, it is erased at power shut down. ROM is dedicated memory typically used to store the *basic input–output system* (BIOS) instructions of a computer or a dedicated component algorithm. The BIOS contains tables and machine subroutines related to the most basic of operations. Variations of these memory forms include *electrically erasable programmable ROM* (EEPROM), which is nonvolatile memory that may be repeatedly reprogrammed, a useful feature for controlling devices. External memory refers to peripheral storage devices, such as hard disk drives, floppy, tape, and zip drives. Access times to external memory are much longer by comparison.

Central Bus

The path for communication between the CPU and memory and peripheral devices is the central bus. As depicted in Figure 7.19, the "central bus" is a collective term for three information paths: the address bus, the data bus, and the control bus. It can be considered as a three-lane highway, one lane for each bus, connecting all devices

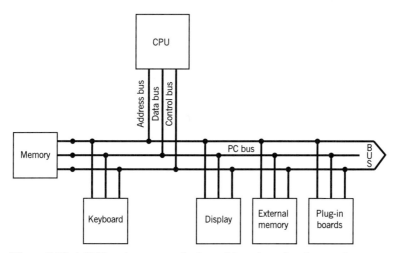

Figure 7.19 A PC bus is composed of an address bus, data bus, and control bus.

that reside along branch roads on the highway. When data or information is to be sent between devices, the address bus will contain the address of the device and the location within that device where the information is to be stored. The data bus will contain the execution instructions or the data being relayed. The control bus sends status information over the central bus. While information is being sent between any two devices connected to the central bus, a busy flag is sent out over the control bus to alert all other devices that the data bus is busy. This allows devices of different speeds to operate over the same lines.

Buffers

Buffers are small amounts of preassigned RAM memory to serve as a sort of holding tank. They are used whenever data must be exchanged between two devices that operate at different speeds. Data are loaded from the source device into the buffer memory, where they are held until extracted by the receiving device. During data acquisition, it is not uncommon to send data to a buffer while awaiting subsequent transfer to a disk for storage. Because transfer from or to a buffer may be slower than the data flow to or from the buffer, it may be necessary to interrupt data acquisition temporarily to avoid buffer overfill. This is done by the use of a status signal sent over the central bus.

7.8   ANALOG INPUT–OUTPUT COMMUNICATION

Data-Acquisition Boards

Analog interfacing with a computer is most often affected by using a DAS I/O board. Typical units are available in the form of an expansion plug-in board or a PCMCIA (Personal Computer Memory Card International Association) card, so the discussion narrows on these devices. A layout of a typical board is given in Figure 7.20. These boards use an expansion slot on the computer to interface with the computer bus. Field wiring from transducers or other analog equipment is usually made to a screw terminal board with subsequent connection directly to the rear of the I/O board. A

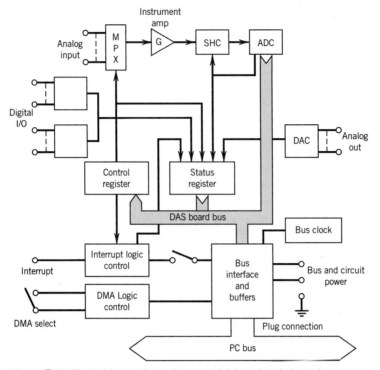

Figure 7.20 Typical layout for a data-acquisition plug-in board.

typical board allows for data transfer both to and from the computer memory by using A/D conversion, D/A conversion, digital I/O, and counter–timer ports. A multipurpose, multichannel high-speed data-acquisition board is shown in Figure 7.21.

The signal flow for plug-in boards is nearly universal. The input lines pass through a multiplexer, which connects through an amplifier and a SHC to an A/D converter. Amplifier gain is set either manually, by a resistor jumper such as in Figure 7.22, or by a software-controlled switch.

For example, the board in Figure 7.21 uses a 16-channel multiplexer and instrument amplifier. With its 12-bit successive approximation A/D converter with an 8–9 ms conversion rate and its 800-ns sample-hold time, sample rates of up to 100,000 Hz are possible. Input signals may be unipolar (e.g., 0–10 V) or bipolar (e.g., ±5 V). For the board shown, this allows input resolution for the 12-bit converter with an $E_{FSR} = 10$ V range of approximately

$$\frac{E_{FSR}}{2^M} = \frac{10 \text{ V}}{2^{12}} = 2.44 \text{ mV}$$

The amplifier permits signal conditioning with gains of $G = 0.5$–1000. This improves the minimum detectable voltage for the 10 V, 12-bit A/D converter to

$$\frac{E_{FSR}}{(G)(2^M)} = \frac{10 \text{ V}}{(1000)(2^{12})} = 2.44 \text{ } \mu\text{V}$$

at maximum gain.

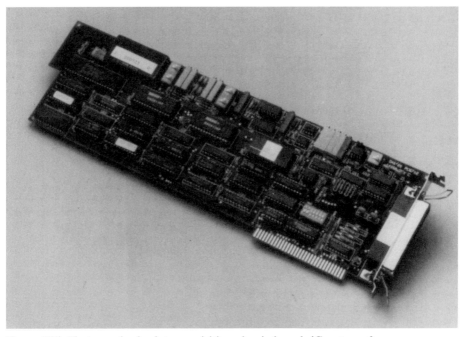

Figure 7.21 Photograph of a data-acquisition plug-in board. (Courtesy of Keithley-Metrabyte.)

Digital I/O can usually be accomplished through these boards for instrument control applications and external triggering. For the board shown, 16-TTL compatible lines are available, giving one 8-bit output and one 8-bit input line.

Single- and Differential-Ended Connections

Analog signal input connections to a DAS board may be single or differential ended. *Single-ended connections* use only one signal line (+ or HIGH) that is measured

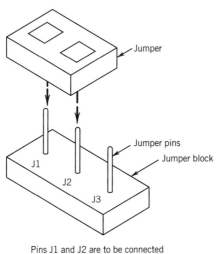

Pins J1 and J2 are to be connected

Figure 7.22 A resistor jumper block.

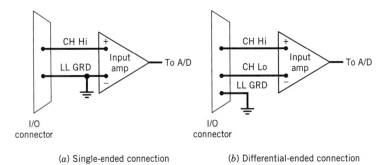

(a) Single-ended connection   (b) Differential-ended connection

Figure 7.23 Single- and differential-ended connections.

relative to ground (GRD), as shown in Figure 7.23. The return line (− or LOW) and ground are connected. There is a common external ground point, usually through the DAS board. Multiple single-ended connections to a DAS board are shown in Figure 7.24. Single-ended connections are suitable only when all of the analog signals can be made relative to the common ground point. There should be no additional local ground at the signal source. Single-ended connecting wires should be short, as they are susceptible to EMI noise.

Why a concern over ground points? Electrical grounds are not all at the same value (See Grounds and Ground Loops, Chapter 6). So if a signal is grounded at two different points, such as at its source and at the DAS board, the grounds could be at different voltage levels. When a grounded source is wired as a single-ended connection, the difference between the source ground voltage and the board ground voltage gives rise to a *common-mode voltage* (CMV). The CMV will combine with the input signal, superimposing interference and noise on the signal. This effect is referred to as a "ground loop." The measured signal will be unreliable.

Differential-ended connections allow the voltage difference between two distinct input signals to be measured. Here the signal input (+ or HIGH) line is paired with a signal return (− or LOW) line, which is isolated from ground (GRD), as shown in Figure 7.23. By the use of twisted wire pairs, the effects of noise are greatly reduced. A

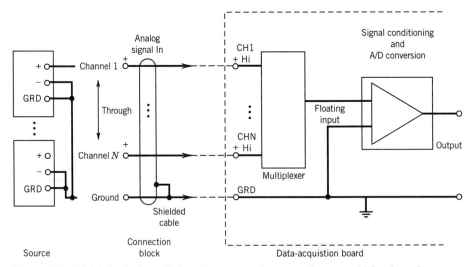

Figure 7.24 Multiple single-ended analog connections to a data-acquisition board.

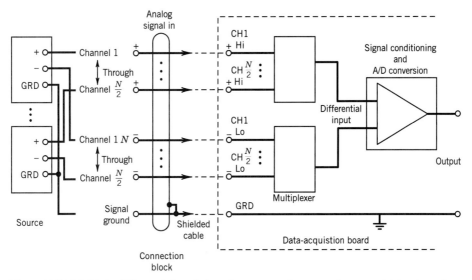

Figure 7.25 Multiple differential-ended analog connections to a data-acquisition board.

differential-ended connection to a DAS board is shown in Figure 7.25. The measured voltage is the voltage difference between these two (+ and −) lines for each channel. Such a connection is also referred to as "floating" because the difference is not measured relative to ground.

When signals from various instruments are connected to a single DAS board, differential-ended connections are usually required. This is also the preferred way to measure low-level signals. However, for low-level measurements, a 10k–100k Ω resistor should be connected between the signal return (− or LOW) line and ground (GRD) at the DAS board. The differential-ended connection is less prone to CMV errors, but be careful—if the measured voltage exceeds the board's common-mode range limit specification, you can damage it!

Special Signal Conditioning Modules

Signal conditioning modules exist for different transducer applications. These connect between the transducer and the data-acquisition card. For example, resistance bridge modules allow the direct interfacing of strain gauges or other resistance sensors through an on-board Wheatstone bridge, such as depicted in Figure 7.26. Temperature modules allow for electronic thermocouple cold junction compensation and can provide for signal linearization for reasonably accurate (down to ±0.5°C) temperature measurements.

Data-Acquisition Triggering

With DAS boards, data acquisition can be triggered by software command, external pulse, or on-board clock. The acquisition mode is sequential, one channel at a time. The channel order of acquisition can be programmed on most boards. Normally, measured signals on adjacent channels are separated by the sample time increment, δt. However, a multichannel SHC placed ahead of the multiplexer allows the simultaneous sampling of several channels.

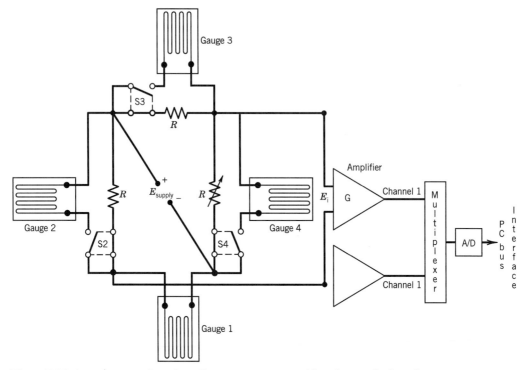

Figure 7.26 A strain-gauge interface. Gauges are connected by wires to the interface.

Data Transfer

Transfer of the data through the analog interface board can take place by software or program control, interrupt mode, or direct memory access (DMA).

In software–program control, a control program instructs the computer when to acquire data. This is also the mode commonly used when the operator controls data acquisition directly from the keyboard.

In the interrupt mode, the analog board sends a request for service along the interface bus to the CPU. This causes the CPU to interrupt its current operations and to implement a service subroutine to access and acquire the information. The CPU controls the data transfer. The interrupt mode is used when data are not being acquired continuously. As an example, computer printers use the interrupt mode to transfer data out of memory onto paper.

In direct memory access the CPU sets up a DMA controller to allow information to flow directly and continuously from the A/D converter into allocated memory. DMA data transfer provides the highest possible sample rates for data acquisition.

7.9 DIGITAL INPUT–OUTPUT COMMUNICATION

Certain standards exist for the manner in which digital information is communicated between digital devices [5, 6]. Serial communication methods transmit data bit by bit. Parallel communication methods transmit data in simultaneous groups of bits, for example, byte by byte. Both methods use a *handshake*, an interface procedure, which controls the data transfer between devices and TTL-level signals. Most lab equipment can be equipped to communicate by at least one of these means, and standards have been defined to codify communications between devices of different manufacturers.

Serial Communications

RS-232C

The RS-232C protocol was initially set up to translate signals between telephone lines and computers via a *modem* (*mo*dulator-*dem*odulator). The original protocol has been adapted as an interface for communication between a computer and a non-modem device, such as a printer, a digital measuring instrument, or another computer. Basically, the protocol allows two-way communication by using two single-ended signal (+) wires, noted as TRANSMIT and RECEIVE, between data-communications equipment (DCE), such as the modem, and data-terminal equipment (DTE), such as the computer. These two signals are analogous to a telephone's mouthpiece and earpiece signals. A signal GROUND wire allows signal return (−) paths. The remaining wires in the original standard are used to access the state of the telephone lines, if needed.

Most PC computers have an RS-232C compatible I/O port. The popularity of this interface is due to the wide range of equipment that can utilize it. Either a 9-pin or 25-pin connector can be used. The full connection protocol is shown in Figure 7.27. Devices may be separated by distances up to ~15 m by using a well-shielded cable. Communications can be half-duplex or full duplex. Half-duplex allows one device to transmit while the other receives. Full-duplex allows for both devices to transmit simultaneously.

The RS-232C is a rather loose standard, and there are no strict guidelines on how to implement the handshake. As such, compatibility problems will sometimes arise. The minimum number of wires required between DTE and DCE equipment is the three-wire connection shown in Figure 7.28. This connects only the TRANSMIT, RECEIVE, and GROUND lines while bypassing the handshake lines. The handshaking lines are jumpered to fool either device into handshaking with itself.

Communication between similar equipment, DTE to DTE or DCE to DCE, requires only the nine lines connected as shown in Figure 7.29, so a 9-pin connector can be used. The 9-pin connector wiring scheme is shown in Figure 7.30.

25–pin Number (DTE)	Description
1	Protective ground
2	Transmitted data (TD)
3	Received data (RD)
4	Request to send (RTS)
5	Clear to send (CTS)
6	Data set ready (DSR)
7	Signal ground (GRD)
8	Data carrier detect (DCD)
15	Transmit signal element timing (TSET)
17	Receive signal element timing (DTR)
20	Data terminal ready (DTR)
22	Ring indicator (RI)

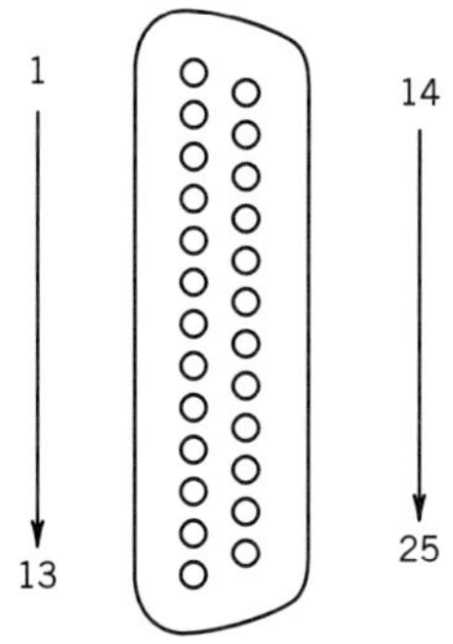

Figure 7.27 Standard RS-232C assignments to a 25-pin connector.

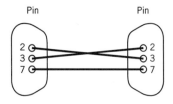

Figure 7.28 Minimum serial connections between DTE and DCE or DTE and DTE equipment.

Serial communication implies that data are sent in successive streams of information, 1 bit at a time. The value of each bit is represented by an analog voltage pulse with a 1 and 0 distinguished by two equal voltages of opposite polarity in the range of 3–25 V. Communication rates are measured in *baud*, which refers to the number of signal pulses per second. A typical transmission is 10 serial bits composed of a start bit followed by a 7-or 8-bit data stream, either 1 or no parity bit, and terminated by 1 or 2 stop bits. The start and stop bits form the serial "handshake" at the beginning and end of each data byte. Asynchronous transmission means that information may be sent at random intervals. Thus the start and stop bits are signals used to initiate and to end the transmission of each byte of data transmitted. The start bit allows for synchronization of the clocks of the two communicating devices. The parity bit allows for limited error checking. *Parity* involves counting the number of 1's in a byte. In one byte of data, there will be an even or odd number of bits with a value of 1. An additional bit added to each byte to make the number of 1 bits a predetermined even or odd number is called a parity bit. The receiving device will count the number of transmitted bits checking for the predetermined even (even parity) or odd (odd parity) number.

Devices that use data buffers will use a software handshaking protocol such as XON/XOFF. With this, the receiving device will transmit an XOFF signal to halt transmission as the buffer nears full and an XON signal when it has emptied and is again ready to receive.

RS-422A/423A/449/485

RS-422A/423A provide recommendations for the electrical interface and RS-449 specifies functional and mechanical characteristics of a serial communication standard. Equipment designed for RS-232C can be used with the new standards with slight modification. Using a 37-pin connector and differential-ended connections to

9–pin
Number   Description

1	Data carrier detect (DCD)
2	Transmitted data (TD)
3	Received data (RD)
4	Data terminal ready (DTR)
5	Signal ground (GRD)
6	Data set ready (DSR)
7	Request to send (RTS)
8	Clear to send (CTS)
9	Ring Indicator (RI)
Shell	Chassis ground

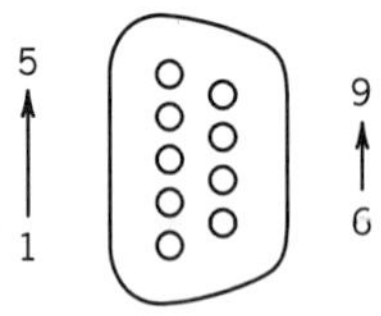

Figure 7.29 Standard serial connections between DTE and DTE or DCE and DCE equipment.

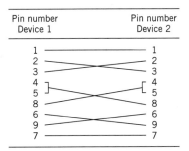

Pin number   Pin number
Device 1   Device 2

Figure 7.30 Nine-wire serial connection between DTE and DTE or DCE and DCE equipment.

reduce noise, they allow for communication at rates up to 2-M baud and distances up to 1000 m. The standard uses a +5 V TTL pulse signal to distinguish between a 1 bit (+2 to +5.5 V) and a 0 bit (−0.6 to +0.8 V).

The RS-485 protocol allows for multidrop (allows up to 32 to 255 devices on one line) operation that is well suited to local area networks (LANs). It communicates in half-duplex along a two-wire bus, which makes it slower than RS-422.

Universal Serial Bus

The *universal serial bus* (USB) is specified to become the industry-standard extension to PC architecture. The intention of this new standard is to permit peripheral expansion for up to 128 devices and to support low to medium transfer rates from 1.5 Mbs up to 12 Mbs. It supports a "hot swap" feature that allows the user to plug in a USB-compatible device and be able to use it without reboot. This bus connects USB devices to a single computer host through a USB root hub.

The USB physical interconnect is a tiered star topology (Figure 7.31). In this setup, the root hub permits one to four attachments, which can be a combination of USB peripheral devices and additional USB hubs. Each successive hub can in turn

Pin number	Description
Data lines	
1	Digital I/O data line 1
2	Digital I/O data line 2
3	Digital I/O data line 3
4	Digital I/O data line 4
13	Digital I/O data line 5
14	Digital I/O data line 6
15	Digital I/O data line 7
16	Digital I/O data line 8
Handshake lines	
6	Data valid (DAV)
7	Not ready for data (NRFD)
8	Not data accepted (NDAC)
Bus management lines	
5	End or identify (EOI)
9	Interface clear (IFC)
10	Service request (SRQ)
11	Attention (ATN)
17	Remote enable (REN)
Ground lines	
12	Shield
18–24	Ground

Figure 7.31 GPIB bus assignments to a 25-pin connector.

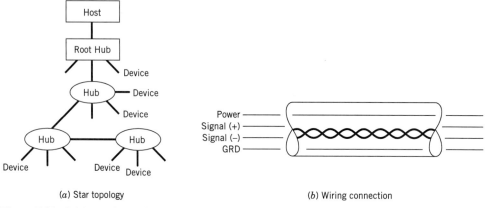

(a) Star topology   (b) Wiring connection

Figure 7.32 Cable configuration for the Universal Serial Bus.

support up to four devices or hubs. The cable length between a device and hub or between two hubs is limited to 5 m. The connecting cable (Figure 7.31) is a four-line wire consisting of a hub power line, two signal lines (+ and −), and a ground (GRD) line.

Parallel Communications

GPIB (IEEE-488)

The *general purpose interface bus* (GPIB) is a high-speed parallel interface originally developed by Hewlett-Packard. The GPIB is usually operated under the IEEE-488 communication standard. The bus allows for the control of other devices through a central controller (e.g., PC computer), and it allows devices to receive or transmit information from or to the controller. This standard is well defined and widely used to interface communication between computers and printers and scientific instrumentation.

The IEEE-488 standard for the GPIB operates from a 16-wire bus with a 24-wire connector (Figure 7.32). A 25-pin connector is standard. The bus is formed by eight data lines plus eight lines for bus management and handshaking (two-way control communication). The additional eight lines are used for grounds and shield. Bit parallel, byte serial communication at data rates up to 1 Mbyte/s are possible, with 1 to 10 kbytes/s most common. Connector lines are limited to a length of roughly 4 m.

The standard requires a controller, a function usually served by the laboratory computer, and permits multidrop operation, allowing up to 15 devices to be attached to the bus at any time. Each device has its own bus address (addresses 1–14). The bus controller (address 0) controls all the bus activities and sequences all communications to and between devices, such as which bus device transmits or receives and when. This is done along the bus management and handshaking lines. The communication along the bus is bidirectional. Normally, ground true TTL logic (≤0.8 V HIGH, ≥2 V LOW) is used.

The IEEE-488 standard specifies the following.

Data bus: This is an 8-bit parallel bus formed by the eight digital I/O data lines (lines 1–4 and 13–16). Data are transmitted as one 8-bit byte at a time. The handshake and bus management lines communicate through the transmission of a HIGH or LOW signal along the line. A HIGH signal asserts a predetermined situation.

Handshake bus: A three-wire handshake is specified. The *Data Valid* (DAV) line (line 6) asserts that data is available on the data bus and is valid. The *Not Ready for Data* (NRFD) line (line 7) is asserted by a device until it is ready to receive data or instructions. The *Not Data Accepted* (NDAC) line (line 8) is asserted by a device until it has accepted a set of data. To illustrate the handshake concept, consider the handshake between a controller (computer) and a measuring instrument. The controller unasserts NRFD, indicating that it is ready to accept data. When ready, the measuring instrument asserts DAV, indicating to the controller that valid data is being sent over the data bus. When the controller has read and accepted the data, it unasserts NDAC.

Bus management: There are five lines reserved to manage the flow of information on the data bus. *The Interface Clear* (IFC) line (line 9) is used by the controller to clear devices, such as during the initial boot. *The Service Request* (SRQ) line (line 10) is used by a device to assert that it is ready to be serviced by the controller, such as when it is ready to transmit data. The *Attention* (ATN) line (line 11) asserts that the data on the data bus are commands from the controller; devices are not to transmit. A low signal allows device messages onto the data line, such as the device data readings. The End or *Identify* (EOI) line (line 5) is used by the transmitting device to indicate the end of a data transmission. With the ATN line asserted, it is used by the controller to poll its devices to determine which device asserted its SRQ. The *Remote Enable* (REN) line (line 17) is used by the controller to place a device in its remote mode. When asserted, the front panels of the device are deactivated and the controller has command over device programming.

The interplay between the controller and the devices on the bus are controlled by software programs. Most scientific devices rely on software drivers; many of these are built around menu-driven programs, which are designed to provide some flexibility in the specific setup or operating parameters of the device. A parallel interface is faster and more efficient at data transfer than a serial interface.

EXAMPLE 7.7

A strain transducer has a static sensitivity of 2.5 V/unit strain ($2.5 \, \text{mV}/\mu\epsilon$) and requires 5 Vdc of power. It is to be connected to a DAS having a ± 5 V, 12-bit A/D converter and its signal measured at 1000 Hz. The transducer signal is to be amplified and filtered. For an expected measurement range of 1–500 $\mu\epsilon$, specify appropriate values for amplifier gain, filter type, and cutoff frequency, and show a signal flow diagram for the connections.

SOLUTION

The signal flow diagram is shown in Figure 7.33. Power is drawn off the data-acquisition board and routed to the transducer. Transducer signal wires are shielded and routed through the amplifier and filter, and they are connected to the data-acquisition board connector block by using twisted pairs to channel 0, as shown. The differential-ended connection at the board will reduce noise.

The amplifier gain is determined by considering the minimum and maximum signal magnitudes expected, and the quantization error of the DAS. The nominal signal magnitude will range from 2.5 μV to 1.25 mV. An amplifier gain of $G = 1000$ will boost this from 2.5 mV to 1.25 V. This lower value is of the order of the quantization

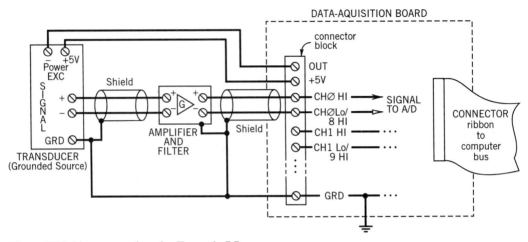

Figure 7.33 Line connections for Example 7.7.

error of the 12-bit converter. A gain of $G = 3000$ will raise the low end of the signal out of quantization noise while keeping the high end out of saturation.

With a sample rate of $f_s = 1000$ Hz, a general purpose anti-alias, low-pass Butterworth filter with 40 dB/decade roll-off and a cutoff frequency set at $f_N = 500$ Hz would meet the task.

EXAMPLE 7.8

The output from an analog device (nominal output impedance of 600 Ω) is input to the 12-bit A/D converter (nominal input impedance of 1 MΩ) of a data-acquisition system. For a 2-V signal, will interstage loading be a problem?

KNOWN

$$Z_1 = 600\,\Omega \qquad E_1 = 2\,\text{V}$$
$$Z_m = 1\,\text{M}\Omega$$

FIND

e_I

SOLUTION

The loading error is given by $e_I = E_1 - E_m$ is the measured voltage. From equation (6.39),

$$e_1 = E_1 \left(1 - \frac{1}{1 + Z_1/Z_m} \right)$$
$$= 1.2\,\text{mV}$$

Although not negligible, the interstage loading error at a 2-V input actually will be less than the quantization error of the 12-bit device.

EXAMPLE 7.9

Data acquisition is to be performed on an analog signal by using a sample rate of 200 Hz with a 12-bit A/D converter that has an input range of -10 to 10 V. The signal contains a 200-Hz component, f_1, with an amplitude of 100 mV. Specify a suitable LC filter that will attenuate the 200-Hz component down to the A/D converter quantization resolution level and will act as an antialias filter for quantization.

KNOWN

$f_s = 200$ Hz $f_1 = 200$ Hz
$M = 12$ $A_1 = 100$ mV
$E_{\mathrm{FSR}} = 20$ V

FIND

Specify a suitable filter

SOLUTION

From equation (7.8) for $f_s = 200$ Hz, f_N is 100 Hz, so an appropriate design for an anti-alias filter would have the properties of $f_c = 100$ Hz and $M\,(100\text{ Hz}) = -3$ dB.
    The A/D converter quantization resolution level is

$$Q = \frac{E_{\mathrm{FSR}}}{2^M} = \frac{20 \text{ V}}{4096} = 4.88 \text{ mV}$$

For a 100-mV signal at 200 Hz, this requires a low-pass Butterworth filter with an attenuation of

$$M(200 \text{ Hz}) = \frac{4.88 \text{ mV}}{100 \text{ mV}} = 0.0488 = -26 \text{ dB}$$
$$= [1 + (f/f_c)^{2k}]^{-1/2}$$

Setting $f_c = 100$ Hz yields $k = 4.3 \approx 5$. Appropriate values for L and C can be set from Table 6.1 and Figure 6.31.

EXAMPLE 7.10

To consider how the GPIB lines might be used, suppose we wish to use a computer to instruct a voltmeter to take a reading and display it. In this case, the computer acts as the controller and transmits commands to the voltmeter, the receiver. With bus addresses set by software to transmit address $= 0$, receive address $= 3$, the process begins with the Attention line, line 11, set HIGH (this halts activity while the bus devices listen to the computer command; here C = Computer; DV = voltmeter).
    C: Initialize all devices: Line 9 (IFC) HIGH then LOW.
    DV: Voltmeter Ready (to receive instructions): Line 7 (NRFD) LOW.
    C: Computer Sends Instructions: Line 5 (EOI) HIGH.
    The computer addresses the voltmeter. The eight parallel data lines become active as instructions are sent to the voltmeter. Instructions are 7- or 8-bit ASCII codes using digital I/O 1–8.
    DV: Voltmeter Accepts: Line 8 (NDAC) LOW.
    The voltmeter indicates it accepts the instructions.

DV: Voltmeter Active: Line 7 (NRFD) HIGH.

The voltmeter takes a reading. The line 7 status informs the computer that the voltmeter is busy so no new instructions should be sent.

DV: Voltmeter Ready: Line 7 (NRFD) LOW. The voltmeter now displays its new reading.

This signals that the voltmeter awaits new instructions. The next instruction might be to send the new reading along the data lines to be stored in computer memory.

7.10   SUMMARY

This chapter has focused on sampling concepts, the interfacing of analog and digital devices, and data-acquisition systems. Despite the advantages of digital systems, they exist in a mostly analog world, and the back-and-forth exchange between an analog and digital signal has limitations. With this in mind, a thorough discussion has been provided on the selection of sample rate and the number of measurements required to reconstruct a continuous process variable from a discrete representation. Equations (7.1) and (7.9) explain the criteria by which a periodic waveform can be accurately represented by such a discrete time series. The limitations resulting from improper sampling and the discrete representation include frequency alias and leakage, and amplitude ambiguity.

The fact that practically any electrical instrument can be interfaced with a data-acquisition system is significant. The working mechanics of A/D and D/A converters is basic to interfacing such analog and digital devices with data-acquisition systems. The limitations imposed by the resolution and range of these devices mandate certain signal conditioning requirements in terms of signal gain and filtering prior to acquisition. These require analog amplifiers and filters, devices discussed previously. The PC-based data-acquisition system is becoming more common in most facets of engineering. Its low cost and significant data-processing power provide for a highly versatile system.

REFERENCES

1. Bendat, J., and Piersol, A., *Random Data: Analysis and Measurement Procedures*, 2d ed., Wiley-Interscience, New York, 1986.
2. Carlson, G. E., *Signal and Linear System Analysis*, 2d ed., Wiley, New York, 1998.
3. Morrison, N., *Introduction to Fourier Analysis*, Wiley, New York, 1994.
4. Vandoren, A., *Data Acquisition Systems*, Reston (a Prentice-Hall company), Reston, VA, 1982.
5. Krutz, R., *Interfacing Techniques in Digital Design: Emphasis on Microprocessors*, Wiley, New York, 1988.
6. Hnatek, E., *A User's Handbook of D/A and A/D Converters*, Krieger, New York, 1988.
7. Parallax, *Basic Stamp Manual*, Parallex, Inc., Rocklin, CA, 1998.

Suggested Reading

Ahmed, H., and Spreadbury, P. J., *Analogue and Digital Electronics for Engineers*, 2d ed., Cambridge University Press, Cambridge, UK, 1984.

Beauchamp, K. G., and C. K. Yuen, *Data Acquisition for Signal Analysis*, Allen & Unwin, London, 1980.

Evans, A. J., *Basic Digital Electronics*, PROMPT Publishers, 1997.

Iotech, Inc., *Signal Conditioning and PC-Based Data Acquisition Handbook*, Iotech, Cleveland, 1998.

Johnson, J. H., *Build Your Own Low-Cost Data Acquisition and Display Devices*, TAB Books, New York, 1993.

Seitzer, D., Pretzl, G., and Hamdy, N., *Electronic Analog-to-Digital Converters: Principles, Circuits, Devices, Testing*, Wiley, New York, 1983.

NOMENCLATURE

e	error	E_i	input voltage [V]
e_Q	quantization error	E_o	output voltage [V]
f	frequency $[t^{-1}]$	E_{FSR}	full-scale analog voltage range
f_a	alias frequency $[t^{-1}]$	G	amplifier gain
f_c	filter cutoff frequency $[t^{-1}]$	I	electric current
f_m	maximum analog signal	M	digital component bit size
	frequency $[t^{-1}]$	$M(f)$	magnitude ratio at frequency, f
f_N	Nyquist frequency $[t^{-1}]$	N	data set size
f_s	sample rate frequency $[t^{-1}]$	$N\delta t$	total digital sample period
k	cascaded filter stage number	R	resistance $[\Omega]$
t	time	Q	A/D converter resolution [V]
$y(t)$	analog signal	δf	frequency resolution of DFT $[t^{-1}]$
$y_1, y(r\,\delta t)$	discrete values of a signal, $y(t)$	δR	change in resistance $[\Omega]$
$\{y(r\,\delta t)\}$	complete discrete time signal	δt	sample time increment $[t]$
	of $y(t)$	τ	time constant $[t]$
E	voltage [V]	$\phi(f)$	phase shift at frequency, f

PROBLEMS

For many of these problems, the use of spreadsheet software or the software files included on the companion disk (*FunSpect*, *DataSpect*, and *Sampling*) will facilitate solution, particularly in creating discrete series and in discrete Fourier analysis.

7.1 Convert the analog voltage, $E(t) = 5\sin 2\pi t$ mV, into a discrete time signal. Specifically, using sample time increments of (a) 0.125 s, (b) 0.30 s, and (c) 0.75 s, plot each series as a function of time over at least one period. Discuss apparent differences between the discrete representations of the analog signal.

7.2 Compute the amplitude spectrum for each of the three discrete signals in Problem 7.1. Discuss apparent differences. Use a data set of 128 points.

7.3 Compute the amplitude spectrum for the discrete time signal that results from sampling the analog signal, $T(t) = 2\sin 4\pi t$ °C, at sample rates of 4 and 8 Hz. Use a data set of 128 points. Discuss and compare your results.

7.4 Determine the alias frequency that results from sampling f_1 at rate f_s:
    a. $f_1 = 60$ Hz; $f_s = 90$ Hz
    b. $f_1 = 1.2$ kHz; $f_s = 2$ kHz
    c. $f_1 = 10$ Hz; $f_s = 6$ Hz
    d. $f_1 = 16$ Hz; $f_s = 8$ Hz

7.5 A particular data-acquisition system is used to convert the analog signal, $E(t) = (\sin 2\pi t + 2\sin 8\pi t)$ V, into a discrete time signal, using a sample rate of 16 Hz. Build the discrete time signal and from that use the Fourier transform to reconstruct the Fourier series.

7.6 Consider the continuous signal found in Example 2.3. What would be an appropriate sample rate and sample period to use in sampling this signal if the resulting discrete series must have a size of $2M$, where M is an integer and the signal is to be filtered at and above 2 Hz?

7.7 Convert the following straight binary numbers to positive integer base-10 numbers: (a) 1 0 1 0, (b) 1 1 1 1 1, (c) 1 0 1 1 1 0 1 1, and (d) 1 1 0 0 0 0 1.

7.8 Convert the following bipolar (ones-complement) binary numbers into integer base-10 numbers: (a) 0 1 1 1, (b) 1 0 0 1, (c) 0 1 1 1 1 1 1 1, and (d) 1 1 1 1 1 1 1 1.

7.9 Convert the following decimal (base 10) numbers into bipolar binary numbers by using a twos-complement code: (a) 10, (b) −10, (c) −247, and (d) 1013.

7.10 A PC does integer arithmetic in twos-complement binary code. How is the largest positive binary number represented in this code for an 8-bit byte? Add one to this number. What base-10 decimal numbers do these represent?

7.11 How is the largest negative binary number represented in twos-complement code for an 8-bit byte? Subtract one from this number. What base-10 decimal numbers do these represent?

7.12 List the possible sources of uncertainty in the dual-slope procedure for A/D conversion. Derive a relationship between the uncertainty in the digital result and the slope of the integration process.

7.13 Compute the resolution and SNR for an M-bit A/D converter having a full-scale range of ± 5 V. Let M be 4, 8, 12, and 16.

7.14 A 12-bit A/D converter having an $E_{FSR} = 5$ V has a relative accuracy of 0.03% full scale. Estimate its quantization error in volts. What is the total possible error expected in volts? What value of relative uncertainty might be used for this device?

7.15 An 8-bit single-ramp A/D converter with $E_{FSR} = 10$ V uses a 2.5-MHz clock and a comparator having a threshold voltage of 1 mV. Estimate the following: (a) the binary output when the input voltage is $E = 6.000$ V and when $E = 6.035$ V; (b) the actual conversion time for the 6.000-V input, and the maximum and average conversion times; and (c) the resolution of the converter.

7.16 Compare the maximum conversion times of a 10-bit successive approximation A/D converter to a dual-slope ramp converter if both use a 1-MHz clock and $E_{ESR} = 10$ V.

7.17 An 8-bit D/A converter produces an output of 3.58 V when straight binary 1 0 1 1 0 0 1 1 is applied. What is the output voltage when 0 1 1 0 0 1 0 0 is applied?

7.18 A 0–10 V, 4-bit successive approximation A/D converter is used to measure an input voltage of 4.9 V.

a. Determine the binary representation and its analog approximation of the input signal. Explain your answer in terms of the quantization error of the A/D converter.

b. If we wanted to ensure that the analog approximation was within 2.5 mV of the actual input voltage, estimate the number of bits required of an A/D converter.

7.19 Discuss the trade-offs between resolution and conversion rate for successive approximation, ramp, and parallel converters.

7.20 A 0–10 V, 10-bit A/D converter displays an output in straight binary code of 1 0 1 0 1 1 0 1 1 1. Estimate the input voltage to within 1 LSB.

7.21 A ± 5 V, 8-bit A/D converter displays an output in twos-complement code of 1 0 1 0 1 0 1 1. Estimate the input voltage to within 1 LSB.

7.22 A dual-slope A/D converter has 12-bit resolution and uses a 10-kHz internal clock. Estimate the conversion time required for an output code equivalent to 2011_{10}.

7.23 How long does it take an 8-bit single ramp A/D converter using a 1-MHz clock to convert the number 173_{10}?

7.24 A successive approximation A/D converter has a full-scale output of 0–10 V and uses an 8-bit register. An input of 6.2 V is applied. Estimate the final register value.

7.25 The voltage from a strain gauge balance scale of 0–5 kg is expected to vary from 0 to 3.50 mV. The signal is to be recorded by using a 12-bit A/D converter having a range of ± 5 V with the weight displayed on a computer monitor. Suggest an appropriate amplifier gain for this situation.

7.26 An aircraft wing will oscillate under wind gusts. The oscillations are expected to be at ~ 2 Hz. Wing mounted strain gauge sensors are connected to a ± 5 V, 12-bit A/D converter and data-acquisition system to measure this. For each test block, 10 s of data are sampled.

a. Suggest an appropriate sample rate. Explain.

b. If the signal oscillates with an amplitude of 2 V, express the signal with a Fourier series.

c. Based on a. and b., sketch a plot of the expected amplitude spectrum. What is the frequency spacing on the abscissa? What is its Nyquist frequency?

7.27 How many data points can be sampled by passing a signal through a 12-bit parallel A/D converter and stored in computer memory, if 8 MB of 32-bit computer memory is available? If the A/D converter uses a 100 MHz clock and acquisition is by DMA, estimate the duration of signal that can be measured.

7.28 Select an appropriate sample rate and data number to acquire with minimal leakage the first five terms of a square wave signal having a fundamental period of 1 s. Select an appropriate cutoff frequency for an anti-alias filter. Hint: approximate the square wave as a Fourier series.

7.29 A triangle wave with a period of 2 s can be expressed by the Fourier series

$$y(t) = \sum [2C_1(1 - \cos n\pi)/n\pi] \cos \pi n t \qquad n = 1, 2, \ldots$$

Specify an appropriate sample rate and data number to acquire the first seven terms with minimal leakage. Select an appropriate cutoff frequency for an antialias filter.

7.30 Using Fourier transform software (or equivalent software), generate the amplitude spectrum for Problem 7.28. Use $C_1 = 1$ V.

7.31 Using Fourier transform software (or equivalent software), generate the amplitude spectrum for Problem 7.29. Use $C_1 = 1$ V.

7.32 Design a low-pass Butterworth filter around a 10 Hz cut-off (-3 dB) frequency. The filter is to pass 95% of signal magnitude at 5 Hz and no more than 90% at 20 Hz. Source and load impedances are 10 Ω.

7.33 A two-stage LC Butterworth filter with $f_c = 100$ Hz is used to filter an analog signal. Determine the attenuation of the filtered analog signal at 10, 50, 75, 200 and 400 Hz.

7.34 A three-stage LC Bessel filter with $f_c = 100$ Hz is used to filter an analog signal. Determine the attenuation of the filtered analog signal at 10, 50, 75, and 200 Hz.

7.35 Design a cascading LC Butterworth low-pass filter that has a magnitude ratio flat to within 3 dB from 0 to 5 kHz but with an attenuation of at least 30 dB for all frequencies at and above 10 kHz.

7.36 A complex periodic analog signal to be sampled at 500 Hz is passed through a low-pass RC Butterworth filter rated at -3 dB at 250 Hz. What is the maximum frequency in the filtered signal for which the amplitude will be affected by a dynamic error of no more than 10%?

7.37 Choose an appropriate cascading low-pass Butterworth filter to remove a 500-Hz component contained within an analog signal that is to be passed through an 8-bit A/D converter having a 10-V range and 200-Hz sample rate. Attenuate the component to within the A/D converter quantization error.

7.38 The voltage output from a J-type thermocouple referenced to 0°C is to be used to measure temperatures of 50–70°C. The output voltages will vary linearly over this range from 2.585 to 3.649 mV.

a. If the thermocouple voltage is input to a 12-bit A/D converter having a ± 5-V range, estimate the percent quantization error in the digital value.

b. If the analog signal can be first passed through an amplifier circuit, compute the amplifier gain required to reduce the quantization error to 5% or less.

c. If the ratio of SNR level in the analog signal is 40 dB, compute the magnitude of the noise after amplification. Discuss the results of b in light of this.

7.39 Specify an appropriate ± 5-V M-bit A/D converter (8- or 12-bit), sample rate (up to 100 Hz) and signal conditioning to convert these analog signals into digital series. Estimate the quantization error and dynamic error resulting from the system specified:

a. $E(t) = 2 \sin 20\pi t$ V
b. $E(t) = 1.5 \sin \pi t + 20 \sin 32\pi t - 3 \sin(60\pi t + \pi/4)$ V
c. $P(t) = -10 \sin 4\pi t + 5 \sin 8\pi t$ kPa; $K = 0.4$ V/kPa

7.40 The following signal is to be sampled by using a 12-bit, ± 5-V data-acquisition board:

$$y(t) = 4 \sin 8\pi t + 2 \sin 20\pi t + 3 \sin 42\pi t \text{ V}$$

Select an appropriate sample rate and sample size that provides minimal spectral leakage.

7.41 Static pressures are to be measured at eight locations under the hood of a NASCAR race car. The pressure transducers to be used have an output span of ± 1 V for an input span of ± 25 cm H_2O. The signals are measured and recorded on a portable DAS, which uses a 10-bit, ± 5-V A/D converter. Pressure has to be resolved to within 0.25 cm H_2O. The dynamic content of the signals is important and has a fundamental period of ~ 0.5 s. The system has 4 MB of memory, powers all instruments, and has 10 min of usable battery life. Suggest an appropriate sample rate, total sample time, and signal conditioning for this application. Sketch a signal flow diagram through the measurement system.

7.42 The transducer in Problem 7.41 has a rated accuracy of 0.25%. Estimate the design stage uncertainty of the transducer–A/D converter system based on known information. Would a 12-bit converter improve things notably? If following measurements the engineer notes a 0.10 cm H_2O uncertainty caused by data scatter alone, what uncertainty dominates the measurement?

Chapter 8

Temperature Measurements

8.1 INTRODUCTION

Temperature is one of the most commonly used and measured engineering variables. Much of our lives is affected by the diurnal and seasonal variations in ambient temperature, but the fundamental, scientific definition of temperature and a scale for the measurement of temperature are not commonly understood. Although temperature is one of the most familiar engineering variables, it is unfortunately not easy to define. This chapter explores the establishment of a practical temperature scale and common methods of temperature measurement. In addition, errors associated with the design and installation of a temperature sensor are discussed.

Historical Background

Early exploration of thermodynamic temperatures was accomplished by a French scientist, Guillaume Amontons (1663–1705). His efforts examined the behavior of a constant volume of air that was subject to temperature changes. The modern liquid-in-glass bulb thermometer traces its origin to Galileo (1565–1642), who attempted to use the volumetric expansion of liquids in tubes as a relative measure of temperature. Unfortunately, this open tube device was actually sensitive to both barometric pressure and temperature changes, and thus could be called a "barothermoscope." A major advance in temperature measurement occurred in 1630 as a result of a seemingly unrelated event: the development of the technology to manufacture capillary glass tubes. These tubes were then used with water and alcohol in a thermometric device resembling the bulb thermometer, and eventually led to the development of a practical temperature measuring instrument.

A temperature scale proposed by Sir Isaac Newton (1642–1727) used the freezing point of water and the armpit temperature of a healthy man as extremes of temperature on a linear 0 to 12 scale. The scale proposed by Gabriel D. Fahrenheit, a German physicist (1686–1736), in 1715 attempted to incorporate body temperature as the median point on a scale having 180 divisions between the freezing point and the boiling point of water. Fahrenheit also successfully used mercury as the liquid in a bulb thermometer, making significant improvements over the attempts of Ismael Boulliau in 1659.

In 1742, the Swedish astronomer Anders Celsius[1] (1701–1744) described a temperature scale that divided the interval between the boiling and freezing points of

---

[1] It is interesting to note that in addition to his work in thermometry, Celsius published significant papers on the aurora borealis and the falling level of the Baltic Sea.

water at 1 atm of pressure into 100 equal parts. The boiling point of water was fixed as 0, and the freezing point of water as 100. Shortly after Celsius death. Carolus Linnaeus (1707–1778) reversed the scale so that the 0 point corresponded to the freezing point of water at 1 atm. Even though this scale may not have been originated by Celsius [1], in 1948 the change from degrees centigrade to degrees Celsius was officially adopted. It is also interesting to note that despite the many practical applications for temperature measurement, a practical temperature scale and measuring devices were initially developed as measurement tools for research. As stated by H. A. Klein in *The Science of Measurement: A Historical Survey*,[2]

> *From the original thermoscopes of Galileo and some of his contemporaries, the measurement of temperature has pursued paths of increasing ingenuity, sophistication and complexity. Yet temperature remains in its innermost essence the average molecular or atomic energy of the least bits making up matter, in their endless dance. Matter without motion is unthinkable. Temperature is the most meaningful physical variable for dealing with the effects of those infinitesimal, incessant internal motions of matter.*

We will begin our discussion of temperature by examining a method for measuring temperature, before attempting a precise definition.

8.2   TEMPERATURE STANDARDS AND DEFINITION

Temperature can be loosely described as the property of an object that describes its hotness or coldness, concepts that are clearly relative. Our experiences indicate that heat transfer tends to equalize temperature; or more precisely, systems that are in thermal communication will eventually have equal temperatures. The zeroth law of thermodynamics states that two systems in thermal equilibrium with a third system are in thermal equilibrium with each other. Although the zeroth law of thermodynamics essentially provides the definition of the equality of temperature, it provides no means for defining a temperature scale.

A *temperature scale* provides for three essential aspects of temperature measurement: (1) the definition of the size of the degree, (2) fixed reference points for establishing known temperatures, and (3) a means for interpolating between these fixed temperature points. These provisions are consistent with the requirements for any standard, as described in Chapter 1. To construct a temperature scale, the three aspects listed above for a temperature scale must be established.

Fixed-Point Temperatures and Interpolation

To begin, consider the definition of the triple point of water as having a value of 0.01 for our temperature scale, as is done for the Celsius scale (0.01°C). This provides for an arbitrary starting point for a temperature scale; in fact, the number value assigned to this temperature could be anything. On the Fahrenheit temperature scale it has a value very close to 32. Consider another fixed point on our temperature scale. Fixed points are typically defined by phase-transition temperatures or the triple point of a pure substance. The point at which pure water boils at one standard atmosphere pressure is an easily reproducible fixed temperature. For our purposes let's assign this fixed point a numerical value of 100.

---

[2]Dover, Mineola, NY, 1988.

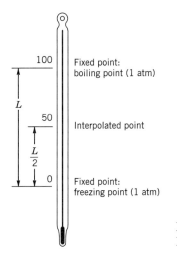

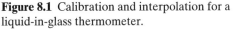

Figure 8.1 Calibration and interpolation for a liquid-in-glass thermometer.

The next problem is to define the size of the degree. Since we have two fixed points on our temperature scale, we can see that the degree is 1/100th of the temperature difference between the ice point and the boiling point of water at atmospheric pressure. Conceptually, this defines a workable scale for the measurement of temperature; however, as yet we have made no provision for interpolating between the two fixed-point temperatures.

The calibration of a temperature measurement device entails not only the establishment of fixed temperature points, but the indication of any temperature between fixed points. The operation of a mercury-in-glass thermometer is based on the thermal expansion of mercury contained in a glass capillary, where the level of the mercury is read as an indication of the temperature. Imagine that we submerged the thermometer in water at the ice point, made a mark on the glass at the height of the column of mercury, and labeled it 0°C, as illustrated in Figure 8.1.

Next we submerged the thermometer in boiling water, and again marked the level of the mercury, this time labeling it 100°C. Clearly we want to be able to measure temperatures other than these two fixed points. How can we determine the appropriate place on the thermometer to mark, say, 50°C?

The process of establishing 50°C without a fixed-point calibration is called interpolation. The simplest option would be to divide the distance on the thermometer between the marks representing 0 and 100 into equally spaced degree divisions. What assumption is implicit in this method of interpolation? It is obvious that we do not have enough information to appropriately divide the interval between 0 and 100 on the thermometer into degrees. Some theory of the behavior of the mercury in the thermometer, or many fixed points for calibration are necessary to resolve our dilemma. Even by the late 18th century, there was no standard for interpolating between fixed points on the temperature scale; the result was that different thermometers indicated different temperatures away from fixed points, sometimes with surprisingly large errors.

Temperature Scales and Standards

At this point, it is necessary to reconcile this arbitrary temperature scale with the idea of thermodynamic and absolute temperature. Thermodynamics defines a

Table 8.1 Temperature Fixed Points as Defined by ITS–90

Defining State	Temperature	
	K	C
Triple point of hydrogen	13.8033	−259.3467
Liquid–vapor equilibrium for hydrogen at 25/76 atm	≈17	≈ −256.15
Liquid–vapor equilibrium for hydrogen at 1 atm	≈20.3	≈ −252.87
Triple point of neon	24.5561	−248.5939
Triple point of oxygen	54.3584	−218.7916
Triple point of argon	83.8058	−189.3442
Triple point of water	273.16	0.01
Solid–liquid equilibrium for gallium at 1 atm	302.9146	29.7646
Solid–liquid equilibrium for tin at 1 atm	505.078	231.928
Solid–liquid equilibrium for zinc at 1 atm	692.677	419.527
Solid–liquid equilibrium for silver at 1 atm	1234.93	961.78
Solid–liquid equilibrium for gold at 1 atm	1337.33	1064.18
Solid–liquid equilibrium for copper at 1 atm	1357.77	1084.62

temperature scale that has an absolute reference, and defines an absolute zero for temperature. For example, this absolute temperature governs the energy behavior of an ideal gas and is used in the ideal gas equation of state. The behavior of real gases at very low pressure may be used as a temperature standard to define a practical measure of temperature that approximates the thermodynamic temperature.

The modern engineering definition of the temperature scale is provided by a standard called the International Temperature Scale of 1990 (ITS–90) [2]. This standard establishes fixed points for temperature and provides standard procedures and devices for interpolating between fixed points. Temperatures established according to ITS–90 do not deviate from the thermodynamic temperature scale by more than the uncertainty in the thermodynamic temperature at the time of adoption of ITS–90. The primary fixed points from ITS–90 are shown in Table 8.1. In addition to these fixed points, other fixed points of secondary importance are available in ITS–90.

Along with the fixed temperature points established by ITS–90, a standard for interpolation between these fixed points is necessary. In the range of interest for most engineering applications, accurate interpolation between fixed points is provided by the variation of resistance with temperature for a platinum wire. For temperatures ranging from 13.8033 to 1234.93 K, ITS–90 establishes a platinum resistance thermometer as the standard interpolating instrument, and establishes interpolating equations that relate temperature to resistance. Above 1234.93 K the temperature is defined in terms of blackbody radiation, without specifying an instrument for interpolation [2]. Prior to the adoption of ITS–90, the high-temperature interpolation standard was based on an optical pyrometer.[3]

In summary, temperature measurement, and a practical temperature scale and standards for fixed points and interpolation have evolved over a period of about two centuries. Present standards for fixed-point temperatures and interpolation allow for practical and accurate measurements of temperature. In the United States, the National Institute of Standards and Technology (NIST) provides for a means to obtain accurately calibrated platinum wire thermometers for use as secondary standards in the calibration of a temperature measuring system to any practical level of uncertainty.

8.3   THERMOMETRY BASED ON THERMAL EXPANSION

Most materials exhibit a change in size with changes in temperature. Since this physical phenomenon is well defined and repeatable, it is useful for temperature measurement. The liquid-in-glass thermometer and the bimetallic thermometer are based on this phenomenon.

Liquid-in-Glass Thermometers

A liquid-in-glass thermometer measures temperature by virtue of the thermal expansion of a liquid. The construction of a liquid-in-glass thermometer is shown in Figure 8.2. The liquid is contained in a glass structure that consists of a bulb and a stem. The bulb serves as a reservoir and provides sufficient fluid for the total volume

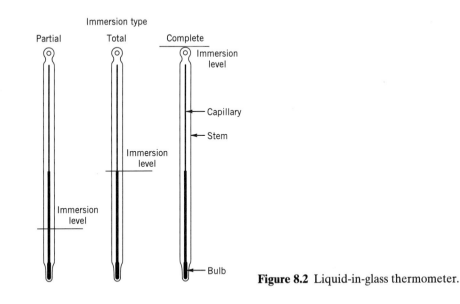

Figure 8.2 Liquid-in-glass thermometer.

---

[3]From 1968 through 1989 the International Practical Temperature Scale of 1968 (IPTS–68) established fixed temperature points and interpolation standards. The differences between IPTS–68 and ITS–90 are documented in [3], which provides the necessary information to assess the importance of changes for instruments calibrated under IPTS–68.

change of the fluid to cause a detectable rise of the liquid in the stem of the thermometer. The stem contains a glass capillary tube, and the level of the liquid in the capillary is an indication of the temperature. The difference in thermal expansion between the liquid and the glass produces a practical change in the level of the liquid in the glass capillary. Principles and practices of temperature measurement using liquid-in-glass thermometers are described in [4].

During calibration, such a thermometer is subject to one of three measuring environments:

1. For a *complete immersion thermometer*, the entire thermometer is immersed in the calibrating temperature environment or fluid.
2. For a *total immersion thermometer*, the thermometer is immersed in the calibrating temperature environment up to the liquid level in the capillary.
3. For a *partial immersion thermometer*, the thermometer is immersed to a predetermined level in the calibrating environment.

For the most accurate temperature measurements, the thermometer should be immersed in the same manner in use as it was during calibration. In practice, it may not be possible to employ the thermometer in exactly the same way as when it was calibrated. In this case, stem corrections can be applied to the temperature reading [5].

Temperature measurements using liquid-in-glass thermometers can provide accuracies to $\pm 0.01°C$ under very carefully controlled conditions; however, such extraneous variables as pressure and changes in bulb volume over time can introduce significant errors in scale calibration. For example, pressure changes increase the indicated temperature by approximately 0.1°C per atmosphere [6]. Practical measurements using liquid-in-glass thermometers typically result in total uncertainties that range from ± 0.2 to $\pm 2°C$, depending upon the specific instrument.

Bimetallic Thermometers

The physical phenomenon employed in a bimetallic temperature sensor is the *differential* thermal expansion of two metals. Figure 8.3 shows the construction and response of a bimetallic sensor to an input signal. The sensor is constructed by bonding two strips of different metals, A and B. The resulting bimetallic strip may be in a variety of shapes, depending upon the particular application. Consider the simple linear construction shown in Figure 8.3. At the assembly temperature, T_1, the bimetallic strip will be straight; however, for temperatures other than T_1 the strip will have a curvature. The physical basis for the relationship between the radius of curvature and temperature is given as

$$r_c \propto \frac{d}{[(C_\alpha)_A - (C_\alpha)_B](T_2 - T_1)} \tag{8.1}$$

where

r_c = radius of curvature
C_α = material thermal expansion coefficient
T = temperature
d = thickness

Bimetallic strips employ one metal having a high coefficient of thermal expansion with another having a low coefficient, providing increased sensitivity. Invar is often

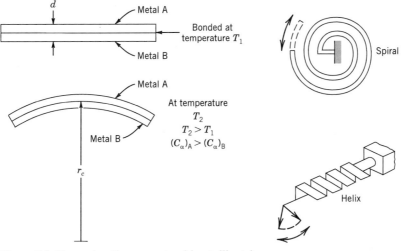

Figure 8.3 Expansion thermometry: bimetallic strip.

used as one of the metals, since for this material $C_\alpha = 1.7 \times 10^{-8}$ m/m °C, as compared to typical values for other metals, such as steels, which range from approximately 2×10^{-5} to 20×10^{-5} m/m °C.

The bimetallic sensor is used in many temperature control systems and is the primary element in most dial thermometers. The geometries shown in Figure 8.3 serve to provide the desired deflection in the bimetallic strip for a given application. Dial thermometers using a bimetallic strip as their sensing element typically provide temperature measurements with uncertainties of ± 1 °C.

8.4 ELECTRICAL RESISTANCE THERMOMETRY

As a result of the physical nature of the conduction of electricity, electrical resistance of a conductor or semiconductor varies with temperature. Using this behavior as the basis for temperature measurement is extremely simple in principle and leads to two basic classes of resistance thermometers: resistance temperature detectors (conductors) and thermistors (semiconductors). Resistance temperature detectors (RTDs) may be formed from a solid metal wire, which exhibits an increase in electrical resistance with temperature. The physical basis for the relationship between resistance and temperature is the temperature dependence of the resistivity of a material, ρ_e. The resistance of a conductor of length l and cross-sectional area A_c may be expressed in terms of the resistivity as

$$R = \frac{\rho_e l}{A_c} \tag{8.2}$$

Thermistors are semiconductor devices that display a very large decrease in resistance as temperature increases. Current manufacturing techniques provide thermistors that are stable and accurate enough to function as temperature sensors.

Resistance Temperature Detectors

In the case of a RTD the sensor is generally constructed by mounting a metal wire on an insulating support structure to eliminate mechanical strains and by encasing

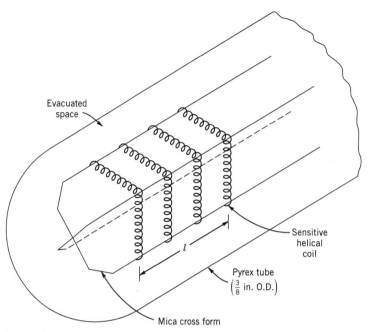

Figure 8.4 Construction of a platinum RTD. (From R. P. Benedict, *Fundamentals of Temperature, Pressure and Flow Measurements*, 3d ed., Wiley, New York, 1984.)

the wire to prevent changes in resistance that are due to influences from the sensor's environment, such as corrosion. Figure 8.4 shows such a typical RTD construction. Mechanical strains change a conductor's resistance and must be eliminated if accurate temperature measurements are to be made. This factor is essential since the resistance changes with mechanical strain are significant, as evidenced by the use of metal wire as sensors for the direct measurement of strain. Such mechanical stresses and resulting strains can be created by thermal expansion. Thus, provision for strain-free expansion of the conductor as its temperature changes is essential in the construction of an RTD.

The relationship between the resistance of a metal conductor and its temperature can be expressed as a polynomial expansion:

$$R = R_0[1 + \alpha(T - T_0) + \beta(T - T_0)^2 + \cdots] \tag{8.3}$$

where R_0 is a reference resistance measured at temperature T_0. The coefficients α, $\beta, \ldots$ are material constants. Figure 8.5 shows the relative relation between resistance and temperature for three common metals. This figure shows that the relationship between temperature and resistance over specific small temperature ranges is linear. This approximation can be expressed as

$$R = R_0[1 + \alpha(T - T_0)] \tag{8.4}$$

where α is the temperature coefficient of resistivity. For example, for platinum conductors the linear approximation is accurate to within $\pm0.3\%$ over the range 0–200°C and $\pm1.2\%$ over the range 200–800°C. Table 8.2 lists a number of temperature coefficients of resistivity, α, for materials at 20°C. With the assumed linear relationship between resistance and temperature, an appropriate value of α should be chosen for the temperature range of interest.

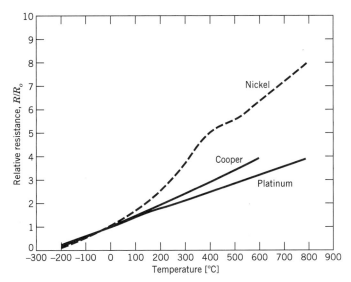

Figure 8.5 Relative resistance of several pure metals (R_0 at 0 °C).

The most common material chosen for the construction of RTDs is platinum. The RTD relies on the change in electrical resistance of a platinum wire to provide a precise measure of temperature, as expressed in equation (8.4). The principle of operation is quite simple: platinum exhibits a predictable and reproducible change in electrical resistance with temperature, which can be calibrated and interpolated to a high degree of accuracy. The linear approximation for the relationship between temperature and resistance is valid over a wide temperature range, and platinum is highly stable. To be suitable for use as a secondary temperature standard, a platinum resistance thermometer should have a value of α not less than 0.003925°C^{-1}. This minimum value is an indication of the purity of the platinum. In general, RTDs may be used for the measurment of temperatures ranging from cryogenic to approximately 650°C. By properly constructing an RTD and correctly measuring its resistance, an uncertainty in temperature measurement of approximately ±0.005°C is possible. Because of this potential for low uncertainties and the

Table 8.2 Temperature Coefficient of Resistivity for Selected Materials at 20°C

Substance	$\alpha[°C^{-1}]$
Aluminum (A1)	0.00429
Carbon (C)	−0.0007
Copper (Cu)	0.0043
Gold (Au)	0.004
Iron (Fe)	0.00651
Lead (Pb)	0.0042
Nickel (Ni)	0.0067
Nichrome	0.00017
Platinum (Pt)	0.003927
Tungsten (W)	0.0048
Thermistors	−0.068 to +0.14

predictable and stable behavior of platinum, the platinum RTD is widely used as a local standard.

Temperature measurements using an RTD are accomplished by measuring the resistance of the platinum wire and relating this resistance to temperature. For a NIST-certified RTD, a table and interpolating equation would be available.

RTD Resistance Measurement

The resistance of an RTD may be measured by a number of means, and the choice of an appropriate resistance measuring device must be made based on the required level of uncertainty in the final temperature measurement. Conventional ohmmeters cause a small current to flow during resistance measurements, creating self-heating in the RTD. An appreciable temperature change of the sensor may be caused by this current, in effect a loading error. This is an important consideration for RTDs.

Bridge circuits are used to measure the resistance of RTDs, to minimize loading errors, and to provide low uncertainties in measured resistance values. Wheatstone bridge circuits are commonly used for these measurements. However, the basic Wheatstone bridge circuit does not compensate for the resistance of the leads in measuring resistance of an RTD, which are a major source of error in electrical resistance thermometers. When greater accuracies are required, three-wire and four-wire bridge circuits can be used.

Figure 8.6(a) shows a three-wire Callender–Griffiths bridge circuit. The lead wires numbered 1, 2, and 3 have resistances r_1, r_2, and r_3, respectively. At balanced conditions.

$$\frac{R_1}{R_2} = \frac{R_3}{R_{\text{RTD}}} \tag{8.5}$$

but with the lead wire resistances included in the circuit analysis.

$$\frac{R_1}{R_2} = \frac{R_3 + r_1}{R_{\text{RTD}} + r_3} \tag{8.6}$$

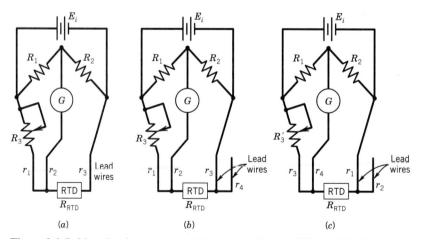

Figure 8.6 Bridge circuits; average of the two readings in (b) and (c) eliminates the effect of lead wire resistances.

and with $R_1 = R_2$, the resistance of the RTD, R_{RTD}, can be found as

$$R_{RTD} = R_3 + r_1 - r_3 \tag{8.7}$$

If $r_1 = r_3$, the effect of these lead wires is eliminated from the determination of the RTD resistance by this bridge circuit. Note that the resistance of lead wire 2 does not contribute to any error in the measurement at balanced conditions, since no current flows through the galvanometer, G.

The four-wire Mueller bridge, as shown in Figure 8.6(b), provides increased compensation for lead wire resistances compared to the Callendar–Griffiths bridge and is used with four-wire RTDs. The four-wire Mueller bridge is typically used when low uncertainties are desired, as in cases in which the RTD is used as a laboratory standard. A circuit analysis of the bridge circuit in the first measurement configuration yields

$$R_{RTD} + r_3 = R_3 + r_1 \tag{8.8}$$

and in the second measurement configuration yields

$$R_{RTD} + r_1 = R'_3 + r_3 \tag{8.9}$$

where R_3 and R'_3 represent the indicated values of resistance in the first and second configurations, respectively. Adding equations (8.8) and (8.9) results in an expression for the resistance of the RTD in terms of the indicated values for the two measurements:

$$R_{RTD} = \frac{R_3 + R'_3}{2} \tag{8.10}$$

With this approach, the effect of variations in lead wire resistances is minimized.

EXAMPLE 8.1

An RTD forms one arm of an equal-arm Wheatstone bridge, as shown in Figure 8.7. The fixed resistances, R_2 and R_3 are equal to 25 Ω. The RTD has a resistance of 25 Ω at a temperature of 0°C and is used to measure a temperature that is steady in time.

The resistance of the RTD over a small temperature range may be expressed as in equation (8.4):

$$R_{RTD} = R_0[1 + \alpha(T - T_0)]$$

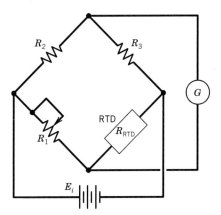

Figure 8.7 RTD Wheatstone bridge arrangement.

Suppose the coefficient of resistance for this RTD is $0.003925°C^{-1}$. A temperature measurement is made by placing the RTD in the measuring environment and balancing the bridge by adjusting R_1. The value of R_1 required to balance the bridge is $37.36\ \Omega$. Determine the temperature of the RTD.

KNOWN

An RTD having a resistance of $25\ \Omega$ at $0°C$ with $\alpha = 0.003925°C^{-1}$ is used to measure a temperature. A value of $R_1 = 37.36\ \Omega$ is required to balance the bridge circuit.

FIND

The temperature of the RTD.

SOLUTION

The resistance of the RTD is measured by balancing the bridge; recall that in a balanced condition.

$$R_{RTD} = R_1 \frac{R_3}{R_2}$$

The resistance of the RTD is found to be $37.36\ \Omega$. With $R_0 = 25\ \Omega$ at $T = 0°C$, and $\alpha = 0.003925°C^{-1}$, equation (8.4) becomes

$$37.36\ \Omega = 25(1 + \alpha T)\ \Omega$$

The temperature of the RTD is found to be $126°C$.

EXAMPLE 8.2

Consider the bridge circuit and RTD of Example 8.1. To select or design a bridge circuit for measuring the resistance of the RTD in this example, the required uncertainty in temperature would be specified. If the required uncertainty in the measured temperature is $\leq 0.5°C$, would a 1% total uncertainty in each of the resistors that make up the bridge be acceptable? Neglect the effects of lead wire resistances for this example.

KNOWN

A required uncertainty in temperature of $\pm 0.5°C$, measured with the RTD and bridge circuit of Example 8.1.

FIND

The uncertainty level for a 1% total uncertainty in each of the resistors that make up the bridge circuit.

ASSUMPTION

All uncertainties are provided and evaluated at the 95% confidence level.

SOLUTION

Perform a design-stage uncertainty analysis. Assuming at the design stage that the total uncertainty in the resistances is 1%, then with initial values of the resistances in the bridge equal to 25 Ω.

$$(u_d)_{R_1} = (u_d)_{R_2} = (u_d)_{R_3} = (0.01)(25) = 0.25 \ \Omega$$

The second power relation is used to estimate the propagation of uncertainty in each resistor to the uncertainty in determining the RTD resistance by

$$u_{RTD} = \sqrt{\left[\frac{\partial R}{\partial R_1}(u_d)_{R_1}\right]^2 + \left[\frac{\partial R}{\partial R_2}(u_d)_{R_2}\right]^2 + \left[\frac{\partial R}{\partial R_3}(u_d)_{R_3}\right]^2}$$

where

$$R = R_{RTD} = \frac{R_1 R_3}{R_2}$$

Then, the design-stage uncertainty in the resistance of the RTD is

$$u_{RTD} = \sqrt{\left[\frac{R_3}{R_2}(u_d)_{R_1}\right]^2 + \left[\frac{-R_1 R_3}{R_2^2}(u_d)_{R_2}\right]^2 + \left[\frac{R_1}{R_2}(u_d)_{R_3}\right]^2}$$

$$u_{RTD} = \sqrt{(1 \times 0.25)^2 + (1 \times -0.25)^2 + (1 \times 0.25)^2}$$

$$= 0.433 \ \Omega$$

To determine the uncertainty in temperature, we know

$$R = R_{RTD} = R_0[1 + \alpha(T - T_0)]$$

and

$$u_T = \sqrt{\left(\frac{\partial T}{\partial R} u_{RTD}\right)}$$

Setting $T_0 = 0°C$ with $R_0 = 25 \ \Omega$ and neglecting uncertainties in T_0, α, and R_0, we have

$$\frac{\partial T}{\partial R} = \frac{1}{\alpha R_0}$$

$$\frac{1}{\alpha R_0} = \frac{1}{(0.003925°C^{-1})(25 \ \Omega)}$$

Then the uncertainty in temperature is

$$u_T = u_{RTD}\left(\frac{\partial T}{\partial R}\right) = \frac{0.433 \ \Omega}{0.098 \ \Omega/°C} = 4.4°C$$

The desired uncertainty in temperature is not achieved with the specified levels of uncertainty in the pertinent variables.

COMMENT

Uncertainty analysis, in this case, would have prevented performing a measurement that would not provide meaningful results.

EXAMPLE 8.3

Suppose the total uncertainty in the bridge resistances of Example 8.1 was reduced to 0.1%. Would the required level of uncertainty in temperature be achieved?

KNOWN

The uncertainty in each of the resistors in the bridge circuit for temperature measurement from Example 8.1 is ±0.1%.

FIND

The resulting uncertainty in temperature.

SOLUTION

The uncertainty analysis from the previous example may be directly applied, with the uncertainty values for the resistances appropriately reduced. The uncertainties for the resistances are reduced from 0.25 to 0.025, yielding

$$u_{RTD} = \pm\sqrt{(1 \times 0.025)^2 + (1 \times -0.025)^2 + (1 \times 0.025)^2}$$
$$= \pm 0.0433 \ \Omega$$

and the resulting uncertainty in temperature is ±0.44°C, which satisfies the design constraint.

COMMENT

This result provides confidence that the effect of the resistors' uncertainties will not cause the uncertainty in temperature to exceed the target value. However, the uncertainty in temperature not only is a result of the uncertainty in the sensor, the RTD, but will depend on other aspects of the measurement system as well. The design-stage uncertainty analysis performed in this example may be viewed as ensuring that the factors considered do not produce a higher than acceptable uncertainty level. Additional sources of uncertainty will exist in any actual measurement employing an RTD.

Practical Considerations

The transient thermal response of typical commercial RTDs is generally quite slow compared to other temperature sensors, and for transient measurements bridge circuits must be operated in a deflection mode. For these reasons, RTDs are not generally chosen for transient temperature measurements. A notable exception is the use of very small platinum wires for temperature measurements in noncorrosive flowing gases. In this application, wires having diameters of the order of 0.01 cm can have frequency responses higher than any other temperature sensor, because of their extremely low thermal capacitance. Obviously, the smallest impact would destroy this sensor. Other resistance sensors in the form of thin metallic films provide fast transient response temperature measurements, often in conjunction with anemometry or heat flux measurements. Such platinum films are constructed by depositing a platinum film onto a substrate and coating the film with a ceramic glass for mechanical protection [7]. Typical film thickness ranges from 1 to 2 μm, with a 10-μm protective coating. Continuous exposure at temperatures of 600°C is possible with this construction. The range of applications for these thin-film sensors is increasing, especially for accuracies of ±0.5 to ±2 °C; practical uses include temperature control circuits for heating systems and cooking devices and surface temperature monitoring on electronic components subject to overheating.

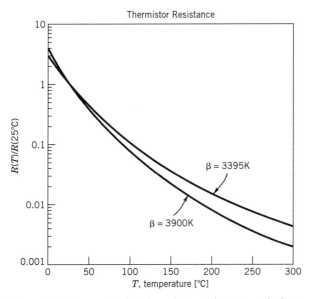

Figure 8.8 Representative thermistor resistance variations with temperature.

Thermistors

Thermistors (from *therm*ally sensitive re*sistors*) are ceramic-like semiconductor devices. The resistance of a typical thermistor decreases rapidly with temperature, which is in contrast to the small increases of resistance with temperature for RTDs. The functional relationship between resistance and temperature for a thermistor is generally assumed to be of the form

$$R = R_0 e^{\beta(1/T - 1/T_0)} \tag{8.11}$$

The parameter β ranges from 3500 to 4600 K, depending on the material, temperature, and individual construction for each sensor, and therefore must be determined for each thermistor. Figure 8.8 shows the variation of resistance with temperature for two common thermistor materials; the ordinate is the ratio of the resistance to the resistance at 25°C. Thermistors, exhibit large resistance changes with temperature in comparison to typical RTDs, as indicated by comparison of Figures 8.5 and 8.8. Equation (8.11) is not accurate over a wide range of temperature, unless β is taken to be a function of temperature; typically the value of β specified by a manufacturer for a sensor is assumed to be constant over a limited temperature range. A simple calibration is possible for determining β as a function of temperature, as illustrated in the circuits shown in Figure 8.9. Other circuits and a more complete discussion of measuring β may be found in the Electronic Industries Association standard Thermistor Definitions and Test Methods [8].

Thermistors are generally used when high sensitivity, ruggedness, or fast response times are required [9]. Thermistors are often encapsulated in glass and thus can be used in corrosive or abrasive environments. The resistance characteristics of the semiconductor material may change at elevated temperatures, and some aging of a thermistor will occur at temperatures above 200°C. The high resistance of a thermistor, compared to that of an RTD, eliminates the problems of lead wire

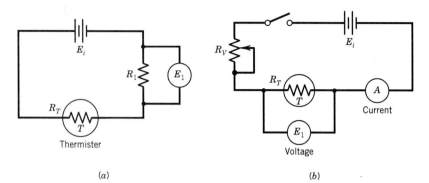

(a)   (b)

Figure 8.9 Circuits for determining β for thermistors. (a) Voltage divider method: $R_T = R_1(E_i/E_1 - 1)$. *Note:* Both R_1 and E_i must be known values. The value of R_1 may be varied to achieve appropriate values of thermistor current. (b) Volt-ammeter method. *Note:* Both current and voltage are measured.

resistance compensation. However, thermistors are not interchangeable, and variations in room temperature resistances for ordinary thermistors having the same nominal characteristics may be as much as 20%.

The zero-power resistance of a thermistor is the resistance value of the thermistor with no flow of electric current. The dissipation constant for a thermistor is defined at a given ambient temperature as

$$\delta = \frac{P}{T - T_\infty} \tag{8.12}$$

where

$$\delta = \text{dissipation constant}$$
$$P = \text{power supplied to thermistor}$$
$$T, T_\infty = \text{thermistor and ambient temperatures}$$

The zero-power resistance should be measured such that a decrease in the current flow to the thermistor will result in not more than a 0.1% change in resistance.

EXAMPLE 8.4

The material constant β is to be determined for a particular thermistor by using the circuit shown in Figure 8.9(a). The thermistor has a resistance of 60 kΩ at 25°C. The reference resistor in the circuit, R_1, has a resistance of 130.5 kΩ. The dissipation constant, δ, is 0.09 mW/°C. The voltage source used for the measurement is constant at 1.564 V. The thermistor is to be used at temperatures ranging from 100 to 150°C. Determine the value of β.

KNOWN

The temperature range of interest is from 100 to 150°C.

$$R_0 = 60,000 \ \Omega, \qquad T_0 = 25°C$$
$$E_i = 1.564 \ V, \qquad \delta = 0.09 \ mW/°C, \qquad R_1 = 130.5 \ k\Omega$$

FIND

The value of β over the temperature range of 100–150°C

SOLUTION

The voltage drop across the fixed resistor is measured for three known values of thermistor temperature. The thermistor temperature is controlled and determined by placing the thermistor in a laboratory oven and measuring the temperature of the oven. For each measured voltage across the reference resistor, the thermistor resistance, R_T, is determined from

$$R_T = R_1 \left(\frac{E_i}{E_1} - 1 \right) \tag{8.13}$$

The results of these measurements are as follows:

Temperature [°C]	R_1 Voltage [V]	R_T [Ω]
100	1.501	5477.4
125	1.531	2812.9
150	1.545	1604.9

Equation (8.11) can be expressed in the form of a linear equation as

$$\ln \frac{R_T}{R_0} = \beta \left(\frac{1}{T} - \frac{1}{T_0} \right) \tag{8.14}$$

When this equation is applied to the measured data, with $R_0 = 60{,}000\ \Omega$, the three data points above yield the following:

$\ln(R_T/R_0)$	$(1/T) - (1/T_0)$ [K^{-1}]	β [K]
−2.394	-6.75×10^{-4}	3546.7
−3.060	-8.43×10^{-4}	3629.9
−3.621	-9.92×10^{-4}	3650.2

COMMENT

These results are for constant β and are based on the behavior described by equation (8.11), over the temperature range from T_0 to the temperature T. The significance of the measured differences in β will be examined further.

The measured values of β in Example 8.4 are different at each value of temperature. If β were truly a temperature independent constant, and these measurements had negligible uncertainty, all three measurements would yield the same value for β. The variation in β may be due to a physical effect of temperature, or may be attributable to the uncertainty in the measured values.

Are the measured differences significant, and, if so, what value of β best represents the behavior of the thermistor over this temperature range? To perform the necessary uncertainty analysis, additional information must be provided concerning the instruments and procedures used in the measurement.

EXAMPLE 8.5

Perform an uncertainty analysis to determine the uncertainty in each measured value of β in Example 8.4, and evaluate a single best estimate of β for this temperature range. The measurement of β involved the measurement of voltages, temperatures, and resistances. For temperature there is a precision error associated with spatial and temporal variations in the oven temperature such that $S_{\bar{T}} = 0.19°C$ for 20 measurements. In addition, based on a manufacturer's specification, there is a known measurement bias limit for temperature of $\pm 0.36°C$ in the thermocouple. All values are given at 95% confidence level.

The bias errors in measuring resistance and voltage are negligible, and estimates of the instrument repeatability based on manufacturer's specifications in the measured values are $\pm 1.5\%$ for resistance and ± 0.002 V for the voltage.

KNOWN

Standard deviation of the means for oven temperature, $S_{\bar{T}} = 0.19°C$, $N = 20$. The remaining values will be treated as bias limits, since no statistical analysis can be performed:

$$(B)_T = \pm 0.36°C$$
$$(B)_R = \pm 1.5\%$$
$$(B)_E = \pm 0.002 \text{ V}$$

FIND

The uncertainty in β at each measured temperature, and a best estimate for β over the measured temperature range.

SOLUTION

Consider the problem of providing a single best estimate of β. One method of estimation might be to average the three measured values. This results in a value of 3609 K. However, since the relationship between (R_T/R_0) and $1/T - 1/T_0$ is expected to be linear, a least-squares fit can be performed on the three data points, and the point $(0, 0)$. The resulting value of β is 3638 K. Is this difference significant, and which value best represents the behavior of the thermistor? To answer these questions, an uncertainty analysis must be performed for β.

For each measured value,

$$\beta = \frac{\ln(R_T/R_0)}{1/T - 1/T_0}$$

Errors in voltage, temperature, and resistance are propagated into the resulting value of β for each measurement.

Consider first the sensitivity indices, θ_i, for each of the variables R_T, R_0, T, and T_0. These may be tabulated by computing the appropriate partial derivatives of β, evaluated at each of the three temperatures, as follows:

T [°C]	θ_{R_T} [K/Ω]	θ_{R_0} [K/Ω]	θ_T	θ_{T_0}
100	−0.270	0.0247	−37.77	59.17
125	−0.422	0.0198	−27.18	48.48
150	−0.628	0.0168	−20.57	41.45

The determination of the uncertainty in β, u_β, requires the uncertainty in the measured value of resistance for the thermistor, u_{R_T}, but R_T is determined from the expression

$$R_T = R_1[(E_i/E_1) - 1]$$

and thus requires an analysis of the uncertainty in the resulting value of R_T, from measured values of R_1, E_i, and E_1. All errors in R_T are treated as bias, yielding

$$(B)_{R_T} = \sqrt{\left[\frac{\partial R_T}{\partial R_1}(B)_{R_1}\right]^2 + \left[\frac{\partial R_T}{\partial E_i}(B)_{E_i}\right]^2 + \left[\frac{\partial R_T}{\partial E_1}(B)_{E_1}\right]^2}$$

To arrive at a representative value, we compute B_{R_T} at 125°C. The uncertainty in R_1 is ±1.5% of 130.5 kΩ, or ±1.96 kΩ. The uncertainty in E_i and E_1 are each ±0.002 V. Using the appropriate values to compute the sensitivity indices, the value of $(B)_{R_T}$ is found as ±247 Ω. (See Problem 8.25).

An uncertainty for β will be determined for each of the measured temperatures. The effect of the measurement bias limits for temperature and resistance is found by combining the individual contributions according to the second power law as

$$B_\beta = \sqrt{[\theta_T(B)_T]^2 + [\theta_{T_0}(B)_{T_0}]^2 + [\theta_{R_T}(B)_{R_T}]^2 + [\theta_{R_0}(B)_{R_0}]^2}$$

where

$(B)_T = ±0.36 \qquad (B)_{R_T} = ±247\ \Omega$

$(B)_{T_0} = ±0.36 \qquad (B)_{R_0} = ±900\ \Omega$

The precision index for β contains contributions only from the statistically determined oven temperature characteristics and is found from

$$P_\beta = \sqrt{(\theta_T S_{\bar{T}})^2 + (\theta_{T_0} S_{\bar{T}_0})^2}$$

where both $S_{\bar{T}}$ and $S_{\bar{T}_0}$ are 0.19, as determined with $N = 20$.

The resulting values of uncertainty in β are found from

$$u_\beta = \sqrt{B_\beta^2 + (t_{19,95} P_\beta)^2}$$

where $t_{19,95}$ is 2.093. At each temperature the uncertainty in β is determined as shown in Table 8.3. The effect of increases in the sensitivity indices, θ_i, on the total uncertainty is to cause increased uncertainty in β as the temperature increases.

Table 8.3 Uncertainties in β

	Uncertainty (95%)		
	Precision	Bias	Total
T	P_β	B_β	u_β
[°C]	[K]	[K]	[K]
100	13.3	74.7	79.7
125	10.6	107.6	109.9
150	8.8	156.7	157.8

The original results of the measured values of β must now be reexamined. The results, from Table 8.3, are plotted as a function of temperature in Figure 8.10, with uncertainty limits on each data point. Clearly, there is no justification for assuming that the measured values indicate a trend of changes with temperature, and it would be appropriate to use either the average value of β or the value determined from the linear least-squares curve fit.

8.5 THERMOELECTRIC TEMPERATURE MEASUREMENT

The most common method of measuring and controlling temperature uses an electrical circuit called a thermocouple. A *thermocouple* consists of two electrical conductors that are made of dissimilar metals and have at least one electrical connection. This electrical connection is referred to as a junction. A thermocouple junction may be created by welding, soldering, or by any method that provides good electrical contact between the two conductors, such as twisting the wires around one another. The output of a thermocouple circuit is a voltage, and there is a definite relationship between this voltage and the temperatures of the junctions that make up the thermocouple circuit. We will examine the causes of this voltage and develop the basis for using thermocouples to make engineering measurements of temperature.

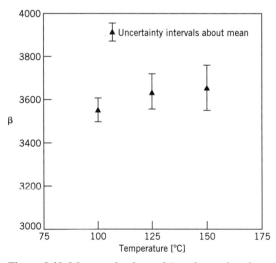

Figure 8.10 Measured values of β and associated uncertainties for three temperatures.

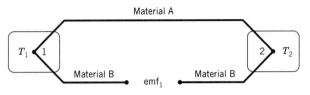

Figure 8.11 Basic thermocouple circuit.

Consider the thermocouple circuit shown in Figure 8.11. The junction labeled 1 is at a temperature T_1 and the junction labeled 2 is at a temperature T_2. If T_1 and T_2 are not equal, a finite open-circuit electric potential, emf_1, will be measured. The magnitude of the potential will depend on the difference in the temperatures and the particular metals the thermocouple circuit contains. A thermocouple junction is the source of an *electromotive force* (emf), which gives rise to the potential difference in a thermocouple circuit. It is the basis for temperature measurement using thermocouples. The circuit shown in Figure 8.11 is the most common form of a thermocouple circuit used for measuring temperature. This thermocouple circuit measures the difference between T_1 and T_2.

Thermoelectric phenomena result from the simultaneous flows of heat and electricity in an electrical conductor. More precisely, these phenomena result from coupled flows of entropy and electricity. The precise mathematical description of the source of these phenomena is provided by irreversible thermodynamics, but such a description of these phenomena is not necessary to use thermocouples to measure temperature. It is our goal to understand the origin of thermoelectric phenomena and the requirements for providing accurate temperature measurements using thermocouples.

In an electrical conductor that is subject to a temperature gradient, there will be both a flow of thermal energy and a flow of electricity. Both of these phenomena are closely tied to the behavior of the free electrons in a metal; it is no coincidence that good electrical conductors are, in general, good thermal conductors. The characteristic behavior of these free electrons in an electrical circuit composed of dissimilar metals results in a useful relationship between temperature and emf. There are three basic phenomena that can occur in a thermocouple circuit: (1) the *Seebeck effect*, (2) the *Peltier effect*, and (3) the *Thomson effect*.

Under ideal measurement conditions, with no loading errors, the emf generated by a thermocouple circuit would be the result of the Seebeck effect only.

Seebeck Effect

The Seebeck effect, named for Thomas Johann Seebeck (1770–1831), refers to the generation of a voltage potential, or emf, in an open thermocouple circuit caused by a difference in temperature between junctions in the circuit. The Seebeck emf can be measured when there is no current flow in the circuit. There is a fixed, reproducible relationship between the emf and the junction temperatures T_1 and T_2 (Figure 8.11). This relationship is expressed by the Seebeck coefficient, α_{AB}, defined as

$$\alpha_{AB} = \left[\frac{\partial(\text{emf})}{\partial T} \right]_{\text{open circuit}} \tag{8.15}$$

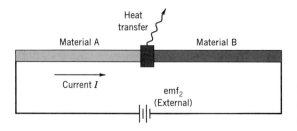

Figure 8.12 Peltier effect caused by current flow across a junction of dissimilar metals.

where A and B refer to the two materials that comprise the thermocouple. Since the Seebeck coefficient specifies the rate of change of voltage with temperature for the materials A and B, it is equal to the static sensitivity of the open-circuit thermocouple.

Peltier Effect

A familiar concept is that of I^2R or joule heating in a conductor through which an electrical current flows. Consider the two conductors having a common junction, shown in Figure 8.12, through which an electrical current, I, flows as a result of an externally applied emf. For any portion of either of the conductors, the energy removal rate required to maintain a constant temperature is I^2R, where R is the resistance to a current flow and is determined by the resistivity and size of the conductor. However, at the junction of the two dissimilar metals the removal of a quantity of energy different than I^2R is required to maintain a constant temperature. The difference in I^2R and the amount of energy generated by the current flowing through the junction is due to the Peltier effect. The Peltier effect is due to the thermodynamically reversible conversion of energy as a current flows across the junction, in contrast to the irreversible dissipation of energy associated with I^2R losses. The Peltier heat is the quantity of heat in addition to the quantity I^2R that must be removed from the junction to maintain the junction at a constant temperature. This amount of energy is proportional to the current flowing through the junction; the proportionality constant is the Peltier coefficient π_{AB}, and the heat transfer required to maintain a constant temperature is

$$Q_\pi = \pi_{AB}I \tag{8.16}$$

caused by the Peltier effect alone. This behavior was discovered by Jean Charles Athanase Peltier (1785–1845) during experiments with Seebeck's thermocouple. He observed that passing a current through a thermocouple circuit having two junctions, as in Figure 8.11, raised the temperature at one junction, while lowering the temperature at the other junction. This effect forms the basis of a device known as a Peltier, or thermoelectric, refrigerator, which provides cooling without moving parts.

Thomson Effect

In addition to the Seebeck effect and the Peltier effect, there is a third phenomenon that occurs in thermoelectric circuits. Consider the conductor shown in Figure 8.13, which is subject to a longitudinal temperature gradient and also to a potential difference, such that there is a flow of current and heat in the conductor. Again, to maintain a constant temperature in the conductor it is found that a quantity of energy different

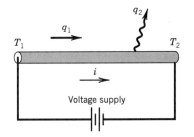

q_1 Energy flow as a result of a temperature gradient
q_2 Heat transfer to maintain constant temperature

Figure 8.13 Thomson effect caused by simultaneous flows of current and heat.

than the joule heat, $I^2 R$, must be removed from the conductor. First noted by William Thomson (1824–1907, Lord Kelvin from 1892) in 1851, this energy is expressed in terms of the Thomson coefficient, σ as

$$Q_\sigma = \sigma I (T_1 - T_2) \tag{8.17}$$

For a thermocouple circuit, all three of these effects may be present and may contribute to the overall emf of the circuit.

Fundamental Thermocouple Laws

The use of thermocouple circuits to measure temperature is based upon observed behaviors of carefully controlled thermocouple materials and circuits. The following laws provide the basis necessary for temperature measurement with thermocouples:

1. Law of Homogeneous Materials. *A thermoelectric current cannot be sustained in a circuit of a single homogeneous material by the application of heat alone, regardless of how it might vary in cross section.* Simply stated, this law requires that at least two materials be used to construct a thermocouple circuit for the purpose of measuring temperature. It is interesting to note that a current may occur in an inhomogeneous wire that is nonuniformly heated; however, this is neither useful nor desirable in a thermocouple.

2. Law of Intermediate Materials. *The algebraic sum of the thermoelectric forces in a circuit composed of any number of dissimilar materials is zero if all of the circuit is at a uniform temperature.* This law allows a material other than the thermocouple materials to be inserted into a thermocouple circuit without changing the output emf of the circuit. As an example, consider the thermocouple circuit shown in Figure 8.14 where the junctions of the measuring device are made of copper and material B is an alloy (not pure copper). The electrical connection between the measuring device and the thermocouple circuit forms yet another thermocouple junction. The law of

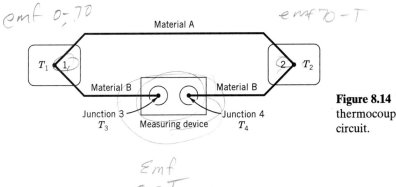

Figure 8.14 Typical thermocouple measuring circuit.

intermediate metals, in this case, provides that the measured emf will be unchanged from the open-circuit emf, which corresponds to the temperature difference between T_1 and T_2, if $T_3 = T_4$. Another practical consequence of this law is that readily available copper extension wires may be used to transmit thermocouple emfs to a measuring device.

3. Law of Successive or Intermediate Temperatures. *If two dissimilar homogeneous materials produce thermal emf$_1$ when the junctions are at* T_1 *and* T_2 *and produce thermal emf$_2$ when the junctions are at* T_2 *and* T_3, *the emf generated when the junctions are at* T_1 *and* T_3 *will be emf$_1$ + emf$_2$.* The law of intermediate temperatures allows a thermocouple calibrated for one reference temperature to be used at another reference temperature.

Basic Temperature Measurement with Thermocouples

The basic thermocouple circuit shown in Figure 8.14 can be used to measure the difference between the two temperatures T_1 and T_2. For practical temperature measurements, one of these junctions becomes a reference junction and is maintained at some known, constant reference temperature. The other junction then becomes the measuring junction, and the emf existing in the circuit for any temperature T_1 provides a direct indication of the temperature of the measuring junction.

Figure 8.15 shows a basic thermocouple measuring system, using a chromel-constantan thermocouple, an ice bath to create a reference temperature, copper extension wires, and a potentiometer to measure the output voltage of the circuit. The law of intermediate materials ensures that neither the potentiometer nor the extension wires will change the emf of the circuit, as long as the connecting junctions are at the same temperature. All that is required to be able to measure temperature with this circuit is to know the relationship between the output emf and the temperature of the measuring junction, for the particular reference temperature. One method of determining this relationship is to calibrate the thermocouple. However, we shall see that for reasonable levels of uncertainty for temperature measurement, standard materials and procedures allow thermocouples to be accurate temperature measuring devices, without the necessity of calibration.

The provisions for a reference junction should provide a temperature that is accurately known, stable, and reproducible. A very common reference junction temperature is provided by the ice point, $0°C$. The creation of a reference junction temperature of $0°C$ is accomplished in either of two basic ways. Prior to the development of an electronic means of creating a reference point in the thermoelectric circuit, an ice bath served to provide the reference junction temperature. An ice bath is typically made by filling a vacuum flask, or Dewar, with finely crushed ice, and adding just enough water to create a transparent slush. Surprising results are often obtained when the temperature of a mixture of ice and water is measured to verify the ice point is achieved. A few ice cubes floating in water does not create a $0°C$ environment! Ice baths can be constructed to provide a reference junction temperature with an uncertainty to within $\pm 0.01°C$.

Electronic reference junctions provide a convenient means of the measurement of temperature without the necessity to construct an ice path. Numerous manufacturers produce commercial temperature measuring devices with built-in reference junction compensation. The electronics generally rely on a thermistor to determine the local environment temperature, as shown in Figure 8.16. Uncertainties for the

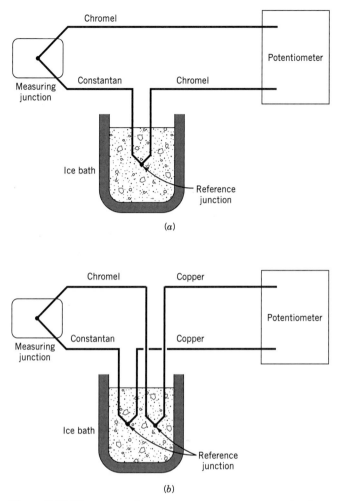

Figure 8.15 Thermocouple temperature measurement circuits.

reference junction temperature in this case are of the order of ±0.1°C, with ±0.5°C
as typical.

Thermocouple Standards

NIST provides specifications for the materials and construction of standard thermocouple circuits for temperature measurement [10]. Many material combinations exist
for thermocouples; these material combinations are identified by the thermocouple

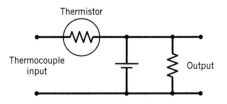

Figure 8.16 Basic thermistor circuit for
thermocouple reference junction compensation.

Table 8.4 Thermocouple Designations

Type	Material Combination		Applications
	Positive	Negative	
E	Chromel(+)	Constantan(−)	Highest sensitivity (<1000°C)
J	Iron(+)	Constantan(−)	Nonoxidizing environment (<760°C)
K	Chromel(+)	Alumel(−)	High temperature (<1372°C)
S	Platinum/ 10% rhodium	Platinum(−)	Long-term stability high temperature (<1768°C)
T	Copper(+)	Constantan(−)	Reducing or vacuum environments (<400°C)

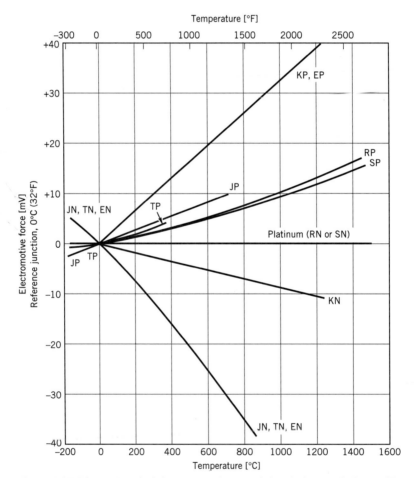

Figure 8.17 Thermal emf of thermocouple materials relative to platinum-67. *Note:* JP indicates the positive leg of a J thermocouple, or iron. (From R. P. Benedict, *Fundamentals of Temperature, Pressure and Flow Measurements*, 3d ed., Wiley, New York, 1984. Reprinted by permission.)

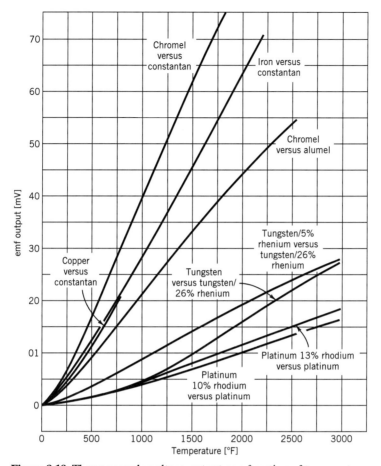

Figure 8.18 Thermocouple voltage output as a function of temperature for some common thermocouple materials. Reference junction is at 0°C. (From R. P. Benedict, *Fundamentals of Temperature, Pressure and Flow Measurements*, 3d ed., Wiley, New York, 1984. Reprinted by permission.)

type and denoted by a letter. Table 8.4 shows the letter designations and the polarity of common thermocouples, along with some basic application information for each type. The choice of a type of thermocouple depends on the temperature range to be measured, the particular application, and the desired uncertainty level.

To determine the emf output of a particular material combination, a thermocouple is formed from a candidate material and a standard platinum alloy to form a thermocouple circuit having a 0°C reference temperature. Figure 8.17 shows the output of various materials in combination with platinum-67. The notation indicates the thermocouple type. The law of intermediate temperatures then allows the emf of any two materials whose emf relative to platinum is known to be determined. Figure 8.18 shows a plot of the emf as a function of temperature for some common thermocouple material combinations. The slope of the curves in this figure corresponds to the static sensitivity of the thermocouple measuring circuit. Choices of a thermocouple type for a particular application should consider the nature of the measuring environment, the temperature range of interest, and the required uncertainty level.

Table 8.5 Standard Thermocouple Compositions[a]

Type	Wire Positive	Wire Negative	Expected Bias Error[b]
S	Platinum	Platinum/ 10% rhodium	±1.5°C or 0.25%
R	Platinum	Platinum/ 13% rhodium	±1.5°C
B	Platinum/ 30% rhodium	Platinum/ 6% rhodium	±0.5%
T	Copper	Constantan	±1.0°C or 0.75%
J	Iron	Constantan	±2.2°C or 0.75%
K	Chromel	Alumel	±2.2°C or 0.75%
E	Chromel	Constantan	±1.7°C or 0.5%

Alloy Designations
Constantan: 55% copper with 45% nickel
Chromel: 90% nickel with 10% chromium
Alumel: 94% nickel with 3% manganese, 2% aluminum, and 1% silicon

[a]From Temperature Measurement ANSI PTC 19.3-1974.
[b]Use greater value; these limits of error do not include installation errors.

Standard Thermocouple Voltage

Table 8.5 provides the standard composition of thermocouple materials, along with standard limits of error for the various material combinations. These limits specify the expected maximum errors resulting from the thermocouple materials. NIST uses high-purity materials to establish the standard value of voltage output for a thermocouple composed of two specific materials. This results in standard tables or equations used to determine temperature from a measured value of emf [10]. An example of such a table is provided in Table 8.6 for an iron–constantan thermocouple, usually referred to as a J-type thermocouple. Table 8.7 provides polynomial equations that relate emf and temperature for standard thermocouples. Because of the widespread need to measure temperature, an industry has grown up to supply high-grade thermocouple wire. Manufacturers can also provide thermocouples having special tolerance limits relative to the NIST standard voltages ranging from ±1.0 °C to perhaps ±0.1 °C. Thermocouples constructed of standard thermocouple wire do not require calibration to provide measurement of temperature with accuracies within the tolerance limits of the thermocouple wire given in Table 8.5.

Thermocouple Voltage Measurement

The Seebeck voltage for a thermocouple circuit is measured with no current flow in the circuit. From our discussion of the Thomson and Peltier effects, it is clear that the emf will be slightly changed from the open-circuit value when there is a current flow in the thermocouple circuit. Therefore, the best method for the measurement of thermocouple voltages is a device that minimizes current flow, such as a potentiometer. A potentiometer has for many years been the laboratory standard for voltage measurement in thermocouple circuits. A potentiometer, as described in Chapter 6, achieves nearly zero loading error by inducing no current to flow at a balanced condition. However, in situations requiring the measurement of a time-varying temperature, a typical potentiometer is not practical. For such applications, voltage measuring devices that have a very high input impedance, such as digital voltmeters, can be

Table 8.6 Thermocouple Reference Table for Type-J Thermocouple[a]

Temperature (°C) measured T_i								Thermocouple emf (mV) oK		
	0	−1	−2	−3	−4	−5	−6	−7	−8	−9
−210	−8.095									
−200	−7.890	−7.912	−7.934	−7.955	−7.976	−7.996	−8.017	−8.037	−8.057	−8.076
−190	−7.659	−7.683	−7.707	−7.731	−7.755	−7.778	−7.801	−7.824	−7.846	−7.868
−180	−7.403	−7.429	−7.456	−7.482	−7.508	−7.534	−7.559	−7.585	−7.610	−7.634
−170	−7.123	−7.152	−7.181	−7.209	−7.237	−7.265	−7.293	−7.321	−7.348	−7.376
−160	−6.821	−6.853	−6.883	−6.914	−6.944	−6.975	−7.005	−7.035	−7.064	−7.094
−150	−6.500	−6.533	−6.566	−6.598	−6.631	−6.663	−6.695	−6.727	−6.759	−6.790
−140	−6.159	−6.194	−6.229	−6.263	−6.298	−6.332	−6.366	−6.400	−6.433	−6.467
−130	−5.801	−5.838	−5.874	−5.910	−5.946	−5.982	−6.018	−6.054	−6.089	−6.124
−120	−5.426	−5.465	−5.503	−5.541	−5.578	−5.616	−5.653	−5.690	−5.727	−5.764
−110	−5.037	−5.076	−5.116	−5.155	−5.194	−5.233	−5.272	−5.311	−5.350	−5.388
−100	−4.633	−4.674	−4.714	−4.755	−4.796	−4.836	−4.877	−4.917	−4.957	−4.997
−90	−4.215	−4.257	−4.300	−4.342	−4.384	−4.425	−4.467	−4.509	−4.550	−4.591
−80	−3.786	−3.829	−3.872	−3.916	−3.959	−4.002	−4.045	−4.088	−4.130	−4.173
−70	−3.344	−3.389	−3.434	−3.478	−3.522	−3.566	−3.610	−3.654	−3.698	−3.742
−60	−2.893	−2.938	−2.984	−3.029	−3.075	−3.120	−3.165	−3.210	−3.255	−3.300
−50	−2.431	−2.478	−2.524	−2.571	−2.617	−2.663	−2.709	−2.755	−2.801	−2.847
−40	−1.961	−2.008	−2.055	−2.103	−2.150	−2.197	−2.244	−2.291	−2.338	−2.385
−30	−1.482	−1.530	−1.578	−1.626	−1.674	−1.722	−1.770	−1.818	−1.865	−1.913
−20	−0.995	−1.044	−1.093	−1.142	−1.190	−1.239	−1.288	−1.336	−1.385	−1.433
−10	−0.501	−0.550	−0.600	−0.650	−0.699	−0.749	−0.798	−0.847	−0.896	−0.946
0	0.000	−0.050	−0.101	−0.151	−0.201	−0.251	−0.301	−0.351	−0.401	−0.451

	0	+1	+2	+3	+4	+5	+6	+7	+8	+9
0	0.000	0.050	0.101	0.151	0.202	0.253	0.303	0.354	0.405	0.451
10	0.507	0.558	0.609	0.660	0.711	0.762	0.814	0.865	0.916	0.968
20	1.019	1.071	1.122	1.174	1.226	1.277	1.329	1.381	1.433	1.485
30	1.537	1.589	1.641	1.693	1.745	1.797	1.849	1.902	1.954	2.006
40	2.059	2.111	2.164	2.216	2.269	2.322	2.374	2.427	2.480	2.532
50	2.585	2.638	2.691	2.744	2.797	2.850	2.903	2.956	3.009	3.062
60	3.116	3.169	3.222	3.275	3.329	3.382	3.436	3.489	3.543	3.596
70	3.650	3.703	3.757	3.810	3.864	3.918	3.971	4.025	4.079	4.133
80	4.187	4.240	4.294	4.348	4.402	4.456	4.510	4.564	4.618	4.672
90	4.726	4.781	4.835	4.889	4.943	4.997	5.052	5.106	5.160	5.215
100	5.269	5.323	5.378	5.432	5.487	5.541	5.595	5.650	5.705	5.759
110	5.814	5.868	5.923	5.977	6.032	6.087	6.141	6.196	6.251	6.306
120	6.360	6.415	6.470	6.525	6.579	6.634	6.689	6.744	6.799	6.854
130	6.909	6.964	7.019	7.074	7.129	7.184	7.239	7.294	7.349	7.404
140	7.459	7.514	7.569	7.624	7.679	7.734	7.789	7.844	7.900	7.955
150	8.010	8.065	8.120	8.175	8.231	8.286	8.341	8.396	8.452	8.507
160	8.562	8.618	8.673	8.728	8.783	8.839	8.894	8.949	9.005	9.060
170	9.115	9.171	9.226	9.282	9.337	9.392	9.448	9.503	9.559	9.614
180	9.669	9.725	9.780	9.836	9.891	9.947	10.002	10.057	10.113	10.168
190	10.224	10.279	10.335	10.390	10.446	10.501	10.557	10.612	10.668	10.723
200	10.779	10.834	10.890	10.945	11.001	11.056	11.112	11.167	11.223	11.278
210	11.334	11.389	11.445	11.501	11.556	11.612	11.667	11.723	11.778	11.834
220	11.889	11.945	12.000	12.056	12.111	12.167	12.222	12.278	12.334	12.389
230	12.445	12.500	12.556	12.611	12.667	12.722	12.778	12.833	12.889	12.944
240	13.000	13.056	13.111	13.167	13.222	13.278	13.333	13.389	13.444	13.500
250	13.555	13.611	13.666	13.722	13.777	13.833	13.888	13.944	13.999	14.055
260	14.110	14.166	14.221	14.277	14.332	14.388	14.443	14.499	14.554	14.609
270	14.665	14.720	14.776	14.831	14.887	14.942	14.998	15.053	15.109	15.164

Table 8.6 (*Continued*)

Temperature (°C)								Thermocouple emf (mV) [a]	
	0	+1	+2	+3	+4	+5	+6	+7	+8	+9
280	15.219	15.275	15.330	15.386	15.441	15.496	15.552	15.607	15.663	15.718
290	15.773	15.829	15.884	15.940	15.995	16.050	16.106	16.161	16.216	16.272
300	16.327	16.383	16.438	16.493	16.549	16.604	16.659	16.715	16.770	16.825
310	16.881	16.936	16.991	17.046	17.102	17.157	17.212	17.268	17.323	17.378
320	17.434	17.489	17.544	17.599	17.655	17.710	17.765	17.820	17.876	17.931
330	17.986	18.041	18.097	18.152	18.207	18.262	18.318	18.373	18.428	18.483
340	18.538	18.594	18.649	18.704	18.759	18.814	18.870	18.925	18.980	19.035
350	19.090	19.146	19.201	19.256	19.311	19.366	19.422	19.477	19.532	19.587
360	19.642	19.697	19.753	19.808	19.863	19.918	19.973	20.028	20.083	20.139
370	20.194	20.249	20.304	20.359	20.414	20.469	20.525	20.580	20.635	20.690
380	20.745	20.800	20.855	20.911	20.966	21.021	21.076	21.131	21.186	21.241
390	21.297	21.352	21.407	21.462	21.517	21.572	21.627	21.683	21.738	21.793
400	21.848	21.903	21.958	22.014	22.069	22.124	22.179	22.234	22.289	22.345
410	22.400	22.455	22.510	22.565	22.620	22.676	22.731	22.786	22.841	22.896
420	22.952	23.007	23.062	23.117	23.172	23.228	23.283	23.338	23.393	23.449
430	23.504	23.559	23.614	23.670	23.725	23.780	23.835	23.891	23.946	24.001
440	24.057	24.112	24.167	24.223	24.278	24.333	24.389	24.444	24.499	24.555
450	24.610	24.665	24.721	24.776	24.832	24.887	24.943	24.998	25.053	25.109
460	25.164	25.220	25.275	25.331	25.386	25.442	25.497	25.553	25.608	25.664
470	25.720	25.775	25.831	25.886	25.942	25.998	26.053	26.109	26.165	26.220
480	26.276	26.332	26.387	26.443	26.499	26.555	26.610	26.666	26.722	26.778
490	26.834	26.889	26.945	27.001	27.057	27.113	27.169	27.225	27.281	27.337
500	27.393	27.449	27.505	27.561	27.617	27.673	27.729	27.785	27.841	27.897
510	27.953	28.010	28.066	28.122	28.178	28.234	28.291	28.347	28.403	28.460
520	28.516	28.572	28.629	28.685	28.741	28.798	28.854	28.911	28.967	29.024
530	29.080	29.137	29.194	29.250	29.307	29.363	29.420	29.477	29.534	29.590
540	29.647	29.704	29.761	29.818	29.874	29.931	29.988	30.045	30.102	30.159
550	30.216	30.273	30.330	30.387	30.444	30.502	30.559	30.616	30.673	30.730
560	30.788	30.845	30.902	30.960	31.017	31.074	31.132	31.189	31.247	31.304
570	31.362	31.419	31.477	31.535	31.592	31.650	31.708	31.766	31.823	31.881
580	31.939	31.997	32.055	32.113	32.171	32.229	32.287	32.345	32.403	32.461
590	32.519	32.577	32.636	32.694	32.752	32.810	32.869	32.927	32.985	33.044
600	33.102	33.161	33.219	33.278	33.337	33.395	33.454	33.513	33.571	33.630
610	33.689	33.748	33.807	33.866	33.925	33.984	34.043	34.102	34.161	34.220
620	34.279	34.338	34.397	34.457	34.516	34.575	34.635	34.694	34.754	34.813
630	34.873	34.932	34.992	35.051	35.111	35.171	35.230	35.290	35.350	35.410
640	35.470	35.530	35.590	35.650	35.710	35.770	35.830	35.890	35.950	36.010
650	36.071	36.131	36.191	36.252	36.312	36.373	36.433	36.494	36.554	36.615
660	36.675	36.736	36.797	36.858	36.918	36.979	37.040	37.101	37.162	37.223
670	37.284	37.345	37.406	37.467	37.528	37.590	37.651	37.712	37.773	37.835
680	37.896	37.958	38.019	38.081	38.142	38.204	38.265	38.327	38.389	38.450
690	38.512	38.574	38.636	38.698	38.760	38.822	38.884	38.946	39.008	39.070
700	39.132	39.194	39.256	39.318	39.381	39.443	39.505	39.568	39.630	39.693
710	39.755	39.818	39.880	39.943	40.005	40.068	40.131	40.193	40.256	40.319
720	40.382	40.445	40.508	40.570	40.633	40.696	40.759	40.822	40.886	40.949
730	41.012	41.075	41.138	41.201	41.265	41.328	41.391	41.455	41.518	41.581
740	41.645	41.708	41.772	41.835	41.899	41.962	42.026	42.090	42.153	42.217
750	42.281	42.344	42.408	42.472	42.536	42.599	42.663	42.727	42.791	42.855
760	42.919	42.983	43.047	43.110	43.174	43.238	43.303	43.367	43.431	43.495

[a] Reference junction at 0°C.

Table 8.7 Reference Functions for Selected Letter Designated Thermocouples

The relationship between emf and temperature is provided in the form of a polynomial in temperature [10]

$$E = \sum_{i=0}^{n} c_i T^i$$

where E is in mV and T is in °C. Constants are provided below.

Thermocouple Type	Temperature Range	Constants
J-type	−210–760°C	$c_0 = 0.000\ 000\ 000\ 0$ $c_1 = 5.038\ 118\ 781\ 5 \times 10^1$ $c_2 = 3.047\ 583\ 693\ 0 \times 10^{-2}$ $c_3 = -8.568\ 106\ 572\ 0 \times 10^{-5}$ $c_4 = 1.322\ 819\ 529\ 5 \times 10^{-7}$ $c_5 = -1.705\ 295\ 833\ 7 \times 10^{-10}$ $c_6 = 2.094\ 809\ 069\ 7 \times 10^{-13}$ $c_7 = -1.253\ 839\ 533\ 6 \times 10^{-16}$ $c_8 = 1.563\ 172\ 569\ 7 \times 10^{-20}$
T-type	−270–0°C	$c_0 = 0.000\ 000\ 000\ 0$ $c_1 = 3.874\ 810\ 636\ 4 \times 10^1$ $c_2 = 4.419\ 443\ 434\ 7 \times 10^{-2}$ $c_3 = 1.184\ 432\ 310\ 5 \times 10^{-4}$ $c_4 = 2.003\ 297\ 355\ 4 \times 10^{-5}$ $c_5 = 9.013\ 801\ 955\ 9 \times 10^{-7}$ $c_6 = 2.265\ 115\ 659\ 3 \times 10^{-8}$ $c_7 = 3.607\ 115\ 420\ 5 \times 10^{-10}$ $c_8 = 3.849\ 393\ 988\ 3 \times 10^{-12}$ $c_9 = 2.821\ 352\ 192\ 5 \times 10^{-14}$ $c_{10} = 1.425\ 159\ 477\ 9 \times 10^{-16}$ $c_{11} = 4.876\ 866\ 228\ 6 \times 10^{-19}$ $c_{12} = 1.079\ 553\ 927\ 0 \times 10^{-21}$ $c_{13} = 1.394\ 502\ 706\ 2 \times 10^{-24}$ $c_{14} = 7.979\ 515\ 392\ 7 \times 10^{-28}$
T-type	0–400°C	$c_0 = 0.000\ 000\ 000\ 0$ $c_1 = 3.874\ 810\ 636\ 4 \times 10^1$ $c_2 = 3.329\ 222\ 788\ 0 \times 10^{-2}$ $c_3 = 2.061\ 824\ 340\ 4 \times 10^{-4}$ $c_4 = -2.188\ 225\ 684\ 6 \times 10^{-6}$ $c_5 = 1.099\ 688\ 092\ 8 \times 10^{-8}$ $c_6 = -3.081\ 575\ 877\ 2 \times 10^{-11}$ $c_7 = 4.547\ 913\ 529\ 0 \times 10^{-14}$ $c_8 = -2.751\ 290\ 167\ 3 \times 10^{-17}$

used with minimal loading error. These devices can also be used in either static or dynamic measuring situations in which the loading error created by the measurement device is acceptable for the particular application. For such needs, high-impedance voltmeters have been incorporated into commercially available temperature indicators, temperature controllers, and digital data acquisition systems.

EXAMPLE 8.6

The thermocouple circuit shown in Figure 8.19 is used to measure the temperature T_1. The thermocouple junction labeled 2 is at a temperature of 0°C, maintained by

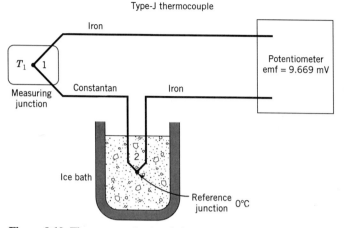

Figure 8.19 Thermocouple circuit for Example 8.6.

an ice-point bath. The voltage output is measured with a potentiometer and is found to be 9.669 mV. What is T_1?

KNOWN

A thermocouple circuit having one junction at 0°C and a second junction at an unknown temperature. The circuit produces an emf of 9.669 mV.

FIND

The temperature T_1.

ASSUMPTION

Thermocouple follows NIST standard.

SOLUTION

Standard thermocouple tables such as Table 8.6 are referenced to 0°C. The temperature of the reference junction for this case is 0°C. Therefore, the temperature corresponding to an output voltage may simply be determined from Table 8.6, in this case as 180°C.

COMMENT

Because of the law of intermediate metals, the junctions formed at the potentiometer do not affect the voltage measured for the thermocouple circuit, and the voltage output reflects accurately the temperature difference between junctions 1 and 2.

EXAMPLE 8.7

Suppose the thermocouple circuit in the previous example (Example 8.6) now has junction 2 maintained at a temperature of 30°C and produces an output voltage of 8.132 mV. What temperature is sensed by the measuring junction?

KNOWN

T_2 is 30°C, and the output emf is 8.132 mV.

ASSUMPTION

Thermocouple follows NIST standard.

FIND

The temperature of the measuring junction.

SOLUTION

By the law of intermediate temperatures the output emf for a thermocouple circuit having two junctions, one at 0°C and the other at T_1, would be the sum of the emfs for a thermocouple circuit between 0 and 30°C and between 30°C and T_1. Thus,

$$\text{emf}_{0-30} + \text{emf}_{30-T_1} = \text{emf}_{0-T_1}$$

This relationship allows the voltage reading from the nonstandard reference temperature to be converted to a 0°C reference temperature by adding $\text{emf}_{0-30} = 1.537$ to the existing reading. This results in an equivalent output voltage, referenced to 0°C as

$$1.537 + 8.132 = 9.669 \, \text{mV}$$

Clearly, this thermocouple is sensing the same temperature as in the previous example, 180°C. This value is determined from Table 8.6.

COMMENT

Note that the effect of raising the reference junction temperature is to lower the output voltage of the thermocouple circuit. Negative values of voltage, as compared with the polarity listed in Table 8.4, indicate that the measured temperature is less than the reference junction temperature.

EXAMPLE 8.8

A J-type thermocouple measures a temperature of 100°C and is referenced to 0°C. The thermocouple is AWG 30 (30-gauge or 0.010-in. wire diameter), and is arranged in a circuit as shown in Figure 8.15(*a*). The length of the thermocouple wire is 10 ft, in order to run from the measurement point to the ice bath and to a potentiometer. The resolution of the potentiometer is 0.005 mV. If the thermocouple wire has a resistance per unit length, as specified by the manufacturer, of 5.6 Ω/ft, estimate the residual current in the thermocouple circuit at balanced conditions.

KNOWN

A potentiometer having a resolution of 0.005 mV is used to measure the emf of a J-type thermocouple that is 10 ft long.

FIND

The residual current in the thermocouple circuit.

SOLUTION

The total resistance of the thermocouple circuit is 56 Ω, for 10 ft of thermocouple wire. The residual current is then found from Ohm's law as

$$I = \frac{E}{R} = \frac{0.005\,\text{mV}}{56\,\Omega} = 8.9 \times 10^{-8}\,\text{A}$$

COMMENT

The loading error caused by this current flow is ~ 0.005 mV/54.3 μV/°C ≈ 0.1°C.

EXAMPLE 8.9

Suppose a high-impedance voltmeter is used in place of the potentiometer in Example 8.8. Determine the minimum input impedance required for the voltmeter that will limit the loading error to the same level as the potentiometer.

KNOWN

Loading error should be less than 8.9×10^{-8} A.

FIND

Input impedance for a voltmeter that would produce the same current flow or loading error.

SOLUTION

At 100°C a J-type thermocouple referenced to 0°C will have a Seebeck voltage of $E_s = 5.269$ mV. At this temperature, the required voltmeter impedance to limit the current flow to 8.9×10^{-8} A is found from Ohm's law:

$$\frac{E_s}{I} = 5.269 \times 10^{-3}\,\text{V}/8.9 \times 10^{-8}\,\text{A}$$
$$= 59.2\,\text{k}\Omega$$

COMMENT

This input impedance is not unusually high for microvoltmeters and indicates that such a voltmeter would be a reasonable choice for the measurement of the thermocouple emf in this situation. As always, the allowable loading error should be determined based on the required uncertainty in the measured temperature.

Multiple-Junction Thermocouple Circuits

A thermocouple circuit composed of two junctions of dissimilar metals produces an open-circuit emf that is related to the temperature difference between the two

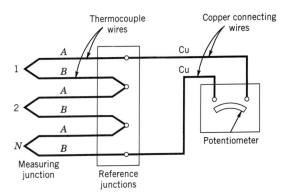

Figure 8.20 Thermopile arrangement. (From R. P. Benedict, *Fundamentals of Temperature, Pressure and Flow Measurements*, 3d ed., Wiley, New York, 1984. Reprinted by permission.)

junctions. More than two junctions can be employed in a thermocouple circuit, and thermocouple circuits can be devised to measure temperature differences or average temperature or to amplify the output voltage of a thermocouple circuit.

Thermopiles

"Thermopile" is a term used to describe a multiple-junction thermocouple circuit that is designed to amplify the output of the circuit. Since thermocouple voltage outputs are typically in the millivolt range, increasing the voltage output may be a key element in reducing the uncertainty in the temperature measurement, or it may be necessary to allow transmission of the thermocouple signal to the recording device. Figure 8.20 shows a thermopile for providing an amplified output signal; in this case the output voltage would be N times the single thermocouple output, where N is the number of measuring junctions in the circuit. The average output voltage corresponds to the average temperature level sensed by the N junctions. This thermopile arrangement can be used to measure a spatially averaged temperature or to measure a single value of temperature. The measurement of a single value of temperature entails considerations of the physical size of a thermopile, as compared to a single thermocouple. In transient measurements, a thermopile may have a more limited frequency range than a single thermocouple, due to its increased thermal capacitance. Thermopiles are particularly useful for reducing the uncertainty in measuring small temperature differences between the measuring and reference junctions. The principle has also been used to generate small amounts of power in spacecraft and to power electric coolers.

Figure 8.21 shows a series arrangement of thermocouple junctions designed to measure the average temperature difference between junctions. This thermocouple circuit could be used in an environment where a uniform temperature was desired. In that case, a voltage output would indicate that a temperature difference existed between two of the thermocouple junctions. It should be noted, however, that a zero voltage output could also occur if the temperature differences that occurred in the circuit summed to a zero emf. Alternatively, junctions $1, 2, \ldots, N$ could be located at one physical location, while junctions $1', 2', \ldots, N'$ could be located at another

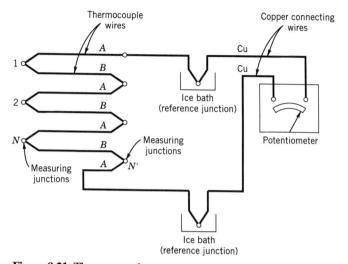

Figure 8.21 Thermocouples arranged to sense temperature differences. (From R. P. Benedict, *Fundamentals of Temperature, Pressure and Flow Measurements*, 3d ed., Wiley, New York, 1984. Reprinted by permission.)

physical location. Applications for such a circuit might be the measurement of heat flux through a solid.

Thermocouples in Parallel

When a spatially averaged temperature is desired, multiple thermocouple junctions can be arranged as shown in Figure 8.22. In such an arrangement of N measuring junctions, a mean emf is produced, given by

$$\overline{\text{emf}} = \frac{1}{N} \sum_{i=1}^{N} (\text{emf})_i \qquad (8.18)$$

The mean emf is indicative of a mean temperature,

$$\overline{T} = \frac{1}{N} \sum_{i=1}^{N} T_i \qquad (8.19)$$

Measuring temperatures by using thermocouples connected to data-acquisition systems is common practice. However, the characteristics of the thermocouple, including the need for a reference or cold junction and the low signal voltages produced, complicate its use. Nevertheless, with a little attention and realistic expectation of achievable accuracy, the systems are quite acceptable for most monitoring and moderate accuracy measurements.

Data Acquisition Considerations

Once the appropriate thermocouple type is selected, attention must be given to the cold junction compensation method. The two connection points of the thermocouple

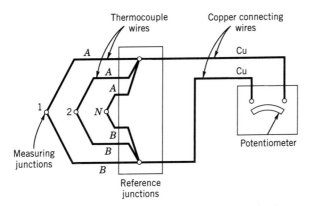

Figure 8.22 Parallel arrangement of thermocouples for sensing the average temperature of the measuring junctions. (From R. P. Benedict, *Fundamentals of Temperature, Pressure and Flow Measurements*, 3d ed., Wiley, New York, 1984. Reprinted by permission.)

to the DAS board form two new thermocouple connections. Use of external cold junction methods between the thermocouple and the board will eliminate this problem, but more frequently, the thermocouple will be connected directly to the board and use built-in electronic cold junction compensation. This is usually accomplished by using a separate thermistor sensor, which measures the temperature at the system connection point to determine the cold junction error, and providing an appropriate bias voltage correction either directly or through software. An important consideration is that the internal correction method has a typical error of the order of 0.5–1.5°C and, as a bias error, this error is directly passed on to the measurement.

These boards also may use internal polynomial interpolation for converting measured voltage into temperature. If not, this can be programmed into the data-reduction scheme by using, for example, the information of Table 8.7. Nonetheless, this introduces a "linearization" error, which is a function of thermocouple material and temperature range and typically specified with the DAS board.

Thermocouples are often used in harsh, industrial environments with significant EMI and rf noise sources. Thermocouple wire pairs should be twisted to reduce noise. Also, a differential-ended connection is preferred between the thermocouple and the DAS board. In this arrangement, though, the thermocouple becomes an isolated voltage source, meaning there is no longer a direct ground path keeping the input within its common mode range. As a consequence, a common complaint is that the measured signal may drift or suddenly jump in the level of its output. Usually, this interference behavior is eliminated by placing a 10-kΩ to 100-kΩ resistor between the low terminal of the input and low-level ground.

Since most DAS boards use A/D converters having a ±5-V full scale, the signal must be conditioned using an amplifier. Usually a gain of 100 to 500 is sufficient. High gain, very low noise amplifiers are important for accurate measurements. Consider a J-type thermocouple using a 12-bit A/D converter with a signal conditioning gain of 100. This allows for a full-scale input range of 100 mV, which is a suitable range

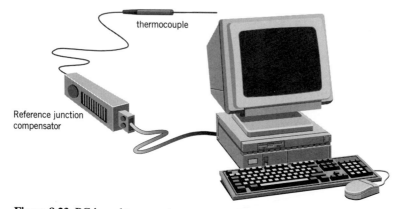

Figure 8.23 PC-based temperature measurement system.

for most measurements. Then, the A/D conversion resolution is

$$\frac{E_{FSR}}{(G)(2^M)} = \frac{10\,V}{(100)(2^{12})} = 24.4\,\mu V/bit$$

A J-type thermocouple has a sensitivity of ~55 μV/°C. Thus, the measurement resolution becomes $(24.4\,\mu V/bit)/(55\,\mu V/°C) = 0.44°C/bit$.

Thermocouples tend to have long time constants relative to the typical sample rate capabilities of general purpose DAS boards. If temperature measurements show greater than expected fluctuations, high-frequency sampling noise is a likely cause. Slowing down the sample rate or using a smoothing filter are simple solutions. The period of averaging should be on the scale of the time constant of the thermocouple.

EXAMPLE 8.10

It is desired to create an off-the-shelf temperature measuring system for a PC-based control application. The proposed temperature measuring system is illustrated schematically in Figure 8.23. The temperature measurement system consists of the following.

- A PC-based data-acquisition system composed of a data-acquisition board, a computer, and appropriate software to allow measurement of analog input voltage signals
- A J-type thermocouple and reference junction compensator. The reference junction compensator serves to provide an output emf from the thermocouple equivalent to the value that would exist between the measuring junction and a reference temperature of 0°C.
- A J-type thermocouple, which is uncalibrated but meets NIST standard limits of error

The system is designed to measure and control a process temperature that varies slowly in time compared to sampling rate of the data acquisition system. The process nominally operates at 185°C.

The following specifications are applicable to the measurement system components.

Component	Characteristics	Accuracy Specifications
Data-acquisition board	Analog voltage input range: 0–0.1 V	12-bit A/D converter accuracy: ±0.01% of reading
Reference junction compensator	J-type compensation range of 0–50°C	±0.5°C over the range 20–36°C ±1°C over the range 36–50°C
Thermocouple (J-type)	Stainless-steel sheathed, ungrounded junction	Accuracy: ±1.0°C based on NIST standard limits of error

The purpose of the measurement system requires that the temperature measurement have a total uncertainty of less than 1.5°C. Based on a design stage uncertainty analysis, does this measurement system meet the overall accuracy requirement?

SOLUTION

The design-stage uncertainty for this measurement system will be determined by expressing the uncertainty of each system component as an equivalent uncertainty in temperature and then combining these design-stage uncertainties.

The 12-bit A/D converter divides the full-scale voltage range into 2^{12} or 4096 equal-sized intervals. Thus, the resolution (quantization error) of the A/D in measuring voltage is

$$\frac{0.1\,\text{V}}{4096\,\text{intervals}} = 0.0244\,\text{mV}$$

The accuracy of the DAS is specified as 0.01% of the reading. A nominal value for the thermocouple voltage must be known or established. In the present case, the nominal process temperature is 185°C, which corresponds to a thermocouple voltage of approximately 10 mV. Thus the calibration uncertainty of the DAS is 0.001 mV.

The contribution to the total uncertainty of the temperature measurement system from the DAS can now be determined. First combine the resolution and calibration uncertainties as

$$u_{\text{DAS}} = \pm\sqrt{(0.0244)^2 + (0.001)^2} = \pm 0.0244\,\text{mV}$$

The relationship between uncertainty in voltage and temperature is provided by the static sensitivity, which can be estimated from Table 8.6 as 0.055 mV/°C. Thus an uncertainty of 0.0244 mV corresponds to an uncertainty in temperature of

$$\frac{0.0244\,\text{mV}}{0.055\,\text{mV/}^\circ\text{C}} = \pm 0.444^\circ\text{C}$$

This uncertainty can now be combined directly with the reference junction uncertainty, and the uncertainty associated with the standard limits of error for the thermocouple, as

$$u_T = \pm\sqrt{(0.44)^2 + (1.0)^2 + (0.5)^2} = \pm 1.2^\circ\text{C}$$

COMMENT

It would be appropriate in many cases to calibrate the thermocouple against a laboratory standard, such as an RTD, which has a calibration traceable to NIST standards. With reasonable expense and care, an uncertainty level in the thermo-couple of $\pm 0.1°C$ can be achieved, compared to the uncalibrated value of $\pm 1°C$. If the thermocouple was calibrated, the resulting uncertainty in the overall measure-ment of temperature is reduced to $\pm 0.67°C$. But even by reducing the uncertainty contribution of the thermocouple by a factor of 10, the system uncertainty would be reduced only by a factor of 2.

8.6 RADIATIVE TEMPERATURE MEASUREMENTS

The measurement of temperature through the detection of thermal radiation presents some unique advantages. The sensor for thermal radiation need not be in contact with the surface to be measured, making this method attractive for a wide variety of applications. The basic operation of a radiation thermometer is predicated upon some knowledge of the radiation characteristics of the surface whose temperature is being measured, relative to the calibration of the thermometer. The spectral charac-teristics of radiative measurements of temperature is beyond the scope of the present discussion; an excellent source for further information is found in [11].

Radiation Fundamentals

Radiation refers to the emission of electromagnetic waves from the surface of an object. This radiation has characteristics of both waves and particles, which leads to a description of the radiation as being composed of photons. The photons generally travel in straight lines from points of emission to another surface, where they are absorbed, reflected, or transmitted. This radiation exists over a large range of wave-lengths that includes X rays, ultraviolet radiation, visible light, and thermal radiation, as shown in Figure 8.24. The thermal radiation emitted from an object is related to its temperature and has wavelengths ranging from approximately 10^{-7} to 10^{-3} m. It is necessary to understand two key aspects of radiative heat transfer in relation to temperature measurements. First, the radiation emitted by an object is proportional

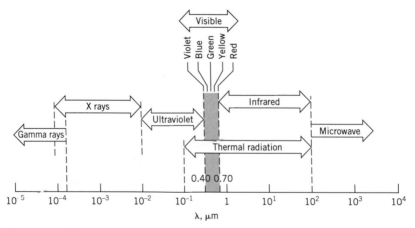

Figure 8.24 Electromagnetic spectrum. (From F. P. Incropera and D. P. DeWitt, *Fundamentals of Heat and Mass Transfer*, 2d ed., Wiley, New York, 1985. Reprinted with permission.)

to the fourth power of its temperature. In the ideal case, this may be expressed as

$$E_b = \sigma T^4 \tag{8.20}$$

where E_b is the flux of energy radiating from an ideal surface, or the blackbody emissive power. The emissive power of a body is the energy emitted per unit area and per unit time. The term "blackbody" implies a surface that absorbs all incident radiation, and as a result emits radiation in an "ideal" manner.

The emissive power is a direct measure of the total radiation emitted by an object. However, energy is emitted by an ideal radiator over a range of wavelengths, and at any given temperature the distribution of the energy emitted as a function of wavelength is unique. Max Planck (1858–1947) developed the basis for the theory of quantum mechanics in 1900 as a result of examining the wavelength distribution of radiation. He proposed the following equation to describe the wavelength distribution of thermal radiation for an ideal or blackbody radiator:

$$E_{b\lambda} = \frac{2\pi h_p c^2}{\lambda^5 [\exp(h_p c / k_B \lambda T) - 1]} \tag{8.21}$$

where

$E_{b\lambda}$ = total emissive power at the wavelength
λ = wavelength
c = speed of light
h_p = Planck's constant = 6.6256×10^{-34} J s
k_B = Boltzmann's constant = 1.3805×10^{-23} J/K

Figure 8.25 is a plot of this wavelength distribution for various temperatures. For the purposes of radiative temperature measurements, it is crucial to note that the maximum energy emission shifts to shorter wavelengths at higher temperatures. Our experiences confirm this behavior through observation of color changes as a surface is heated.

Consider an electrical heating element, as can be found in an electric oven. With no electric current flow through the element, it appears almost black, its room-temperature color. With a current flow, the element temperature rises and it appears to change color to a dull red, and perhaps to a reddish orange. If its temperature continued to increase, eventually it would appear white. This change in color signifies a shift in the maximum intensity of the emitted radiation to shorter wavelengths, out of the infrared and into the visible. There is also an increase in total emitted energy. The Planck distribution provides a basis for the measurement of temperature through color comparison.

Radiation Detectors

Radiative energy flux can be detected in a sensor by two basic techniques. The detector is subject to radiant energy from the source whose temperature is to be measured. The first technique involves a thermal detector in which absorbed radiative energy elevates the detector temperature, as shown in Figure 8.26. These thermal detectors are certainly the oldest sensors for radiation, and the first such detector can probably be credited to Sir William Herschel, who verified the presence of infrared radiation using a thermometer and a prism. The equilibrium temperature of the

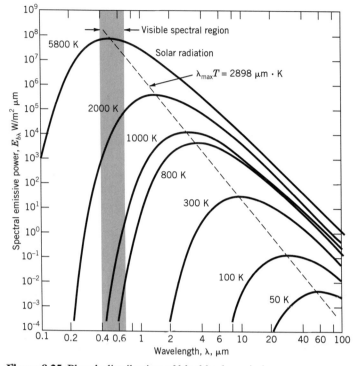

Figure 8.25 Planck distribution of blackbody emissive power as a function of wavelength. (From F. P. Incropera and D. P. DeWitt, *Fundamentals of Heat and Mass Transfer*, 2d ed., Wiley, New York, 1985. Reprinted by permission.)

detector is a direct measure of the amount of radiation absorbed. The resulting rise in temperature must then be measured. Thermopile detectors provide a thermoelectric power resulting from a change in temperature. A thermistor can also be used as the detector and results in a change in resistance with temperature.

A second basic type of detector relies on the interaction of a photon with an electron, resulting in an electric current. In a photomultiplier tube, the emitted electrons are accelerated and used to create an amplified current, which is measured. Photovoltaic cells may be employed as radiation detectors. The photovoltaic effect results from the generation of a potential across a p-n junction in a semiconductor when it is subject to a flux of photons. Electron-hole pairs are formed if the incident photon has an energy level of sufficient magnitude. This process will result in the direct conversion of radiation into electrical energy and results in high sensitivity

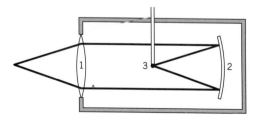

Figure 8.26 Schematic of a basic radiometer: 1, lens; 2, focusing mirror; 3, detector (thermopile or thermistor).

and a fast response time when used as a detector. In general, the photon detectors tend to be spectrally selective, so that the relative sensitivity of the detector tends to change with the wavelength of the measured radiation.

Many considerations enter into the choice of a detector for radiative measurements. If time response is important, photon detectors are significantly faster than thermopile or thermistor detectors, and therefore have a much wider frequency response. Photodetectors saturate, whereas thermopile sensors may slowly change their characteristics over time. Some instruments have variations of sensitivity with the incident angle of the incoming radiation; this factor may be important for solar insolation measurements. Other considerations include wavelength sensitivity, cost, and allowable operating temperatures.

Radiative Temperature Measurements

Commercially applicable radiation thermometers vary widely in their complexity and the accuracy of the resultant measurements. We will consider only the basic techniques that allow measurement of temperature. Perhaps the simplest form, a radiometer measures a source temperature by measuring the voltage output from a thermopile detector. A schematic of such a device is shown in Figure 8.26. The increase in temperature of the thermopile is a direct indication of the temperature of the radiation source. One application of this principle is in the measurement of total solar radiation incident upon a surface. Figure 8.27 shows a schematic of a pyranometer, used to measure global solar irradiance. It would have a hemispherical field of view, and it measures both the direct or beam radiation, and diffuse radiation. The diffuse and beam components of radiation can be separated by shading the pyranometer from the direct solar radiation, thereby measuring the diffuse component.

Optical pyrometry identifies the temperature of a surface by its color, or more precisely the color of the radiation it emits. A schematic of an optical pyrometer is shown in Figure 8.28. A standard lamp is calibrated so that the current flow through its filament is controlled and calibrated in terms of the filament temperature. Comparison is made optically between the color of this filament and the surface of the object whose temperature is being measured. The comparator can be the human eye. Uncertainties in the measurement may be reduced by appropriately filtering the incoming light. Corrections must be applied for surface emissivity associated with the measured radiation; uncertainties vary with the skill of the user, and generally are of the order of 5°C. Replacing the human eye with a different detector extends the range of useful temperature measurement and improves the precision.

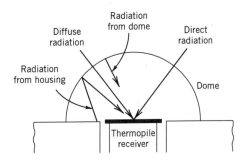

Figure 8.27 Pyranometer construction.

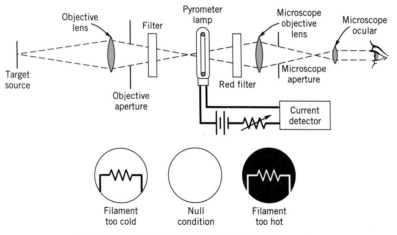

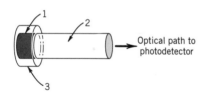

Filament too cold   Null condition   Filament too hot

Appearance of lamp filament in eyepiece of optical pyrometer.

Figure 8.28 Schematic diagram of a disappearing filament optical pyrometer.

The major advantage of an optical pyrometer lies in its ability to measure high temperatures remotely. For example, it could be used to measure the temperature of a furnace without having any sensor in the furnace itself. For many applications this provides a safe and economical means of measuring high temperatures.

Optical Fiber Thermometers

The optical fiber thermometer is based on the creation of an ideal radiator that is optically coupled to a fiber optic transmission system [12, 13], as shown in Figure 8.29. The temperature sensor in this system is a thin, single-crystal aluminum oxide (sapphire) fiber; a metallic coating on the tip of the fiber forms a blackbody radiating cavity, which radiates directly along the sapphire crystal fiber. The single-crystal sapphire fiber is necessary because of the high-temperature operation of the thermometer. The operating range of this thermometer is 300–1900°C. Signal transmission is accomplished by using standard, low-temperature fiber optics. A specific wavelength band of the transmitted radiation is detected and measured, and these raw data are reduced to yield the temperature of the blackbody sensor.

The absence of electrical signals associated with the sensor signal provides excellent immunity from electromagnetic and rf interference. The measurement system has superior frequency response and sensitivity. The system has been employed for measurement in combustion applications. Temperature resolution of 0.0001°C or better is possible.

Optical path to photodetector

Figure 8.29 Optical fiber thermometer: 1, blackbody cavity (iridium film); 2, sapphire fiber (single crystal); 3, protective coating (Al_2O_3).

8.7   PHYSICAL ERRORS IN TEMPERATURE MEASUREMENT

In general, errors in temperature measurement derive from two fundamental sources. The first source of errors derives from uncertain information about the temperature of the sensor itself. Such uncertainties can result from random interpolation errors, calibration bias errors, or a host of other error sources. However, errors in temperature measurement could occur even if the temperature of the sensor was measured exactly. In such cases, the probe does not sense accurately the temperature it was intended to measure.

A list of typical errors associated with the use of temperature sensors is provided in Table 8.8. Precision errors in temperature measurements are a result of resolution limits of measuring and recording equipment, time variations in extraneous variables, and other sources of random error. Thermocouples have some characteristics that can lead to both bias and precision errors, such as the effect of extension wires and connectors. Another major source of error for thermocouples involves the accuracy of the reference junction. Ground loops can lead to spurious readings, especially when thermocouple outputs are amplified for control or data acquisition purposes. As with any measurement system, calibration of the entire measurement system, in place if possible, is the best means of identifying error sources and reducing the resulting measurement uncertainty to acceptable limits.

Insertion Errors

This discussion will focus on ensuring that a sensor output accurately represents the temperature it is intended to measure. For example, suppose it is desired to measure the outdoor temperature. This measurement could employ a large dial thermometer, which might be placed on a football field or a tennis court in the direct sunlight, and

Table 8.8 Measuring Errors Associated with Temperature Sensors

Precision Errors

1. Imprecision of readings
2. Time and spatial variations

Bias Errors

1. Insertion errors, heating or cooling of junctions
   a. Conduction errors
   b. Radiation errors
   c. Recovery errors
2. Effects of plugs and extension wires
   a. Non-isothermal connections
   b. Loading errors
3. Ignorance of materials or material changes during measurements
   a. Aging following calibration
   b. Annealing effects
   c. Cold work hardening
4. Ground loops
5. Magnetic field effects
6. Galvanic error
7. Reference junction inaccuracies

assumed to represent "the temperature," perhaps as high as 50°C (120°F). But what temperature is being indicated by this thermometer? Certainly the thermometer is not measuring the air temperature, nor is it measuring the temperature of the field or the court. The thermometer is subject to the very sources of error we wish to describe and analyze. The thermometer, very simply, indicates its own temperature! The temperature of the thermometer is the thermodynamic equilibrium temperature that results from the radiant energy from the sun, convective exchange with the air, and conduction heat transfer with the surface on which it is resting. Considering the fact that these thermometers typically have a glass cover, which acts as a solar collector, it is very likely that the thermometer temperature is significantly higher than the air temperature.

The physical mechanisms that may cause a temperature probe to indicate a temperature different than that intended include conduction, radiation, and recovery errors. In any real measurement system, their effects are coupled, and therefore should not be considered independently. However, for simplicity, each of these error sources will be considered separately. Our purpose is to provide only approximate analyses of the errors, and not to provide predictive techniques for correcting measured temperatures. These analyses should be used for selection and design of temperature measuring systems. The goal of the measurement engineer should be to minimize these errors, as far as is possible, through the careful installation and design of temperature probes.

EXAMPLE 8.11

An effective method of evaluating data-acquisition and reduction errors associated with the use of multiple temperature sensors within a test facility is to provide a known temperature point at which the sensor outputs can be compared. Suppose the outputs from M similar thermocouple sensors (e.g., all T-type) are to be measured and stored on an M-channel data-acquisition system. Each sensor is referenced to the same reference junction temperature (e.g., ice point) and operated in the normal manner. The sensors are exposed to a known and uniform temperature. N (say 30) readings for each of the M thermocouples are recorded. What information can be obtained from the data?

KNOWN

$M(j = 1, 2, \ldots, M)$ thermocouples
$N(i = 1, 2, \ldots, N)$ readings measured for each thermocouple

SOLUTION

The mean value for all readings of the ith thermocouple is given as

$$\bar{T}_j = \frac{1}{N} \sum_{i=1}^{N} T_{ij}$$

The pooled mean for all the thermocouples is given as

$$\langle \bar{T} \rangle = \frac{1}{M} \sum_{j=1}^{M} \bar{T}_j$$

The difference between the pooled mean temperature and the known temperature would provide an estimate of the bias limit that can be expected from any channel during data acquisition. In contrast, the differences between each $\bar{T}_j$ and $\langle \bar{T} \rangle$ must reflect the precision among the M channels. The precision index for the data-acquisition and reduction instrumentation system is then

$$\langle S_T \rangle = \sqrt{\frac{\sum\limits_{j=1}^{M}\sum\limits_{i=1}^{N}(T_{ij} - \bar{T}_j)^2}{M(N-1)}}$$

with degrees of freedom, $\nu = M(N-1)$

COMMENT

Elemental errors accounted for in these estimates include the following:

- reference junction precision errors
- precision errors in the known temperature
- data-acquisition system precision errors
- extension cable and connecting plug bias errors
- thermocouple emf–T correlation bias errors

These estimates would not include instrument bias errors or any probe insertion errors resulting from subsequent measurements.

Conduction Errors

Errors that result from conduction heat transfer between the measuring environment and the ambient are often called *immersion errors*. Consider the temperature probe shown in Figure 8.30. In many circumstances, a temperature probe extends from the measuring environment through a wall into the ambient environment, where indicating or recording systems are located. The probe and the electrical leads form a path for the conduction of energy from the measuring environment to the ambient. The fundamental nature of the error created by conduction in measured temperatures can be illustrated by the model of a temperature probe shown in Figure 8.31. The essential physics of immersion errors associated with conduction can be discerned by modeling the temperature probe as a fin. Suppose we assume that the measured temperature is higher than the ambient temperature. If we consider a differential element of the fin at steady state, as shown in Figure 8.31(b), there is energy conducted along the fin and transferred by convection from the surface. The surface area for

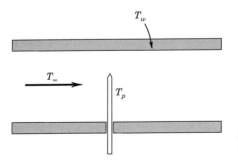

Figure 8.30 Temperature probe inserted into a measuring environment.

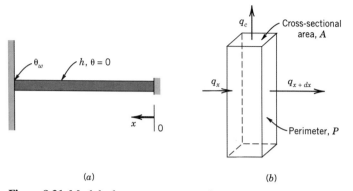

Figure 8.31 Model of a temperature probe as a one-dimensional fin.

convection is $P\,dx$, where P is the perimeter or circumference. Applying the first law of thermodynamics to this differential element yields

$$q_{x+dx} - q_x = hP\,dx[T(x) - T_\infty] \tag{8.22}$$

where h is the convection coefficient. If q is expanded in a Taylor series about the point x, and the substitutions

$$\theta = T - T_\infty \qquad q = -kA\frac{dT}{dx} \qquad m = \sqrt{\frac{hP}{kA}} \tag{8.23}$$

are made, then the governing differential equation becomes

$$\frac{d^2\theta}{dx^2} - m^2\theta = 0 \tag{8.24}$$

Here k is the effective thermal conductivity of the temperature probe. The solution to this differential equation for the boundary conditions that the wall has a temperature T_w, or a normalized value $\theta_w = T_w - T_\infty$, and the end of the fin is small in surface area, is

$$\frac{\theta(x)}{\theta_w} = \frac{\cosh mx}{\cosh mL} \tag{8.25}$$

The point $x = 0$ is the location where the temperature is being assumed to be measured, and therefore the solution is evaluated at $x = 0$ as

$$\frac{\theta(0)}{\theta_w} = \frac{T(0) - T_\infty}{T_w - T_\infty} = \frac{1}{\cosh mL} \tag{8.26}$$

From this analysis, the error that is due to conduction, e_C, can be estimated. An ideal sensor would indicate the fluid temperature, T_∞; therefore, if the sensor temperature is $T_p = T(0)$, then the conduction error is

$$e_C = T_p - T_\infty = \frac{T_w - T_\infty}{\cosh mL} \tag{8.27}$$

Probe Design

The purpose of the preceeding analysis is to gain some physical understanding of ways to minimize conduction errors (not to correct inaccurate measurements). The

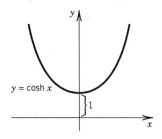

Figure 8.32 Behavior of the hyperbolic cosine.

behavior of this solution is such that the ideal temperature probe would have $T_p = T_\infty$, or $\theta(0) = 0$, implying that $e_C = 0$. Equation (8.26) shows that a value of $\theta(0)$ different from zero results from a nonzero value of θ_w, and a finite value of $\cosh(mL)$. The difference between the fluid temperature being measured and the wall temperature should be as small as possible; clearly, this implies that the wall should be insulated to minimize this temperature difference, and the resulting conduction error.

The term $\cosh(mL)$ should be as large as possible. The behavior of the cosh function is shown in Figure 8.32. Since the hyperbolic cosine monotonically increases for increasing values of the argument, the goal of a probe design should be to maximize the value of the product mL, or $(hP/kA)^{1/2}L$. In general, the thermal conductivity of a temperature probe and the convection coefficient are not design parameters. Thus, two important conclusions are that the probe should be as small in diameter as possible and should be inserted as far as possible into the measuring environment, away from the bounding surface, producing a large value of L. A small diameter increases the ratio of the perimeter, P, to the cross-sectional area, A. For a circular cross section, this ratio is $4/D$, where D is the diameter. A good rule of thumb based on equation (8.27) is to have an $L/D > 50$ for negligible conduction error.

Although this analysis clearly indicates the fundamental aspects of conduction errors in temperature measurements, it does not provide the capability to correct measured temperatures for conduction errors. Conduction errors should be minimized through appropriate design and installation of temperature probes. Usually, the physical situation is sufficiently complex to preclude accurate mathematical description of the measurement errors. Additional information on modeling conduction errors may be found in [14].

Radiation Errors

Consider a temperature probe used to measure a gas temperature. In the presence of significant radiation heat transfer, the equilibrium temperature of a temperature probe may be different than the fluid temperature being measured. Because radiation heat transfer is proportional to the fourth power of temperature, the importance of radiation effects increase as the absolute temperature of the measuring environment increases. The error caused by radiation can be estimated by considering steady-state thermodynamic equilibrium conditions for a temperature sensor. Consider the case in which energy is transferred to or from a sensor by convection from the environment and from/to the sensor by radiation to a body at a different temperature, such as a pipe or furnace wall. For this analysis, conduction will be neglected. A first law analysis of a system containing the probe, at steady-state conditions, yields

$$q_v + q_r = 0 \tag{8.28}$$

where

q_v = convective heat transfer
q_r = radiative heat transfer

The heat-transfer components can then be expressed in terms of the appropriate fundamental relations as

$$q_v = hA_s(T_\infty - T) \qquad q_r = \sigma\epsilon(T_w^4 - T^4) \tag{8.29}$$

Assuming that the surroundings may be treated as a blackbody, the first law for a system consisting of the temperature probe is

$$hA_s(T_\infty - T_p) = FA_s\epsilon\sigma(T_p^4 - T_w^4) \tag{8.30}$$

A temperature probe is generally small compared to its surroundings, which justifies the assumption that the surroundings may be treated as black. The radiation error, e_r, is estimated by

$$e_r = (T_p - T_\infty) = \frac{F\epsilon\sigma}{h}(T_w^4 - T_p^4) \tag{8.31}$$

where

σ = the Stefan–Boltzmann constant ($\sigma = 5.669 \times 10^{-8}$ W/m² K⁴)
ϵ = emissivity of the sensor
F = radiation view factor
T_p = probe temperature
T_w = temperature of the surroundings
T_∞ = fluid temperature being measured

Again, if the sensor is small compared to the scale of the surroundings, the view factor from the sensor to the surroundings may be taken as 1.

EXAMPLE 8.12

A typical situation in which radiation would be important occurs in measuring the temperature of a furnace. Figure 8.33 shows a small temperature probe for which conduction errors are negligible, which is placed in a high-temperature enclosure, where the fluid temperature is T_∞ and the walls of the enclosure are at T_w. Convection

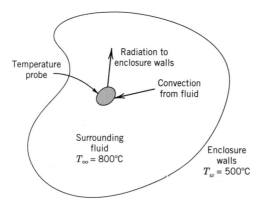

Figure 8.33 Analysis of a temperature probe in a radiative and convective environment.

and radiation are assumed to be the only contributing heat transfer modes at steady state. Develop an expression for the equilibrium temperature of the probe. Also determine the equilibrium temperature of the probe and the radiation error in the case in which $T_\infty = 800°C$, $T_w = 500°C$, and the emissivity of the probe is 0.8. The convective heat transfer coefficient is 100 W/m² °C.

KNOWN

Temperatures $T_\infty = 800°C$ and $T_w = 500°C$, with $h = 100$ W/m² °C. The emissivity of the probe is 0.8.

FIND

An expression for the equilibrium temperature of the probe, and the resulting probe temperature for the stated conditions.

ASSUMPTIONS

The surroundings may be treated as black, and conduction heat transfer is neglected.

SOLUTION

The probe is modeled as a small spherical body within the enclosed furnace; the radiation view factor from the probe to the furnace is 1.0. At steady state, an equilibrium temperature may be found from an energy balance. The first law for a system consisting of the temperature probe from equation (8.30) is

$$h(T_\infty - T_p) + \sigma\epsilon(T_w^4 - T_p^4) = 0 \tag{8.32}$$

The probe is attempting to measure T_∞. Equation (8.32) may be solved by trial and error to yield an equilibrium temperature for the probe of $T_p = 642.5°C$. The radiation error in this case is $-157.5°C$ [from equation (8.31)]. This result indicates that radiative heat transfer can create significant measurement errors at elevated temperatures.

COMMENT

Equation (8.31) does not allow direct solution for the radiation error, since it contains the probe temperature. A trial-and-error approach, as followed here, provides the simplest solution.

Radiation shielding is a key concept in controlling radiative heat transfer; shielding for radiation is analogous to insulation to reduce conduction heat transfer. A *radiation shield* is an opaque surface interposed between a temperature sensor and its radiative surroundings so as to reduce electromagnetic wave interchange. In principle, the shield attains an equilibrium temperature closer to the fluid temperature than the surroundings. Because the probe can no longer "see" the surroundings, with the radiation shield in place, the probe temperature is closer to the fluid temperature. Additional information on radiation error in temperature measurements may be found in [14]. The following example serves to demonstrate radiation errors and the effect of shielding.

EXAMPLE 8.13

Consider again Example 8.12, where the oven is maintained at a temperature of 800°C. Because of energy losses the walls of the oven are cooler, having a temperature of 500°C. For the present case, consider the temperature probe as a small spherical object located in the oven, having no thermal conduction path to the ambient. (All energy exchange is through convection and radiation). Under these conditions, the probe temperature is 642.5°C, as found in Example 8.12.

Suppose a radiation shield is placed between the temperature probe and the walls of the furnace, which blocks the path for radiative energy transfer. Examine the effect of adding a radiation shield on the probe temperature.

KNOWN

A radiation shield is added to a temperature probe in an environment with $T_\infty = 800°C$ and $T_w = 500°C$.

FIND

The radiation error in the presence of the shield.

ASSUMPTIONS

The radiation shield completely surrounds the probe, and the surroundings may be treated as a blackbody.

SOLUTION

The shield equilibrium temperature is higher than the wall temperature by virtue of convection with the fluid. As a result, the probe "sees" a higher temperature surface, and the probe temperature is closer to the fluid temperature, resulting in less measurement error.

For a single radiation shield placed so that it completely surrounds the probe, which is small compared to the size of the enclosure and has an emissivity of 1, the equilibrium temperature of the shield can be determined from equation (8.30). The temperature of the shield is found by trial and error solution to be 628°C. Because the sensor now "sees" the shield, rather than the wall, the temperature measured by the probe will be 697°C, which is also determined by trial and error from equation (8.30).

One shield with an emissivity of 1 provides for an improvement over the case of no shields, but a better choice of the surface characteristics of the shield material can result in much better performance. If the shield has an emissivity of 0.1, the shield temperature rises to 756°C, and the probe temperature to 771°C.

COMMENT

Shielding provides improved temperature measurements by reducing radiative heat transfer. Another area for improvement in this temperature measurement could be the elevation of the wall temperature through insulation.

This discussion of radiation shielding serves to demonstrate the usefulness of shielding as a means of improving temperature measurements in radiative environments. As with conduction errors, the development should be used to guide the design and installation of temperature sensors, not to correct measured temperatures. Further information on radiation errors may be found in [5].

Recovery Errors in Temperature Measurement

The kinetic energy of a gas moving at high velocity can be converted to sensible energy by reversibly and adiabatically bringing the flow to rest at a point. The temperature resulting from this process is called the *stagnation* or *total temperature*, T_t. In contrast, the *static temperature* of the gas, T_∞, is the temperature that would be measured by an instrument moving at the local stream velocity. From a molecular point of view, the static temperature measures the magnitude of the random kinetic energy of the molecules that comprise the gas, while the stagnation temperature includes both the directed and random components of kinetic energy. Generally, the engineer would be content with knowledge of either temperature, but in high-speed gas flows neither temperature is indicated by the sensor.

For negligible changes in potential energy, and in the absence of heat transfer or work, the energy equation for a flow may be written

$$h_1 + \frac{U^2}{2g_c} = h_2 \qquad (8.33)$$

where state 2 refers to the stagnation condition, and state 1 to a condition where the gas is flowing with the velocity U. Assuming ideal gas behavior, the enthalpy difference $h_2 - h_1$ may be expressed as $c_p(T_2 - T_1)$, or in terms of static and stagnation temperatures

$$\frac{U^2}{2g_c} = c_p(T_t - T_\infty) \qquad (8.34)$$

The term $U^2/2g_c c_p$ is called the *dynamic temperature*.

What implication does this have for the measurement of temperature in a flowing gas stream? The physical nature of gases at normal pressures and temperatures is such that the velocity of the gas on a solid surface is zero, because of the effects of viscosity. Thus, when a temperature probe is placed in a moving fluid, the fluid is brought to rest on the surface of the probe. However, this process may not be thermodynamically reversible, and the gas in question may not behave as an ideal gas. Deceleration of the flow by the probe converts some portion of the directed kinetic energy of the flow to thermal energy, and it elevates the temperature of the probe above the static temperature of the gas. The fraction of the kinetic energy recovered as thermal energy is called the *recovery factor*, r, defined as

$$r \equiv \frac{T_p - T_\infty}{U^2/2g_c c_p} \qquad (8.35)$$

where T_p represents the equilibrium temperature of the stationary (with respect to the flow) real temperature probe. In general, r may be a function of the velocity of the flow, or more precisely, the Mach number and Reynolds number of the flow, and the shape and orientation of the temperature probe. For thermocouple junctions of round wire, Moffat [15] reports values for r of

$$0.68 \pm 0.07\,(95\%) \qquad \text{for wires normal to the flow}$$
$$0.86 \pm 0.09\,(95\%) \qquad \text{for wires parallel to the flow}$$

These recovery factor values tend to be constant at velocities for which temperature errors are significant, usually flows where the Mach number is greater than 0.1. For thermocouples having a welded junction, a spherical weld bead significantly larger than the wire diameter tends to a value of the recovery factor of 0.75, for the wires parallel or normal to the flow. The relationships between temperature and velocity for temperature probes with known recovery factors are

$$T_p = T_\infty + \frac{rU^2}{2g_c c_p} \tag{8.36}$$

or in terms of the *recovery error*, e_U,

$$e_U = T_\infty - T_p = -\frac{rU^2}{2g_c c_p} \tag{8.37}$$

The probe temperature is related to the stagnation temperature by

$$T_p = T_t - \frac{(1-r)U^2}{2g_c c_p} \tag{8.38}$$

Fundamentally, in liquids the stagnation and static temperatures are essentially equal [5], and the recovery error may generally be taken as zero. In any case, high-velocity flows are rarely encountered in liquids.

EXAMPLE 8.14

A temperature probe having a recovery factor of 0.86 is to be used to measure a flow of air at velocities up to the sonic velocity, at a pressure of 1 atm and a static temperature of 30°C. Calculate the value of the recovery error in the temperature measurement as the velocity of the air flow increases, from 0 to the speed of sound, using equation (8.37).

KNOWN

$r = 0.86$ $p_\infty = 1$ atm abs $= 101$ kPa abs
$M \leq 1$ $T_\infty = 30°C = 303$ K

FIND

The recovery error as a function of air velocity.

ASSUMPTION

Air behaves as an ideal gas.

SOLUTION

Assuming that air behaves as an ideal gas, the speed of sound is expressed as

$$a = \sqrt{kRTg_c} \tag{8.39}$$

For air at 101 kPa and 303 K, with $R = 0.287$ kJ/kg K, the speed of sound is approximately 349 m/s.

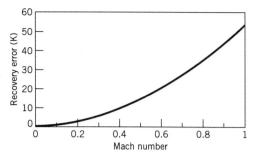

Figure 8.34 Behavior of recovery error as a
function of Mach number.

The Mach number is defined as the ratio of the flow velocity to the speed of
sound. Figure 8.34 shows the recovery error as a function of Mach number for this
temperature probe based on equations (8.37) and (8.39).

COMMENT

Typically, the static and total temperatures of the flowing fluid stream are to be
determined from the measured probe temperature. In this case a second independent
measurement of the velocity is necessary.

8.8 SUMMARY

Temperature is a fundamentally important quantity in science and engineering, both
in concept and practice. As such, temperature is one of the most widely measured engineering variables, providing the basis for a variety of control and safety systems. This
chapter provides the basis for the selection and installation of temperature sensors.

Temperature is defined for practical purposes through the establishment of a
temperature scale, such as the Kelvin scale, that encompasses fixed reference points
and interpolation standards. The International Temperature Scale of 1990 is the
accepted standard for establishing a universal means of temperature measurement.

The two most common methods of temperature measurement employ thermocouples and resistance temperature detectors. Standards for the construction and
use of these temperature measuring devices have been established and provide the
basis for selection and installation of commercially available sensors and measuring
systems.

Installation effects on the accuracy of temperature measurements are a direct
result of the influence of radiation, conduction, and convection heat transfer on the
equilibrium temperature of a temperature sensor. The installation of a temperature probe into a measuring environment can be accomplished in such a way as to
minimize the uncertainty in the resulting temperature measurement.

REFERENCES

1. Patterson, E. C., Eponyms: why celsius? *American Scientist* 77(4):413, 1989.
2. Committee Report, The International Temperature Scale of 1990, *Metrologia* 27(3), 1990. (The text
   of the superseded International Practical Temperature Scale of 1968 appears as an appendix in the
   National Bureau of Standards monograph 124. An amended version was adopted in 1975, and the
   English text published: *Metrologia* 12:7–17, 1976.)

3. Quinn, T. J., News from BIPM, *Metrologia* 26:69–74, 1989.

4. *Temperature Measurement*, American Society of Mechanical Engineers PTC 19.3, 1998.

5. Benedict, R. P., *Fundamentals of Temperature, Pressure and Flow Measurements*, 3d ed., Wiley, New York, 1984.

6. McGee, T. D., *Principles and Methods of Temperature Measurement*, Wiley-Interscience, New York, 1988.

7. Diehl, W., Thin-film PRTD, *Measurements and Control*, December: 155–159, 1982.

8. Thermistor definitions and test methods, *ANSI/Electronic Industries Standard 275-A-1971*, May 1999.

9. *NTC Thermistors*, Vol.1, 1993. Thermometrics, Inc., Location, 1993.

10. Burns, G. W., Scroger, M. G., and Strouse, G. F., Temperature-electromotive force reference functions and tables for the letter-designated thermocouple types based on the ITS–90, *NIST Monograph* 175, April 1993 (supersedes NBS Monograph 125).

11. Dewitt, D. P., and Nutter, G. D., *Theory and Practice of Radiation Thermometry*, Wiley-Interscience, New York, 1988.

12. Dils, R. R., High-temperature optical fiber thermometer, *Journal of Applied Physics*, 54(3):1198–1201, 1983.

13. Optical fiber thermometer, *Measurements and Control*, April 1987.

14. Sparrow, E. M., Error estimates in temperature measurement, in Eckert, E. R. G., and Goldstein, R. J. (eds.), *Measurements in Heat Transfer*, 2d ed., Hemisphere, Washington, DC, 1976.

15. Moffat, R. J., Gas temperature measurements, in *Temperature—Its Measurement and Control in Science and Industry*, Vol. 3, Part 2, Reinhold, New York, 1962.

NOMENCLATURE

c	speed of light in a vacuum $[l\,t^{-1}]$	P	precision index; perimeter $[l]$
c_p	specific heat $[l^2\,t^{-2}/°]$	Q	heat transfer $[m\,l^2\,t^{-3}]$
d	thickness of bimetallic strip $[l]$	R	resistance $[\Omega]$
e_C	conduction temperature error $[°]$	R	gas constant $[l^2\,t^{-2}/°]$
e_U	recovery temperature error $[°]$	R_0	reference resistance $[\Omega]$
e_r	radiation temperature error $[°]$	R_T	thermistor resistance $[\Omega]$
emf	electromotive force	$S_{\bar{x}}$	standard deviation of the means
h	convective heat transfer		for the variable x
	coefficient $[m\,t^{-3}/°]$	T	temperature $[°]$
k	thermal conductivity $[m\,l\,t^{-3}/°]$	T_0	reference temperature $[°]$
k	ratio of specific heats (equation	T_p	probe temperature $[°]$
	for the speed of sound)	T_w	wall or boundary temperature $[°]$
l	length $[l]$	T_∞	fluid temperature $[°]$
m	$\sqrt{h\,p/kA}$ [equation (7.15)] $[l^{-2}]$	U	fluid velocity $[l\,t^{-1}]$
q	heat flux $[m\,t^3]$	α	temperature coefficient of resistivity
r	recovery factor		$[\Omega/°]$
r_c	radius of curvature $[l]$	α_{AB}	Seebeck coefficient
u	uncertainty	β	material constant for thermistor
u_d	design-stage uncertainty		resistance $[°]$
A_c	cross-sectional area $[l^2]$	β	constant in polynomial expansion
A_N	surface area $[l^2]$	π_{AB}	Peltier coefficient $[m\,l^2\,t^{-3}\,A^{-1}]$
B	bias limit	σ	Thomson coefficient $[m\,l^2\,t^{-3}\,A^{-1}]$
C_α	coefficient of thermal expansion $[l\,l^{-1}/°]$	σ	Stefan–Boltzmann constant for radiation
D	diameter $[l]$		$[m\,t^{-3}(°)^4]$
E_b	blackbody emissive power $[m\,t^{-3}]$	θ	nondimensional temperature
$E_{b\lambda}$	spectral emissive power $[m\,t^{-3}]$	θ_x	sensitivity index for variable x
E_i	input voltage $[V]$		(uncertainty analysis)
E_1	voltage drop across R_1 $[V]$	ρ_e	resistivity $[\Omega\,l]$
F	radiation view factor	δ	thermistor dissipation constant
I	current $[A]$		$[m\,l^2\,t^{-3}/°]$
L	length of temperature probe $[l]$	ϵ	emissivity

PROBLEMS

8.1 Define and discuss the significance of the following terms, as they apply to temperature and temperature measurements:

    a. temperature scale
    b. temperature standards
    c. fixed points
    d. interpolation

8.2 Fixed temperature points in the International Temperature Scale are phase equilibrium states for a variety of pure substances. Discuss the conditions necessary within an experimental apparatus to accurately reproduce these fixed temperature points. How would elevation, weather, and material purity affect the uncertainty in these fixed points?

8.3 Calculate the resistance of a platinum wire that is 2 m in length and has a diameter of 0.1 cm. The resistivity of platinum at $25°C$ is 9.83×10^{-6} Ω cm. What implications does this result have for the construction of a resistance thermometer using platinum?

8.4 An RTD forms one arm of a Wheatstone bridge, as shown in Figure 8.35. The RTD is used to measure a constant temperature, with the bridge operated in a balanced mode. The RTD has a resistance of $25\,\Omega$ at a temperature of $0°C$, and a thermal coefficient of resistance, $\alpha = 0.003925°C^{-1}$. The value of the variable resistance R_1 must be set to $41.485\,\Omega$ to balance the bridge circuit, with the RTD in thermal equilibrium with the measuring environment.

    a. Determine the temperature of the RTD.
    b. Compare this circuit to the equal-arm bridge in Example 8.2. Which circuit provides the greater static sensitivity?

8.5 A thermistor is placed in a $100°C$ environment, and its resistance measured as $20,000\,\Omega$. The material constant, β, for this thermistor is $3650°C$. If the thermistor is then used to measure a particular temperature, and its resistance is measured as $500\,\Omega$, determine the thermistor temperature.

8.6 Estimate the required level of uncertainty in the measurement of resistance for a platinum RTD if the RTD is to serve as a local standard for the calibration of a temperature measurement system for an uncertainty of $\pm0.005°C$. Assume $R(0°C) = 100\,\Omega$.

8.7 Define and discuss the following terms related to thermocouple circuits:

    a. thermocouple junction
    b. thermocouple laws

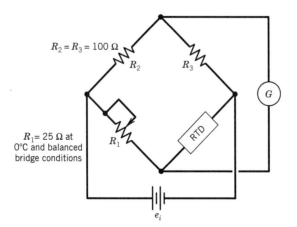

Figure 8.35 Wheatstone bridge circuit for Problem 8.4.

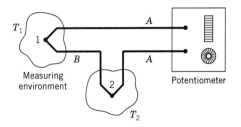

Figure 8.36 Thermocouple circuit for Problems 8.8–8.10: N1, measuring junction; 2, reference junction.

c. reference junction
d. Peltier effect
e. Seebeck coefficient

8.8 The thermocouple circuit in Figure 8.36 represents a J-type thermocouple with the reference junction having $T_2 = 0°C$. The output emf is 13.777 mV. What is the temperature of the measuring junction, T_1?

8.9 The thermocouple circuit in Figure 8.36 represents a J-type thermocouple. The circuit produces an emf of 15 mV for $T_1 = 750°C$. What is T_2?

8.10 The thermocouple circuit in Figure 8.36 is composed of copper and constantan, and has an output voltage of 6 mV for $T_1 = 200°C$. What is T_2?

8.11 a. The thermocouple shown in Figure 8.37(a) yields an output voltage of 7.947 mV. What is the temperature of the measuring junction?
b. The ice bath that maintains the reference junction temperature melts, allowing the reference junction to reach a temperature of 25°C. If the measuring junction of part a remains at the same temperature, what voltage would be measured by the potentiometer?
c. Copper extension leads are installed as shown in Figure 8.37(b). For an output voltage of 7.947 mV, what is the temperature of the measuring junction?

8.12 A J-type thermocouple referenced to 70°F has a measured output emf of 2.878 mV. What is the temperature of the measuring junction?

8.13 A J-type thermocouple referenced to 0°C indicates 4.115 mV. What is the temperature of the measuring junction?

8.14 A temperature measurement requires an uncertainty of ±2°C at a temperature of 200°C. A standard T-type thermocouple is to be used with a readout device that provides electronic ice-point reference junction compensation, and has a stated accuracy of ±0.5°C with 0.1°C resolution. Determine if the uncertainty constraint is met at the design stage.

8.15 A temperature difference of 3.0°C is measured by using a thermopile having three pairs of measuring junctions, arranged as shown in Figure 8.21.

a. Determine the output of the thermopile for J-type thermocouple wire, if all pairs of junctions sense the 3.0°C temperature difference. The average temperature of the junctions is 80°C.
b. If the thermopile is constructed of wire that has a maximum emf variation from the NIST standard values of ±0.8%, and the voltage measuring capabilities in the system are such that the uncertainty is ±0.0005 V, perform an uncertainty analysis to estimate the uncertainty in the measured temperature difference at the design stage.

8.16 Complete the following table for a J-type thermocouple:

Temperature [°C]		emf
Measured	Reference	[mV]
100	0	
	0	−0.5
100	50	
	50	2.5

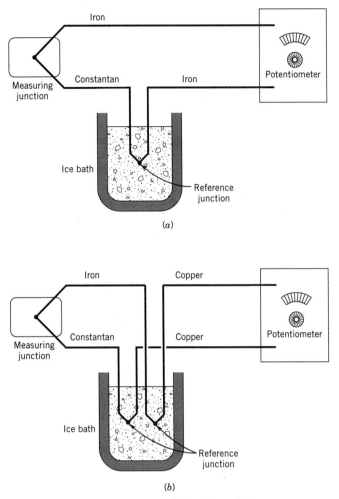

Figure 8.37 Schematic diagram for Problem 8.11.

8.17 A J-type thermopile is constructed as shown in Figure 8.20, to measure a single temperature. For a four-junction thermopile, referenced to 0°C, what would be the emf produced at a temperature of 125°C? If a voltage measuring device was available that had a total uncertainty of ±0.0001 V, how many junctions would be required in the thermopile to reduce the uncertainty in the measured temperature to 0.1°C?

8.18 You are employed as a heating, ventilating, and air conditioning engineer. Your task is to decide where in a residence to place a thermostat, and how it is to be mounted on the wall. A thermostat contains a bimetallic temperature measuring device that serves as the sensor for the control logic of the heating and air conditioning system for the house. Consider the heating season. When the temperature of the sensor falls 1°C below the set-point temperature of the thermostat, the furnace is activated; when the temperature rises 1°C above the set point, the furnace is turned off. Discuss where the thermostat should be placed in the house, what factors could cause the temperature of the sensor in the thermostat to be different from the air temperature, and possible causes of discomfort for the occupants of the house. How does the thermal capacitance of the temperature sensor affect the operation of the thermostat? Why are thermostats typically set 5°C higher in the air conditioning season?

8.19 A J-type thermocouple for use at temperatures between 0 and 100°C was calibrated at the steam point in a device called a hypsometer. A hypsometer creates a constant temperature environment at the saturation temperature of water, at the local barometric pressure. The steam-point temperature is strongly affected by barometric pressure variations. Atmospheric pressure on the day of this calibration was 30.1 in Hg. The steam-point temperature as a function of barometric pressure may be expressed as

$$T_{st} = 212 + 50.422 \left(\frac{p}{p_0} - 1 \right) - 20.95 \left(\frac{p}{p_0} - 1 \right)^2 \quad [^\circ F]$$

where $p_0 = 29.921$ in Hg. At the steam point, the emf produced by the thermocouple, referenced to $0^\circ C$, is measured as 5.310 mV. Construct a calibration curve for this thermocouple by plotting the difference between the thermocouple reference table value and the measured value ($emf_{ref} - emf_{meas}$) versus temperature. What is this difference at $0^\circ C$? Suggest a means for measuring temperatures between 0 and $100^\circ C$ using this calibration, and estimate the contribution to the total uncertainty.

8.20 A J-type thermocouple is calibrated against an RTD standard within $\pm 0.01^\circ C$ between 0 and $200^\circ C$. The emf is measured with a potentiometer having 0.001-mV resolution and less than 0.015-mV bias. The reference junction temperature is provided by an ice bath. The calibration procedure yields the following results:

$T_{RTD} [^\circ C]$	0.00	20.50	40.00	60.43	80.25	100.65
emf [mV]	0.010	1.038	2.096	3.207	4.231	5.336

a. Determine a polynomial to describe the relation between the temperature and thermocouple emf.
b. Estimate the uncertainty in temperature using this thermocouple and potentiometer.
c. Suppose the thermocouple is connected to a digital temperature indicator having a resolution of $0.1^\circ C$ and better than $0.3^\circ C$ accuracy. Estimate the uncertainty in indicated temperature.

8.21 A beaded thermocouple is placed in a duct in a moving gas stream having a velocity of 200 ft/s. The thermocouple indicates a temperature of $1400^\circ R$.

a. Determine the true static temperature of the fluid, based on correcting the reading for velocity errors. Take the specific heat of the fluid to be 0.6 Btu/lb$_m$R, and the recovery factor to be 0.22.
b. Estimate the error in the thermocouple reading caused by radiation if the walls of the duct are at $1200^\circ R$. The view factor from the probe to the duct walls is 1, the convective heat transfer coefficient, h, is 30 Btu/h ft^2 $^\circ R$, and the emissivity of the temperature probe is 1.

8.22 It is desired to measure the static temperature of the air outside of an aircraft flying at 20,000 ft with a speed of 300 miles per hour, or 438.3 ft/s. A temperature probe is used that has a recovery factor, r, of 0.75. If the static temperature of the air is $413^\circ R$, and the specific heat is 0.24 Btu/lb$_m$R, what is the temperature indicated by the probe? Local atmospheric pressure at 20,000 ft. is approximately 970 lb/ft^2, which results in an air density of 0.0442 lb$_m$/ft^3. Discuss additional factors that might affect the accuracy of the static temperature reading.

8.23 Consider the typical construction of a sheathed thermocouple, as shown in Figure 8.38. Analysis of this geometry to determine conduction errors in temperature measurement is difficult.

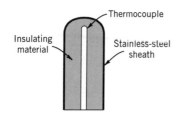

Insulating material

Thermocouple

Stainless-steel sheath

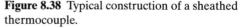

Figure 8.38 Typical construction of a sheathed thermocouple.

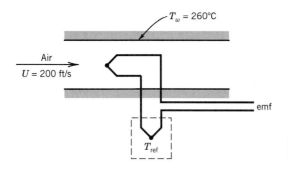

Figure 8.39 Schematic diagram for Problem 8.24.

Suggest a method for placing a realistic upper limit on the conduction error for such a probe, for a specified immersion depth into a convective environment.

8.24 An iron–constantan thermocouple is placed in a moving air stream in a square duct, as shown in Figure 8.39. The thermocouple reference junction is maintained at 80°C. The emf output from the thermocouple is 18 mV.

a. Determine the thermocouple junction temperature.
b. By considering recovery and radiation errors, estimate the total error in the indicated temperature (i.e., the error if we assumed the junction temperature represented the air temperature). Discuss whether this estimate of the error is conservative, and why or why not. The heat transfer coefficient may be taken as 400 W/m²°C.

Air Properties	Thermocouple Properties
$c_p = 1$ kJ/kg K	$r = 0.7$
$u = 61$ m/s	$\varepsilon = 0.25$

8.25 In Example 8.5, an uncertainty value for R_T was determined at 125°C as $B_{R_T} = \pm 247\,\Omega$. Show that this value is correct by performing an uncertainty analysis on R_T. In addition, determine the value of B_{R_T} at the temperatures 150 and 100°C. What error is introduced into the uncertainty analysis for β by using the value of B_{R_T} at 125°C?

8.26 The thermocouple circuit shown in Figure 8.40 measures the temperature T_1. The potentiometer limits of error are given as follows.

Limits of error: ±0.05% of reading +15 μV at 25°C
Resolution: 5 μV

Give a best estimate for the temperature T_1, if the output emf is 9 mV.

8.27 A concentration of salt of 600 ppm in tap water will cause a 0.05°C change in the freezing point of water. For an ice bath prepared using tap water and ice cubes from tap water at a local laboratory, and having 1500 ppm of salts, determine the error in the ice-point reference. Upon repeated measurements, is this error manifested as a bias or precision error? Explain.

8.28 A platinum RTD ($\alpha = 0.00392°C^{-1}$) is to be calibrated in a fixed-point environment. The probe itself is used in a balanced mode with a Wheatstone bridge, as shown in Figure 8.41. The

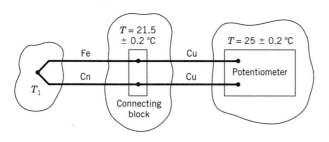

Figure 8.40 Thermocouple circuit for Problem 8.26.

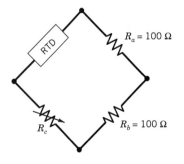

Figure 8.41 Bridge circuit for Problem 8.28.

bridge resistances are known to an uncertainty of $\pm 0.001\ \Omega$ (95%). At $0°C$ the bridge balances when $R_c = 100.000\ \Omega$. At $100°C$ the bridge balances when $R_c = 139.200\ \Omega$.

a. Find the RTD resistance corresponding to 0 and $100°C$, and the uncertainty in each value.
b. Calculate the uncertainty in determining a temperature using this RTD-bridge system for a measured temperature that results in $R_c = 300\ \Omega$. Assume $u_\alpha = \pm 1 \times 10^{-5}$ (95%).

8.29 A T-type thermopile is used to measure temperature difference across insulation in the ceiling of a residence in an energy monitoring program. The temperature difference across the insulation is used to calculate energy loss through the ceiling from the relationship

$$Q = kA_c(\Delta T/L)$$

where

$A_c = $ ceiling area $= 15\ m^2$
$k = $ insulation thermal conductivity $= 0.4\ W/m\,°C$
$L = $ insulation thickness $= 0.25\ m$
$\Delta T = $ temperature difference $= 5°C$
$Q = $ heat loss (W)

The value of the temperature difference is expected to be $5°C$, and the thermocouple emf is measured with an uncertainty of ± 0.04 mV. Determine the required number of thermopile junctions to yield an uncertainty in Q of $\pm 5\%$, assuming the uncertainty in all other variables may be neglected.

8.30 A T-type thermocouple referenced to $0°C$ is used to measure the temperature of boiling water. What is the emf of this circuit at $100°C$?

8.31 A T-type thermocouple referenced to $0°C$ develops an output emf of 1.2 mV. What is the temperature sensed by the thermocouple?

8.32 A temperature measurement system consists of a digital voltmeter and a T-type thermocouple. The thermocouple leads are connected directly to the voltmeter, which is placed in an air-conditioned space at $25°C$. The output emf from the thermocouple is 10 mV. What is the measuring junction temperature?

Chapter 9

Pressure and Velocity Measurements

9.1 INTRODUCTION

In this chapter, methods to measure the pressure within a stationary and a moving fluid are presented. Instruments and procedures for establishing known values of pressure for calibration purposes, as well as various types of transducers for pressure measurement, are discussed. Three well-established methods for the measurement of the local velocity within a moving fluid are also discussed. Practical considerations, including common error sources, for pressure and velocity measurements are presented.

9.2 PRESSURE CONCEPTS

Pressure represents a contact force per unit area. It acts inward and normal to the surface of any physical boundary that a fluid contacts. A basic understanding of the origin of a pressure involves the consideration of the forces acting between the fluid molecules and the solid boundaries containing the fluid. For example, consider the measurement of pressure at the wall of a vessel containing a perfect gas. As a molecule with some amount of kinetic energy collides with the solid boundary, it will rebound off in a different direction. From Newton's second law, we know that the change in linear momentum of the molecule produces an equal but opposite force on the boundary. It is the net effect of these collisions that yields the pressure sensed at the boundary surface. Factors that affect the magnitude or frequency of the collisions, such as fluid temperature and fluid density, will affect the pressure. In fact, this reasoning is the basis of the kinetic theory from which the ideal gas equation of state may be derived.

A pressure scale can be related to molecular activity, since a lack of any molecular activity must form the limit of absolute zero pressure. A pure vacuum, which contains no molecules, would form the primary standard for absolute zero pressure. As shown in Figure 9.1, the absolute pressure scale is quantified relative to this absolute zero pressure. The pressure under standard atmospheric conditions is defined [1] as 1.01320×10^5 Pa absolute (1 Pa $= 1 \text{ N/m}^2$). This is equivalent to

- 101.32 kPa absolute
- 1 atm absolute

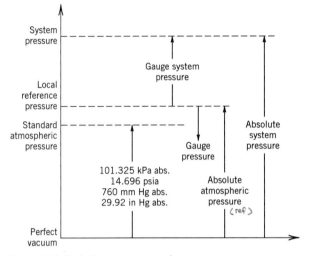

Figure 9.1 Relative pressure scales.

- 14.696 lb/in.^{2} absolute (psia)
- 1.013 bar absolute

Also indicated in Figure 9.1 is a gauge pressure scale. The gauge pressure scale is measured relative to some absolute reference pressure that is defined in a manner convenient to the measurement. The relation between an *absolute pressure*, p_{abs}, and its corresponding *gauge pressure*, p_g, is given by

$$p_g = p_{abs} - p_0 \tag{9.1}$$

where p_0 is a reference pressure. A commonly used reference pressure is the local absolute atmospheric pressure existing during the measurement. For example, if we choose the reference pressure as 1 atm absolute, then the pressure at one atmosphere is either 1 atm absolute or 0 atm. As a convention, an absolute pressure will always be noted as such by using the term "absolute" or the abbreviations "abs" or "a" following its units. Absolute pressure will be a positive number. Gauge pressure can be positive or negative depending on the value of measured pressure relative to the reference pressure. A *differential pressure*, such as $p_1 - p_2$, is a relative measure and cannot be written as an absolute pressure.

Pressure can also be described in terms of the pressure exerted on a surface submerged in a column of fluid at a depth, h, as depicted in Figure 9.2. From hydrostatics, the pressure at any depth within a fluid of specific weight γ can be written as

$$p_{abs}(h) = p_0(h_0) + \gamma h \tag{9.2}$$

In equation (9.2), p_0 is determined at and h measured from an arbitrary datum line, h_0. The fluid specific weight is given by $\gamma = \rho g/g_c$. When equation (9.2) is rearranged, the equivalent head of fluid of depth, h, becomes

$$h = (p_{abs} - p_0)/\gamma \tag{9.3}$$

The equivalent pressure head at 1 standard atmosphere is defined to be

$$760\,\text{mm Hg abs} = 760\,\text{Torr abs} = 1\,\text{atm abs}$$
$$= 10{,}350.8\,\text{mm H}_2\text{O abs}$$
$$= 29.92\,\text{in. Hg abs}$$
$$= 407.513\,\text{in. H}_2\text{O abs}$$

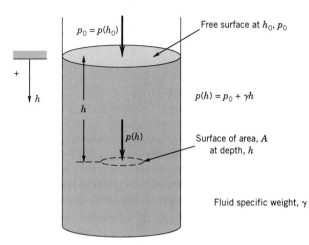

Figure 9.2 Hydrostatic equivalent pressure head and pressure.

The standard is based on mercury with a density of 0.0135951 kg/cm³ at 0°C and water at 0.000998207 kg/cm³ at 20°C [1]. $\rho_{Hg} = 13595.1 \ kg/m^3 \ at \ 0°C$

$\rho_{H_2O} = 998.207 \ kg/m^3 \ at \ 20°C.$

EXAMPLE 9.1

Determine the absolute and gauge pressures and the equivalent pressure head at a depth of 10 m below the free surface of a pool of water at 20°C.

KNOWN

$h = 10$ m, where $h = 0$ is the free surface
$T = 20°C$
$\rho_{H_2O} = 998.207$ kg/m³
Specific gravity of mercury, $S_{Hg} = 13.57 = \dfrac{\rho_{Hg}}{\rho_{H_2O}}$ where   Specific Weight $= \gamma_{Hg} = \dfrac{\rho g}{g_c}$

ASSUMPTIONS

Water density constant
$p_0(h = 0) = 1.0132 \times 10^5$ N/m² abs

FIND

p_{abs}, p_g, and h

SOLUTION

The absolute pressure can be determined directly from equation (9.2). Using the pressure at the free surface as the reference pressure and the datum line for h_0, the absolute pressure must be

$$p_{abs} = 1.0132 \times 10^5 \text{ N/m}^2 + \frac{(997.4 \text{ kg/m}^3)(9.8 \text{ m/s}^2)(10 \text{ m})}{1 \text{ kg m/N s}^2}$$

$$= 1.9906 \times 10^5 \text{ N/m}^2 \text{ abs}$$

$$= 199.06 \text{ kPa abs} = 1.96 \text{ atm abs} = 28.80 \text{ lb/in.}^2 \text{ abs} = 1.99 \text{ bar abs.}$$

The gauge pressure is found from equation (9.1) to be

$$p_g = p_{abs} - p_0 = \gamma h$$
$$= 9.7745 \times 10^4 \text{ N/m}^2$$
$$= 97.7 \text{ kPa} = 0.96 \text{ atm} = 14.11 \text{ lb/in.}^2 = 0.97 \text{ bar}$$

For example, then, the pressure could be stated as

$$p = 1.96 \text{ atm abs} = 0.96 \text{ atm}$$

The pressure in terms of equivalent head is found from equation (9.3):

$$h_{abs} = \frac{1.9906 \times 10^5 \text{ N/m}^2 \text{ abs}}{(997.4 \text{ kg/m}^3)(9.8 \text{ m/s}^2)}$$
$$= 20.36 \text{ m H}_2\text{O abs} = 1.50 \text{ m Hg abs}$$

or in terms of gauge pressure relative to 760 mm Hg abs:

$$h = \frac{(1.9906 \times 10^5) - (1.0132 \times 10^5) \text{ N/m}^2}{(998.207 \text{ kg/m}^3)(9.8 \text{ m/s}^2)(1 \text{ N s}^2/\text{kg m})}$$
$$= 10 \text{ m H}_2\text{O} = 0.73 \text{ m Hg}$$

9.3 PRESSURE REFERENCE INSTRUMENTS

The units of pressure are defined by the standards of the fundamental dimensions of mass, length, and time. However, pressure transducers are generally calibrated by comparison against certain reference instruments from which pressure is determined through methods intrinsic to the reference instrument operating principle. In this section, several basic reference instruments are discussed that are used to establish pressure for the purposes of calibration by comparison, as well as for general measurement.

McLeod Gauge

The McLeod gauge [2] is a pressure-measuring instrument and laboratory reference standard used to establish gas pressures in the subatmospheric range of 1 mm Hg abs down to 0.1 μm Hg abs. One variation of this instrument is sketched in Figure 9.3(a), in which the gauge is connected directly to the low-pressure source. The glass tubing is arranged so that a sample of the gas at the low pressure to be determined can be trapped by inverting the gauge from the sensing position in Figure 9.3(a) to that of the measuring position of Figure 9.3(b). In this way, the trapped gas within the capillary is isothermally compressed by a rising column of mercury. Boyle's law is then used to relate the two pressures on either side of the mercury to the distance of travel of the mercury within the capillary. Mercury is the preferred working fluid because of its high density and very low vapor pressure.

At the equilibrium and measuring position in Figure 9.3(b), the capillary pressure, p_2, is related to the unknown gas pressure to be determined, p_1, by

$$p_2 = p_1(V_1/V_2)$$

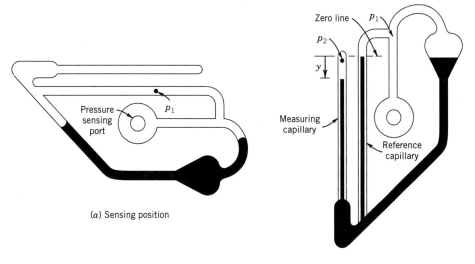

(a) Sensing position

(b) Indicating position

Figure 9.3 McLeod gauge.

where V_1 is the gas volume of the gauge in Figure 9.3(a) (a constant for a gauge at any pressure) and V_2 is the capillary volume in Figure 9.3(b). But $V_2 = Ay$, where A is the known cross-sectional area of the capillary and y is the vertical length of the capillary occupied by the gas. Letting γ be the specific weight of the mercury, the difference in pressures is related by

$$p_2 - p_1 = \gamma y$$

such that the unknown gas pressure is just a function of y:

$$p_1 = \gamma Ay^2/(V_1 - Ay) \tag{9.4}$$

In practice, a commercial McLeod gauge will have the capillary etched and calibrated to indicate pressure, p_1, or equivalent head, p_1/γ, directly.

The McLeod gauge indication generally does not require correction. The reference stem offsets capillary forces acting in the measuring capillary. Instrument bias error will be of the order of 0.5% (95%) at 1 μm Hg abs and increasing to 3% (95%) at 0.1 μm Hg abs.

Barometer

A barometer consists of an initially evacuated tube that is closed on one end. The open end is inverted and immersed within a liquid-filled reservoir as shown in the illustration of the Fortin barometer in Figure 9.4. The reservoir is open to atmospheric pressure, which forces the liquid to rise up the tube. From equations (9.2) and (9.3), the resulting height of the liquid column above the reservoir free surface is a measure of the absolute atmospheric pressure in equivalent head [equation (9.3)]. Evangelista Torricelli (1608–1647), a colleague of Galileo, can be credited with developing and interpreting the working principles of the barometer in 1644.

As Figure 9.4 shows, the closed end of the tube will be at the vapor pressure of the liquid at room temperature. Mercury is the most common liquid used because of its very low vapor pressure. Even so, the barometer will have to be

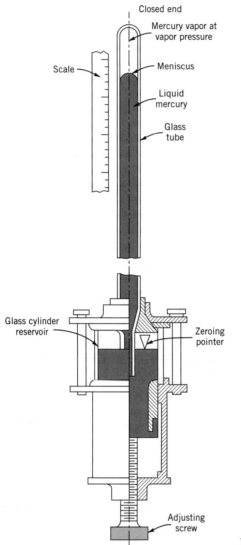

Figure 9.4 Fortin barometer.

corrected for temperature effects on the indicated pressure, for temperature and altitude effects on the weight of mercury, and for deviations from standard gravity (9.80665 m/s² or 32.17405 ft/s²). Correction curves are usually provided by instrument manufacturers.

Barometers are used as local standards for measuring atmospheric pressure. Under standard conditions for pressure temperature and gravity, the mercury will rise 760 mm (29.92 in.) above the reservoir surface. The weather services always report a barometric pressure that has been corrected to sea-level elevation.

Manometers

A manometer is an instrument that utilizes the hydrostatic relationship between pressure and the hydrostatic equivalent head of fluid. Several variations in manometer design are available, but all are basically pressure differential instruments used

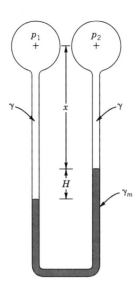

Figure 9.5 U-tube manometer.

to measure differential pressures ranging on an order of magnitude of between 0.001 mm of manometer fluid to several meters.

The U-tube manometer in Figure 9.5 consists of a transparent tube filled with an indicating liquid of specific weight, γ_m. This forms two free surfaces of the manometer liquid. The difference in pressures p_1 and p_2 applied across the two free surfaces brings about a deflection, H, in the level of the manometer liquid. For a fluid of specific weight γ the hydrostatic equation can be applied to the manometer of Figure 9.5 as

$$p_1 = p_2 + \gamma x + \gamma_m H - \gamma(H + x)$$

which yields the relation between the manometer deflection and applied differential pressure,

$$p_1 - p_2 = (\gamma_m - \gamma)H \tag{9.5}$$

By relating an input of differential pressure with an output of manometer deflection, equation (9.5) indicates that the static sensitivity of the U-tube manometer is given by

$$K = 1/(\gamma_m - \gamma)$$

For the maximum manometer sensitivity, then, liquids giving as small a value of $(\gamma_m - \gamma)$ as possible should be used as manometeric fluids. From a practical standpoint, however, the manometer fluid must typically have a greater specific weight than, and must not be soluble with, the working fluid. The manometer fluid should be selected to provide a deflection that is measurable yet not so great that it becomes awkward to observe.

A variation in the U-tube manometer is the micromanometer shown in Figure 9.6. These special purpose instruments are used to measure very small differential pressures, down to 0.005 mm H_2O (0.0002 in. H_2O). In the micromanometer, the manometer reservoir is moved up or down until the level of the manometer fluid within the reservoir is at the same level as a set mark within a magnifying sight glass. At that point the manometer meniscus will be at the set mark, and this serves as a reference position. Changes in pressure bring about fluid displacement so that the reservoir must be moved up or down to bring the meniscus back to the set mark. The

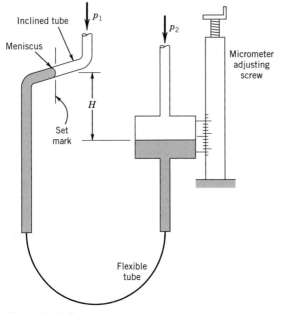

Figure 9.6 Micromanometer.

amount of this repositioning is equal to the change in equivalent pressure head. The position of the reservoir is controlled by a micrometer or other calibrated displacement measuring device so that relative changes in pressure can be measured with a high resolution.

The inclined tube manometer is also used to measure small changes in pressure. It is essentially a U-tube manometer with one tube leg inclined at an angle, typically 10–30° relative to the horizontal. As indicated in Figure 9.7, a change in pressure equivalent to a deflection of height H in a U-tube manometer would bring about a change in position of the meniscus in the inclined leg of $L = H/\sin\theta$. This provides increased sensitivity over the conventional U-tube by a factor of $1/\sin\theta$.

A number of elemental errors affect the instrument uncertainty of all types of manometers. These include scale and alignment errors, zero error, temperature error, gravity error, and capillary and meniscus errors. The specific weight of the manometer fluid will vary with temperature but can be corrected. For example, mercury has a

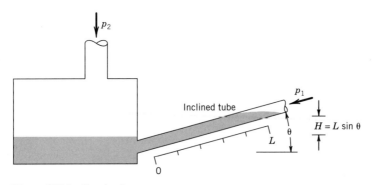

Figure 9.7 Inclined tube manometer.

temperature dependence approximated by

$$\gamma_{Hg} = \frac{848.707}{1 + 0.000101(T - 32)} \text{ [lb/ft}^3\text{]} = \frac{133.084}{1 + 0.00006T} \text{ [N/m}^3\text{]}$$

with T in °F or °C, respectively. A gravity correction for elevation, z, and latitude, ϕ, can be applied to correct for gravity error effects by using the dimensionless correction,

$$e_1 = -(2.637 \times 10^{-3} \cos 2\phi + 9.6 \times 10^{-8} z + 5 \times 10^{-5}) \qquad (9.6a)$$

$$= -(2.637 \times 10^{-3} \cos 2\phi + 2.9 \times 10^{-8} z + 5 \times 10^{-5}) \qquad (9.6b)$$

where ϕ is in degrees and z is in feet for equation (9.6a) and meters in equation (9.6b). Surface tension effects between the manometer and measured fluids give rise to capillary forces on the manometer column and lead to the development of a meniscus. Although the actual effect varies with purity of the manometer liquid, these effects can be minimized by using manometer tube bores of greater than ~6 mm (0.25 in.). In general, the instrument uncertainty in measuring pressure can be as low as 0.02–0.2% of the reading.

EXAMPLE 9.2

An inclined manometer with indicating leg at 30° is to be used at 20°C to measure a gas pressure of nominal magnitude of 100 N/m^2 relative to ambient. "Unity" oil ($S = 1$) is to be used. The specific weight of the oil is 9770 ± 0.5% N/m^2 (95%) at 20°C, the angle of inclination can be set to within 1° using a bubble level, and the manometer resolution is 1 mm with a manometer zero error equal to its interpolation error. Estimate the uncertainty in indicated differential pressure at the design stage, $\gamma_{gas} = 11.5 \pm 0.5\%$ N/m^3 (95%).

KNOWN

$p \approx 100$ N/m^2

Manometer resolution: 1 mm

Zero error: 0.5 mm

$\theta = 30 \pm 1°$ (95% assumed)

$\gamma_m = 9770 \pm 0.5\%$ N/m^2 (95%)

ASSUMPTIONS

Temperature and capillary effects in manometer and gravity error in the specific weights of the fluids are negligible.

FIND

Design stage uncertainty, u_d

SOLUTION

The relation between pressure and manometer deflection is given by equation (9.5) with $H = L \sin \theta$:

$$p_1 - p_2 = L(\gamma_m - \gamma) \sin \theta$$

where p_2 is ambient pressure. Hence, the indicated gauge pressure can be written as

$$p = L(\gamma_m - \gamma) \sin \theta$$

For a nominal pressure of 100 N/m^2, the nominal manometer deflection L would be

$$L = \frac{p}{(\gamma_m - \gamma) \sin \theta} \approx 21 \text{ mm}$$

For the design-stage analysis, $p = f(\gamma, \gamma_m, L, \theta)$, so that the uncertainty in pressure, p, is estimated by

$$(u_d)_p = \pm \sqrt{\left[\frac{\partial p}{\partial \gamma_m}(u_d)_{\gamma_m}\right]^2 + \left[\frac{\partial p}{\partial L}(u_d)_L\right]^2 + \left[\frac{\partial p}{\partial \theta}(u_d)_\theta\right]^2}$$

where the uncertainty in γ is assumed to be negligible ($\gamma_m \gg \gamma$). At assumed 95% confidence levels, the manometer specific weight uncertainty and angle uncertainty are set by judgment at

$$(u_d)_{\gamma_m} = (9770 \text{ N/m}^2)(0.005) \approx 49 \text{ N/m}^2$$
$$(u_d)_\theta = 0.0175 \text{ rad}$$

The uncertainty in estimating the pressure from the indicated deflection is due both to the manometer resolution and zero point bias, where the instrument error u_c is taken as the zero point bias, that is the ability to zero the manometer.

$$(u_d)_L = \sqrt{(u_0)^2 + (u_c)^2}$$
$$(u_d)_L = \sqrt{(0.5 \text{ mm})^2 + (0.5 \text{ mm})^2} = 0.7 \text{ mm}$$

This gives a design-stage uncertainty of,

$$(u_d)_p = \pm \sqrt{(0.26)^2 + (3.42)^2 + (3.10)^2}$$
$$= \pm 4.6 \text{ N/m}^2 \quad (95\%)$$

COMMENT

At a 30° inclination and for this pressure, the uncertainty in pressure is affected almost equally by the instrument inclination and deflection uncertainties. As the inclination approaches a vertical orientation, that is, the U-tube manometer, inclination uncertainty becomes less important. However, for a U-tube manometer, the deflection is reduced to less than 11 mm, a 50% reduction in manometer sensitivity, with an associated design-stage uncertainty of 6.8 N/m^2.

Deadweight Testers

The deadweight tester is used as a laboratory standard for the calibration of pressure-measuring devices over the pressure range from 70 to 7×10^7 N/m^2 (0.01 to 10,000 psi). This device determines pressure directly through its fundamental definition of a force per unit area. A deadweight tester, such as that shown in Figure 9.8, consists of an internal chamber filled with a liquid, and a close-fitting piston and cylinder. Chamber

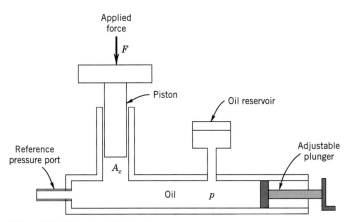

Figure 9.8 Deadweight tester.

pressure is produced by the compression of the liquid, usually oil, by the adjustable plunger. This pressure acts on the end of the carefully machined piston. A static equilibrium will exist when the external pressure exerted by the piston on the fluid balances with the chamber pressure. This external piston pressure is created by a downward force acting over the equivalent area, A_e, of the piston. The weight of the piston plus the additional weight of calibrated masses are used to provide this external force, F. At static equlibrium the piston will float, and the chamber pressure can be deduced as

$$p = \frac{F}{A_e} + \sum \text{errors} \tag{9.7}$$

A pressure-measuring device, such as a pressure transducer, can be connected to the reference port and calibrated by comparison to the chamber pressure.

The instrument uncertainty in the chamber pressure using a deadweight tester can be as low as 0.05–0.01% of the reading. A number of elemental errors contribute, including air buoyancy effects, variations in local gravity, uncertainty in the known mass of the piston and added masses, shear effects, thermal expansion of the piston area, and elastic deformation of the piston.

An indicated pressure, p_i can be corrected for gravity effects, e_1 from equation (9.6a) or (9.6b) and air buoyancy effects, e_2, by

$$p = p_i(1 + e_1 + e_2) \tag{9.8}$$

where

$$e_2 = -(\gamma_{air}/\gamma_{masses}) \tag{9.9}$$

The tester fluid lubricates the piston and the piston will be partially supported by the shear forces of the oil between the cylinder and the piston. Error is introduced which will vary inversely with the tester fluid viscosity [3]. Hence, high-viscosity oils are preferred. In a typical tester, this error is less than 0.01% of the reading. At high pressures, elastic deformation of the piston will affect the actual piston area. For the above two reasons, the effective area is usually based on the average of the piston and cylinder diameters, another source of uncertainty in the measurement.

EXAMPLE 9.3

A deadweight tester indicates 100.00 lb/in.^{2}, or 100.00 psi, at 70°F in Clemson, SC ($\phi = 34°$, $z = 841$ ft). Manufacturer specifications for the effective piston area were stated at 72°F so that thermal expansion effects remain negligible. Assume $\gamma_{air} = 0.076$ lb/ft^3 and $\gamma_{mass} = 496$ lb/ft^3. Correct for known errors.

KNOWN

$p_i = 100.00$ psi
$z = 841$ ft
$\phi = 34°$

ASSUMPTION

Bias corrections for altitude and latitude apply.

FIND

p

SOLUTION

From equation (9.9), the indicated pressure can be corrected for buoyancy effects as

$$e_2 = -(\gamma_{air}/\gamma_{masses}) = -0.000154$$

and for gravity effects by equation (9.6a)

$$e_1 = [2.637 \times 10^{-3} \cos(2 \times 34) + 9.6 \times 10^{-8} \times 841 + 5 \times 10^{-5}]$$
$$= -0.001119$$

From equation (9.8), the corrected pressure becomes

$$p = 100 \times (1 - 0.000154 - 0.001119) \text{ lb/in.}^2 = 99.87 \text{ lb/in.}^2$$

COMMENT

This amounts to the correction of an indicated signal for known bias errors.

9.4   PRESSURE TRANSDUCERS

A pressure transducer converts a measured pressure into a mechanical or electrical signal. The transducer is actually a hybrid sensor–transducer. The primary sensor is usually an elastic element that deforms or deflects under pressure. Several common elastic elements used, shown in Figure 9.9, include the Bourdon tube, bellows, capsule, and diaphragm. A secondary transducer element converts the elastic element deflection into a readily measurable signal such as an electrical voltage or mechanical rotation of a pointer. There are many methods available to perform this secondary function, but electrical transducers will require additional external signal conditioning equipment and external power supplies to drive their electrical output signals.

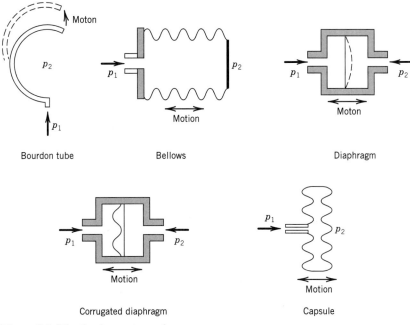

Figure 9.9 Elastic elements used as pressure sensors.

Pressure transducers are subject to some or all of the following elemental errors: resolution, zero shift error, linearity error, sensitivity error, hysteresis, and drift caused by environmental temperature changes. Electrical transducers are also subject to loading errors between the transducer output and its indicating device (Chapter 6). This error increases transducer nonlinearity over its operating range. A voltage follower (Chapter 6) can be inserted at the output of the transducer to isolate transducer load.

Bourdon Tube

The Bourdon tube is a curved metal tube having an elliptical cross section that mechanically deforms under pressure. It is used as the primary sensor in a large class of pressure gauges. In practice, one tube end is held fixed and the input pressure applied internally. A pressure difference between the outside of the tube and the inside of the tube will bring about tube deformation and a deflection of the tube free end. This action of the tube under pressure can be likened to the action of a deflated balloon that is inflated slightly. The magnitude of the deflection is proportional to the magnitude of the pressure difference. Several variations exist, such as the C shape (Figure 9.9), the spiral, and the twisted tube. The exterior of the tube is usually open to atmosphere (hence, the origin of the term "gauge" pressure referring to pressure referenced to atmospheric pressure), but in some variations the tube may be placed within a sealed housing and the tube exterior exposed to some other reference pressure.

The Bourdon tube mechanical dial gauge is perhaps the most commonly used pressure transducer. A typical design is shown in Figure 9.10, in which the secondary element is a mechanical linkage that converts the tube displacement into a rotation of a pointer. The instrument has a range in which pressure will be linearly related

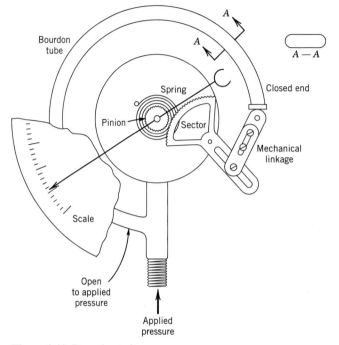

Figure 9.10 Bourdon tube pressure gauge.

to pointer rotation range of the instrument and this is usually the range specified by its manufacturer. Designs exist that can be used for low or high pressures. Various gauges exist that are suitable within a pressure range of 10^4–10^9 Pa (0.1–100,000 psi). The best Bourdon tube gauges have instrument uncertainties as low as 0.1% of the full-scale deflection of the gauge, but values of 0.5–2% are more common.

Bellows and Capsule

A bellows sensing element is a thin-walled, flexible metal tube formed into deep convolutions and sealed at one end (Figure 9.9). One end is held fixed and pressure is applied internally. A difference between the internal and external pressures will cause the bellows to change length. The bellows is housed within a chamber that can be sealed and evacuated for absolute measurements, vented through a reference pressure port for differential measurements, or opened to atmosphere for gauge pressure measurements. A capsule sensing element is a thin-walled, flexible metal tube similar to the bellows but tending to have a wider diameter and shorter length (Figure 9.9).

A mechanical linkage is used to convert the translational displacement of the bellows or capsule sensors into a measurable form. A common secondary transducer is the sliding arm potentiometer (voltage-divider; Chapter 6) found in the potentiometric pressure transducer shown in Figure 9.11. Another type uses a linear variable displacement transducer (LVDT; Chapter 12) to measure bellows or capsule displacement. The LVDT design has a high sensitivity and is commonly found in pressure transducers rated for low pressures and small pressure ranges, such as zero to several inches of water absolute, gauge or differential.

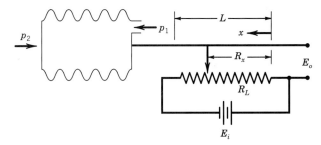

Figure 9.11 Potentiometric pressure transducer.

EXAMPLE 9.4

An engineer finds a desirable potentiometric bellows gauge with a stated range of 0–700 kPa in a catalog. It uses a sliding contact potentiometer having a terminal resistance of 50 Ω –10 kΩ ± 10% over full scale. The following manufacturer information is available:

Calibration linearity:	<0.5% full scale
Repeatability:	±0.1% full scale
Excitation voltage:	5–10 V dc

A voltmeter having an input impedance of 100 kΩ is available to measure output. For an applied transducer excitation voltage of 5 V, estimate the expected instrument uncertainty in measuring a 350-kPa pressure.

KNOWN

$E_i = 5$ V
$R_L = 50\ \Omega$ –10 kΩ ± 10% (95% assumed)

ASSUMPTIONS

Manufacturer specifications are reasonably accurate.

FIND

u_c

SOLUTION

The electrical arrangement of a potentiometric gauge is a voltage divider circuit (Chapter 6). The circuit is illustrated in Figure 9.12. For an infinite measuring instrument input impedance and assuming that the terminal resistance is at its minimum at $p = 0$ kPa and at its maximum at $p = 700$ kPa, then

$$E_o(p) = \frac{R_L(p)\,E_i}{10\ \text{k}\Omega}$$

so that $E_o(0) = 0.0250$ V, $E_o(350) = 2.4875$ V, and $E_o(700) = 5$ V. The output is nearly linear, so the static sensitivity will be approximately

$$K = \frac{E_o(700) - E_o(0)}{700 - 0\ \text{kPa}} = 7.11\ \mu\text{V/Pa}$$

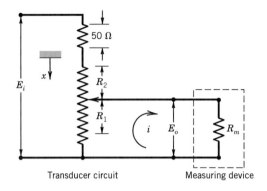

Figure 9.12 Potentiometer transducer circuit for Example 9.4.

Transducer circuit   Measuring device

The transducer instrument uncertainty will be affected by the elemental errors caused by linearity, e_L, repeatability, e_R, and loading, e_I. For a full-scale voltage of 5 V, the specifications indicate

$$e_L = 5 \text{ V} \times 0.005 = 0.0250 \text{ V} \qquad e_R = 5 \text{ V} \times 0.001 = 0.0050 \text{ V}$$

Loading error will depend on the ratio of terminal resistance to measuring instrument input impedance, R_m. Using equation (6.35) with Figure 6.16, at 350 kPa, $R_1 = 4975 \, \Omega$, $R_2 = 5025 \, \Omega$, and $R_m = 100,000 \, \Omega$. So with interstage loading, the output voltage becomes

$$E_o(350) = 2.4268 \text{ V}$$

The error caused by loading is $e_I = 2.4875 - 2.4268 = 0.0607$ V.

At 350 kPa, the instrument uncertainty, u_c, is estimated from the three elemental errors to be

$$u_c = \left[\sum_{i=1}^{3} e_i^2 \right]^{1/2}$$

$$= \sqrt{0.0250^2 + 0.0050^2 + 0.0607^2} = 0.0652 \text{ V}$$

For a static sensitivity of 7.11 μV/Pa, this equates to

$$u_c = 9.17 \text{ kPa} \quad (95\%)$$

Diaphragms

An effective primary element is a diaphragm, which is a thin elastic circular plate supported about its circumference. The action of a diaphragm within a pressure transducer is similar to the action of a trampoline, and a pressure differential across the surface of the diaphragm acts to deform it. The magnitude of the deformation is proportional to the pressure difference. Both membrane and corrugated designs are used. Membranes are made of metal or nonmetallic material, such as plastic or neoprene. The material chosen depends on the pressure range anticipated and the fluid in contact with it. Corrugated diaphragms contain a number of corrugations that serve to increase diaphragm stiffness and to increase the diaphragm effective surface area.

Pressure transducers that use a diaphragm sensor are well suited for either static or dynamic pressure measurements. They have good linearity and resolution over their useful range. An advantage of the diaphragm sensor is that the very low mass and relative stiffness of the thin diaphragm gives the sensor a very high natural frequency with a small damping ratio. Hence, these transducers can have a very wide frequency response and very short 90% rise and settling times. The natural frequency of a circular diaphragm can be estimated by [4]

$$\omega_n = 64.15 \sqrt{\frac{E_m t^2 g_c}{12(1 - v_p^2) r^4 \rho}} \qquad (9.10)$$

where E_m is the bulk modulus [psi or N/m^2], t the thickness [in. or m], r the radius [in. or m], ρ the material density [lb/in.^{3} or kg/m^3], and v_p Poisson's ratio for the diaphragm material with $g_c = 386$ lb$_m$ in./lb s^2 = 1 kg m/N s^2. The maximum elastic deflection of a uniformly loaded, circular diaphragm supported about its circumference occurs at its center and can be estimated by

$$y_{max} = \frac{3(p_1 - p_2)(1 - v_p^2) r^4}{16 E_m t^3} \qquad (9.11)$$

provided that the deflection does not exceed one-third the diaphragm thickness. Diaphragms should be selected so as to not exceed this maximum deflection over the anticipated operating range.

Various secondary elements are available to translate this displacement of the diaphragm into a measurable signal. Several methods are discussed as follows.

Strain Gauge Elements

The most common method for converting diaphragm displacement into a measurable signal is to sense the strain induced on the diaphragm surface as it is displaced. Strain gauges, devices whose measurable resistance is proportional to their sensed strain (Chapter 11), can be bonded directly onto the diaphragm or onto a deforming element (such as a thin beam) attached to the diaphragm so as to deform with the diaphragm and sense strain. Strain gauge resistance is reasonably linear over a wide range of strain and can be directly related to the diaphragm sensed pressure [5]. A diaphragm transducer using strain gauge detection is depicted in Figure 9.13.

The use of semiconductor technology in pressure transducer construction has led to the development of a variety of very fast, very small, highly sensitive strain gauge diaphragm transducers. Silicone piezoresistive strain gauges can be diffused into a single crystal of silicone wafer, which forms the diaphragm. Semiconductor strain gauges have a sensitivity that is 50 times greater than conventional metallic strain gauges. In addition, because the piezoresistive gauges are integral to the diaphragm, they are relatively immune to the thermoelastic strains prevalent in conventional metallic strain gauge–diaphragm constructions. Furthermore, a silicone diaphragm will not creep with age (as will a metallic gauge), thus minimizing calibration drift over time. However, gauge failure is catastrophic and silicone is not well suited to wet environments.

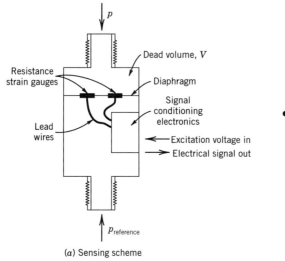

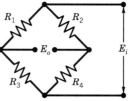

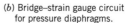

(b) Bridge–strain gauge circuit for pressure diaphragms.

(a) Sensing scheme

Figure 9.13 Diaphragm pressure transducer.

Capacitance Elements

When one or more fixed metal plates are placed directly above or below a metallic diaphragm, a capacitor is created that forms an effective secondary element. Such a transducer using this method is depicted in Figure 9.14. It is known as a capacitance transducer. The capacitance, C, developed between two parallel plates separated by a distance, t, is determined by

$$C = c\epsilon A/t \qquad (9.12)$$

where ϵ is the dielectric constant of the material between the plates (for air, $\epsilon = 1$), A is the overlapping area of the two plates, and c is the proportionality constant given by 0.225 (when A is measured in [in.^{2}] and t in [in.]) or 0.0885 (when A is measured in [cm^2] and t in [cm]). Displacement of the diaphragm changes the average gap separation. In the circuit shown, the measured voltage will be essentially linear with

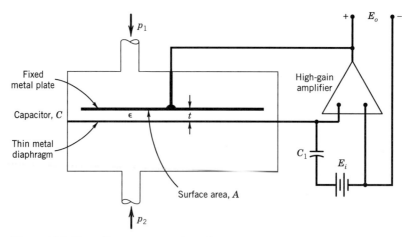

Figure 9.14 Capacitance pressure transducer.

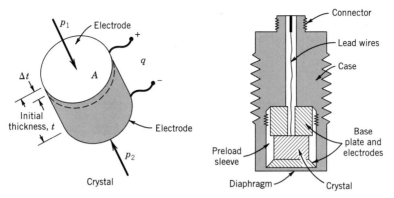

Figure 9.15 Piezoelectric pressure transducer.

developed capacitance

$$E_o = (C_1/C)E_i \tag{9.13}$$

and pressure can be inferred. The capacitance pressure transducer possesses the attractive features of other diaphragm transducers, including small size and a very wide operating range. However, it is sensitive to temperature changes and has a relatively high impedance output.

Piezoelectric Crystal Elements

Piezoelectric crystals form effective secondary elements for dynamic (transient) pressure measurements. Under the action of compression, tension, or shear, a piezoelectric crystal will deform and develop a surface charge, q, which is proportional to the force acting to bring about the deformation. In a piezoelectric pressure transducer, a preloaded crystal is mounted to the diaphragm sensor as indicated in Figure 9.15. Pressure acts normal to the crystal axis and changes the crystal thickness, t, by a small amount Δt. This sets up a charge

$$q = K_q p A$$

where p is the pressure acting over the electrode area A and K_q is the crystal charge sensitivity, a material property. The voltage developed across the electrodes is given by

$$E_o = q/C$$

where C is the capacitance of the crystal–electrode combination. The capacitance can be obtained from equation 9.12 to yield

$$E_o = K_q pt/c\epsilon = K_E pt \tag{9.14}$$

where K_E is the voltage sensitivity of the transducer. The crystal sensitivities for quartz, the most common material used, are $K_q = 2.2 \times 10^{-9}$ C/N and $K_E = 0.055$ V m/N. A charge amplifier (Chapter 6) is used to convert charge to voltage.

9.5 PRESSURE TRANSDUCER CALIBRATION

Static Calibration

The static calibration of a pressure transducer can be accomplished by direct comparison against any of the pressure reference instruments discussed (Section 9.3) or a certified reference transducer. For the low-pressure range, the manometric instruments along with the laboratory barometer serve as convenient and easy working standards. For the high-pressure range, the deadweight tester is a desirable pressure reference standard.

Dynamic Calibration

The rise time and frequency response of a pressure transducer are found by dynamic calibration. As discussed in Chapter 3, the rise time of an instrument is found through a step change in input. The frequency response is found through the application of periodic input signals.

An electrical switching valve is useful for creating a step change in pressure, but the mechanical lag of the valve limits its use to transducers having an expected rise time of 100 ms or more. Faster applications require a shock tube calibration. As shown in Figure 9.16, the *shock tube* consists of a long pipe separated into two chambers by a thin diaphragm. The transducer is mounted into the pipe wall of one chamber at pressure p_1. The pressure in the other chamber is raised from p_1 to p_2. A mechanically controlled needle is used to burst the diaphragm on command. Upon bursting, the pressure differential causes a pressure shock wave to move down the low pressure chamber. A shock wave has a thickness of the order of 1 μm and moves at the speed of sound, a. So as the shock passes the transducer, the transducer experiences a change in pressure from p_1 to p over a time $t = d/a$, where d is the diameter of the transducer pressure port and pressure p is found from

$$p = p_1\left[1 + (2k/k+1)\left(M_1^2 - 1\right)\right] \qquad (9.15)$$

where k is the gas-specific heat ratio and M_1 is the Mach number at pressure p_1. The velocity of the shock wave can also be deduced from the output of fast-acting

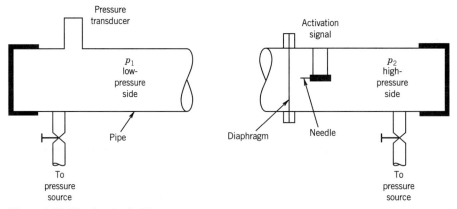

Figure 9.16 Shock tube facility.

pressure sensors mounted in the shock tube wall. Typical values of t are of the order of 1–10 μs. From the transducer output record its rise time is calculated.

A reciprocating piston within a cylinder is a common means to generate a sinusoidal variation in pressure for frequency response calibration. The piston can be driven by a variable speed motor and its displacement measured by a fast responding transducer, such as an LVDT (Chapter 12). Under properly controlled conditions (Example 1.2), the actual pressure variation can be estimated from the piston displacement. An encased loudspeaker or an acoustically resonant enclosure can also be used as frequency drivers, as can an oscillating flow control valve.

EXAMPLE 9.5

A common method to estimate the data-acquisition and reduction errors present in the pressure-measuring instrumentation of a test rig is to apply a series of replication tests on the M calibrated pressure transducers used on that rig (when the number of transducers installed is large, a few transducers can be selected at random and tested). In such a test each transducer is interconnected through a common manifold to the output of a single deadweight tester (or suitable standard) so that each transducer is exposed to exactly the same static pressure. Each transducer is operated at its normal excitation voltage. The test proceeds as follows: record applied test pressure (from the deadweight tester), record each transducer output N (say, $N \geq 10$) times, vent manifold to local atmosphere, close manifold to reapply test pressure, repeat procedure at least K (say, $K \geq 5$) times. What information is available in such test data?

KNOWN

$M(j = 1, 2, \ldots, M)$ transducers
$N(i = 1, 2, \ldots, N)$ repetitions at a test pressure
$K(k = 1, 2, \ldots, K)$ replications of test pressure

SOLUTION

The mean value for each transducer for any replication is given by the mean of the repetitions for that transducer,

$$\bar{p}_{jk} = \frac{1}{N} \sum_{i=1}^{N} p_{ijk}$$

The mean value for the replications of the jth transducer is given by

$$\langle \bar{p}_j \rangle = \frac{1}{K} \sum_{k=1}^{K} \bar{p}_{jk}$$

The difference between the pooled mean pressure and the known applied pressure would provide an estimate of the bias limit to be expected from any transducer during data acquisition. In contrast, differences between $\bar{p}_{jk}$ and $\langle \bar{p}_j \rangle$ must be due to the precision error in the data-acquisition and reduction procedure. This precision error is estimated by considering the variation of the test pressure mean, $\bar{p}_{jk}$, about the pooled mean, $\langle \bar{p}_j \rangle$, for each transducer:

$$\langle S_j \rangle = \sqrt{\frac{\sum\limits_{k=1}^{K} (\bar{p}_{jk} - \langle \bar{p}_j \rangle)^2}{K - 1}}$$

The pooled standard deviation of the mean for M transducers provides the estimate of the precision index of the data-acquisition and reduction procedure:

$$\langle S_{\bar{p}} \rangle = \sqrt{\frac{\sum\limits_{j=1}^{M} \sum\limits_{k=1}^{K} (\bar{p}_{jk} - \langle p_j \rangle)^2}{M(K - 1)}}$$

with degrees of freedom, $\nu = M(K - 1)$.

COMMENT

The statistical estimate contains the effects on precision caused by

- pressure standard (applied pressure repeatability)
- pressure transducer repeatability (repetition and replication)
- excitation voltages
- recording system

It does not contain the effects of instrument calibration errors, large deviations in environmental conditions, pressure tap design errors, or dynamic pressure effects. Bias limits for these must be set by the engineer based on other information.

Note that if the above procedure were repeated over a range of known pressures, then this would allow instrument calibration errors and data-reduction curve fit errors to be entered into the analysis.

9.6 PRESSURE MEASUREMENTS IN MOVING FLUIDS

Pressure measurements in moving fluids deserve special consideration. Consider the flow over the bluff body shown in Figure 9.17. Assume that the upstream flow is uniform and steady. Points along the two streamlines labeled as A and B are to be

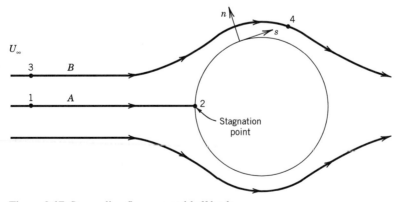

Figure 9.17 Streamline flow over a bluff body.

studied. Along streamline A, the flow moves with a velocity, U_1, such as at point 1 upstream of the body. As the flow approaches point 2 it must slow down and finally stop at the front end of the body. Point 2 is known as the stagnation point and streamline A the stagnation streamline for this flow. Along streamline B, the velocity at point 3 will be U_3 and because the upstream flow is considered to be uniform it follows that $U_1 = U_3$. As the flow along B approaches the body, it is deflected around the body. From conservation of mass principles, $U_4 > U_3$. Application of conservation of energy between points 1 and 2 and between 3 and 4 yields

$$p_1 + \rho U_1^2/2g_c = p_2 + \rho U_2^2/2g_c$$
$$p_3 + \rho U_3^2/2g_c = p_4 + \rho U_4^2/2g_c \qquad (9.16)$$

However, because point 2 is the stagnation point, $U_2 = 0$, and

$$p_2 = p_t = p_1 + \rho U_1^2/2g_c \qquad (9.17)$$

Hence, it follows that $p_2 > p_1$ by an amount equal to $U_1^2/2g_c$, an amount equivalent to the kinetic energy per unit mass of the flow as it moves along the streamline. If the flow is brought to rest in an isentropic manner (i.e., no energy is lost through irreversible processes such as through a transfer of heat[1]), this translational kinetic energy will be transferred completely into p_2. The value of p_2 is known as the stagnation or the total pressure and will be noted as p_t. The *total pressure* can be determined by bringing the flow to rest at a point in an isentropic manner.

The pressures at 1, 3, and 4 are known as the stream or static pressures[2] of the flow. Because the flow is uniform, $U_1 = U_3$, so that $p_3 = p_1$. The static pressure and velocity at points 1 and 3 are known as the freestream pressure and freestream velocity. However, as the flow accelerates around the body its velocity increases such that, from equation (9.16), $p_4 \neq p_3$. The pressure, such as at point 4, is called a local static pressure. The *static pressure* is that pressure sensed by a fluid particle as it moves with the same velocity as the local flow.

Total Pressure Measurement

In practice, the total pressure is measured using an impact probe, such as those depicted in Figure 9.18. A small hole in the impact probe is aligned with the flow so as to cause the flow to come to rest at the hole. The sensed pressure is transferred through the impact probe to a pressure transducer or other pressure sensing device such as a manometer. Alignment with the flow is somewhat critical, although the probes in Figure 9.18(*a*) and 9.18(*b*) are relatively insensitive (within ~1% error in indicated reading) to misalignment within a ±7° angle. A special type of impact probe shown in Figure 9.18(*c*), known as a Kiel probe, uses a shroud around the impact port. The effect of the shroud is to force the local flow to align itself with the shroud axis so as to impact directly onto the impact port. This effectively decreases total pressure misalignment sensitivity to negligible levels up to approximately ±40°.

---

[1]This is a realistic assumption for subsonic flows. In supersonic flows, the assumption will not be valid across a shock wave.

[2]The term "static pressure" is somewhat of a misnomer in moving fluids, but its use here conforms to common expression. The term "stream pressure" is more appropriate.

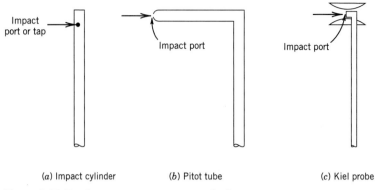

(a) Impact cylinder   (b) Pitot tube   (c) Kiel probe

Figure 9.18 Total pressure measurement devices.

Static Pressure Measurement

As defined, the static pressure is more awkward to measure than the total pressure since it requires the measurement probe to physically move through the fluid at the flow velocity. However, consider the coordinate system, s-n, fixed to the body as shown in Figure 9.17. As the flow passes over any real object, a boundary layer will be formed. Within a boundary layer, the pressure gradient in the direction normal to the streamwise direction s will be $dp/dn \approx 0$ [6]. This suggests that the local value of static pressure can be sensed in a direction that is oriented normal to the flow streamline.

Within ducted flows, static pressure is sensed by wall taps, small holes drilled into the duct wall perpendicular to the flow direction at the measurement point. The tap is fitted with a hose or tube, which is connected to a pressure gauge or transducer. A recommended design for a wall tap is shown in Figure 9.19. The tap hole diameter is typically between 1 and 10% of the pipe diameter, with the smaller size preferred [5].

Curvature of the flow streamlines introduces acceleration forces that change the local velocity. From equation (9.16) and Figure 9.17, we see that curvature of the flow streamlines must affect the local static pressure, since $p_1 = p_3$ but $p_3 \neq p_4$. The freestream static pressure given by p_1 and p_3 can be measured only where there is no curvature of the streamlines. So if pressure at the wall is to be measured, the wall taps should not disturb the flow in any way, for a disturbance would cause streamline curvature. The tap should be square with the wall, with no drilling burrs [8].

A pressure probe cap inserted into the flow for the purpose of measuring static pressure should be a streamlined design to minimize the disturbance of the flow. It

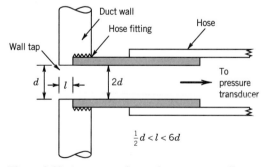

Figure 9.19 Anatomy of a static pressure wall tap.

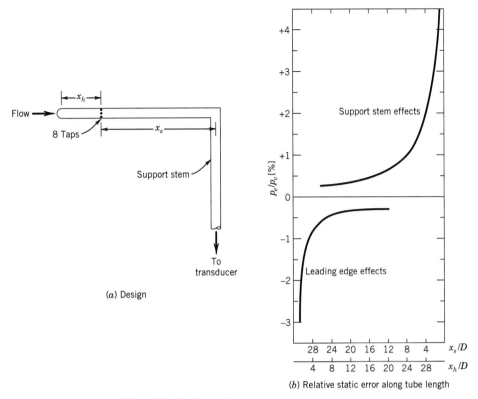

Figure 9.20 Improved Prandtl tube for static pressure.

should be physically small so as not to cause more than a negligible increase in velocity in the vicinity of measurement. The static pressure sensing port should be located well downstream of the leading edge of the probe so as to allow the streamlines to realign themselves parallel with the probe. Such a probe design, the improved Prandtl tube, is shown in Figure 9.20(a).

A Prandtl tube probe consists of eight holes arranged about the probe circumference and positioned 8–16 probe diameters downstream of the probe leading edge and 16 probe diameters upstream of its support stem. These positions are chosen to minimize static pressure error caused by the disturbance to the flow streamlines that is due to the probe's leading edge and stem. This is illustrated in Figure 9.20(b), where the relative static error, $p_e/p_v = (p_i - p)/(1/2\,\rho U^2)$, as a function of tap location along the probe body is plotted with p_i the indicated pressure. Real viscous effects around the static probe cause a slight discrepancy between the actual static pressure and the indicated static pressure. To account for this a correction factor, C_0, is used with $p = C_0 p_i$ where $0.99 < C_0 < 0.995$. A pressure transducer or manometer is connected to the probe stem to measure the sensed pressure.

9.7 DESIGN AND INSTALLATION: TRANSMISSION EFFECTS

The size of the pressure tap diameter and the length of tubing between a pressure tap and the connecting transducer form a pressure-measuring system that can have dynamic response characteristics very different from the pressure transducer itself.

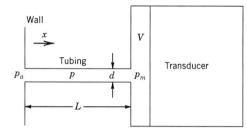

Figure 9.21 Wall tap to pressure transducer connection.

In the discussions to follow it is assumed that the transducer has frequency response and rise time characteristics that exceed those of the entire pressure measuring system (tubing plus transducer). Consider the configuration depicted in Figure 9.21 in which a rigid tube of length L and diameter d is used to connect a pressure tap to a pressure transducer of internal dead volume V (e.g., Figure 9.13). Initially, we can assume that under static conditions the input pressure at the tap will be indicated by the pressure transducer, but if the pressure tap is exposed to a time-dependent pressure, $p_a(t)$, the response behavior of the tubing will cause the system output from the transducer, $p(t)$, to lag the input. By considering the one-dimensional pressure forces acting on a lumped mass of fluid within the connecting tube, a model for the pressure system can be developed. Be aware that this model is based on several very simplifying assumptions and should be used only as a guide in the design of a time-dependent pressure measurement system, not as an accurate indicator of exact system performance or as a correction method.

Gases

Compressibility of a gas can be described through the fluid bulk modulus of elasticity, E_m. Pressure changes will act on the fluid in an effort to move it back and forth by a distance x within the tube. Consider a free body of a volume of fluid within the tube (Figure 9.22). At any instant, we can expect the fluid to be acted upon by the driving pressure force, $p_a \pi d^2/4$, a damping force that is due to fluid shear forces, $8\pi\mu L\dot{x}$, and a compression-restoring force, $\pi^2 E_m d^2 x/16V$. Summing the forces in Newton's

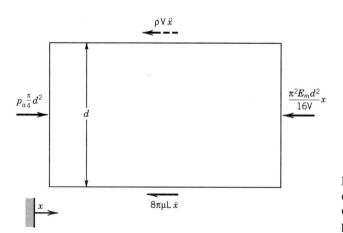

Figure 9.22 Free-body diagram of forces acting on a fluid volume in a pressure transmission line.

second law yields the response equation

$$\frac{4L\rho V}{\pi E_m d^2}\ddot{p}_m + \frac{128\,\mu LV}{\pi E_m d^4}\dot{p}_m + p_m = p_a(t) \tag{9.18}$$

in which p_m is the measured pressure and p_a the applied pressure. With the use of equation (3.13), this gives

$$\omega_n = \frac{d\sqrt{\pi E_m/LV\rho}}{2} \tag{9.19}$$

$$\zeta = \frac{32\mu\sqrt{VL/\pi E_m\rho}}{d^3} \tag{9.20}$$

If we define $a = \sqrt{kRTg_c}$ as the acoustic wave speed of a perfect gas, where T is the absolute temperature of the gas, the natural frequency and damping ratio of the system become

$$\omega_n = \frac{d\sqrt{\pi a^2/LV}}{2} \tag{9.21}$$

$$\zeta = \frac{32\mu\sqrt{VL/\pi}}{a\rho d^3} \tag{9.22}$$

When the tube volume, $V_t = \pi d^2 L/4 \gg V$, then the entire pressure-measuring system will behave in a manner similar to an operating organ pipe. A series of standing pressure waves occur that alter the model to the extent that a better predictor is given by [9]

$$\omega_n = \frac{a}{L(0.5 + 4V/V_t)^{1/2}} \tag{9.23}$$

$$\zeta = \frac{16\mu L\sqrt{0.5 + 4V/V_t}}{d^2\rho a} \tag{9.24}$$

Note that in either case larger diameters and shorter tubes improve pressure system response.

EXAMPLE 9.6

A pressure transducer with a natural frequency of 100 kHz is connected to a 0.10-in. static wall pressure tap using a 0.10-in i.d. rigid tube that is 5 in. long. The transducer has a dead volume of 1 in.^{3}. Determine the pressure transmission system magnitude ratio response to fluctuating pressures of air at 72°F if the fluctuations are ~1 atm abs mean pressure, $R_{air} = 53.3$ ft-lb/lb$_m$ °R, and $\mu = 4 \times 10^{-7}$ lb s/ft^2.

KNOWN

$L = 5$ in. $k = 1.4$
$d = 0.1$ in. $T = 72°F = 532°R$
$V = 1$ in.^{3} $\rho = p/RT = 0.075$ lb$_m$/ft^3

ASSUMPTION

Air behaves as a perfect gas.

FIND

Find $M(\omega)$

SOLUTION

The magnitude ratio is given by equation (3.18) as

$$M(\omega) = \frac{1}{\{[1 - (\omega/\omega_n)^2]^2 + [2\zeta(\omega/\omega_n)]^2\}^{1/2}}$$

We need to find ω_n and ζ and solve $M(\omega)$ for various frequencies.
For this geometry,

$$V_t = \frac{\pi d^2 L}{4} = 0.04 \text{ in.}^3$$

Therefore, $V_t \ll V$. Using $a = \sqrt{kRTg_c} = 1130$ ft/s, then from equation (9.21)

$$\omega_n = \frac{d\sqrt{\pi a^2/LV}}{2}$$
$$= 537.4 \text{ rad/s}$$

and from equation (9.22)

$$\zeta = \frac{32\mu\sqrt{VL/\pi}}{a\rho d^3}$$
$$= 0.08$$

The system is lightly damped. Computation of $M(\omega)$ yields the following representative values:

ω [rad/s]	$M(\omega)$
63	1.01
315	1.50
535	6.27
3150	0.03

Liquids

In liquid flows, sudden pressure changes will be transported through the pressure system more readily. Still the analysis leading to equation (9.18) will hold. However, a momentum corection factor or equivalent mass is introduced to account for the inertial effects not included in a one-dimensional analysis (e.g., [10]). This in effect increases the inertial force by 1.33 to yield

$$\omega_n = \frac{d\sqrt{3\pi E_m/L\rho V}}{4} \tag{9.25}$$

$$\zeta = \frac{16\mu\sqrt{3VL/\pi E_m\rho}}{d^3} \tag{9.26}$$

Heavily Damped Systems

In systems in which equations (9.22), (9.24), or (9.26) indicate a damping ratio greater than 1.5, the system can be considered as heavily damped. The behavior of the pressure-measuring system will closely follow that of a first-order system. A typical pressure transducer will have a rated compliance, C_{vp}, which is a measure of the transducer volume change relative to an applied pressure change. The response of the first-order system is indicated through its time constant, which can be approximated by [11].

$$\tau = \frac{128 \, \mu L C_{vp}}{\pi d^4} \tag{9.27}$$

An important aspect of equation (9.27) is that the time constant is proportional to $L/d^4 \propto (L/d)^2 (1/V_t)$. Thus, long and small diameter connecting tubes will result in relatively sluggish response to changes in pressure.

EXAMPLE 9.7 : A Test Case

An engineer wishes to measure aerodynamic downforce on a race stock car as it moves on a track. The pressure difference between the top and bottom surfaces of the car is responsible for this downward force, which allows the tires to adhere better at high speeds. Pressure is measured by using surface (wall) taps connected by 5-mm i.d. tubing to ± 25.4 cm H_2O, 0–5 V capacitance pressure transducers (accuracy: 0.25%), such as in Figures 9.14, 9.19, and 9.21. Transducer output is measured by a portable data-acquisition system using a 12-bit, 5-V A/D converter (accuracy: 2 LSB). In testing at 180 kph, the maximum average pressure found is on the rear deck (rear window–trunk lid) and is approximately 8 cm H_2O ± 0.10 cm H_2O (95%); the precision uncertainty is due to pressure fluctuations. The engineer is asked two questions: (1) Will increasing the resolution of the A/D conversion system help accuracy? (2) How accurately can rear downforce be measured on the track?

SOLUTION

Force is pressure acting over an area. Aerodynamic downforce is estimated by integrating the pressure over the surface area of the car. An approximation is to measure the average pressure acting on discrete effective areas on the car surface.

   If we ignore installation errors and ambient influences and just look at direct measurement errors, the measurement system errors and data-precision errors are important. The resolution of the transducer and A/D system is limited to

$$Q_{transducer} = 50.8 \text{ cm } H_2O/5 \text{ V} = 10.16 \text{ cm } H_2O/V$$
$$Q_{A/D} = 5 \text{ V}/2^{12} = 0.00122 \text{ V/digit}$$

Or, the total measuring system resolution is $Q = 0.00122 \times 10.16 = 0.0124$ cm H_2O. This yields a $u_o = 0.0124$ cm H_2O. However, improving the DAS system to a 16-bit system improves Q and u_o to 0.008 cm H_2O.

   Pressure fluctuations occur as the flow fluctuates around the car. We can measure pressure at a moderate sample rate (25 Hz), use a smoothing filter, and average the readings. Statistical variations from these average values yield a precision uncertainty, u_{pavr}. The precision errors vary with position on the car as a result of differences in the flow.

If we look at the rear deck of the car:

$$u_{c\ \text{transducer}} \approx (0.0025)(8\ \text{cm H}_2\text{O}) = 0.02\ \text{cm H}_2\text{O}$$

$$u_{c\ \text{A/D}} = (2\ \text{bits})(0.0124\ \text{cm H}_2\text{O/bit}) = 0.025\ \text{cm H}_2\text{O}$$

or

$$(u_d)_p = \pm[(0.02)^2 + (0.025)^2 + (0.0124)^2]^{1/2} = \pm0.034\ \text{cm H}_2\text{O}\ (95\%).$$

Hence from design-stage analysis, it is clear that the higher resolution cannot improve the measurement. The instrument errors dominate the design-stage uncertainty.

The effective area on the rear deck can be estimated from carefully controlled measurements within a wind tunnel with the car at various angles to the wind. A highly accurate balance scale is used to measure downforce. From this, the rear effective area is found. A typical result is

$$A_{\text{eff}} = 10832 \pm 147\ \text{cm}^2\ (95\%)$$

The downforce is simply $F_D = pA_{\text{eff}}$, so the percent uncertainty is found by

$$u_F/F_D = \pm[(u_p/p)^2 + (u_A/A_{\text{eff}})^2] = 0.014\ \text{ or }\ 1.4\%\ (95\%)$$

Experience shows that on-track effects contribute to raising this number to $\sim2\%$.

COMMENT

Here we see that the dominant uncertainty is due to the flow process and not the instrumentation. Improving the uncertainty requires going to alternate methods to estimate the downforce.

9.8   FLUID VELOCITY MEASURING SYSTEMS

Velocity measuring systems are used when information about a moving fluid within a localized portion of the flow is needed. Desirable information can consist of the mean velocity, as well as any of the dynamic components of the velocity. Dynamic components are found in pulsating or oscillating flows, or in turbulent flows. For most general engineering applications, information about the mean flow velocity is usually sufficient. The dynamic velocity information is often sought during applied and basic fluid mechanics research and development, such as in attempting to study airplane wing response to air turbulence, a complex periodic waveform as the wing sees it.

In general, the instantaneous velocity can be written as

$$U(t) = \bar{U} + u \tag{9.28}$$

where $\bar{U}$ is the mean velocity and u is the time-dependent dynamic component of the velocity. The instantaneous velocity can also be expressed in terms of a Fourier series

$$U(t) = \bar{U} + \sum C_i \sin(\omega_i t + \phi_i') \tag{9.29}$$

so that the mean velocity and the amplitude and frequency information concerning the dynamic velocity component can be found through a Fourier analysis of the time-dependent velocity signal.

Pitot–Static Pressure Probe

For a steady, incompressible, isentropic flow, equation (9.16) can be written at any arbitrary point x in the flow field as

$$p_t = p_x + \frac{1}{2g_c}\rho U_x^2 \qquad (9.30)$$

or as

$$p_v = p_t - p_x = \frac{1}{2g_c}\rho U_x^2 \qquad (9.31)$$

Here p_v, the difference between the total and static pressures at any point in the flow, is called the *dynamic pressure*. The determination of the dynamic pressure of a moving fluid at point x would provide a method for the estimation of the local velocity existing at point x. From equation (9.30),

$$U_x = \sqrt{\frac{2g_c p_v}{\rho}} = \sqrt{\frac{2g_c(p_t - p_x)}{\rho}} \qquad (9.32)$$

In practice, equation (9.32) is utilized through a device known as a pitot–static pressure probe. Such an instrument has an outward appearance similar to that of an improved Prandtl static pressure probe [Figure 9.20(a)] except that the pitot–static probe contains an interior pressure tube attached to an impact port at the leading edge of the probe, as shown in Figure 9.23. This creates two coaxial internal cavities within the probe, one exposed to the total pressure and the second exposed to the static pressure. The two pressures are typically measured using a differential pressure transducer so as to indicate p_v directly.

The pitot–static pressure probe is relatively insensitive to misalignment over the yaw angle range of $\pm 15°$. When possible, the probe can be rotated until a maximum signal is measured, a condition that is indicative of alignment with the mean flow direction. However, the probes have a lower velocity limit of use that is brought about by strong viscous effects in the entry regions of the pressure ports. In general, viscous effects should not be a concern, provided that the Reynolds number based on the probe radius $\mathrm{Re}_r = Ur/\nu > 500$, where ν is the kinematic viscosity of the fluid. For $10 < \mathrm{Re}_r < 500$, a correction to the dynamic pressure should be applied, $p_v = C_v p_i$,

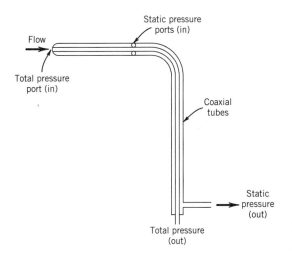

Figure 9.23 Pitot–static pressure probe.

where

$$C_v = 1 + (4/\text{Re}_r) \tag{9.33}$$

and p_i is the dynamic pressure indicated by the probe. However, even with this correction the uncertainty (bias limit) in measured dynamic pressure will be of the order of 40% at $\text{Re}_r \approx 10$ but decreases to 1% at $\text{Re}_r \approx 500$.

In high-speed gas flows, compressibility effects near the probe leading edge necessitate a closer inspection of the governing equation for a pitot–static pressure probe. Recalling equation (8.34), which states the energy balance for a perfect gas between the freestream and a stagnation point.

$$\frac{U^2}{2g_c} = c_p(T_t - T) \tag{8.34}$$

For an isentropic process, the relationship between temperature and pressure can be stated as

$$\frac{T}{T_t} = \left(\frac{p}{p_t}\right)^{(k-1)/k} \tag{9.34}$$

where k is the ratio of specific heats for the gas, $k = c_p/c_v$. The Mach number of a moving fluid relates its local velocity to the local speed of sound,

$$M = U/a \tag{9.35}$$

where the acoustic wave speed, or speed of sound, is defined for a perfect gas as

$$a = \sqrt{kRTg_c} \tag{9.36}$$

where T is the absolute temperature of the gas. Combining equations (8.28) with (9.34)–(9.36) and using a binomial expansion yields the relationship between total pressure and static pressure in a moving compressible flow,

$$p_v = p_t - p = (1/2g_c)\rho\, U^2[1 + M^2/4 + (2 - k)M^4/24 + \cdots] \tag{9.37}$$

It is apparent that equation (9.37) reduces to equation (9.31) when $M \ll 1$. The error in the estimate of p_v based on use of equation (9.31) relative to the true dynamic pressure becomes significant for $M > 0.2$ as shown in Figure 9.24.

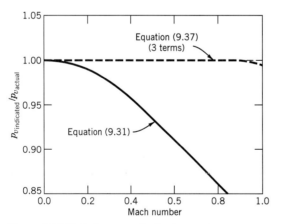

Figure 9.24 Relative error in the dynamic pressure between using equations (9.31) and (9.37) at increasing flow speeds.

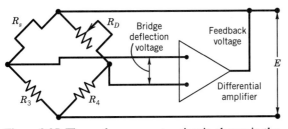

Figure 9.25 Thermal anemometer circuit, shown in the constant resistance mode.

For $M > 1$, the local velocity can be estimated by the Rayleigh relation

$$U = \sqrt{2g_c[k/(k-1)](p/\rho)[(p_t/p)^{(k-1)/k} - 1]}$$ (9.38)

where p and p_t must be measured by independent means.

Thermal Anemometry

The rate at which energy, $\dot{Q}$, is transferred between a warm body at T_s and a cooler moving fluid at T_f is proportional both to the temperature difference between them and to the thermal conductance of the heat transfer path, hA. This thermal conductance increases with fluid velocity, thereby increasing the rate of heat transfer at any given temperature difference. Hence, a relationship between the rate of heat transfer and velocity will exist, and this forms the working basis of a *thermal anemometer*.

A thermal anemometer utilizes a sensor, a metallic RTD element, which makes up one active leg of a Wheatstone bridge circuit, as indicated in Figure 9.25. The resistance-temperature relation for such a sensor was shown in Chapter 8 to be well represented by

$$R_s = R_0[1 + \alpha(T_s - T_0)]$$ (9.39)

so that sensor temperature can be inferred through a resistance measurement. A current is passed through the sensor to heat it to some desired temperature above that of the host fluid. The relationship between the rate of heat transfer from the sensor and the cooling fluid velocity is given by King's law [12]:

$$\dot{Q} = I^2 R_s = A + BU^n$$ (9.40)

where A and B are constants that depend on the fluid and sensor physical properties and operating temperatures, and n is a constant that depends on sensor dimensions [13]. Typically, $0.45 \leq n \leq 0.52$ [14]. A, B, and n are found through calibration.

Two types of sensors are common: the hot-wire and the hot-film sensors. As shown in Figure 9.26, the hot-wire sensor consists of a tungsten or platinum wire ranging from 1 to 4 mm in length and from 1.5 to 15 μm in diameter. The wire is supported between two rigid needles that protrude from a ceramic tube that houses the lead wires. A hot-film sensor usually consists of a thin (2 μm) platinum or gold film deposited onto a glass substrate and covered with a high thermal conductivity coating. The coating acts to electrically insulate the film and offers some mechanical protection. Hot wires are generally used in electrically nonconducting fluids, whereas hot films are used both in nonconducting and conducting fluids and where a more rugged sensor is needed.

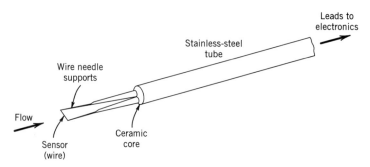

Figure 9.26 Schematic diagram of a hot-wire probe.

Two anemometer bridge operating modes are possible: (1) constant current and (2) constant resistance. In constant current operation, a fixed current is passed through the sensor to heat it. The sensor resistance, and therefore its temperature, are permitted to vary with the rate of heat transfer between the sensor and its environment. Bridge deflection voltage provides a measure of the cooling velocity. The more common mode of operation for velocity measurements is constant resistance. In constant resistance operation, the sensor resistance, and therefore its temperature, is originally set by adjustment of the bridge balance. The sensor resistance is then maintained constant by using a differential feedback amplifier to sense small changes in bridge balance, which would be equivalent to sensing changes in the sensor set point resistance. The feedback amplifier rapidly readjusts the bridge applied voltage, thereby adjusting the sensor current to bring the sensor back to its set point resistance and corresponding temperature. Since the current through the sensor will vary with changes in the velocity, the instantaneous power $(I^2 R_s)$ required to maintain this constant temperature is equivalent to the instantaneous rate of heat transfer from the sensor $(\dot{Q})$. In terms of the instantaneous applied bridge voltage, E, required to maintain a constant sensor resistance, the velocity is found by the correlation

$$E^2 = C + DU^n \tag{9.41}$$

where constants C, D, and n are found by calibration under a fixed sensor and fluid temperature condition. An electronic or digital linearizing scheme is usually employed to condition the signal by performing the transformation

$$E_1 = K\left(\frac{E^2 - C}{D}\right)^{1/n} \tag{9.42}$$

such that the measured output from the linearizer, E_1, is

$$E_1 = KU \tag{9.43}$$

where K is found through a static calibration.

For simple mean velocity measurements the thermal anemometer is a simple device to use. It has a better usable sensitivity than the pitot–static tube at lower velocities. Multiple velocity components can be measured by using multiple sensors, each sensor aligned differently to the mean flow direction and operated by independent anemometer circuits [13,15]. In highly turbulent flows with fluctuations of $\sqrt{\overline{u^2}} \geq 0.1\overline{U}$, signal interpretation can become complicated but has been well investigated (e.g., [15]). Low-frequency fluid temperature fluctuations can be compensated for by placing resistor R_3 directly adjacent to the sensor and exposed to the flow.

An extensive bibliography of thermal anemometry theory and signal interpretation exists [16].

In constant temperature mode using a fast responding differential feedback amplifier, a hot-wire system can attain a frequency response that is flat up to 100,000 Hz, which makes it particularly useful in fluid mechanics turbulence research. However, less expensive and more rugged systems are commonly used for industrial flow monitoring, where a fast dynamic response is desirable. An upper frequency limit on a cylindrical sensor of diameter d is brought about by the natural oscillation in the flow immediately downstream of a body that vibrates the sensor. The frequency of this oscillation, known as the Strouhal frequency, occurs at approximately

$$f \approx 0.22\,[\overline{U}/(2\pi d)]   (10^2 < \mathrm{Re}_d < 10^7) \tag{9.44}$$

The heated sensor will warm the fluid within its proximity. Under flowing conditions this will not cause any measurable problems as long as the condition

$$\mathrm{Re}_d \geq Gr^{1/3} \tag{9.45}$$

is met, where $\mathrm{Re}_d = Ud/v$, $Gr = d^3 g\beta(T_s - T_f)/v^2$, and β is the coefficient of thermal expansion of the fluid. Equation (9.45) ensures that the inertial forces of the moving fluid dominate over the buoyant forces brought on by the heated sensor. For air, this forms a lower velocity limit of the order of 2 ft/s (0.6 m/s).

Doppler Anemometry

The Doppler effect describes the phenomenon experienced by an observer whereby the frequency of light or sound waves emitted from a source that is traveling away from or toward the observer will be shifted from its original value. Most readers are familiar with the change in pitch of a train heard by an observer as the train changes from approaching to receding. Any radiant energy wave, such as a sound or light wave, will experience a Doppler effect. The effect was recognized and modeled by Johann Doppler (1803–1853). The observed shift in frequency, called the Doppler shift, is directly related to the speed of the emitter relative to the observer. To an independent observer, the frequency of emission is perceived to be higher than actual if the emitter is moving toward the observer and lower if moving away, since the arrival of the emission at the observer location will be affected by the relative velocity of the emission source.

Doppler anemometry refers to a class of techniques that utilize the Doppler effect to measure local velocity in a moving fluid. In these techniques, the emission source and the observer remain stationary. However, small scattering particles suspended in the moving fluid can be used to generate the Doppler effect. The emission source is a coherent narrow incident wave. Either acoustic waves or light waves are used.

When a laser beam is used as the incident wave source, the velocity measuring device is called a *laser Doppler anemometer* (LDA). The first practical LDA system was discussed by Yeh and Cummins in 1964 [17]. A laser beam provides a ready emission source that is monochromatic and remains coherent over long distances. As a moving particle suspended in the fluid passes through the laser beam, it scatters light in all directions. An observer viewing this encounter between the particle and the beam will perceive the scattered light at a frequency, f_s:

$$f_s = f_i \pm f_D \tag{9.46}$$

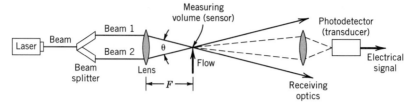

Figure 9.27 LDA, shown in the dual-beam mode of operation.

where f_i is the frequency of the incident laser beam and f_D is the Doppler shift. Using visible light, an incident laser beam frequency will be of the order of 10^{14} Hz. For most engineering applications, the velocities are such that the Doppler shift frequency, f_D, will be of the order of 10^3–10^7 Hz. Such a small shift in the incident frequency can be difficult to detect in a practical instrument. An operating mode that overcomes this difficulty is the dual-beam mode shown in Figure 9.27. In this mode, a single laser beam is divided into two coherent beams of equal intensity using an optical beam splitter. These incident beams are passed through a focusing lens, which focuses the beams to a point in the flow. The focal point forms the effective measuring volume (sensor) of the instrument. Particles suspended in and moving with the fluid will scatter light as they pass through the beams. The frequency of the scattered light will be that given by equation (9.46) everywhere but at the measuring volume. There, the two beams cross and the incident information from the two beams mixes, a process known as optical heterodyne. The outcome of this mixing is a separation of the incident frequency from the Doppler frequency. A stationary observer, such as an optical photodiode, focused on the measuring volume will see two distinct frequencies, the Doppler shift frequency and the unshifted incident frequency, instead of their sum. It is a simple matter to separate the much smaller Doppler frequency from the incident frequency by filtering.

For the setup shown in Figure 9.27, the velocity is related directly to the Doppler shift by

$$U = \frac{\lambda}{2 \sin \theta/2} f_D = d_f f_D \tag{9.47}$$

where the component of the velocity measured is that which is in the plane of and bisector to the crossing beams. In theory, by using beams of different color or polarization, different velocity components can be measured simultaneously. However, the dependence of the lens focal length on color will cause a small displacement between the different measuring volumes formed by the different colors. For most applications this can be corrected. The LDA technique requires no direct calibration beyond determination of the parameters in d_f and the ability to measure f_D.

In the dual-beam mode, the output from the photodiode transducer is a current of a magnitude proportional to the square of the amplitude of the scattered light seen and of a frequency equal to f_D. This effect is seen as a Doppler "burst," shown in the typical oscilloscope trace of Figure 9.28. The Doppler burst is the frequency signal created by a particle moving through the measuring volume. If the instantaneous velocity of a dynamic flow varies with time, the Doppler shift from successive scatters will vary with time. This time-dependent frequency information can be extracted by any of a variety of processing equipment that can interpret the signal current, including frequency trackers, counters, and burst analyzers. The most common is the burst analyzer.

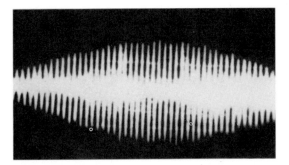

Figure 9.28 Oscilloscope trace of a photodiode output, showing the Doppler frequency from a single particle moving through the measuring volume.

Burst analyzers extract Doppler frequency information by performing a Fourier analysis (see Chapter 2) on the input signal. This is done by first discretizing the photodetector analog signal at a high sample rate and then analyzing the signal. The analysis can work in one of two ways. In the first approach, the analyzer performs the FFT continuously on small blocks of data from which the Doppler frequency is directly determined. In the second mode, the sampled signal is correlated with itself through a mathematical transformation of the form

$$R_j = \sum_{i=1}^{n} x(i)x(i+j) \tag{9.48}$$

where i refers to the sample value at time t and j refers to the sample value at time delay Δt. This operation improves the SNR of the signal; the frequency is determined from the correlation function and the sample rate. From equation (9.47), the Doppler frequency can be converted to a velocity and output. The acquisition and analysis occur rapidly so that the signal appears nearly continuous in time and with only a short time lag. In contrast, frequency counters count the number of zero crossings in the periodic Doppler signal and are suited to flows containing relatively few particles and a high SNR.

All methods output a voltage that is proportional to the instantaneous velocity, which makes their signal easy to process, or a digital output path to a digital computer for signal analysis. At very low light levels and very few scattering particles, the signal level to noise level can be very low. In such cases, photon correlation techniques are successful [18, 19].

EXAMPLE 9.8

A laser Doppler anemometer made up of a He–Ne laser ($\lambda = 632.8$ nm) is used to measure the velocity of water at a point in a flow. A 150-mm lens having $\theta = 11°$ is used to operate the LDA in a dual-beam mode. If an average Doppler frequency of 1.41 MHz is measured, estimate the velocity of water.

KNOWN

$\bar{f}_D = 1.41$ MHz $F = 150$ mm
$\lambda = 632.8$ nm $\theta = 11°$

ASSUMPTION

Scattering particles follow the water exactly.

FIND

$\overline{U}$

SOLUTION

Using equation (9.47),

$$
\begin{aligned}
\overline{U} &= \frac{\lambda}{2 \sin \theta/2} \bar{f}_D \\
&= \left(\frac{632.8 \times 10^{-9} \text{ m}}{2 \sin 5.5°} \right) (1.41 \times 10^6 \text{ Hz}) \\
&= 4.655 \text{ m/s}
\end{aligned}
$$

Selection of Velocity Measuring Methods

The selection of the best velocity measuring system for a particular application will depend on a number of factors that should be weighted accordingly. These factors include the required spatial resolution, the required velocity range, the sensitivity to velocity changes only, the required need to quantify dynamic velocity, the acceptable probe blockage of flow, the ability to be used in hostile environments, the calibration requirements, and low cost and ease of use. When used under appropriate conditions, the uncertainty in velocity determined by any of the discussed methods can be as low as 1% of the measured velocity, although under special conditions LDA methods can have an uncertainty 1 order of magnitude lower [21].

Pitot–Static Pressure Methods

The pressure probe methods are best suited for the determination of mean velocity in fluids of constant density. Relative to other methods, they are the simplest and cheapest method available to measure velocity at a point. Probe blockage of the flow is not a problem in large ducts and away from walls. Fluid particulate will block the impact ports, but aspirating models are available for such situations. They are subject to mean flow misalignment errors. They require no calibration and are well used in the field and laboratory alike.

Thermal Anemometers

Thermal anemometers are best suited for use in clean fluids of constant temperature and density. They offer a small spatial resolution and possess a high-frequency response, making them well suited for measuring dynamic velocities. However, signal interpretation in strongly dynamic flows can be complicated [15, 22]. Hot-film sensors are less fragile and less susceptible to contamination than hot-wire sensors. Probe blockage is not significant in large ducts and away from walls. Thermal anemometers are 180° directionally insensitive, an important factor in flows that may contain flow reversal regions. An industrial-grade system can be built rather inexpensively.

The thermal anemometer is usually calibrated against either pressure probes or an LDA.

Laser Doppler Anemometers

The LDA is a relatively expensive and technically complicated point velocity measuring technique that is suitable for most types of flows but is well suited to hostile, combusting, or highly dynamic flow environments. It can offer good frequency response, small spatial resolution, no probe blockage, and simple signal interpretation, but it requires optical access and the presence of scattering particles. The method measures the velocity of particles suspended in the moving fluid, not the fluid velocity. Thus careful planning is required in particle selection to ensure that the particle velocities represent the fluid velocity exactly. The size and concentration of the particles will govern the system frequency response [23, 24].

9.9 SUMMARY

Several reference pressure instruments have been presented that form the working standards for pressure transducer calibration. Pressure transducers convert sensed pressure into an output form that is readily quantifiable. These transducers come in many forms but tend to operate on either hydrostatic principles, expansion techniques, or force-displacement methods.

In moving fluids, special care must be taken in the measurement of the pressure to delineate between static and total pressure. Methods for the separate measurement of static and total pressure or for the measurement of the dynamic pressure are readily available and well documented.

The measurement of the local velocity within a moving fluid can be accomplished in a number of ways. Specifically, dynamic pressure, thermal anemometry, and Doppler anemometry methods have been presented. As discussed, each method offers advantages over the other, and the best technique must be carefully weighed against the needs and constraints of a particular application.

REFERENCES

1. Brombacher, W. G., Johnson, D. P., and Cross, J. L., *NBS Monograph* 8, 1960.
2. McLeod, H., *Philos. Mag.*, 48, 1874.
3. Sweeney, R. J., *Measurement Techniques in Mechanical Engineering*, Wiley, New York, 1953.
4. Hetenyi, M., ed., *Handbook of Experimental Stress Analysis*, Wiley, New York, 1950.
5. Way, S., Bending of circular plates with large deflection. *Transactions of the ASME*, 56; 1934.
6. Schlicting, H., *Boundary Layer Theory*, McGraw-Hill, New York, 1969.
7. Franklin, R. E., and Wallace, J. M., Absolute measurements of static-hole error using flush transducers, *Journal of Fluid Mechanics*, 42, 1970.
8. Rayle, R. E., Influence of static pressure measurements, ASME Paper No. 59-A-234, 1959.
9. Iberall, A. S., Attenuation of oscillatory pressures in instrument lines, *Transactions of the ASME*, 72, 1950.
10. Streeter, V. L., and Wylie, E. B., *Fluid Mechanics*, 7th ed., McGraw-Hill, New York, 1979.
11. Doebelin, E. O., *Measurement Systems: Application and Design*, 2d ed., McGraw-Hill, New York, 1975.
12. King, L. V., On the convection from small cylinders in a stream of fluid: Determination of the convection constants of small platinum wires with application to hot-wire anemometry, *Proceedings of the Royal Society, London*, 90, 1914.

13. Hinze, J. O., *Turbulence*, McGraw-Hill, New York, 1959.
14. Collis, D. C., and M. J. Williams, Two-dimensional convection from heated wires at low Reynolds numbers, *Journal of Fluid Mechanics*, 6, 1959.
15. Rodi, W., A new method for analyzing hot-wire signals in a highly turbulent flow and its evaluation in a round jet, *DISA Information*, 17, Dantek Electronics, Denmark, 1975 (see also Bruun, H. H., Interpretation of X-wire signals, *DISA Information*, 18, Dantek Electronics, Denmark, 1975.)
16. Freymuth, P., A bibliography of thermal anemometry, *TSI Quarterly*, 4, 1978.
17. Yeh, Y., and H. Cummins, Localized fluid flow measurement with a He–Ne laser spectrometer. *Applied Physic Letters*, 4, 1964.
18. *Photon Correlation and Light Beating Spectroscopy*, Cummins H. Z. and Pike, E. R., eds., *Proc. NATO ASI*, Plenum, New York, 1973.
19. Durst, F., Melling, A., and Whitelaw, J. H., *Principles and Practice of Laser Doppler Anemometry*, Academic, New York, 1976.
20. Goldstein, R. J., ed., *Fluid Mechanics Measurements*, Hemisphere, New York, 1983.
21. Goldstein, R. J., and Kried, D. K., Measurement of laminar flow development in a square duct using a laser doppler flowmeter, *Journal of Applied Mechanics*, 34, 1967.
22. Yavuzkurt, S., A guide to uncertainty analysis of hot-wire data. *Transactions of the ASME, Journal of Fluids Engineering*, 106, 1984.
23. Maxwell, B. R., and Seaholtz, R. G., Velocity lag of solid particles in oscillating gases and in gases passing through normal shock waves. NASA-TN-D-7490, 1974.
24. Dring, R. P., and Suo, M. Particle trajectories in swirling flows, *Transactions of the ASME, Journal of Fluids Engineering*, 104, 1982.

NOMENCLATURE

d	diameter $[l]$	E_m	bulk modulus of elasticity $[ml^{-1}t^{-2}]$
e_l	elemental errors	Gr	Grashof number
h	depth $[l]$	H	manometer deflection $[l]$
h_0	reference depth $(h=0)$ $[l]$	K	static sensitivity
k	ratio of specific heats	K_q	charge sensitivity $[C\,m^{-1}l^{-1}t^2]$
p	pressure $[ml^{-1}t^{-2}]$	K_E	voltage sensitivity $[V\,m^{-1}t^2]$
p_a	applied pressure $[ml^{-1}t^{-2}]$	L	manometer riselength;
p_{abs}	absolute pressure $[ml^{-1}t^{-2}]$		transmission length $[L]$
p_e	relative static pressure error $[ml^{-1}t^{-2}]$	M	Mach number
p_i	indicated pressure $[ml^{-1}t^{-2}]$	Re_d	Reynolds number, $\mathrm{Re}_d = Ud/v$
p_m	measured pressure $[ml^{-1}t^{-2}]$	S	specific gravity
p_o	reference pressure $[ml^{-1}t^{-2}]$	U	velocity $[l\,t^{-1}]$
p_t	total or stagnation pressure $[ml^{-1}t^{-2}]$	V	volume $[l^3]$
p_v	dynamic pressure $[ml^{-1}t^{-2}]$	γ	specific weight $[ml^{-2}t^{-2}]$
q	charge	ϵ	dielectric constant
r	radius $[l]$	λ	wavelength $[l]$
t	thickness $[l]$	ρ	density $[ml^{-3}]$
u	uncertainty	τ	time constant $[t]$
u_d	design-stage uncertainty	ϕ	latitude
y	displacement $[l]$	ω_n	natural frequency $[t^{-1}]$
z	altitude $[l]$	μ	absolute viscosity $[m\,t^{-1}l^{-1}]$
C	capacitance [C]	v	kinematic viscosity $[l^2/t]$
E	voltage [V]	v_p	Poisson ratio

PROBLEMS

9.1 Convert the following absolute pressures to gauge pressure units of N/m^2:

a. 10.8 psia

b. 1.75 bars abs

c. 30.36 inches H_2O abs

d. 791 mm Hg abs

9.2 Convert the following gauge pressures into absolute pressure relative to one standard atmosphere:

a. −0.55 psi

b. 100 mm Hg

c. 98.6 kPa

d. 7.62 cm H_2O

9.3 A water-filled manometer is used to measure the pressure in an air-filled tank. One leg of the manometer is open to atmosphere. For a measured manometer deflection of 250 cm H_2O, determine the tank static pressure. Barometric pressure is 101.3 kPa abs.

9.4 A deadweight tester is used to provide a standard reference pressure for the calibration of a pressure transducer. A combination of 25.3 kg$_f$ of 7.62-cm.-diameter stainless-steel disks is found to be necessary to balance the tester piston against its internal pressure. For an effective piston area of 5.065 cm^2 and a piston weight of 5.35 kg$_f$, determine the standard reference pressure in: bars, N/m^2, and Pa abs. Barometric pressure is 770 mm Hg abs, elevation is 20 m, and latitude is 42°.

9.5 An inclined tube manometer indicates a change in pressure of 5.6 cm H_2O when switched from a null mode (both legs at atmospheric pressure) to deflection mode (one leg measuring, one leg at atmospheric pressure). For an inclination of 30° relative to horizontal, determine the pressure change indicated.

9.6 Show that the static sensitivity of an inclined tube manometer is a factor of $1/\sin\theta$ higher than for a U-tube manometer.

9.7 Determine the static sensitivity of an inclined tube manometer set at an angle of 30°. The manometer tube measures the pressure difference of air and uses mercury as its fluid.

9.8 Show that the instrument uncertainty in the inclined tube manometer of Example 9.2 increases to 6.8 N/m^2 as θ goes to 90°.

9.9 Determine the maximum deflection and the natural frequency of a 0.1-in.-thick diaphragm made of steel ($E_m = 30$ Mpsi, $v_p = 0.32$, $\rho = 0.28$ lb$_m$/in.^{3}), if the diaphragm must have a diameter of 0.75 in. Determine its differential pressure limit.

9.10 A strain gauge–diaphragm pressure transducer (accuracy: <0.1% reading) is subjected to a pressure differential of 10 kPa. If the output is measured by using a voltmeter having a resolution of 10 mV and an accuracy of better than 0.1% of the reading, estimate the uncertainty in pressure at the design stage. How does this change at 100 and 1000 kPa?

9.11 Select a practical manometeric fluid to measure pressures up to 68,950 Pa of an inert gas ($\gamma = 10.4$ N/m^3), if water ($\gamma = 9790$ N/m^3), oil ($S = 0.82$), and mercury ($S = 13.57$) are available.

9.12 An air pressure in the range 200–400 N/m^2 is to be measured relative to atmosphere by using a U-tube manometer with mercury ($S = 13.57$). Manometer resolution will be 1 mm with a zero error of 0.5 mm. Estimate the design-stage uncertainty in gauge pressure based on the manometer indication at 20°C. Would an inclined manometer ($\theta = 30°$) be a better choice if the inclination can be set to within 0.5°?

9.13 Plot the design-stage uncertainty in estimating a nominal pressure of 10,000 N/m^2 by using an inclined manometer (resolution: 1 mm; zero error: 0.5 mm) with water at 20°C for inclination angles of 10–90° (using 10° increments). The inclination angle can be set to within 1°.

9.14 A capacitance pressure transducer, such as shown in Figure 9.14, uses a C_1 of $0.01 \pm 0.005 \mu f$ and an excitation voltage of $5 \pm 1\%$ V. The plates have an overlap area of 8 ± 0.01 mm^2 and are separated by an air gap of 1.5 ± 0.1 mm. If the plates move apart by 0.2 mm, estimate the change in capacitance and the output voltage.

9.15 A diaphragm pressure transducer is calibrated against a pressure standard that has been certified by NIST [accuracy: within ± 0.5 psi (95%)]. Both the standard and pressure transducer output a voltage signal, which is to be measured by a voltmeter (accuracy: ±10 mV; resolution: 1 mV). A calibration curve fit yields $p = 0.564 + 24.0E \pm 1$ psi (95%) based on six points over the range 0–100 psi. When the transducer is installed for its intended purpose, installation effects are estimated to affect pressure up to ±0.5 psi. Estimate the uncertainty associated with a pressure measurement by using the installed transducer–voltmeter system.

9.16 A diaphragm pressure transducer has a water-cooled sensor for high-temperature environments. Its manufacturer claims it to have a rise time of 10 ms, a ringing frequency of 200 Hz, and damping ratio of 0.8.

a. Describe a test plan to verify the manufacturer's specifications.
b. Would this transducer have a suitable frequency response to measure the pressure variations in a typical four-cylinder car? Show your reasoning.

9.17 Find the natural frequency of a 1 mm thick, 6-mm-diameter steel diaphragm to be used for high-frequency pressure measurements. What would be the maximum operating pressure difference that could be applied? What is the effect of a larger diameter for this application?

9.18 The pressure fluctuations in a pipe filled with air at 20°C are to be measured by using a static wall tap, rigid connecting tubing, and a diaphragm pressure transducer. The transducer has a natural frequency of 100,000 Hz. For a tap and tubing diameter of 3.5 mm, a tube length of 0.25 m, and a transducer dead volume of 1600 mm^3, estimate the resonance frequency of the system. What is the maximum frequency that this system can measure with no more than a 10% dynamic error? Plot the frequency response of the system.

9.19 Estimate the sensitivity of a pitot–static tube pressure signal to the velocity that it senses.

9.20 A pitot–static pressure probe inserted within a large duct indicates a differential pressure of 20.3 cm H_2O. Determine the velocity measured.

9.21 A pitot–static tube is placed in a flow of 20°C air at the centerline of a round duct. The pressure difference is sensed by a differential piezoelectric pressure transducer–charge amplifier system whose voltage is noted by a voltmeter (accuracy: ±10 mV; resolution: 1 mV). The transducer calibration ($N = 30$) is given by

$$p = 0.205 + 0.950E \text{ [V]} \pm 0.002 \text{ N/m}^2 \quad (95\%)$$

Three replications for a desired operating condition give the following data:

Run	N	E [V]	S_E [V]
1	21	2.439	0.010
2	21	2.354	0.009
3	21	2.473	0.012

Estimate the flow velocity and its uncertainty.

9.22 A tall pitot–static tube is mounted through and 1 m above the roof of a performance car such that it senses the freestream. Estimate the static, stagnation, and dynamic pressure sensed at 325 kph, if (a) the car is moving along a long straight section of road, and (b) the car is stationary within an open circuit wind tunnel where the flow is blown over the car (Hint: $p_t = 0$ gauge).

9.23 Wall pressure taps (e.g., Figures 9.19 and 9.21) are often used to sense surface pressure and are connected to transducers by connecting tubing. Two race engineers discuss the preferred diameter of the tubing to measure pressure changes on a car as it moves along a track. The tubing length may be up to 2 m. Engineer A suggests very small 2-mm-diameter tubing to reduce air volume so to improve response time. Engineer B disagrees and suggests 5-mm tubing to balance air friction with air volume to improve response time. Offer your opinion and its basis. (Hint: Look at length to diameter effects.)

9.24 A system similar to that described in Example 9.7 is used to measure surface pressures on a car during a closed-circuit wind tunnel test. Large pressure data sets are taken. Estimate the overall uncertainty of the measurements using the 12-bit A/D converter.

Typical Wind Tunnel Measurements:
Stock Race Car at 180 kmph

Position	Pressure (cm H_2O)	
	$|\bar{P}_{avr}|$	S_P
hood	0.8	0.025
roof	3.3	0.0025
rear deck	8.0	0.05

9.25 Pressure is measured 20 times at random time intervals over the course of a test run on a gas turbine compressor section under fixed operating conditions. This procedure is duplicated at each of four measuring positions separated by 90° in the compressor's cross-plane. The results are

	Station			
	1	2	3	4
$\bar{p}$ (MN/m^2)	153	142	161	157
S_p (MN/m^2)	7	9	9	7

What would be the significance of pooling the data in determining the mean pressure here? What information could be found by comparing the pooled mean value to the local mean values? Do this. What new information would replications provide?

9.26 Determine the resolution of a manometer required to measure the velocity of air from 5 to 50 m/s by using a pitot–static tube and a manometric fluid of mercury ($S = 13.57$) to a zero-order uncertainty of 5% and 1%.

9.27 A long cylinder is placed into a wind tunnel and aligned perpendicular to an oncoming freestream. Static wall pressure taps are located circumferentially about the centerline of the cylinder at 45° increments with 0° at the impact (stagnation) position. Each tap is connected to a separate manometer referenced to atmosphere. A pitot–static tube indicates an upstream dynamic pressure of 20.3 cm H_2O, which is used to determine the freestream velocity. The following static pressures are measured:

Tap (deg)	p (cm H_2O)	Tap (deg)	p (cm H_2O)
0	0.0	135	23.1
45	41.4	180	23.9
90	81.3		

Compute the local velocities around the cylinder if the total pressure in the flow remains constant; $p_{atm} = 101.3$ kPa abs, $T_{atm} = 16°C$.

9.28 A pitot–static tube is used as a working standard to calibrate a hot-wire anemometer in 20°C air. If dynamic pressure is measured by using a water-filled micromanometer, determine the smallest manometer deflection for which the pitot–static tube can be considered as accurate without correction for viscous effects.

9.29 For the thermal anemometer in Figures 9.25 and 9.26, determine the decade resistance setting required to set a platinum sensor at 40°C above ambient if the sensor ambient resistance is $110\,\Omega$ and $R_3 = 500\,\Omega$ and $R_4 = 500\,\Omega$. $\alpha = 0.00395°C^{-1}$.

9.30 Determine the static sensitivity of the output from a constant resistance anemometer as a function of velocity. Is it more sensitive at high or at low velocities?

9.31 An LDA setup in a dual-beam mode uses a 600-mm focal length lens ($\theta = 5.5°$) and an argon–ion laser ($\lambda = 514.4$ nm). Compute the Doppler shift frequency expected at 1, 10 and 100 m/s. Repeat for a 300-mm lens ($\theta = 7.3°$).

9.32 A set of 5000 measurements of velocity at a point in a flow using a dual-beam LDA give the following results: $\overline{U} = 21.37$ m/s and $S_U = 0.43$ m/s. If the Doppler shift can be measured with an uncertainty of better than 0.9%, the optical angle can be measured to within 0.25°, and the laser can be tuned to $\lambda = 623.8 \pm 0.5\%$ nm, determine the best estimate of the velocity.

Chapter 10

Flow Measurements

10.1 INTRODUCTION

Flow rate can be expressed in terms of a volume per unit time, known as the *volume flow rate*, or as a mass per unit time, known as the *mass flow rate*. Not only is this quantity useful in flow metering, but many engineering systems require flow measurement information for proper process control. For example, in heat exchange processes the rate at which energy can be removed from or added to a moving fluid is directly proportional to the mass flow rate of the fluid. Properly selected and designed flow measurement equipment can be a very high engineering priority because from the flow of water into our homes to the flow of petroleum at the oil well, flow measurements are vitally linked to the economy.

This chapter discusses some of the most common and accepted methods for flow quantification. Size, accuracy, cost, pressure drop, pressure losses, and compatibility with the fluid are important engineering design considerations for flow metering devices. All methods have both desirable and undesirable features that necessitate some compromise in the selection of the best method for the particular application, and most of the more important of these considerations are included in this chapter. Inherent uncertainties in fluid properties, such as density, viscosity, or specific heat, can affect the accuracy of a flow measurement made with some metering methods. Novel techniques that preclude knowledge of fluid properties are being introduced for use in the more demanding of these applications. The chapter objective is to present both an overview of basic flow metering techniques for proper meter selection, as well as those design considerations important in the integration of a flow metering system with the process system it will meter.

10.2 HISTORICAL COMMENTS

Their importance in engineering systems gives flow measurement methods a rich history. The earliest available accounts of flow metering were recorded by Hero of Alexandria (ca. 150 B.C.), who proposed a scheme to regulate water flow by using a siphon pipe attached to a constant head reservoir. The early Romans developed elaborate water systems to supply public baths and private homes. In fact, Sextus Frontinius (A.D. 40–103), Commissioner of Water Works for Rome, prepared a treatise on Roman methods of water distribution. Evidence suggests that the Romans understood that a relation between volume flow rate and pipe flow area existed, although the role of velocity in flow rate was not recognized. Weirs were used to

389

regulate bulk flow through aqueducts, and the cross-sectional area of terra cotta pipe was used to meter supplies to individual buildings.

Following a number of experiments conducted using olive oil and water, Leonardo da Vinci (1452–1519) first proposed the continuity principle: that area, velocity, and flow rate were related. However, most of his writings were lost until centuries later, and Benedetto Castelli (ca. 1577–1644), a student of Galileo, has been credited in some texts with developing the same steady, incompressible continuity concepts in his day. Isaac Newton (1642–1727), Daniel Bernoulli (1700–1782), and Leonhard Euler (1707–1783) built the mathematical and physical bases on which modern flow meters would later be developed. By the 19th century, the concepts of continuity, energy, and momentum were sufficiently understood for practical exploitation. Relations between flow rate and pressure losses were developed that would permit the tabulation of the hydraulic coefficients necessary for the quantitative engineering design of many modern flow meters.

10.3   FLOW RATE CONCEPTS

The flow rate through a pipeline, duct, or other flow system can be described by use of a control volume, a judiciously selected volume in space through which a fluid flows. The amount of fluid that passes through this volume in a given period of time will determine the flow rate. A geometrical boundary of a control volume is called a control surface. Such a control volume is shown in Figure 10.1, which consists here of a defined volume within a pipe.

The velocity of a fluid at a point can be described by the use of a three-dimensional velocity vector given here in cylindrical coordinates by

$$\mathbf{U} = \mathbf{U}(x, r, \theta) = u\hat{e}_x + v\hat{e}_r + w\hat{e}_\theta$$

where u, v, and w are the scalar velocity magnitudes and $\hat{e}_x$, $\hat{e}_r$ and $\hat{e}_\theta$ are unit vectors in each of the component directions, x, r, and θ, respectively.

The amount of fluid of density ρ that passes through the control volume of volume V at any instant in time depends on the amount of fluid that crosses the control surfaces. This can be expressed by examination of mass flow into, out of, and remaining within the control volume (CV) at any instant. Conservation of mass demands that the rate at which mass accumulates within the control volume plus the net flow of mass through a control volume that physically crosses any of its control

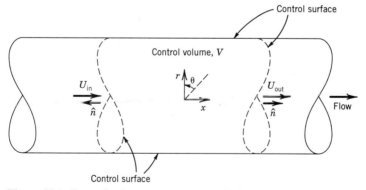

Figure 10.1 Control volume concept as applied to flow through a pipe.

surfaces (CSs) be zero. This is expressed by

$$\frac{\partial}{\partial t} \iiint_{CV} \rho \, dV + \oiint_{CS} \rho \, \mathbf{U} \cdot \hat{n} \, dA = 0 \tag{10.1}$$

where $\hat{n}$ is the outward normal from a control surface of area A.

A steady flow situation exists when the sum of the mass flow across all control surfaces *into* the control volume equals the sum of the mass flow across all the control surfaces *out of* the control volume. For steady flows, equation (10.1) can be simplified to

$$\dot{m}_{in} = \dot{m}_{out} \tag{10.2}$$

where $\dot{m}$ is defined as the mass flow rate through any area A

$$\dot{m} = \iint_{A} \rho \mathbf{U} \cdot \hat{n} \, dA$$

If the average mass flux through the control surface, $\overline{\rho U}$, is known then equation (10.2) becomes

$$\dot{m} = (\overline{\rho U}) A \tag{10.3}$$

where $\overline{U}$ is the average velocity over the control surface. Mass flow rate has the dimensions of mass per unit of time [e.g., units of lb_m/s, kg/s, etc.].

As a general rule, isothermal flows of liquids can be considered incompressible (i.e., constant density). This can also be assumed for isothermal gas flows that move at speeds of less than 0.3 times the speed of sound in that fluid. In these flows, equation (10.3) can be further reduced to

$$Q_{in} = Q_{out} \tag{10.4}$$

where Q is defined as the volume flow rate

$$Q = \iint_{A} \mathbf{U} \cdot \hat{n} \, dA$$

For example, the velocity at a control surface located at axial position x along a pipe will be described by the single component $u(r, \theta)$. In practical terms, knowledge of $u(r, \theta)$ at x in a steady, incompressible flow would be sufficient to estimate the average velocity and yield the volume flow rate through the control surface. In a pipe of circular cross section, the volume flow rate at position x is found by

$$Q = \int_{0}^{r_1} \int_{0}^{2\pi} u(r, \theta) r \, d\theta \, dr \tag{10.5}$$

where r_1 is the pipe radius.

If the average velocity, $\overline{U}$, over a control surface is known, then the volume flow rate can be found by

$$Q = \overline{U} A \tag{10.6}$$

The importance of the preceding analysis is it indicates that methods for determining steady flow rate must depend on the use of techniques that are sensitive either to the average mass flux, $\overline{\rho U}$, to estimate mass flow rate or to the average velocity, $\overline{U}$, to estimate volume flow rate. There are many direct or indirect methods to do either of these.

The flow through a pipe or duct can be characterized as being laminar, turbulent, or something in between called transitional. In flow measurements this flow character can be established through the nondimensional Reynolds number, defined by

$$\mathrm{Re}_{d_1} = \frac{\bar{U}d_1}{\nu} = \frac{4Q}{\pi d_1 \nu} \tag{10.7}$$

where ν is the fluid kinematic viscosity and d_1 is the diameter for circular pipes or the hydraulic diameter for noncircular pipes, $4r_H$, where r_H is defined as the wetted area divided by the wetted perimeter. In pipes, turbulent flows are encountered when $\mathrm{Re}_{d_1} > 4000$ and laminar flows are encountered when $\mathrm{Re}_{d_1} < 2000$. In flow system design, operation in the transitional flow regime should be avoided.

10.4 VOLUME FLOW RATE THROUGH VELOCITY DETERMINATION

The direct implementation of equation (10.5) for estimating the volume flow rate through a duct requires measurement of the velocity at points along several cross sections of a flow control surface. Methods for determining the velocity at a point include any of those previously discussed in Chapter 9. This procedure is most often used for the one-time verification or calibration of system flow rates. For example, the procedure has been used in ventilation system setup and problem diagnosis, where the installation of an in-line flow meter is uncommon since it is not needed in regular operation.

In using this technique in circular pipes, a number of discrete measuring positions are chosen along m flow cross sections (radii) spaced at $360°/m$ apart, such as shown in Figure 10.2. A velocity probe is traversed along each flow cross section with readings taken at each measurement position. There are several options as to the selection of measuring positions [2, 13]. A simple method is to divide the flow area into smaller equal areas with measurements made at the centroid of each of the smaller areas. This latter method is also useful for similar measurements in rectangular ducts. Regardless of the option selected, the average flow rate is estimated along each cross section traversed by using equation (10.5), and the pooled mean of

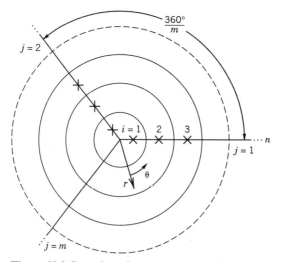

Figure 10.2 Location of n measurements along m radial lines in a pipe.

the flow rates for the m cross sections is determined to yield the best estimate of the duct flow rate. It is important that the flow rate remain fixed during each traverse to minimize temporal errors during data acquisition. Example 10.1 illustrates this method for estimating volume flow rate.

EXAMPLE 10.1

A steady flow of air at 70°F passes through a 10-in.-i.d. circular pipe. A velocity-measuring probe is traversed along three cross-sectional lines ($j = 1, 2, 3$) of the pipe and measurements are made at four positions ($i = 1, 2, 3, 4$) along each traverse line. The locations for each measurement are selected at the centroids of equally spaced areal increments as indicated in the following table [2, 13]. Determine the volume flow rate in the pipe.

Radial Location,		U_{ij} [ft/s]		
i	r/r_1	Line 1 ($j=1$)	Line 2 ($j=2$)	Line 3 ($j=3$)
1	0.3536	8.71	8.62	8.78
2	0.6124	6.26	6.31	6.20
3	0.7906	3.69	3.74	3.79
4	0.9354	1.24	1.20	1.28

KNOWN

$U_{ij}(r/r_1)$; $i = 1, 2, 3, 4$, $j = 1, 2, 3$
$d_1 = 10$ in. $\left(A = \pi d_1^2/4 = 0.54 \text{ ft}^2\right)$

ASSUMPTIONS

Constant and steady pipe flow during all measurements
Incompressible flow

FIND

$\langle \bar{Q} \rangle$

SOLUTION

The flow rate is found by integration of the velocity profile across the duct along each line and subsequent averaging of the three values. For discrete velocity data, equation (10.5) is written along each line, $j = 1, 2, 3$, as

$$Q_j = 2\pi \int_0^{r_1} ur \, dr \approx 2\pi \sum_{i=1}^{4} U_{ij} \, r \, \Delta r$$

where Δr is the radial distance separating each position of measurement. This can be further simplified since the velocities are located at positions that make up centroids of equal areas:

$$Q_j = \frac{A}{4} \sum_{i=1}^{4} U_{ij} \qquad j = 1, 2, 3$$

Then, the mean flow rate along each line of traverse is

$$Q_1 = 2.71 \text{ ft}^3/\text{s} \qquad Q_2 = 2.71 \text{ ft}^3/\text{s} \qquad Q_3 = 2.73 \text{ ft}^3/\text{s}$$

The estimate of the average pipe flow rate is found from the pooled mean of the individual flow rates

$$\langle \bar{Q} \rangle = \frac{1}{3} \sum_{j=1}^{3} Q_j$$

so,

$$\langle \bar{Q} \rangle = 2.72 \text{ ft}^3/\text{s} \quad (0.077 \text{ m}^3/\text{s})$$

10.5   PRESSURE DIFFERENTIAL METERS

The operating principle of a pressure differential meter is based on the relationship between volume flow rate and the pressure drop $\Delta p = p_1 - p_2$, along the flow path,

$$Q \approx (p_1 - p_2)^n$$

where the value of n equals one for laminar flow occurring between the pressure measurement locations and n equals one-half in fully turbulent flow.

An intentional reduction in flow area will cause a measurable local pressure drop across the flow path. The reduced flow area causes a local increase in velocity. So the pressure drop is in part due to the so-called Bernoulli effect, the inverse relationship between local velocity and pressure, as well as flow energy losses. The class of pressure differential meters that use such methods is commonly called *obstruction meters*. A special class of pressure differential meters alter the flow in such a manner as to force the flow to become laminar ($\text{Re}_{d_1} < 2000$) between two locations of pressure measurement. This class is known as *laminar flow meters*.

Obstruction Meters

Three common obstruction meters are the *orifice plate*, the *venturi*, and the *flow nozzle*. Flow area profiles of each are shown in Figure 10.3. These meters are usually inserted in-line with a pipe. This class of meters operates by using similar physical

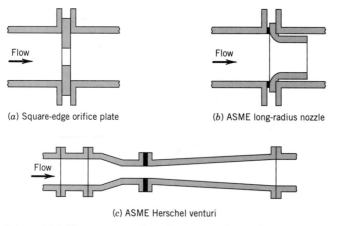

(a) Square-edge orifice plate   (b) ASME long-radius nozzle

(c) ASME Herschel venturi

Figure 10.3 Flow area profiles of common obstruction meters.

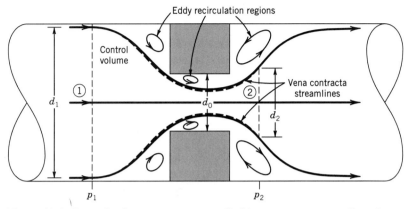

Figure 10.4 Control volume concept as applied between two streamlines for flow through an obstruction meter.

reasonings to relate volume flow rate to pressure drop. Referring to Figure 10.4, consider the energy equation written between two control surfaces for an incompressible fluid flow through the arbitrary control volume shown. It is assumed that (1) no external energy in the form of heat is added to the flow, (2) there is no shaft work done within the control volume, and that the flow is (3) steady and (4) one dimensional. This yields

$$\frac{p_1}{\gamma} + \frac{\bar{U}_1^2}{2g} = \frac{p_2}{\gamma} + \frac{\bar{U}_2^2}{2g} + h_{L_{1-2}} \tag{10.8}$$

where $h_{L_{1-2}}$ denotes the head losses occurring as a result of frictional effects between control surfaces 1 and 2. From conservation of mass [equations (10.4) and (10.6)],

$$\bar{U}_1 = \bar{U}_2 A_2 / A_1$$

Substituting $\bar{U}_1$ into equation (10.8) and rearranging yields the incompressible volume flow rate,

$$Q_I = \bar{U}_2 A_2 = \frac{A_2}{[1 - (A_2/A_1)^2]^{1/2}} \sqrt{\frac{2(p_1 - p_2)}{\rho} + 2gh_{L_{1-2}}} \tag{10.9}$$

where the subscript I is used only to emphasize that equation (10.9) applies to an incompressible flow. Later we drop the subscript.

When the flow area changes abruptly, the effective flow area immediately downstream of the alteration will not necessarily be the same as the pipe flow area. This is due to the vena contracta effect, originally investigated by Jean Borda (1733–1799) and illustrated in Figure 10.4. This effect is brought about by an inability of a fluid to expand immediately upon encountering an area expansion as a result of the inertia of each fluid particle. This forms a central core flow bounded by regions of slower moving recirculating eddies. As a consequence, the pressure sensed with pipe wall taps located within the vena contracta region will correspond to the higher moving velocity within the vena contracta of unknown flow area, A_2. The unknown vena contracta area will be accounted for by introducing a contraction coefficient C_c, where

$C_c = A_2/A_0$, into equation (10.9). This yields

$$Q_I = \frac{C_c A_0}{[1 - (C_c A_0/A_1)^2]^{1/2}} \sqrt{\frac{2\Delta p}{\rho} + 2gh_{L_{1-2}}} \tag{10.10}$$

The frictional head losses can be incorporated into a friction coefficient, C_f, such that equation (10.10) becomes

$$Q_I = \frac{C_f C_c A_0}{[1 - (C_c A_0/A_1)^2]^{1/2}} \sqrt{\frac{2\Delta p}{\rho}} \tag{10.11}$$

For convenience, the coefficients are factored out of equation (10.11) and replaced by a single coefficient known as the *discharge coefficient*, C, and equation (10.11) becomes

$$Q_I = CEA \sqrt{\frac{2\Delta p}{\rho}} \tag{10.12}$$

where E, known as the velocity of approach factor, is defined by

$$E \equiv \frac{1}{[1 - (A_0/A_1)^2]^{1/2}} = \frac{1}{(1 - \beta^4)^{1/2}} \tag{10.13}$$

with $\beta \equiv d_0/d_1$. In engineering handbooks, the product CE is often represented by the *flow coefficient*, K_0.

The discharge coefficient can be defined as the ratio of the actual flow rate through a meter to the ideal flow rate possible for the pressure drop measured. As shown, it is derived from frictional effects and vena contracta effects. Both effects alter the flow rate from ideal flow concepts. Because of its nature, C will depend on the flow Reynolds number and the β ratio, d_0/d_1, for each particular obstruction flow meter design. Because the magnitude of the vena contracta and head loss effects must vary along the length of a meter, the flow rate estimate based on pressure drop is very sensitive to pressure tap location, and consistency in tap placement is imperative for correct operation [2].

Compressibility Effects

In compressible gas flows, compressibility effects change the value of the discharge coefficient. Rather than modify C, it is customary to introduce the compressible adiabatic *expansion factor*, Y, defined as the ratio of the actual compressible volume flow rate, Q, divided by the incompressible flow rate Q_I. Combining with equation (10.12) yields

$$Q = YQ_I = CEAY\sqrt{2\Delta p/\rho_1} \tag{10.14}$$

Equation (10.14) represents a general form of the working equation for obstruction meter volume flow rate determination.

The value for the expansion factor, Y, depends on several values: the β ratio, the particular gas specific heat ratio, k, and the relative pressure ratio across the meter, $(p_1 - p_2)/p_1$, for a particular meter type. As a general rule, compressibility effects become important when $(p_1 - p_2)/p_1 \geq 0.1$. Note that when $Y \approx 1$, equations (10.12) and (10.14) become identical.

Standards

The flow behavior of the orifice plate, venturi, and flow nozzle has been studied to such an extent that these meters are used extensively without calibration. Values for the flow coefficients and expansion factors of these common obstruction meters have been tabulated and are available in standard flow handbooks along with standardized construction, installation, and operation techniques [2–4, 12]. Nonstandard installation requires in-line calibration.

Orifice Meter

An orifice meter consists of a circular plate, containing a hole (orifice), which is inserted into a pipe such that the orifice is concentric with the pipe inside diameter (i.d.). Several variations in the orifice design exist, but the square-edged orifice, shown in Figure 10.5, is common. Installation is simplified by housing the orifice plate

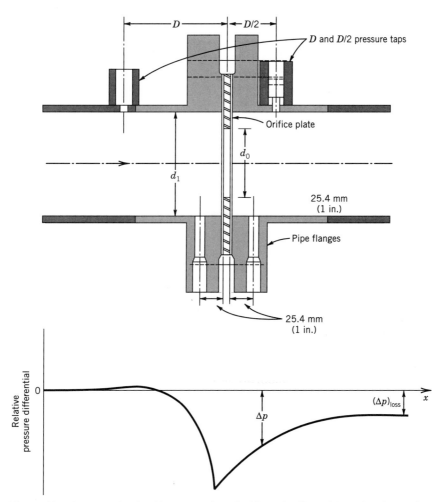

Figure 10.5 Square-edged orifice meter installed in a pipeline with optional one pipe and one-half pipe diameters, and flange pressure taps shown. Relative flow pressure drop along the pipe axis is shown.

between two pipe flanges. With this technique an orifice plate is interchangeable with others of different β value. The simplicity of the installation and orifice design allows for a range of β values to be maintained on hand at modest expense. Rudimentary versions of the orifice plate have existed for several centuries. Both Torricelli and Newton used orifice plates to study the relation between pressure head and efflux from reservoirs, although neither ever got the discharge coefficients quite right [5].

For an orifice plate, equation (10.14) is used with values of A and β being based on the orifice hole diameter. The exact location of the pressure taps is crucial when tabulated values for flow coefficient and expansion factor are used. Standard pressure tap locations include (1) flange taps where pressure tap centers are located 25.4 mm (1 in.) upstream and 25.4 mm (1 in.) downstream of the nearest orifice face, and (2) taps located one pipe diameter upstream and one-half diameter downstream of the upstream orifice face. Nonstandard tap locations require on-site meter calibration.

Values for the flow coefficient as functions of β and Re_{d_1} and for the expansion factor as functions of k and $(p_1 - p_2)/p_1$ for a square-edged orifice plate are given in Figures 10.6 and 10.7 based on the use of flange taps. Uncertainty bias limits for discharge coefficient [12] are ~0.6% of C for $0.2 \leq \beta \leq 0.6$ and β% of C for all $\beta > 0.6$. Bias limits for the expansion factor are ~$[4(p_1 - p_2)/p_1]$% of Y. Although the orifice plate represents a relatively inexpensive flow meter choice that provides an easily measurable pressure drop, it introduces a large permanent pressure loss, $(\Delta p)_{loss}$, into the flow system. The magnitude of pressure drop is illustrated in Figure 10.5 with the pressure loss estimated from Figure 10.8.

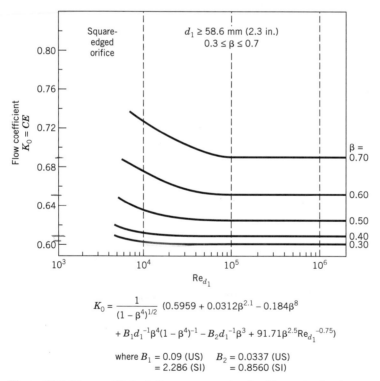

$$K_0 = \frac{1}{(1-\beta^4)^{1/2}} \left(0.5959 + 0.0312\beta^{2.1} - 0.184\beta^8 \right.$$

$$\left. + B_1 d_1^{-1}\beta^4(1-\beta^4)^{-1} - B_2 d_1^{-1}\beta^3 + 91.71\beta^{2.5}Re_{d_1}^{-0.75}\right)$$

where B_1 = 0.09 (US) B_2 = 0.0337 (US)
        = 2.286 (SI)   = 0.8560 (SI)

Figure 10.6 Flow coefficients for a square-edged orifice meter having flange pressure taps (compiled from data in [2]).

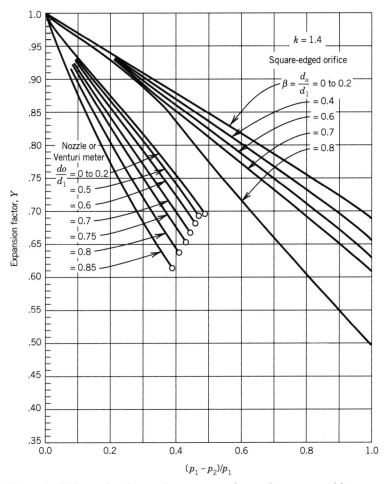

Figure 10.7 Expansion factors for common obstruction meters with $k = c_p/c_v = 1.4$. (Courtesy of the American Society of Mechanical Engineers, New York; compiled and reprinted from [2].1

Venturi Meter

A venturi meter consists of a smooth converging contraction to a narrow throat followed by a shallow diverging section, as shown in Figure 10.9. The standard venturi can utilize either a 15° or 7° divergent section. The meter is installed between two flanges intended for this purpose. Pressure is sensed between a location upstream of the throat and a location at the throat such that equation (10.14) is used with values for both A and β based on the throat diameter.

The quality of a venturi meter ranges from cast to precision machined units. The discharge coefficient varies little for pipe diameters above 7.6 cm (3 in.). For $2 \times 10^5 \leq \mathrm{Re}_{d_1} \leq 2 \times 10^6$ and $0.4 \leq \beta \leq 0.75$, values for discharge coefficient of $0.984 \pm 0.7\%$ bias (95%) for cast and $0.995 \pm 1\%$ bias (95%) for machined units should be used [2, 12]. Values for expansion factor are given in Figure 10.7 and have a bias limit of $\sim[(4 + 100\beta^2)(p_1 - p_2)/p_1]\%$ of Y [12]. Although representing a relatively expensive initial cost, as much as fivefold that of an orifice plate, Figure 10.8 demonstrates that a venturi meter shows a much smaller permanent pressure loss

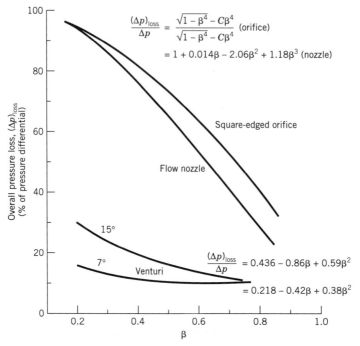

$$\frac{(\Delta p)_{loss}}{\Delta p} = \frac{\sqrt{1-\beta^4}-C\beta^4}{\sqrt{1-\beta^4}-C\beta^4} \text{ (orifice)}$$

$$= 1 + 0.014\beta - 2.06\beta^2 + 1.18\beta^3 \text{ (nozzle)}$$

Square-edged orifice

Flow nozzle

15°

7°

Venturi

$$\frac{(\Delta p)_{loss}}{\Delta p} = 0.436 - 0.86\beta + 0.59\beta^2$$

$$= 0.218 - 0.42\beta + 0.38\beta^2$$

Figure 10.8 The permanent pressure loss associated with flow through common obstruction meters. (Courtesy of the American Society of Mechanical Engineers, New York; compiled and reprinted from [2].)

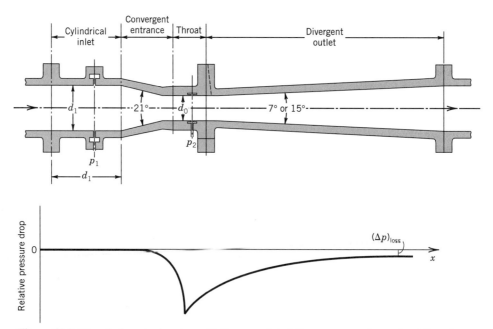

Figure 10.9 Herschel venturi meter with the associated flow pressure drop along its axis.

for a given installation. This translates into much lower system operating costs for the pump or blower used to move the flow.

The modern venturi meter was first proposed by Clemens Herschel (1842–1930). Herschel's design was based on his understanding of the principles developed by several men, most notably those of Daniel Bernoulli. However, he cited the studies of contraction–expansion angles and their corresponding resistance losses by Giovanni Venturi (1746–1822) and later those by James Francis (1815–1892) as instrumental to his design of a practical flow meter.

Flow Nozzles

A flow nozzle consists of a gradual contraction to a narrow throat. It needs less installation space than a venturi meter and has ~80% of the initial cost. A common form for the nozzle is the ASME long radius nozzle, in which the nozzle contraction is that of the quadrant of an ellipse with major axis aligned with the flow axis, as shown in Figure 10.10. The nozzle is typically installed in line, but can also be used at

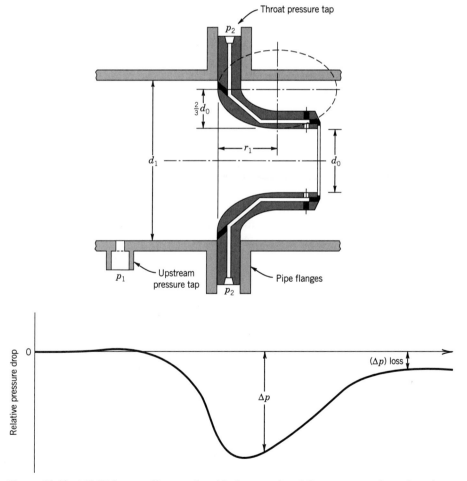

Figure 10.10 ASME long-radius nozzle with the associated flow pressure drop along its axis.

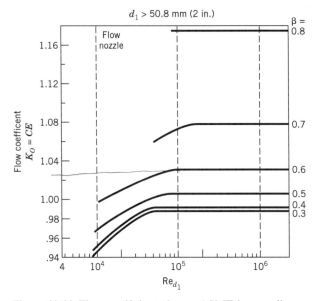

Figure 10.11 Flow coefficients for an ASME long-radius nozzle with a throat pressure tap (compiled from [2]).

the inlet to and the outlet from a plenum or reservoir or at the outlet of a pipe. Pressure taps are usually located at one pipe diameter upstream of the nozzle inlet and at the nozzle throat by using either wall or throat taps. The flow rate is determined from equation (10.14) with values for A and β based on the throat diameter. Typical values for the flow coefficient and expansion factor are given in Figures 10.11 and 10.7. Bias limits for the discharge coefficient are $\sim 2\%$ of C and for the expansion factor are $\sim [2(p_1 - p_2)/p_1]\%$ of Y [12]. Because it lacks the gradual divergent section of a venturi, the permanent loss associated with the nozzle is larger (see Figure 10.8) for the same pressure drop.

The idea of using a nozzle as a flow meter was first proposed in 1891 by John Ripley Freeman (1855–1932), an inspector and engineer employed by a factory fire insurance firm. His work required tedious tests to quantify pressure losses in pipes, hoses, and fittings. He noted a consistent relationship between pressure drop and flow rate through fire nozzles.

EXAMPLE 10.2

A U-tube manometer filled with manometer fluid (of specific gravity S_m) is used to measure the pressure drop across an obstruction meter. A fluid of specific gravity, S, flows through the meter. Determine a relationship between the meter flow rate and the measured manometer deflection, H.

KNOWN

Fluid (of specific gravity S and specific weight γ)
Manometer fluid (of specific gravity S_m and specific weight γ_m)

ASSUMPTION

Density of fluids remain constant

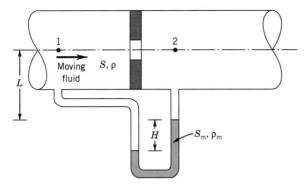

Figure 10.12 Manometer of Example 10.2.

FIND

$Q = f(H)$

SOLUTION

Equation (10.14) provides a relationship between flow rate and a flow pressure drop. From Figure 10.12 and hydrostatic principles, the pressure drop measured by the manometer is

$$p_1 - p_2 = \gamma_m H - \gamma H = H(\gamma_m - \gamma) = \gamma H[(S_m/S) - 1]$$

Combining this equation with equation (10.14) yields the working equation when pressure drop is based on the equivalent pressure head

$$Q = CEYA\sqrt{2gH[(S_m/S) - 1]} \tag{10.15}$$

EXAMPLE 10.3

A 10-cm-diameter square-edged orifice plate is used to meter the steady flow of 16°C water through an 20-cm pipe. Flange taps are used and the pressure drop measured is 50 cm Hg. Determine the pipe flow rate. The specific gravity of mercury is 13.5.

KNOWN

$d_1 = 20$ cm
$H = 50$ cm Hg
$d_0 = 10$ cm

Properties (properties are from Appendix B)

$\mu = 1.08 \times 10^{-3}$ N s/m^2
$\rho = 999$ kg/m^3

ASSUMPTIONS

Steady flow
Incompressible flow ($Y = 1$)

FIND

Q

SOLUTION

Use of an orifice plate is governed by equation (10.14), which requires knowledge of E and C. Since the beta ratio for this installation is

$$\beta = d_0/d_1 = 0.5$$

the velocity of approach factor can be calculated from equation (10.13):

$$E = \frac{1}{(1 - \beta^4)^{1/2}} = 1.0328$$

Because $C = f(\text{Re}_{d_1}, \beta)$ the discharge coefficient, C, requires information about the flow Reynolds number. But from equation (10.7) with $\nu = \mu/\rho$,

$$\text{Re}_{d_1} = Ud_1/\nu = 4Q/\pi d\nu$$

Without information concerning either Q or U, C cannot be determined explicitly and a trial-and-error approach must be undertaken: Guess a value for C and iterate. A good guess is to choose a value for C at a high value of Re_{d_1}. This is the flat region of Figure 10.6. Most real flow systems operate in this region, so fewer iterations may be required if we start there. Using this as a guide, guess from Figure 10.6 a value of $K_0 = CE \approx 0.625$.

Based on the manometer deflection and equation (10.15) (see Example 10.2),

$$Q = CEYA \sqrt{2gH[(S_m/S) - 1]}$$
$$= 0.055 \text{ m}^3/\text{s}$$

But this value resulted from a guessed value for C, so we must test to determine if that guessed value was correct. For this Q,

$$\text{Re}_{d_1} = 4Q/\pi d\nu = 3.5 \times 10^5$$

From Figure 10.6, at this Reynolds number, $K_0 \approx 0.625$, so we conclude that $Q = 0.055 \text{ m}^3/\text{s}$.

EXAMPLE 10.4

Air flows at $20°C$ through a 6-cm pipe. A square-edged orifice plate with $\beta = 0.4$ is chosen to meter the flow rate. A pressure drop of 250 cm H_2O is measured at flange taps with an upstream pressure of 93.7 kPa abs. Find the flow rate.

KNOWN

$d_1 = 6$ cm $p_1 = 93.7$ kPa abs
$\beta = 0.4$ $T_1 = 20°C = 293$ K
$H = 250$ cm H_2O

Properties (Appendix B)

Air: $\nu = 1.53 \times 10^{-5}$ m²/s
Water: $\rho_{H_2O} = 999$ kg/m³

ASSUMPTIONS

Steady air flow
Ideal gas ($p = \rho RT$)

FIND

Q

SOLUTION

An orifice is governed by equation (10.14) that will require information about E, C, and Y. From the given information, the orifice area, $A_{d_0} = 4.52 \times 10^{-4}$ m^2, and the velocity of approach factor, $E = 1.013$, are found. The air density is found from the ideal gas equation of state:

$$\rho_1 = \frac{p_1}{RT_1} = \frac{(93{,}700 \text{ N/m}^2)}{(287 \text{ N m/kg K})(293 \text{ K})}$$

$$= 1.114 \text{ kg/m}^3$$

The pressure drop is found from equation (9.3) to be $p_1 - p_2 = \rho_{H_2O} g H = 24{,}500 \text{ N/m}^2$.

Since the pressure ratio for this gas flow, $(p_1 - p_2)/p_1$, is 0.26, the air cannot be treated as incompressible. Using $k = 1.4$ (air) and a pressure ratio of 0.26, Figure 10.7 indicates that $Y \approx 0.92$.

As in the previous example, the discharge coefficient cannot be found explicitly unless the flow rate is known, since $C = f(\text{Re}_{d_1}, \beta)$. So a trial-and-error approach is used. From Figure 10.6, guess a value for $K_0 = CE \approx 0.61$ or $C = 0.60$. Then,

$$Q = CEYA \sqrt{\frac{2\Delta p}{\rho_1}} = 0.60 \times 1.013 \times 0.92 \times 4.52 \times 10^{-4} \text{ m}^2$$

$$\times \sqrt{\frac{(2)24{,}500 \text{ N/m}^2}{1.114 \text{ kg/m}^3}}$$

$$= 0.053 \text{ m}^3/\text{s}$$

Now check to determine whether the guessed value for C is correct. For this flow rate, $\text{Re}_{d_1} = 4Q/\pi d_1 \nu = 7.35 \times 10^4$ and from Figure 10.6, $K_0 \approx 0.61$ as assumed. The flow rate through the orifice is taken to be 0.053 m^3/s.

Sonic Nozzles

Sonic nozzles are used to meter and to control the flow rate of compressible gases [6]. They may take the form of any of the previously described obstruction meters. If the gas flow rate through an obstruction meter becomes sufficiently high, the sonic condition will be achieved at the meter throat. At the sonic condition, the gas velocity will equal the acoustic wave speed (speed of sound) of the gas. At that point the throat is considered to be choked and the mass flow rate through the throat will be at a maximum for the given inlet conditions regardless of any further increase in pressure drop across the meter. The theoretical basis for such a meter stems from the early work of Bernoulli, Venturi, and St. Venant (1797–1886). In 1866, Julius Weisbach (1806–1871) developed a direct relation between pressure drop and a maximum mass flow rate.

For a perfect gas undergoing an isentropic process, the pressure drop corresponding to the onset of the choked flow condition at the meter minimum area, the meter throat, is given by the *critical pressure ratio*:

$$\frac{p_0}{p_1} = \left(\frac{2}{k+1}\right)^{k/(k-1)} \tag{10.16}$$

where p_0 is the throat pressure. A pressure ratio at or below critical results in choked flow at the meter throat.

The steady-state energy equation written for a perfect gas is given by

$$c_p T_1 + \frac{\bar{U}_1^2}{2g_c} = c_p T_0 + \frac{\bar{U}_0^2}{2g_c} \tag{10.17}$$

where c_p is the constant pressure specific heat, which is assumed constant. Combining equations (10.6), (10.16), and (10.17) with the ideal gas equation of state yields the mass flow at and below the critical pressure ratio:

$$\dot{m}_{critical} = \rho_1 A \sqrt{2RT_1} \sqrt{\frac{k}{k+1}\left(\frac{2}{k+1}\right)^{2/(k-1)}} \tag{10.18}$$

where k refers to the specific heat ratio of the gas. Equation (10.18) provides a measure of the ideal mass flow rate for a perfect gas. As with all obstruction meters, this ideal rate must be modified using a discharge coefficient to account for losses. However, the ideal and actual flow rates tend to differ by no more than 3%. When calibrations cannot be run, a $C = 0.99 \pm 2\%$ bias (95%) should be assumed [2].

The sonic nozzle provides a very convenient method to meter and to regulate a gas flow. The judicious selection of throat diameter can establish any desired fluid flow rate provided that the flow is sonic at the throat. This capability makes the sonic nozzle attractive as a local calibration standard for gases. Both large pressure drops and system pressure losses must be tolerated with the technique, although low loss venturi designs will minimize losses [2]. Special orifice plate designs exist for metering at very low flow rates [2].

EXAMPLE 10.5

A flow nozzle is to be used at choked conditions to regulate the flow of nitrogen N_2 at 1.3 kg/s through a 6-cm-i.d. pipe. The pipe is pressurized at 690 kPa abs and gas flows at 20°C. Determine the maximum β ratio nozzle that can be used, $R_{N_2} = 297$ N m/kg K.

KNOWN

$N_2(k = 1.4)$ $p_1 = 690$ kPa abs $\dot{m} = 1.3$ kg/s
$d_1 = 6$ cm $T_1 = 293$ K

ASSUMPTIONS

Perfect gas
Steady, choked flow
$C = 0.99$

FIND

Find $\beta_{max} = d_{0,max}/d_1$

SOLUTION

Equation (10.18) can be rewritten in terms of nozzle throat area, A:

$$A_{max} = \frac{\dot{m}}{\rho_1(2RT)^{1/2}\{k/(k+1)[2/(k+1)]^{2/(k-1)}\}^{1/2}}$$
$$= 8.41 \times 10^{-4} \text{ m}^2$$

Since $A = \pi d_0^2/4$, this yields a maximum nozzle throat diameter, d_0:

$$d_{0,max} = 3.27 \text{ cm}.$$

Hence, $\beta = d_0/d_1 \leq 0.545$ is required to maintain the specified mass flow rate.

Obstruction Meter Selection

Selection between obstruction meter types depends on a number of factors that require some engineering compromises. Primary considerations include meter placement, overall pressure loss, accuracy, and overall costs.

It is very important to provide sufficient upstream and downstream pipe lengths on each side of a flow meter to provide for proper flow development. The flow development dissipates swirl, promotes a symmetric velocity distribution, and allows for proper pressure recovery downstream of the meter. For most situations, these installation effects can be minimized by placement of the flow meter in adherence to the considerations outlined in Figure 10.13. However, the engineer needs to be wary that installation effects are difficult to predict, particularly downstream of elbows (out-of-plane double elbow turns are notoriously difficult), and Figure 10.13 is to be used as a guide only [11, 12]. Even with recommended lengths, an additional 0.5% uncertainty should be added to the uncertainty in the discharge coefficient to account for swirl effects [12]. Unorthodox installations will require meter calibration to determine flow coefficients or use of special correction procedures [11, 12]. The physical size of a meter may become an important consideration in itself, since in large-diameter pipelines, venturi, and nozzle meters can take up a considerable length of pipe.

The unrecoverable overall pressure loss, $(\Delta p)_{loss}$, associated with a flow meter depends on β ratio and flow rate. These losses must be overcome by the system prime mover (e.g., a pump driven by a motor) in addition to any other pipe system losses and must be considered during prime mover sizing. In established systems, flow meters must be chosen on the basis of pressure loss that the prime mover can accommodate and still maintain a desired system flow rate. The power, $\dot{W}$, required to overcome any loss in a system is given by

$$\dot{W} = \rho g h_{L_{1-2}}\frac{Q}{\eta g_c} = Q\frac{(\Delta p)_{loss}}{\eta} \tag{10.19}$$

where η is the prime mover overall efficiency.

The actual cost of any meter depends on the initial capital cost of the meter, the costs of installing the meter (including system down time), calibration costs, and the added capital and operating costs associated with flow meter pressure losses to

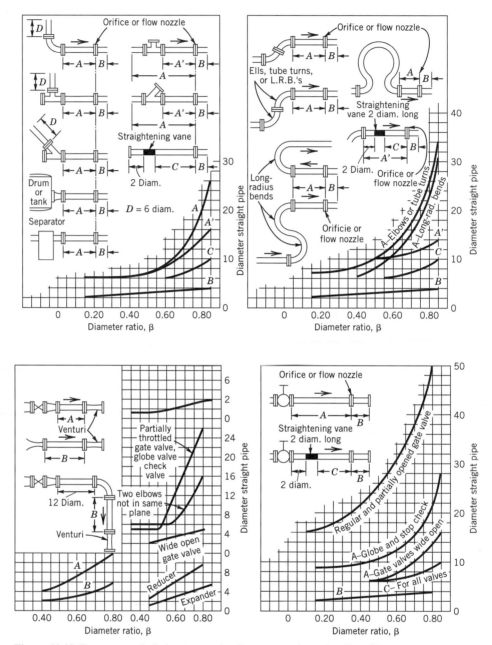

Figure 10.13 Recommended placements for flow meters in a pipe line. (Courtesy of American the Society of Mechanical Engineers, New York; reprinted from [2].)

the system. Indirect costs may include product loss caused by the meter flow rate uncertainty.

The ability of an obstruction flow meter to accurately estimate the volume flow rate in a pipeline depends on both the method used to calculate flow rate and factors inherent to the meter. If standard tabulated values for meter coefficients are used, factors that contribute to the overall uncertainty of the measurement enter through the following data-acquisition elemental errors: (1) actual β ratio precision and pipe eccentricity, (2) pressure measurement position precision, (3) temperature effects leading to relative thermal expansion of components, and (4) upstream flow profile. These errors are in addition to the errors inherent in the coefficients themselves and errors associated with the estimation of the upstream fluid density [2]. Direct in situ calibration of any flow meter installation can reduce these contributions to overall uncertainty. For example, the actual β ratio value, pressure tap locations, and other installation effects will be included in a system calibration and accommodated in the actual computed flow coefficient.

The selection of any meter should consider whether the system into which the meter is to be installed will be used at more than one flow rate. If so, considerations as to meter performance and its effect on system performance over the entire anticipated flow rate range should be included. The range over which any one meter can be used is known as the meter *turndown*.

EXAMPLE 10.6

For the orifice meter in Example 10.3 having a β = 0.5 and a pressure drop of 50 cm Hg, calculate the permanent pressure loss caused by the meter.

KNOWN

$H = 50$ cm Hg or $p_1 - p_2 = 66.5$ kPa
$\beta = 0.5$

ASSUMPTION

Steady flow

FIND

$(\Delta p)_{loss}$

SOLUTION

For a properly designed and installed orifice meter, Figure 10.8 indicates that the permanent pressure loss of the meter will be ~75% of the pressure drop across the meter. Hence,

$$(\Delta p)_{loss} = 0.75 \times 66.5 \text{ kPa}$$
$$= 49.9 \text{ kPa or 38 cm Hg}$$

COMMENT

In comparison, for $Q = 0.055$ m^3/s a typical venturi (say 15° outlet) with a β = 0.5 would provide a pressure drop equivalent to 19.6 cm Hg with a permanent loss of only 3.1 cm Hg. A prime mover (such as the pump here) would need to be able to add additional energy to the flow to make up for the meter permanent pressure loss while providing enough power to sustain the desired flow rate. A smaller pump would be needed if the venturi were used.

EXAMPLE 10.7

For the orifice of Example 10.6, estimate the operating costs required to overcome these permanent losses if electricity is available at $0.08/kW h, the pump is used 6000 h/year and the pump-motor efficiency is 60%.

KNOWN

Water at $16°C$ $(\Delta p)_{loss} = 49.9$ kPa
$Q = 0.055$ m^3/s $\eta = 0.60$

ASSUMPTIONS

Steady flow
Meter installed according to standards [2]

FIND

Annual cost from $(\Delta p)_{loss}$ alone

SOLUTION

We can determine the pump power required to overcome the permanent pressure losses in the system resulting from the use of the orifice meter by equation (10.19):

$$\dot{W} = Q\frac{\Delta p_{loss}}{\eta}$$
$$= \frac{(0.055 \text{ m}^3/\text{s})(49,900 \text{ N/m}^2)(3600 \text{ s/h})}{0.60} = 4530 \text{ W} = 4.53 \text{ kW}$$

The additional annual pump operating cost required to overcome this is

$$\text{Cost} = (4.53 \text{ kW})(6000 \text{ h/year})(\$0.08/\text{kW h}) = \$2175/\text{year}$$

COMMENT

In comparison, the venturi discussed in the Comment to Example 10.6 would cost about \$180/year to operate. Such a venturi meter might cost 50 times more than the orifice plate installed, but the venturi meter will pay for itself in energy savings in ~2.5 years and allow a smaller pump to be used.

EXAMPLE 10.8

An ASME long-radius nozzle ($\beta = 0.5$) is to be installed into a horizontal section of 12-in. (schedule 40) pipe. A long, straight length of pipe exists just downstream of a fully open gate valve but upstream of an in-plane $90°$ elbow. Determine the minimum lengths of straight unobstructed piping that should exist upstream and downstream of the meter.

KNOWN

$d_1 = 11.94$ in. (30.33 cm)
$\beta = 0.5$
Installation layout

ASSUMPTION

Steady flow

FIND

Meter placement

SOLUTION

From Figure 10.13, the meter should be placed a minimum of eight pipe diameters ($A=8$) downstream of the valve and a minimum of three pipe diameters ($B=3$) upstream of any flow fitting. Note that since a typical flow nozzle has a length of about 1.5 pipe diameters, this installation will require a straight pipe section of at least 12.5 pipe diameters ($8+3+1.5$).

COMMENT

A general rule for initial planning is to use a 10-diameter upstream length before and a four diameter downstream length after the meter pressure taps.

Laminar Flow Elements

The relationship between flow rate and pressure drop across a length of pipe of diameter, d_1, and containing a laminar flow follows a linear relationship. The energy equation for such a case is written as

$$\frac{p_1}{\gamma} + \frac{\overline{U}_1^2}{2g} = \frac{p_2}{\gamma} + \frac{\overline{U}_2^2}{2g} + h_{L_{1-2}} \tag{10.20}$$

The head loss term, $h_{L_{1-2}}$, can be estimated by the Darcy–Weisbach relation:

$$h_{L_{1-2}} = f\frac{L}{d_1}\frac{\overline{U}_1^2}{2g} \tag{10.21}$$

where L is the distance between pressure taps and f is the friction factor, which for a laminar flow is simply

$$f = \frac{64}{\mathrm{Re}_{d_1}} \tag{10.22}$$

From conservation of mass, equation (10.6),

$$Q = \overline{U}_1\frac{\pi d_1^2}{4} = \overline{U}_2\frac{\pi d_1^2}{4}$$

which for a constant diameter conduit requires $\overline{U}_1 = \overline{U}_2$. This yields

$$p_1 - p_2 = \gamma h_{L_{1-2}} \tag{10.23}$$

Accordingly, by substitution into the Darcy–Weisbach relation using $\overline{U} = 4Q/\pi d_1^2$ yields

$$Q = \frac{\pi d_1^4}{128\mu}\frac{p_1 - p_2}{L} \qquad \mathrm{Re}_{d_1} < 2000 \tag{10.24}$$

Equation (10.24) is fundamental to the operation of a laminar flow meter. It reveals that volume flow rate is linear with pressure drop in a laminar pipe flow. This relation was first demonstrated by Jean Poiseulle (1799–1869), who conducted meticulous tests documenting the resistance of flow through capillary tubes.

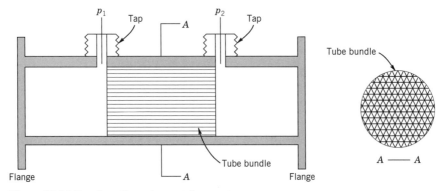

Figure 10.14 Laminar flow element flow meter.

The simplest type of laminar flow meter consists of two pressure taps separated by a length of piping.[1] However, the Reynolds number constraint for laminar flow restricts the size of pipe diameter that can be used or, alternatively, the flow rate to which equation (10.24) can be applied for a given fluid. This limitation is overcome in commercial units through the use of laminar flow elements (Figure 10.14), which consist of a bundle of small-diameter tubes, or some proprietary design of geometric passages, placed in parallel. The strategy of a laminar flow element is to divide up the flow by passing it through the tube bundle so as to reduce the flow rate per tube such that the individual Reynolds number in each tube remains below 2000. Pressure drop is measured between the entrance and the exit of the laminar flow element. Because of the additional entrance and exit losses associated with the laminar flow element, a flow coefficient is used to modify equation (10.24) where the coefficient must be determined by calibration. No standard tables for these coefficients exist because of the wide possible variation in flow meter design, but commercial units come supplied with individual calibration charts. The coefficient will be essentially constant over the useful meter range.

For any given meter there will exist a flow rate above which laminar flow will no longer exist in the elements. Consequently, the application of any particular meter will be limited to some maximum flow rate. Various meter sizes and designs are available to accommodate user needs. Turndowns up to 100:1 are available.

Laminar flow elements offer some distinct advantages over other pressure differential meters. These include (1) a high sensitivity even at extremely low flow rates, (2) the ability to measure pipe system flows in either meter direction, (3) a wide usable flow range, and (4) the ability to indicate an average flow rate in pulsating flows. The overall uncertainty bias limit in flow rate determination can be as low as 0.25% of the flow rate with these meters. However, these meters are very susceptible to clogs, restricting their use to clean fluids. All of the pressure drop measured remains a system pressure loss.

10.6 INSERTION VOLUME FLOW METERS

Dozens of volume flow meter types based on a number of different principles have been proposed, developed, and sold commercially. The largest group of meters are based on some phenomenon that is actually sensitive to the average velocity across a

---

[1] Alternately, if flow rate can be measured or is known, equation (10.24) provides the basis for a capillary tube viscometer.

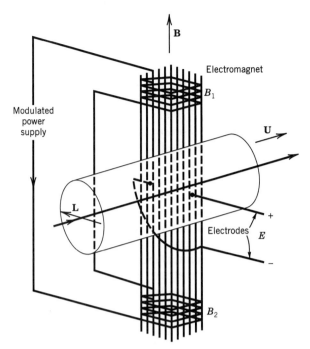

Figure 10.15 Electromagnetic principle as applied to a working flow meter.

control surface of known area. Several of these designs are included in the discussion that follows.

Electromagnetic Flow Meters

The operating principle of an *electromagnetic flow meter* [7] is based on the fundamental principle that an emf of electric potential, E, is induced in a conductor of length, $\mathbf{L}$, which moves with a velocity, $\mathbf{U}$, through a magnetic field of magnetic flux, $\mathbf{B}$. This physical behavior was first recorded by Michael Faraday (1791–1867). In principle, we write

$$E = \mathbf{U} \times \mathbf{B} \cdot \mathbf{L} \tag{10.25}$$

A practical utilization of equation (10.25) is shown in Figure 10.15. From equation (10.25) the magnitude of E is affected by

$$E = \overline{U} BL \sin\alpha = f(\overline{U})$$

where α is the angle between the mean velocity vector and the magnetic flux vector, usually at 90°. In general, electrodes are embedded in the pipe wall in a diametrical plane that is normal to the known magnetic field. As the flow moves through the magnetic field, the induced electric potential is detected and measured by the electrodes, which are separated by the length, L. The average magnitude of the velocity, $\overline{U}$, across the pipe is thus inferred through the measured emf. The flow rate is found by

$$Q = \frac{\overline{U} \pi d_1^2}{4} = \frac{E}{BL} \frac{\pi d_1^2}{4} = K_1 E \tag{10.26}$$

The value of L is of the order of the pipe diameter, the exact value depending on the meter construction and magnetic flux lines. The static sensitivity K_1 is a meter constant found by calibration and supplied by a manufacturer. The relationship between flow rate and measured potential is linear.

The electromagnetic flow meter comes commercially as a packaged flow device, which is installed directly in line and connected to an external electronic output unit. Units are available using either permanent magnets, called dc units, or variable flux strength electromagnets, called ac units. The magnetic flux strength of an ac unit can be increased on site for a strong signal at low flow rates of low conductivity fluids such as water. Special designs include a flow sensor unit that actually can clamp over (not in-line with) a nonmagnetic pipe, a design favored to monitor blood flow rate through major arteries during surgery.

The electromagnetic flow meter has a very low pressure loss associated with its use as a result of its open tube, no obstruction design, and it is suitable for installations that can tolerate only a small pressure drop. This absence of internal parts is very attractive for metering corrosive and "dirty" fluids. The operating principle is independent of fluid density and viscosity, responding only to average velocity, and there is no difficulty with measurements in either laminar or turbulent flows, provided that the velocity profile is reasonably symmetrical. It can be used in either steady or pulsatile flows, providing either time-averaged or instantaneous data in the latter. Data-acquisition errors down to ±0.25% of the measured flow rate can be attained, although values between ±1 and 5% are more common for these meters. The use of any meter is limited to fluids having a threshold value of electrical conductivity, the actual value of which depends on a particular meter's design, but fluids with values as low as 0.1 μsieman/cm have been metered. The addition of salts to a fluid will increase its conductivity.

Vortex Shedding Meters

Nearly everyone has observed an oscillating street sign or heard the "singing" of power lines on a windy day. These are but a couple of the effects induced by vortex shedding from bluff-shaped bodies, a natural phenomenon in which alternating vortices are shed in the wake of the body at a frequency that will depend on the velocity of the flow past the body. The vortices formed on opposite sides of the body are carried downstream in the body's wake forming a "vortex street," each vortex having an opposite sign of rotation. This behavior is seen in Figure 10.16, a photograph that

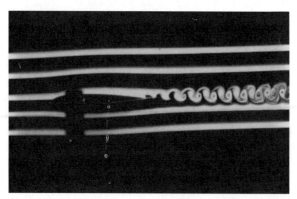

Figure 10.16 Smoke lines in this photograph reveal the vortex shedding behind a streamlined wing-shaped body in a moving flow.

captures the vortex shedding downstream of a section of an aircraft wing. The existence of the vortex street was deduced theoretically by the aerodynamicist Theodore von Karman (1881–1963), although da Vinci appears to have been the first to actually record the phenomenon [1]. The alternating vortices are manifestations of an oscillating pressure field about the body, a field that exerts an equal but opposite force on the body.

The vortex shedding phenomenon is used to sense average velocity in pipe flows in a *vortex flow meter*. The basic relationship between shedding frequency, ω, and average velocity, $\overline{U}$, for a given shape is given by the Strouhal number,

$$St = \omega d / \overline{U} \qquad (10.27)$$

where d is a characteristic length for the body, such as diameter for a cylinder. In general, the Strouhal number is a function of Reynolds number, but various geometrical shapes, known as shedders, can be used to produce a stable vortex flow that has a *constant* Strouhal number over a broad range of flow Reynolds number. Examples are given in Table 10.1. The quality and strength of the shedding can be improved by manipulation of the design of the tail end of the body and by providing a slightly concave upstream body face that traps the stagnation streamline at a point giving way to a stable oscillation. Accordingly, there are a number of proprietary designs in existence. In general, abrupt edges on the shedder restrict the dependence

Table 10.1 Shedder Shape and Strouhal Number

Cross Section	Strouhal Number[a]
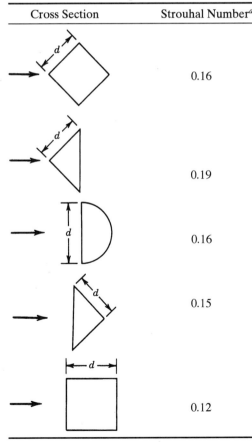	0.16
	0.19
	0.16
	0.15
	0.12

[a] For Reynolds number $Re_d \geq 10^4$.

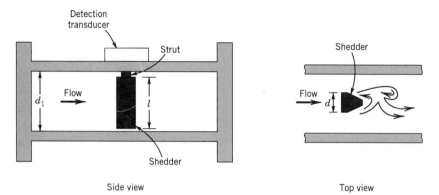

Figure 10.17 Vortex shedding flow meter. Different shedder shapes are available.

on Reynolds number by fixing the flow separation points. A typical design is shown in Figure 10.17. The shedder spans the pipe, $d/l \approx 0.3$, to provide an average pipe velocity with $d/d_1 \approx 0.3$ for strong, stable vortex strength.

For a constant Strouhal number, the flow rate can be deduced from continuity based on measured shedding frequency. With $K_1 = \pi d^3/4(St)$, equation (10.27) becomes

$$Q = K_1 \omega \tag{10.28}$$

where K_1 is a constant value of static sensitivity provided by the manufacturer and depends on the shedder design. Shedding frequency can be measured in many ways. The shedder strut can be instrumented to detect the force oscillation by using strut-mounted strain gauges, for example, or to detect the pressure oscillations by using a piezoelectric crystal.

The lower flow rate limit on vortex meters appears to be at Reynolds numbers near 10,000, below which the Strouhal number varies nonlinearly with flow rate and shedding becomes unstable regardless of shedder design. This can be a problem in metering high-viscosity pipe flows ($\mu > 20$ cp). The upper flow bound is limited only by the onset of cavitation in liquids and by the onset of compressibility effects in gases at Mach numbers exceeding 0.2. Property variations affect meter performance only indirectly. Density variations affect the strength of the shed vortex and this places a lower limit on fluid density which is based on the sensitivity of the vortex shedding detection equipment. Viscosity affects the operating Reynolds number. Otherwise, within bounds, the meter is insensitive to property variations.

The meter has no moving parts and relatively low pressure losses compared to obstruction meters. A single meter can operate over a flow range of up to 20:1 above its minimum with a linearity in K_1 of $\pm 0.5\%$. In contrast, equation (10.28) shows the shedding frequency is inversely related to the pipe diameter cubed. This suggests that meter resolution can drop quickly in large diameter pipes, placing a limit on the maximum meter size that can be used effectively.

Rotameters

The rotameter remains a widely used insertion meter for flow rate indication. As depicted in Figure 10.18, the meter consists of a float within a vertical tube, tapered to an increasing cross-sectional area at its outlet. Flow entering through the bottom passes over the float, which is free to move. The basic principle of the device is based

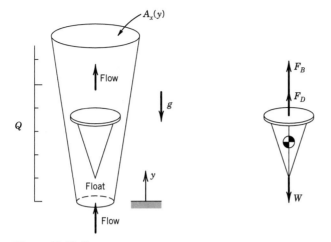

Figure 10.18 Rotameter.

on the simple balance between the drag force, F_D, and the weight, W, and buoyancy forces, F_B, acting on the float in the moving fluid. It is the drag force on the float that varies with the average velocity over the float.

The force balance in the vertical direction y yields

$$\Sigma F_y = 0 = -F_D + W - F_B \tag{10.29}$$

or with $F_D = (1/2)C_D \rho \bar{U}^2 A_x$, $W = \rho_b V_b$, and $F_B = \rho V_b$,

$$C_D(1/2)\rho A_x \bar{U}^2 = g(\rho_b - \rho)V_b \tag{10.30}$$

where

ρ_b = density of float
ρ = density of fluid
C_D = drag coefficient of the float; $C_D = f(\text{Re})$
A_x = tube cross-sectional area
$\bar{U}$ = average velocity past the float
V_b = volume of float

In operation, the float will rise to some position within the tube at which such a force balance exists. The height of this position increases with flow velocity and, hence, flow rate. This flow rate is found by

$$Q = \bar{U} A_a(y) = |C_D K_1|^{1/2} A_a(y) \tag{10.31}$$

where $A_a(y)$ is the annular area between the float and the tube, which depends on the height of the float in the tube, and K_1 is a constant depending on the meter design and fluid in use. Since annular area is a function of float position within the vertical tube, the float's vertical position gives a direct measure of flow rate that can be read from a graduated scale, electronically sensed with an optical cell, or detected magnetically. Floats with sharp edges are less sensitive to fluid viscosity changes with temperature. Rotameters are used in noncritical applications in which accuracy is not of prime concern. Uncertainty bias limits of $\pm 2\%$ of flow rate are common as is a turndown of 10:1.

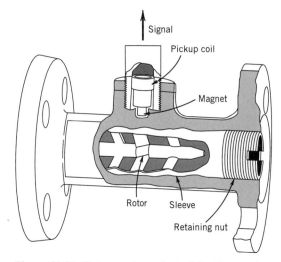

Figure 10.19 Cutaway view of a turbine flow meter.
(Courtesy of Schlumberger Industries,
Measurements Division, Greenwood, SC.)

Turbine Meters

Turbine meters make use of angular momentum principles to meter flow rate. In a typical design (Figure 10.19), a rotor is encased within a bored housing through which the fluid to be metered is passed. The housing contains flanges for direct in-sertion into a pipeline. The exchange of momentum between the flow and the rotor turns the rotor at a rotational speed that is proportional to the flow rate. Rotor rotation can be measured in a number of ways. For example, a reluctance pickup coil can sense the passage of magnetic rotor blades producing a pulse train signal at a frequency that is directly related to rotational speed.

The rotor angular velocity, η, will depend on the average flow velocity and fluid viscosity, v, through the meter bore of diameter, d_1. Dimensionless analysis of these parameters [8] suggests

$$Q/\eta D^3 = f(\eta D^2/v) \qquad (10.32)$$

In practice, there will exist a region in which the rotor angular velocity will vary linearly with flow rate.

Turbine meters offer a low-pressure drop and very good accuracy. Uncertainty bias limits in flow rate down to $\pm 0.25\%$ are typical, with turndown of 20:1. They are exceptionally repeatable, making them good candidates for local flow rate standards. However, their use must be restricted to clean fluids because of possible fouling of their rotating parts. Temperature changes affect fluid viscosity, a property to which a turbine meter rotational speed is sensitive. Some compensation for viscosity variations can be made electronically [9]. The turbine meter is very susceptible to installation errors caused by pipe flow swirl [11] and a careful selection of installation position is suggested.

Positive-Displacement Meters

Positive-displacement meters contain mechanical elements that define a known volume. The free-moving elements are displaced by the action of the moving fluid. A counting mechanism counts the number of element displacements to provide a direct reading of volume of fluid passed through the meter. Usually used as volume meters, volume per unit time can be discerned in conjunction with a timer. These units are typically installed where applications demand ruggedness and high accuracy in steady flow metering. This metering method is common to water, gasoline, and natural gas meters.

Wobble meters contain a disk that is seated in a chamber. The flow of fluid through the chamber causes the disk to oscillate so that a known volume of fluid moves through the chamber with each oscillation. The disk is connected to a counter that records each oscillation as volume displaced. Wobble meters are used for domestic water applications where they must have an uncertainty of no greater than 1% of actual delivery. Frequently found on oil trucks and at the gasoline pump, *rotating vane meters* use rotating cups or vanes that move about an annular opening displacing a known volume with each rotation. Uncertainty can be as low as ±0.3% of actual delivery. Because of the good accuracy of these meters, they are often used as local standards to calibrate volume flow meters.

10.7 MASS FLOW METERS

There are many situations in which the mass flow rate is the quantity of interest. Direct conversion from measured volume flow rate to actual mass flow rate requires knowledge of the density of the metered fluid under the exact conditions of measurement. For many applications, one can assume that both the errors in the tabulated values for fluid density and the fluid density variations within a volume meter are small enough to ignore. However, not all fluids have readily established equations of state to provide a known value of density (e.g., petroleum products, polymers, and cocoa butter). When volume flow methods are used, uncertainty in the density increases the overall uncertainty in mass flow rate above the uncertainty in volume flow rate. Therefore, it is not uncommon to use density measuring devices just upstream of the volume flow measurement to improve the uncertainty in the mass flow determination in critical applications. The direct measurement of mass flow rate is desirable since it eliminates the uncertainties associated with estimating or measuring actual density.

The difference between a meter that is sensitive to Q as opposed to $\dot{m}$ is not trivial. Prior to the 1970s, reliable commercial mass flow meters with sufficient accuracy to circumvent volume flow rate corrections were generally not available, even though the basic principles and implementation schemes for such meters had been understood in theory for several decades. U.S. patents dating back to the 1940s record schemes for using heat transfer, Coriolis forces, and momentum methods to infer mass flow rate directly.

Thermal Flow Meter

The rate at which energy, $\dot{E}$, must be input to a flowing fluid to raise the temperature of the fluid some desired amount between two control surfaces is directly related to the mass flow rate by

$$\dot{E} = \dot{m} c_p \, \Delta T \tag{10.33}$$

where c_p is the fluid specific heat. Methods to utilize this effect to directly measure mass flow rate incorporate an inline meter having some means to input energy to the fluid over the meter length. The passing of a current through an immersed filament is a common method. Fluid temperatures are measured at the upstream and the downstream locations of the meter. This type of meter is quite easy to use and appears reliable. In fact, in the 1980s, the technique was adapted for use in automobile fuel injection systems to provide an exact air–fuel mixture to the engine cylinders despite short-term altitude, barometric, and seasonal environmental temperature changes.

A major consideration involved with the use of this type of meter concerns the assumption that c_p is known for the fluid and remains constant over the length of the meter. For common gases, such as air, this assumption is quite valid at reasonable temperatures and pressures. Flow rate turndown of up to 100:1 is possible with uncertainties down to $\pm 0.5\%$ of flow rate with very little pressure drop. But the assumptions become restrictive for liquids and for many gases in which c_p either may be a strong function of temperature or may simply not be well established.

A second type of thermal mass flow meter is actually a velocity-sensing meter and thermal sensor together in one direct insertion unit. The meter uses both hot-film anemometry methods to sense fluid velocity through a conduit of known diameter and an adjacent RTD sensor for temperature measurement. For sensor and fluid temperatures, T_s and T_f, respectively, mass flow rate is inferred from the correlation

$$\dot{E} = [C + B(\rho \overline{U})^{1/n}](T_s - T_f) \tag{10.34}$$

where C, B, and n, are constants that depend on fluid properties [10] and are determined through calibration. In a scheme to reduce the fluid property sensitivity of the meter, the RTD may be used as an adjacent resistor leg of the anemometer Wheatstone bridge circuit to provide a temperature-compensated velocity output over a wide range of fluid temperatures, with excellent repeatability ($\sigma \approx 0.25\%$). Gas velocities of up to 12,000 ft/m and flow rate turndown of 50:1 are possible with uncertainties down to $\pm 2\%$ of the flow rate and very little pressure drop. A series of these mass flow probes can be combined to span across large ducts or chimneys to provide better averaging information.

Coriolis Flow Meter

The term "Coriolis flow meter" refers to the family of insertion meters that meter mass flow rate by inducing a Coriolis acceleration on the flowing fluid and measuring the resulting developed force. The developed force is directly related to the mass flow rate independent of the fluid properties. The Coriolis effect was proposed by Gaspard de Coriolis (1792–1843) following his studies of accelerations in rotating systems. A large number of methods to utilize this effect have been proposed since the first U.S. patent for a Coriolis effect meter was issued in 1947.

Commercially available units for metering of liquids include a scheme in which the pipe flow is diverted from the main pipe and divided between two bent, parallel, adjacent tubes of equal diameter, such as shown in Figure 10.20. The tubes themselves are mechanically driven in a relative out-of-phase sinusoidal oscillation by an electromagnetic driver. In general, a fluid particle passing through the meter tube that is rotating (as a result of the oscillating tube) relative to the fixed pipe experiences an acceleration at any arbitrary position S. The total acceleration at S, $\ddot{\mathbf{r}}$, is composed of several components (see Figure 10.21),

$$\ddot{\mathbf{r}} = \ddot{\mathbf{R}}_{O'} + \dot{\boldsymbol{\omega}} \times \mathbf{r}_{S/O'} + \boldsymbol{\omega} \times \boldsymbol{\omega} \times \mathbf{r}_{S/O'} + \ddot{\mathbf{r}}_{S/O'} + 2\boldsymbol{\omega} \times \dot{\mathbf{r}}_{S/O'} \tag{10.35}$$

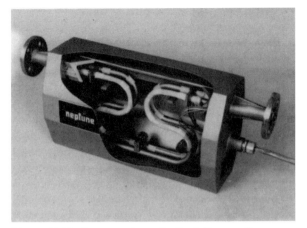

Figure 10.20 Cutaway view of a Coriolis mass flow meter. (Courtesy of Schlumberger Industries, Measurements Division, Greenwood, SC.)

where

$$\ddot{\mathbf{R}}_{O'} = \text{translation acceleration of rotating origin}$$
$$O' \text{ relative to fixed origin } O$$
$$\boldsymbol{\omega} \times \boldsymbol{\omega} \times \mathbf{r}_{S/O'} = \text{centripetal acceleration of } S \text{ relative to } O'$$
$$\dot{\boldsymbol{\omega}} \times \mathbf{r}_{S/O'} = \text{tangential acceleration of } S \text{ relative to } O'$$
$$\ddot{\mathbf{r}}_{S/O'} = \text{translational acceleration of } S \text{ relative to } O'$$
$$2\boldsymbol{\omega} \times \dot{\mathbf{r}}_{S/O'} = \text{Coriolis acceleration at } S \text{ relative to } O'$$

with ω the angular velocity of the tube relative to O'. In these meters, the tubes are rotated but not translated, so that the translational accelerations are zero. A fluid particle will experience forces that are due to the remaining accelerations that cause equal and opposite reactions on the meter tube walls. The Coriolis acceleration acts in a plane perpendicular to the tube axes and develops a force gradient that creates a twisting motion or oscillating rotation about the tube plane.

The utilization of the Coriolis force depends on the shape of the meter tube. However, the basic principle is illustrated in Figure 10.22. Rather than rotating the

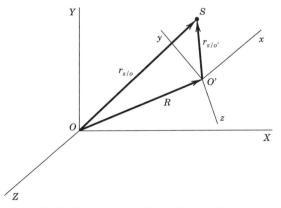

Figure 10.21 Fixed and rotating reference frames.

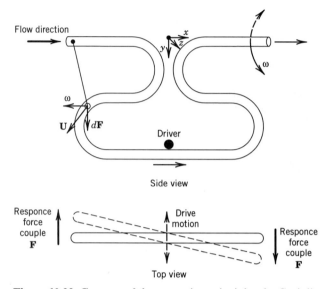

Figure 10.22 Concept of the operating principle of a Coriolis mass flow meter.

tubes a complete 360° about the pipe axis, the meter tubes are vibrated continuously at a drive frequency, ω, with amplitude displacement, z, about the pipe axis. This eliminates rotational seal problems. The driving frequency is selected at the tube resonant frequency that places the driven tube into what is called a limit cycle, a continuous, single-frequency oscillation. This is essentially a self-sustaining tuning fork since the meter will naturally respond to any disturbance at this frequency with a minimum in input energy. As a fluid particle of elemental mass, dm, flows through a section of the flow meter, it experiences the Coriolis acceleration, $2\boldsymbol{\omega} \times \dot{\mathbf{r}}_{S/O'}$, and an inertial Coriolis force

$$d\mathbf{F} = (2\boldsymbol{\omega}_c \times \dot{\mathbf{r}}_{S/O'})\, dm \qquad (10.36)$$

acting in the z direction. For the flow meter design of Figure 10.22, as the particle travels along the meter the direction of the velocity vector changes. This results, using the right-hand rule, in a change in direction of vector force, $d\mathbf{F}$. The resultant forces experienced by the tube are of equal magnitude but opposite sign of those experienced by the particle. Each tube segment senses a corresponding differential torque, $d\mathbf{T}$, a rotation about the y axis at a frequency $\boldsymbol{\omega}_c$.

$$d\mathbf{T} = \mathbf{x}' \times d\mathbf{F} = \mathbf{x}' \times dm(2\boldsymbol{\omega}_c \times \dot{\mathbf{r}}_{S/O'}) \qquad (10.37)$$

where x' refers to the x distance between the elemental mass and the y axis, $\mathbf{x}' = x'\hat{e}_x$. The total torque experienced by each tube is found by integration along the total path length of the tube, L,

$$\mathrm{T} = \int_0^L d\mathrm{T} \qquad (10.38)$$

A differential element of fluid will have the mass, $dm = \rho A\, dl$, for an elemental cross section of fluid, A, of differential length, dl, and of density, ρ. If this mass moves with an average velocity, $\overline{U}$, then the differential mass can be written as

$$dm = \rho A\, dl = \dot{m}(dl/\overline{U}) \qquad (10.39)$$

The Coriolis cross-product can be expressed as

$$\boldsymbol{\omega}_c \times \dot{\mathbf{r}}_{S/O'} = (2\,\omega_c\,\dot{r}_{S/O'}\,\sin\theta)\hat{e}_z = (2\,\omega_c\,\overline{U}\,\sin\theta)\hat{e}_z \qquad (10.40)$$

where θ is the angle between the Coriolis rotation and the velocity vector. Then,

$$\dot{m} = \frac{T}{2\int_0^L(\rho x'\,\omega_c\,\sin\theta)\,dl}\hat{e}_y \qquad (10.41)$$

Since the velocity direction changes by $180°$, the Coriolis developed torque will act in opposite directions on each side of the tube.[2] The meter tubes will twist about the y-axis of the tube at an angle δ. For small angles of rotation the twist angle is related to torque by

$$\delta = k_S\,T = \text{constant} \times \dot{m} \qquad (10.42)$$

where k_S is related to the stiffness of the tube. The objective becomes to measure the twist angle, which can be accomplished in many ways. For example, by driving the two meter tubes $180°$ out of phase, the relative phase at any time between the two tubes will be directly related to the mass flow rate. The exact relationship is linear over a wide flow range and determined by calibration with any fluid.

The tangential and the centripetal accelerations remain of minor consequence as a result of the tube stiffness in the directions in which they act. However, at high mass flow rates they can excite modes of vibration in addition to the driving mode, ω, and response mode, ω_c. In doing so, they affect the meter linearity and zero error (drift), which affects mass flow uncertainty and meter turndown. Essentially, the magnitude of these undesirable effects are inherent to the particular meter shape and are controlled, if necessary, through tube-stiffening members.

Another interesting problem occurs mostly in small meters where the tube mass approaches the mass of the fluid in the tube. At flow rates that correspond to flow transition from a laminar to turbulent regime, the driving frequency can excite the flow instabilities responsible for the flow transition. The fluid and tube can go out of phase, reducing the response amplitude and its corresponding torque. This affects the meter's calibration linearity, but a good design can contain this effect to within 0.5% of the meter reading.

The meter principle is unaffected by changing fluid properties but temperature changes will affect the overall meter stiffness, an effect that can be compensated for electronically. A very desirable feature is an apparent insensitivity to installation position. Commercially available Coriolis flow meters can measure flow rate with uncertainty bias limits as low as $\pm0.10\%$ of mass flow rate, with a turndown of approximately 20:1. The meter is also used as an effective densitometer.

10.8   FLOW METER CALIBRATION AND STANDARDS

A fundamental primary standard for flow rate does not exist, but a number of calibration procedures do exist in lieu of this. The general procedure for the calibration of most in-line flow meters requires the establishment of steady flow in a calibration flow loop and the determination of the volume or mass of flowing fluid passing through the flow meter in an accurately determined time interval. Flow loop systems in which the flow rate is established by some means so as to calibrate a

---

[2]It is not difficult to envision a design in which the velocity does not reverse direction along the flow path but ω does.

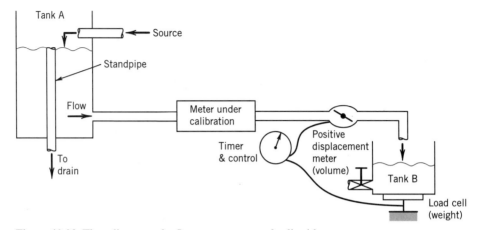

Figure 10.23 Flow diagram of a flow meter prover for liquids.

flow meter are known as *provers*. Several methods to establish the flow rate are discussed.

In liquids, variations of the "catch-and-weigh" technique are often employed in flow provers. One variation of the technique consists of a calibration loop with catch tank as depicted in Figure 10.23. Tank A is a large tank from which fluid is pumped back to a constant head reservoir, which supplies the loop with a steady flow. Tank B is the catch-and-weigh tank into which liquid can be diverted for an accurately determined period of time. The liquid volume is measured, either directly using a positive displacement meter or other means, or indirectly through its weight, and flow rate deduced through time. The ability to determine the volume and the uncertainty in the initial and final time of the event are fundamental limitations to the accuracy of this technique. Neglecting installation effects, the ultimate limits of uncertainty (at 95%) in flow rate of liquids are of the order of 0.03%, a number based on $u_m \approx 0.02\%$, $u_\rho \approx 0.02\%$, and $u_t \approx 0.01\%$.

Flow meter calibration by velocity profile determination is particularly effective for in situ calibration in both liquids and gases, provided that the gas velocity does not exceed ~70% of the sonic velocity. Velocity traverses at any cross-sectional location some 20–40 pipe diameters downstream of any pipe fitting in a long section of straight pipe are preferred.

Comparison calibration by insertion of a local standard flow meter into a calibration loop is another common means of establishing the flow rate through a system. Other flow meters can be installed in tandem with the standard and directly calibrated against it. Such a method has the advantage of being useful with either liquids or gases. Turbine and vortex meters and Coriolis mass flow meters have consistent, highly accurate calibration curves and are often used as local standards. Other provers use an accurate positive-displacement meter to determine flow volume over time. Of course, these standards must be periodically recalibrated. In the United States, NIST maintains flow meter calibration facilities for this purpose. However, installation effects in the end-user facility will not be accounted for in an NIST calibration. This last point accounts for much of the uncertainty in a flow meter calibration. Lastly, a sonic nozzle can also be used as a standard to establish the flow rate of a gas in a comparison calibration. In any of these methods, the calibration uncertainty is limited by the standard used, installation effects, and the inherent limitations of the flow meter calibrated.

10.9 SUMMARY

Flow quantification has been an important engineering task for well over two millennia. In this chapter, methods to determine the volume rate of flow and the mass rate of flow were presented. The engineering decision involving the selection of a particular meter was found to depend on a number of constraining factors. In general, flow rate can be determined to within about ±0.25% of actual flow rate with the best of present technology. However, new methods may push these limits even further, provided that calibration standards can be developed to document method accuracies and the effects of installation.

REFERENCES

1. *Leonardo da Vinci: Del Moto e Misura Dell'Aqua*, Carusi E., and Favaro, A., eds., Zanichelli, Bologna, 1923.
2. *Fluid Meters, Theory and Applications*, 6th ed., American Society of Mechanical Engineers, New York, 1971.
3. *Flow of Fluids through Valves and Fittings*, Technical Report No. 410, Crane Co., Chicago, 1969.
4. *ASHRAE Fundamentals*, American Society of Heating, Refrigeration and Air Conditioning Engineers, Atlanta, CA 1997.
5. Rouse, H., and Ince, S., *History of Hydraulics*, Dover, New York, 1957.
6. Arnberg, B. T., A review of critical flowmeters for gas flow measurements, *Transactions of the ASME*, 84:447, 1962.
7. Shercliff, J. A., *Theory of Electromagnetic Flow Measurement*, Cambridge University Press, New York, 1962.
8. Hochreiter, H. M., Dimensionless correlation of coefficients of turbine-type flowmeters, *Transactions of the ASME*, October: 1363, 1958.
9. Lee, W. F., and Karlby, H., A study of viscosity effects and its compensation on turbine flow meters, *Journal of Basic Engineering*, September: 717–728, 1960.
10. Hinze, J. O., *Turbulence*, McGraw-Hill, New York, 1953.
11. Mattingly, G., Fluid measurements: standards, calibrations and traceabilities, Proc. ASME/AIChE National Heat Transfer Conference, Philadelphia, PA, 1989.
12. *Measurement of Fluid Flow in Pipes Using Orifice, Nozzle and Venturi*, ASME Standard MFC-3M-1985, American Society of Mechanical Engineers, New York, 1985.
13. *Fan Engineering*, 8th ed., Buffalo Forge Co., Buffalo, New York, 1983.

NOMENCLATURE

c_p, c_v	specific heats $[l^2\, t^{-2/\circ}]$		r	radial coordinate $[l]$
d	diameter $[l]$		r_1	pipe radius $[l]$
d_0	flow meter throat or minimum diameter $[l]$		r_H	hydraulic radius $[l]$
			A	area $[l^2]$
d_1	upstream pipe diameter $[l]$		A_0	area based on d_0 $[l^2]$
d_2	diameter (see Figure 9.4) $[l^2]$		A_1	area based on d_1 $[l^2]$
f	friction factor		A_2	area based on d_2 $[l^2]$
g	gravitational acceleration constant $[l\, t^{-2}]$		$\mathbf{B}$	magnetic field flux vector
			C	discharge coefficient
$h_{L_{1-2}}$	energy head loss between points 1 and 2 $[l]$		C_D	drag coefficient
			E	velocity of approach factor $[v]$
k	ratio of specific heats, c_p/c_v		Gr	Grashof number
$\dot{m}$	mass flow rate $[m\, t^{-1}]$		H	manometer deflection $[l]$
p	pressure $[m\, l^{-1}\, t^{-2}]$		K_0	flow coefficient
$p_1 - p_2, \Delta p$	pressure drop $[m\, l^{-1}\, t^{-2}]$		K_1	flow meter constant
$(\Delta p)_{\text{loss}}$	permanent pressure loss $[m\, l^{-1}\, t^{-2}]$		L	length $[l]$
			Q	volume flow rate $[l^3\, t^{-1}]$

R gas constant $[l^2\, t^{-2}\,°]$

Re_{d_1} Reynolds number; $Re = Ud_1/\nu$

S specific gravity

T temperature $[°]$;
torque $[m\, l^2\, t^{-2}]$

U velocity $[l\, t^{-2}]$

V volume $[l^3]$

Y expansion factor

β diameter ratio

δ twist angle

ν kinematic viscosity $[l^2\, t^{-1}]$

η rotational speed $[t^{-1}]$

ρ density $[m\, l^{-3}]$

ω circular frequency $[t^{-1}]$;
angular velocity $[t^{-1}]$

σ standard deviation

PROBLEMS

10.1 Determine the average mass flow rate of 5°C air at 1 bar abs through a 5-cm-i.d. pipe whose velocity profile is found to be symmetric and described by

$$U(r) = 25[1 - (r/r_1)^2]\ \text{cm/s}$$

10.2 A 10-cm-i.d. pipe of flowing 10°C air is traversed along three radial lines with measurements taken at five equidistant stations along each radial. Determine the best estimate of the pipe flow rate.

Location	Radial r(cm)	Line 1	Line 2	Line 3
1	1.0	25.31	24.75	25.10
2	3.0	22.48	22.20	22.68
3	5.0	21.66	21.53	21.79
4	7.0	15.24	13.20	14.28
5	9.0	5.12	6.72	5.35

Column header spanning Line 1–3: $U(r)$ [cm/s]

10.3 A manometer is used in a 5.1-cm water line to measure the pressure drop across a flow meter. Mercury is used as the manometric fluid ($S = 13.57$). If the manometer deflection is 10.16 cm Hg, determine the pressure drop in N/m^2.

10.4 A Bourdon tube pressure gauge indicates a 69-kPa drop across an orifice meter located in a 25.4-cm pipe containing flowing air at 32°C. Find the equivalent pressure head in cm H_2O.

10.5 For a square-edged orifice plate using flange taps to meter the flow of O_2 at 60°F through a 3-in.-i.d. pipe, the upstream pressure is 100 lb/in.^{2} and the downstream pressure is 76 lb/in.^{2}. If the orifice diameter is 1.5 in., can this flow be treated as incompressible for the purposes of estimating the flow rate? Demonstrate your reasoning. $R_{O_2} = 48.3$ ft-lb/lb$_m$ °R

10.6 Find the discharge coefficient of a 5-cm-diameter, square-edged orifice plate using flange taps and located in a 15-cm pipe if the Reynolds number is 250,000.

10.7 At what flow rate of 20°C water through a 10-cm-i.d. pipe would the discharge coefficient of a square-edged orifice plate ($\beta = 0.4$) become essentially independent of Reynolds number?

10.8 Estimate the flange tap pressure drop across a square-edged orifice plate ($\beta = 0.5$) within a 12-cm-i.d. pipe if 25°C water flows at 50 l/s. Compute the permanent pressure loss associated with the plate.

10.9 Determine the flow rate of 38°C air through a 6-cm pipe if a flow nozzle with a 3-cm throat is used, the pressure drop measured is 75 cm H_2O, and the upstream pressure is 94.4 kPa abs.

10.10 A square-edged orifice ($\beta = 0.5$) is used to meter N_2 at 520°R through a 4-in. pipe. If the upstream pressure is 20 psia and the downstream pressure is 15 psia, determine the flow rate through the pipe. Flange taps are used. $R_{N_2} = 55.13$ ft-lb/lb$_m$ °R

10.11 Size a suitable orifice plate to meter water at 20°C flowing through a 38-cm pipe if the nominal flow rate expected is 200 kg/s and the pressure drop must not exceed 15 cm Hg.

10.12 A venturi meter is to be used to meter the flow of 15°C water through a 10-cm pipe. For a maximum differential pressure of 76 cm H_2O and a nominal 0.5 m^3/min flow rate, select a suitable throat size.

10.13 For 120 ft^3/m in of 60°F water flowing through a 6-in. pipe, size a suitable orifice, venturi, and nozzle flow meter if the maximum pressure drop cannot exceed 20 in. Hg. Estimate the permanent losses associated with each meter. Compare the annual operating cost associated with each if power cost is $0.10/kW h, a 60% efficient pump motor is used, and the meter is operated 6000 h/year.

10.14 Estimate the flow rate of water through a 15-cm-i.d. pipe that contains an ASME long-radius nozzle ($\beta = 0.6$) if the pressure drop across the nozzle is 25.4 cm Hg. The water temperature is 27°C.

10.15 Select a location for a 9-cm square-edged orifice along a 4-m straight run of 18-cm pipe if the orifice must be situated downstream of a 90° elbow. The orifice is to be operated uncalibrated under ASME codes.

10.16 An orifice ($\beta = 0.4$) is used as a sonic nozzle to meter a 40°C air flow in a 5-cm pipe. If upstream pressure is 695 mm Hg abs and downstream pressure is 330 mm Hg abs, estimate the mass flow rate through the meter; $R = 287$ N m/kg K, $k = 1.4$.

10.17 Compute the flow rate of 20°C air through a 0.5-m square-edged orifice plate that is situated in a 1.0-m-i.d. pipe. The pressure drop across the plate is 90 mm H_2O and the upstream pressure is 2 atm.

10.18 An ASME long-radius nozzle ($\beta = 0.5$) is to be used to meter the flow of 20°C water through a 8-cm-i.d. pipe. If flow rates will range from 0.6 to 1.6 l/s, select the range required of a single pressure transducer if that transducer will be used to measure pressure drop over the flow rate range. If the type of transducer to be used has a typical error of 0.25% of full scale, estimate the uncertainty in flow rate at the design stage.

10.19 A square-edged orifice plate is selected to meter the flow of water through a 2.3 in.-i.d. pipe over the range 10–50 gal/min at 60°F. It is desired to operate the orifice in the range where C is independent of the Reynolds number for all flow rates. A pressure transducer having an accuracy to within 0.5% of reading is to be used. Select a suitable orifice plate and estimate the design-stage uncertainty in measured flow rate at 10, 25, and 50 gal/min. Assume flange pressure taps and reasonable values for the bias limits on pertinent parameters.

10.20 Estimate the error contribution to the uncertainty in flow rate caused by the effect of the relative humidity of air on air density. Consider the case of using an orifice plate to meter air flow if the relative humidity of the air can vary 10–80% but a density equivalent to 45% relative humidity is used in the computations.

10.21 For Problem 10.20, suppose the air flow rate is 17 m^3/h at 20°C through a 6-cm-i.d. pipe, a square-edged orifice ($\beta = 0.4$) is used with flange taps, and the pressure drop can be measured to within ± 0.5% of reading for all pressures above 5 cm H_2O by using a manometer. If basic dimensions are maintained to within 0.1 mm, estimate a design-stage uncertainty in the flow rate. Use $p_1 = 96.5$ kPa abs, $R = 287$ N m/kg K.

10.22 It is desired to regulate the flow of air through an air sampling device using a sonic nozzle. A flow rate of 45 ft^3/m relative to 70°F is to be maintained. For an ASME long-radius nozzle, determine the downstream pressure required to choke the nozzle if air is supplied at 14.1 psia. Select a suitable nozzle throat diameter; $k = 1.4$, $R = 53.3$ ft-lb/lb_m °R.

10.23 An orifice meter is to be installed between flanges in a 10-cm diameter pipe. It is desired that the meter develop a 100-mm Hg pressure head at 2 m^3/min of 20°C water using flange taps. Specify an appropriate size (d_o) for the orifice meter; $S_{Hg} = 13.6$.

10.24 An ASME long-radius nozzle is used to meter the flow of 20°C water through a 20-cm diameter pipe. The operating flow rate expected is between 5000 cm^3/s and 50,000 cm^3/s. For $\beta = 0.5$, specify the input range required of a pressure transducer used to measure the expected pressure drop. Estimate the permanent pressure loss associated with this nozzle. If a catalog is available, select an appropriate pressure transducer from its listings based on input range.

10.25 Select a pressure transducer from a catalog to meet the needs of Problem 10.24 based on input and output range. If its output is to be measured by using a data-acquisition system that uses a ±5 V, 12-bit A/D converter, estimate the percent relative quantization error at the low and high flow rates. Do these numbers seem appropriate? If not, how can they be improved?

10.26 A vortex flow meter uses a shedder having a Strouhal number of 0.20. Estimate the mean duct velocity if the shedding frequency indicated is 77 Hz and the shedder characteristic length is 1.27 cm.

10.27 A thermal mass flow meter is used to meter 30°C air flow through a 2-cm-i.d. pipe. If 25 W of power are required to maintain a 1°C temperature rise across the meter, estimate the mass flow rate through the meter. Clearly state any assumptions; $c_p = 1.006$ kJ/kg K.

10.28 When used with air (or any perfect gas), a sonic nozzle can be used to regulate volume flow rate. However, uncertainty arises because of variations in density brought on by local atmospheric pressure and temperature changes. For a range of 101.3 ± 7 kPa abs and $10 \pm 5°C$, estimate the error in regulating a 1.4 m^3/min flow rate at critical pressure ratio as a result of these variations alone.

10.29 Estimate an uncertainty in the determined flow rate in Example 10.4, assuming that dimensions are known to within 0.025 mm, that pressure is known to within 0.25 cm H$_2$O, and that the pressure drop shows a standard deviation of 0.5 cm H$_2$O in 20 readings. The upstream pressure is constant; all assumptions should be justified and reasonable.

Chapter 11

Strain Measurement

11.1 INTRODUCTION

The design of load-carrying components for machines and structures requires information about the distribution of forces within the particular component. Proper design of devices, such as shafts, pressure vessels, and support structures, must consider capacity and allowable deflections. Mechanics of materials provides a basis for predicting these essential characteristics of a mechanical design and provides the fundamental understanding of the behavior of load-carrying parts. However, theoretical analysis is often not sufficient and experimental measurements are required to achieve a final design.

Our interest in this chapter is the measurement of physical displacements in engineering components. Engineering designs are based on a safe level of stress within a material. In an object that is subject to loads, forces within the material act to balance the external load.

As a simple example, consider a slender rod that is placed in uniaxial tension, as shown in Figure 11.1. If the rod is sectioned at B—B, a force within the material at B—B is necessary to maintain static equilibrium for the sectioned rod. Such a force within the rod, per unit area, is called *stress*. Design criteria are based on stress levels within a part. In most cases stress cannot be measured directly. But the length of the rod in Figure 11.1 will change when the load is applied, and such changes in length or shape of a material can be measured. The stress is calculated from these measured deflections. Before we can proceed to develop techniques for these measurements, we will briefly review the relationship between deflections and stress.

11.2 STRESS AND STRAIN

The experimental analysis of stress is accomplished by measuring the deformation of a part under load and inferring the existing state of stress from the measured deflections. Again, consider the rod in Figure 11.1. If the rod has a cross-sectional area of A_c, and the load is applied only along the axis of the rod, the normal stress is defined as

$$\sigma_a = F_N / A_c \tag{11.1}$$

where A_c is the cross-sectional area and F_N is the tension force applied to the rod, normal to the area A_c. The ratio of the change in length of the rod (which results from applying the load) to the original length is the axial strain, defined as

$$\varepsilon_a = \delta L / L \tag{11.2}$$

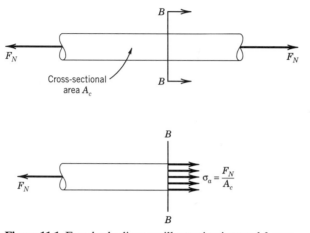

Figure 11.1 Free-body diagram illustrating internal forces for a rod in uniaxial tension.

where ε_a is the average strain over the length L, δL is the change in length, and L is the original unloaded length. For most engineering materials, strain is a small quantity; strain is usually reported in units of 10^{-6} in./in. or 10^{-6} m/m. These units are equivalent to a dimensionless unit called a microstrain ($\mu\varepsilon$).

Stress–strain diagrams are very important in understanding the behavior of a material under load. Figure 11.2 is such a diagram for mild steel (a ductile material). For loads less than that required to permanently deform the material, most engineering materials display a linear relationship between stress and strain. The range of stress over which this linear relationship holds is called the elastic region. The relationship between uniaxial stress and strain for this elastic behavior is expressed as

$$\sigma_a = E_m \varepsilon_a \tag{11.3}$$

where E_m is the modulus of elasticity, or Young's modulus, and the relationship is

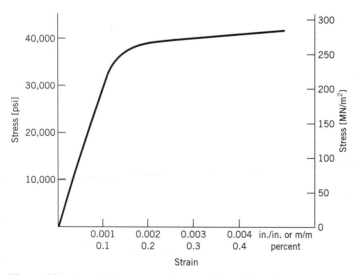

Figure 11.2 A typical stress–strain curve for mild steel.

called Hooke's law. Hooke's law applies only over the range of applied stress where the relationship between stress and strain is linear. Different materials respond in a variety of ways to loads beyond the linear range, largely depending on whether the material is ductile or brittle. For almost all engineering components, stress levels are designed to remain well below the elastic limit of the material; thus, a direct linear relationship may be established between stress and strain. Under this assumption, Hooke's law forms the basis for experimental stress analysis, through the measurement of strain.

Consider the elongation of the rod shown in Figure 11.1, which occurs as a result of the load F_N. As the rod is stretched in the axial direction, the cross-sectional area must decrease, since the total mass (or volume for constant density) must be conserved. Similarly, if the rod were compressed in the axial direction, the cross-sectional area would increase. This change in cross-sectional area is most conveniently expressed in terms of a lateral (transverse) strain. For a circular rod, the lateral strain is defined as the change in the diameter divided by the original diameter. In the elastic range, there is a constant rate of change in the lateral strain as the axial strain increases. In the same sense that the modulus of elasticity is a property of a given material, the ratio of lateral strain to axial strain is also a material property. This property is called Poisson's ratio, defined as

$$v_p = \frac{|\text{lateral strain}|}{|\text{axial strain}|} = \frac{\varepsilon_L}{\varepsilon_a} \tag{11.4}$$

Engineering components are seldom subject to one-dimensional axial loading. The relationship between stress and strain must be generalized to a multidimensional case. Consider a two-dimensional geometry, as shown in Figure 11.3, subject to tensile loads in both the x and y directions, resulting in normal stresses σ_x and σ_y. In this case, for a biaxial state of stress the stresses and strains are

$$\varepsilon_y = \frac{\sigma_y}{E_m} - v_p \frac{\sigma_x}{E_m} \qquad \varepsilon_x = \frac{\sigma_x}{E_m} - v_p \frac{\sigma_y}{E_m}$$

$$\sigma_x = \frac{E_m(\varepsilon_x + v_p\varepsilon_y)}{1 - v_p^2} \qquad \sigma_y = \frac{E_m(\varepsilon_y + v_p\varepsilon_x)}{1 - v_p^2} \tag{11.5}$$

$$\tau_{xy} = G\gamma_{xy}$$

In this case, all of the stress and strain components lie in the same plane. The state

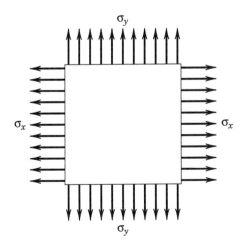

Figure 11.3 Biaxial state of stress.

of stress in the elastic condition for a material is similarly related to the strain in a complete three-dimensional situation [1, 2]. Since stress and strain are related, it is possible to determine stress from measured strains under appropriate conditions. However, strain measurements are made at the surface of an engineering component. The measurement yields information about the state of stress on the surface of the part. The analysis of measured strains requires application of the relationship between the complete state of stress and strain at a surface. Such analysis of strain data is described in [3]. Our emphasis in this chapter will be on techniques for the measurement of strain.

11.3   RESISTANCE STRAIN GAUGES

The measurement of the small displacements that occur in a material or object under mechanical load determines the strain. Strain can be measured by methods as simple as observing the change in the distance between two scribe marks on the surface of a load-carrying member, or as advanced as optical holography. In any case, the ideal sensor for the measurement of strain would (1) have good spatial resolution, implying that the sensor would measure strain at a point; (2) be unaffected by changes in ambient conditions; and (3) have a high-frequency response for dynamic (time-based) strain measurements. A sensor that closely meets these characteristics is the *bonded resistance strain gauge.*

The resistance of a strain gauge changes when it is deformed, and this is easily related to the local strain. Both metallic and semiconductor materials experience this change in electrical resistance when they are subjected to a strain. The amount that the resistance changes depends on how the gauge is deformed, the materials it is made of, and the design of the gauge. Gauges can be made quite small for good resolution and with a low mass to provide a high-frequency response. With some ingenuity, ambient effects can be minimized or eliminated.

In an 1856 publication in the *Proceedings of the Royal Society* in England, Lord Kelvin (William Thomson) [4] laid the foundations for understanding the changes in electrical resistance that metals undergo when subjected to loads, which eventually led to the strain gauge concept. Two individuals began the modern development of strain measurement in the late 1930s—Edward Simmons at the California Institute of Technology and Arthur Ruge at the Massachusetts Institute of Technology. Their development of the bonded metallic wire strain gauge lead to commercially available strain gauges. The resistance strain gauge also forms the basis for a variety of other transducers, such as load cells, pressure transducers, and torque meters.

Metallic Gauges

To understand how metallic strain gauges work, consider a conductor having a uniform cross-sectional area, A_c and a length, L, made of a material having a resistivity, ρ_e. For this electrical conductor, the resistance, R, is given by

$$R = \rho_e L / A_c \tag{11.6}$$

If the conductor is subjected to a normal stress along the axis of the wire, the cross-sectional area and the length will change, resulting in a change in the total electrical resistance, R. The total change in R is due to several effects, as illustrated in the total differential:

$$dR = \frac{A_c(\rho_e \, dL + L \, d\rho_e) - \rho_e L \, dA_c}{A_c^2} \qquad (11.7)$$

which may be expressed in terms of Poisson's ratio as

$$\frac{dR}{R} = \frac{dL}{L}(1 + 2v_p) + \frac{d\rho_e}{\rho_e} \qquad (11.8)$$

Hence, the changes in resistance are caused by two basic effects: the change in geometry as the length and cross-sectional area change, and the change in the value of the resistivity, ρ_e. The dependence of resistivity on mechanical strain is called piezoresistance and may be expressed in terms of a *piezoresistance coefficient*, π_1, defined by

$$\pi_1 = \frac{1}{E_m} \frac{d\rho_e/\rho_e}{dL_1/L} \qquad (11.9)$$

With this definition, the change in resistance may be expressed as the ratio

$$\frac{dR/R}{dL/L} = 1 + 2v_p + \pi_1 E_m \qquad (11.10)$$

EXAMPLE 11.1

Determine the total resistance of a copper wire having a diameter of 1 mm and a length of 5 cm. The resistivity of copper is $1.7 \times 10^{-8} \, \Omega$ m.

KNOWN

$D = 1$ mm
$L = 5$ cm
$\rho_e = 1.7 \times 10^{-8} \, \Omega$ m

FIND

The total electrical resistance.

SOLUTION

The resistance may be calculated from equation (11.6) as

$$R = \rho_e L / A_c$$

where

$$A_c = \frac{\pi}{4} D^2 = \frac{\pi}{4}(1 \times 10^{-3})^2 = 7.85 \times 10^{-7} \text{m}^2$$

The resistance is then

$$R = \frac{(1.7 \times 10^{-8} \, \Omega \, \text{m})(5 \times 10^{-2} \, \text{m})}{7.85 \times 10^{-7} \, \text{m}^2} = 1.08 \times 10^{-3} \, \Omega$$

COMMENT

If the material were nickel instead of copper, for the same diameter and length of wire, what would the resistance be? The resistivity of nickel is 7.8×10^{-8} Ω m, which results in a resistance of 5×10^{-3} Ω.

EXAMPLE 11.2

A very common material for the construction of strain gauges is the alloy constantan (55% copper with 45% nickel), having a resistivity of 49×10^{-8} Ω m. A typical strain gauge might have a resistance of 120 Ω. What length of constantan wire of diameter 0.025 mm would yield a resistance of 120 Ω?

KNOWN

The resistivity of constantan is 49×10^{-8} Ω m.

FIND

The length of constantan wire needed to produce a total resistance of 120 Ω.

SOLUTION

From equation (11.6), we may solve for the length, which yields in this case

$$L = \frac{RA_c}{\rho_e} = \frac{(120\,\Omega)(4.91 \times 10^{-10}\ \text{m}^2)}{49 \times 10^{-8}\ \Omega\,\text{m}} = 0.12\ \text{m}$$

The wire would need to be 12 cm in length to achieve a resistance of 120 Ω.

As shown by Example 11.2, a single straight conductor is normally not practical for strain measurement with meaningful resolution. Instead, a simple solution is to bend the wire conductor so that several lengths of wire are oriented along the axis of the strain gauge, as shown in Figure 11.4.

Figure 11.5 illustrates the construction of a typical metallic-foil bonded strain gauge. Such a strain gauge consists of a metallic foil pattern that is formed in a

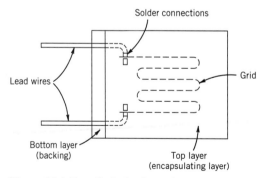

Figure 11.4 Detail of a basic strain gauge construction. (Courtesy of Micro-Measurements Division, Measurements Group, Inc., Raleigh, NC.)

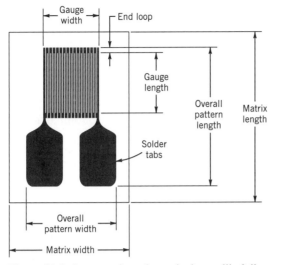

Figure 11.5 Construction of a typical metallic foil strain gauge. (Courtesy of Micro-Measurements Division, Measurements Group, Inc., Raleigh, NC.)

manner similar to the process used to produce printed circuits. This photoetched metallic foil pattern is mounted on a plastic backing material. The *gauge length*, as illustrated in Figure 11.5, is an important specification for a particular application. Since strain is usually measured at the location on a component where the stress is a maximum and the stress gradients are high, the strain gauge averages the measured strain over the gauge length. Since the maximum strain is the quantity of interest, errors can result from improper choice of a gauge length [5].

The variety of conditions encountered in particular applications require special construction and mounting techniques, including design variations in the backing material, the grid configuration, bonding techniques, and total gauge electrical resistance. The adhesives used in the bonding process and the mounting techniques for a particular gauge and manufacturer will vary according to the specific application. However, there are some fundamental aspects that are common to all bonded resistance gauges.

The strain gauge backing serves several important functions. It electrically isolates the metallic gauge from the test specimen and transmits the applied strain to the sensor. A bonded resistance strain gauge must be appropriately mounted to the specimen for which the strain is to be measured. The backing provides the surface used for bonding with an appropriate adhesive. Backing materials are available that are useful over temperatures that range from −270 to 290°C.

The adhesive bond serves as a mechanical and thermal coupling between the metallic gauge and the test specimen. As such the strength of the adhesive should be sufficient to accurately transmit the strain experienced by the test specimen, and should have thermal conduction and expansion characteristics suitable for the application. If the adhesive shrinks or expands during the curing process, apparent strain can be created in the gauge. A wide array of adhesives are available for bonding strain gauges to a test specimen. Among these are epoxies, cellulose nitrate cement, and ceramic-based cements.

If a strain gauge is placed in a state of stress in which there are both normal and lateral (transverse) components of strain, the total change in resistance of the gauge will be a result of both the axial and lateral components of strain. Strictly speaking, the relationship for change in resistance expressed in equation (11.8) is true only for a single conductor in a state of uniaxial tension or compression. The goal of the design of electrical resistance strain gauges is to create a sensor that is sensitive only to strain along one axis, and therefore has zero sensitivity to lateral and shearing strains. Actual strain gauges are sensitive to lateral strains to some degree; for most gauges the sensitivity to shearing strains may be neglected.

Gauge Factor

The change in resistance of a strain gauge is normally expressed in terms of an empirically determined parameter called the gauge factor, GF. For a particular strain gauge, the gauge factor is supplied by the manufacturer. The gauge factor is expressed as

$$ GF \equiv \frac{\delta R/R}{\delta L/L} = \frac{\delta R/R}{\varepsilon_a} \tag{11.11} $$

Relating this definition to equation (11.10), we see that the gauge factor is dependent on the Poisson ratio for the gauge material and its piezoresistivity. For metallic strain gauges, the Poisson ratio is approximately 0.3 and the resulting gauge factor is ~2.

The gauge factor represents the total change in resistance for a strain gauge, under a calibration loading condition. The calibration loading condition generally creates a biaxial strain field, and the lateral sensitivity of the gauge influences the measured result. Strictly speaking then, the sensitivity to normal strain of the material used in the gauge and the gauge factor are not the same. Generally gauge factors are measured in a biaxial strain field, which results from the deflection of a beam having a value of Poisson's ratio of 0.285. Thus, for any other strain field there is an error in strain indication that is due to the lateral sensitivity of the strain gauge. The following equation expresses the percentage error that is due to lateral sensitivity for a strain gauge mounted on any material, at any orientation in the strain field:

$$ e_L = \frac{K_t(\varepsilon_L/\varepsilon_a + v_{po})}{1 - v_{po}K_t} \times 100 \tag{11.12} $$

where

$\varepsilon_a, \varepsilon_L$ = axial and lateral strains, respectively (with respect the axis of the gauge)

v_{po} = Poisson's ratio of the material on which the manufacturer measured GF (usually 0.285 for steel)

e_L = error as a percentage of axial strain (with respect to the axis of the gauge)

K_t = lateral (transverse) sensitivity of the strain gauge

Typical values of the lateral sensitivity for commercial strain gauges range from 0.05 to −0.19. Figure 11.6 shows a plot of the percentage error for a strain gauge as a function of the ratio of lateral loading to axial loading and the lateral sensitivity. It is possible to correct for the lateral sensitivity effects [6].

Semiconductor Strain Gauges

When subjected to a load, a semiconductor material exhibits a change in resistance and therefore can be used for the measurement of strain. Silicon crystals are the

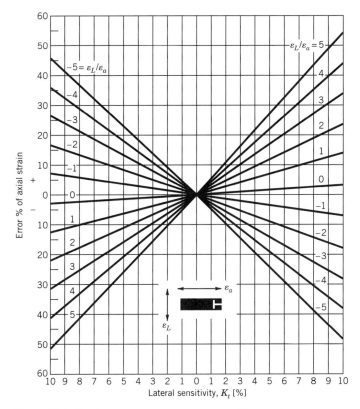

Figure 11.6 Strain measurement error caused by strain gauge lateral (transverse) sensitivity. (Courtesy of Measurements Group, Inc., Raleigh, NC.)

basic material for semiconductor strain gauges; the crystals are sliced into very thin sections to form strain gauges. Because of the high piezoresistance coefficient, the semiconductor gauge exhibits a very high gauge factor, as high as 200 for some gauges. These gauges also exhibit higher resistance, longer fatigue life, and lower hysteresis under some conditions than metallic gauges. However, the output of the semiconductor strain gauge is nonlinear with strain, and the strain sensitivity or gauge factor may be markedly dependent on temperature.

In general, materials exhibit a change in resistivity with strain, characterized by the piezoresistance coefficient, π_1. For a semiconductor, this change in resistivity with strain can be very large. The effect of applied stress is to change the number and the mobility of the charge carriers within the material, thus causing large changes in the resistivity. Resistivity is a direct measure of the charge carrier density. The semiconductor crystals from which strain gauges are constructed contain a certain amount of impurities. These impurities control the number and mobility of the charge carriers within the gauge, and thus allow control of the gauge characteristics during the manufacturing process. Semiconductor materials for strain gauge applications have resistivities ranging from 0.001 to 1.0 Ω cm. Semiconductor strain gauges may have a relatively high or low density of charge carriers [3, 7]. Semiconductor strain gauges made of materials having a relatively high density of charge carriers ($\approx 10^{20}$ carriers/cm^3) exhibit little variation of their gauge factor with strain or temperature. In contrast, for the case in which the crystal contains a low number of

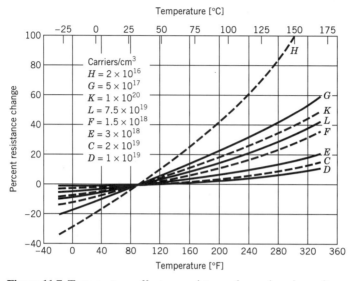

Figure 11.7 Temperature effect on resistance for various impurity concentrations for P-type semiconductors (reference resistance at 27.2°C). (Courtesy of Kulite Semiconductor Products, Inc., location).

charge carriers ($<10^{17}$ carriers/cm^3), the gauge factor may be approximated as

$$\mathrm{GF} = \frac{T_0}{T}\mathrm{GF}_0 + C_1\left(\frac{T_0}{T}\right)^2 \varepsilon \tag{11.13}$$

where GF_0 is the gauge factor at the reference temperature T_0, under conditions of zero strain [8], and C_1 is a constant for a particular gauge. The behavior of a high-resistivity P-type semiconductor is shown in Figure 11.7.

Because of the capability for producing small gauge lengths, silicon semiconductor strain gauge technology provides for the construction of very small transducers. These transducers may, for example, be used for the measurement of pressure and by virtue of their size would allow measurement of strain at higher frequencies than previously was possible. However, silicon diaphragm pressure transducers require special procedures for measuring in liquid environments. Semiconductor strain gauges are somewhat limited in the maximum strain that they can measure, approximately 5000 $\mu\varepsilon$ for tension but higher in compression.

Semiconductor strain gauges find their primary application in the construction of transducers, such as load cells and pressure transducers. Because of the possibility of an inherent sensitivity to temperature, careful consideration must be made for each application to provide appropriate temperature compensation or correction. Temperature effects can result in zero drift during the duration of a measurement.

11.4 STRAIN GAUGE ELECTRICAL CIRCUITS

A Wheatstone bridge is generally used to detect the small changes in resistance that are the output of a strain gauge measurement circuit. A typical strain gauge measuring installation on a steel specimen will have a sensitivity of 10^{-6} Ω/(kN m^2). As such,

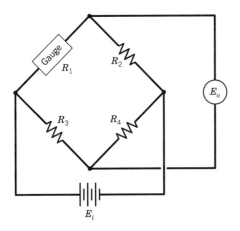

Figure 11.8 Basic strain gauge Wheatstone bridge circuit.

a high-sensitivity device, such as a Wheatstone bridge, is desirable for measuring resistance changes for strain gauges. The fundamental relationships for the analysis of such bridge circuits are discussed in Chapter 6. Their application to strain gauge circuits will be discussed in this section. Equipment is commercially available that can measure changes in gauge resistance of less than 0.0005 Ω(0.000001 $\mu\varepsilon$).

A simple strain gauge Wheatstone bridge circuit is shown in Figure 11.8. The bridge output under these conditions is given by equation (6.15):

$$E_o + \delta E_o = E_i \frac{(R_1 + \delta R)R_4 - R_3 R_2}{(R_1 + \delta R + R_2)(R_3 + R_4)} \tag{6.15}$$

E_o is the bridge output at initial conditions, δE_o is the bridge deflection and δR is the strain gauge resistance change. Consider the case where all the fixed resistors and the strain gauge resistance are initially equal, and the bridge is balanced such that $E_o = 0$. If the strain gauge is then subjected to a state of strain, the change in the output voltage, δE_o, of equation (6.15) reduces to

$$\frac{\delta E_o}{E_i} = \frac{\delta R/R}{4 + 2(\delta R/R)} \approx \frac{\delta R/R}{4} \tag{11.14}$$

The simplified form of equation (11.14) is suitable for all but those measurements that demand the highest accuracy and remains valid for values of $\delta R/R \ll 1$. Using the relationship from equation (11.11) that $\delta R/R = \mathrm{GF}\varepsilon$,

$$\frac{\delta E_o}{E_i} = \frac{\mathrm{GF}\varepsilon}{4 + 2\mathrm{GF}\varepsilon} \approx \frac{\mathrm{GF}\varepsilon}{4} \tag{11.15}$$

Equations (11.14) and (11.15) yield two practical equations for strain gauge measurements.

The Wheatstone bridge has several distinct advantages for use with electrical resistance strain gauges. The bridge may be balanced by changing the resistance of one arm of the bridge. Therefore, once the gauge is mounted in place on the test specimen under a condition of zero loading, the output from the bridge may be zeroed. Two schemes for circuits to accomplish this balancing are shown in Figure 11.9. Shunt balancing provides the best arrangement for strain gauge applications, since the changes in resistance for a strain gauge are small. Also, the strategic placement

Circuit arrangement for shunt balance

Differential shunt balance arrangement

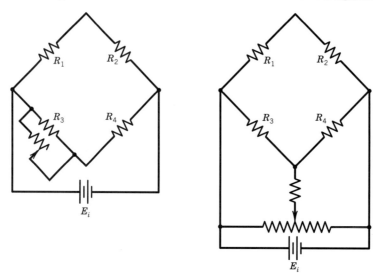

Figure 11.9 Balancing schemes for bridge circuits.

of multiple gauges in a Wheatstone bridge can both increase the bridge output and cancel out certain ambient effects and unwanted components of strain as discussed in the next two sections.

EXAMPLE 11.3

A strain gauge, having a gauge factor of 2, is mounted on a rectangular steel bar ($E_m = 200 \times 10^6$ kN/m^2), as shown in Figure 11.10. The bar is 3 cm wide and 1 cm high, and is subjected to a tensile force of 30 kN. Determine the resistance change of the strain gauge, if the resistance of the gauge was 120 Ω in the absence of the axial load.

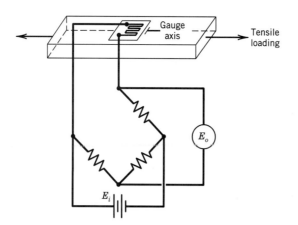

Figure 11.10 Strain gauge circuit subject to uniaxial tension.

KNOWN

$$GF = 2 \qquad E_m = 200 \times 10^6 \text{ kN/m}^2 \quad F_N = 30 \text{ kN}$$
$$R = 120 \, \Omega \qquad A_c = 0.03 \text{ m} \times 0.01 \text{ m}$$

FIND

The resistance change of the strain gauge for a tensile force of 30 kN.

SOLUTION

The stress in the bar under this loading condition is

$$\sigma_a = \frac{F_N}{A_c} = \frac{30 \text{ kN}}{(0.03 \text{ m})(0.01 \text{ m})} = 1 \times 10^5 \text{ kN/m}^2$$

and the resulting strain is

$$\varepsilon_a = \frac{\sigma_N}{E_m} = \frac{1 \times 10^5 \text{ kN/m}^2}{200 \times 10^6 \text{ kN/m}^2} = 5 \times 10^{-4} \text{ m/m} \qquad (11.16)$$

For strain along the axis of the strain gauge, the change in resistance from equation (11.11) is

$$\delta R / R = \varepsilon GF$$

or

$$\delta R = R \varepsilon GF = (120 \, \Omega)(5 \times 10^{-4})(2.0)$$
$$= 0.12 \, \Omega$$

EXAMPLE 11.4

Suppose the strain gauge described in Example 11.3 is to be connected to a measurement device capable of determining a change in resistance with a stated accuracy to ±0.005 Ω (95%). This stated accuracy includes a resolution of 0.001 Ω. What uncertainty in stress would result using this resistance measurement device?

KNOWN

A stress is to be inferred from a strain measurement by using a strain gauge having a gauge factor of 2 and a zero load resistance of 120 Ω. The measurement of resistance has a stated accuracy to ±0.005 Ω (95%).

FIND

The design-stage uncertainty in stress.

SOLUTION

The design-stage uncertainty in stress, $(u_d)_\sigma$, is given by

$$(u_d)_\sigma = \frac{\partial \sigma}{\partial (\delta R)} (u_d)_{\delta R}$$

with

$$\sigma = \varepsilon E_m = \frac{\delta R/R}{GF} E_m$$

Then with

$$\frac{\partial \sigma}{\partial(\delta R)} = \frac{E_m}{R(GF)}$$

we can express the uncertainty as

$$(u_d)_\sigma = \frac{E_m}{R(GF)}(u_d)_{\delta R} = \frac{200 \times 10^6 \text{ kN/m}^2}{120\,\Omega(2.0)}(0.005\,\Omega)$$

This results in a design-stage uncertainty in stress of $(u_d)_\sigma = 4.17 \times 10^3$ kN/m^2 (95%) or ~2.4% of the expected stress.

11.5 PRACTICAL CONSIDERATIONS FOR STRAIN MEASUREMENT

This section will describe some characteristics of strain gauge applications that allow practical implementation of strain measurement.

The Multiple Gauge Bridge

The output from a bridge circuit can be increased by the appropriate use of more than one active strain gauge. This increase can be related to a *bridge constant* as illustrated next. In addition, the use of multiple gauges can be used to compensate for unwanted effects, such as temperature or specific strain components. Consider the case when all four resistances in the bridge circuit of Figure 11.8 represent active strain gauges. In general, the bridge output is given by

$$E_o = E_i \left[\frac{R_1}{R_1 + R_2} - \frac{R_3}{R_3 + R_4} \right] \tag{11.17}$$

The strain gauges R_1, R_2, R_3, and R_4 are assumed initially to be in a state of zero strain. If these gauges are now subjected to strains such that the resistances change by dR_i, where $i = 1, 2, 3$, and 4, then the change in the bridge output voltage can be expressed as

$$dE_o = \sum_{i=1}^{4} \frac{\partial E_o}{\partial R_i} dR_i \tag{11.18}$$

Evaluating the appropriate partial derivatives from equation (11.17) yields

$$dE_o = E_i \left[\frac{R_2\, dR_1 - R_1\, dR_2}{(R_1 + R_2)^2} + \frac{R_3\, dR_4 - R_4\, dR_3}{(R_3 + R_4)^2} \right] \tag{11.19}$$

Then from equations (11.2) and (11.11), $dR_i = R_i \varepsilon_i GF_i$, and the value of dE_o can be determined. Assuming $dR_i \ll R_i$, the resulting change in output voltage δE_o may now be expressed as δE_o where

$$\delta E_o = E_i \left[\frac{R_1 R_2}{(R_1 + R_2)^2}(\varepsilon_1 GF_1 - \varepsilon_2 GF_2) + \frac{R_3 R_4}{(R_3 + R_4)^2}(\varepsilon_4 GF_4 - \varepsilon_3 GF_3) \right] \tag{11.20}$$

If $R_1 = R_2 = R_3 = R_4$, then

$$\frac{\delta E_o}{E_i} = \frac{1}{4}(\varepsilon_1 GF_1 - \varepsilon_2 GF_2 + \varepsilon_4 GF_4 - \varepsilon_3 GF_3) \qquad (11.21)$$

It is possible to purchase matched sets of strain gauges for a particular application, so that $GF_1 = GF_2 = GF_3 = GF_4$, and

$$\frac{\delta E_o}{E_i} = \frac{GF}{4}(\varepsilon_1 - \varepsilon_2 + \varepsilon_4 - \varepsilon_3) \qquad (11.22)$$

Equation (11.22) is important and forms the basic working equation for a strain gauge–bridge circuit using multiple gauges (compare this equation with 11.15).

Equation (11.22) shows that for a bridge containing one or more active strain gauges, equal strains on opposite bridge arms sum, whereas equal strains on adjacent arms of the bridge cancel. These characteristics can be used to increase the output of the bridge, to provide temperature compensation, or to cancel unwanted components of strain. Practical means of achieving these desirable characteristics will be explored further, after the concept of the bridge constant is developed.

Bridge Constant

Commonly used strain gauge–bridge arrangements may be characterized by a *bridge constant*, κ, defined as the ratio of the actual bridge output to that output of a single gauge sensing the maximum strain, ε_{max} (assuming the remaining bridge resistances remain fixed). The output for a single gauge experiencing the maximum strain may be expressed

$$\frac{\delta R}{R} = \varepsilon_{max} GF \qquad (11.23)$$

So that, again for a single gauge,

$$\frac{\delta E_o}{E_i} \cong \frac{\varepsilon_{max} GF}{4} \qquad (11.24)$$

The bridge constant, κ, is found from the ratio of the actual bridge output given by equation (11.22) to the output for a single gauge given by equation (11.24). When more than one gauge is used in the bridge circuit, equation (11.15) becomes

$$\frac{\delta E_o}{E_i} = \frac{\kappa \delta R / R}{4 + 2\delta R / R} = \frac{\kappa GF\varepsilon}{4 + 2GF\varepsilon} \approx \frac{\kappa GF\varepsilon}{4} \qquad (11.25)$$

The simplified form of equation (11.25) is suitable for all but those measurements demanding the highest accuracy and remains valid for values of $\delta R / R \ll 1$.

The bridge constant concept is illustrated in Example 11.5.

EXAMPLE 11.5

Determine the bridge constant for two strain gauges mounted on a member, as shown in Figure 11.11. The member is subject to uniaxial tension, which produces an axial strain ε_a and a lateral strain $\varepsilon_L = -\nu_p \varepsilon_a$. Assume that all the resistances in Figure 11.11 are initially equal, and therefore the bridge is initially balanced. Let $(GF)_1 = (GF)_2$.

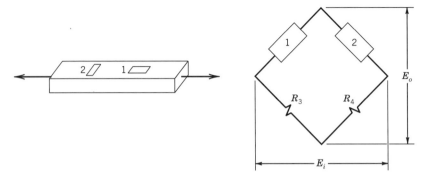

Figure 11.11 Bridge circuit with two arms active; strain gauge installation for increased sensitivity.

KNOWN

Strain gauge installation shown in Figure 11.11.

FIND

The bridge constant for this installation.

ASSUMPTION

The change in the strain gauge resistances are small compared to the initial resistance (see explanation as follows).

SOLUTION

If the gauges are mounted so that gauge 1 is aligned with the axial tension and gauge 2 is mounted transversely on the member, the output of the bridge will be greater than for gauge 1 alone. Since strain gauges generally experience small resistance changes, it is often convenient to develop approximate relationships for specific bridge circuit arrangements and strain gauge installations under this assumption.

The changes in resistance for the gauges may be expressed

$$\frac{\delta R_1}{R_1} = \varepsilon_a(\text{GF})_1 \tag{11.26}$$

and

$$\frac{\delta R_2}{R_2} = -\nu_p \varepsilon_a(\text{GF})_2 = -\nu_p \frac{\delta R_1}{R_1} \tag{11.27}$$

With only one gauge active, equation (11.14) would be applicable. If $\delta R_1 / R_1 \ll 1$,

$$\frac{\delta E_o}{E_i} = \frac{\delta R_1 / R_1}{4 + 2(\delta R_1 / R_1)} \approx \frac{\delta R_1 / R_1}{4} \tag{11.14}$$

But with both gauges installed and active, as shown in Figure 11.11, the output of the bridge is determined from an analysis of the bridge response, which results in

$$\frac{\delta E_o}{E_i} = \frac{(\delta R_1 / R_1)(1 + \nu_p)}{4 + 2(\delta R_1 / R_1)(1 + \nu_p)}$$

In practical applications, it is most often the case that changes in resistance are small in comparison to the resistance values; thus

$$\frac{\delta E_o}{E_i} = \frac{(\delta R_1 / R_1)(1 + \nu_p)}{4} \tag{11.28}$$

Therefore, the bridge constant is the ratio of equation (11.28) to (11.14)

$$\frac{(\delta R_1 / R_1)(1 + \nu_p)/4}{(\delta R_1 / R_1)/4} \tag{11.29}$$

or in the case where $\delta R_1 / \delta R_1 \ll 1$,

$$\kappa = 1 + \nu_p$$

Comparing equation (11.28) to (11.14) shows that the use of two gauges oriented as described has increased the output from the bridge by a factor of $1 + \nu_p$ over that of using a single gauge.

Apparent Strain and Temperature Compensation

Apparent strain is manifested as any change in gauge resistance that is not due to the component of strain being measured. Techniques for accomplishing temperature compensation, eliminating certain components of strain, and increasing the value of the bridge constant can be devised by examining more closely equation (11.22). The bridge constant is influenced by (1) the location of strain gauges on the test specimen and (2) the gauge connection positions in the bridge circuit. The combined effect of these two factors is determined by examining the existing strain field and using equation (11.22) to determine the resulting bridge output.

Let us look at how a component of strain can be removed (compensation) from the measured signal. Consider a beam having a rectangular cross section and subject to the loading condition shown in Figure 11.12, where the beam is subject to an axial load F_N and a bending moment, M. The stress distribution in this cross section is given by

$$\sigma_x = -12My/bh^3 + F_N/bh \tag{11.30}$$

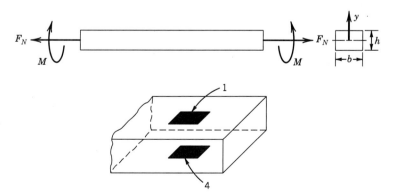

Figure 11.12 Strain gauge installation for bending compensation.

To remove the effects of bending strain, identical strain gauges are mounted to the top and bottom of the beam as shown in Figure 11.12, and they are connected to bridge locations 1 and 4 (opposite bridge arms). The gauges experience equal but opposite bending strains [see equation (11.30)], and both strain gauges are subject to the same axial strain caused by F_N. The bridge output under these conditions is

$$\frac{\delta E_o}{E_i} = \frac{GF}{4}(\varepsilon_1 + \varepsilon_4) \tag{11.31}$$

where $\varepsilon_1 = \varepsilon_{a_1} + \varepsilon_{b_1}$ and $\varepsilon_4 = \varepsilon_{a_4} - \varepsilon_{b_4}$, with subscripts a and b referring to axial and bending strain, respectively. Hence, the bending strains cancel but the axial strains will sum, giving

$$\frac{\delta E_o}{E_i} = \frac{GF}{2}\varepsilon_a \tag{11.32}$$

For a single gauge experiencing the maximum strain,

$$\frac{\delta E_o}{E_i} = \frac{GF}{4}\varepsilon_a \tag{11.33}$$

equations (11.32) with (11.33) differ by a value of 2, which is the bridge constant ($\kappa = 2$) for the arrangement in Figure 11.12. So, this arrangement compensates for the bending strain.

A guide for some practical bridge–gauge configurations is provided in Table 11.1.

Differential thermal expansion between the gauge and the specimen on which it is mounted creates an apparent strain in the strain gauge. Thus, temperature sensitivity of strain gauges is a result of both the changes in resistance caused by temperature changes in the gauge itself, and the strain experienced by the gauge caused by differential thermal expansion between the gauge and the material on which it is mounted. Using gauges of identical alloy composition as the specimen would minimize this latter effect. However, even keeping the specimen at a constant temperature may not be enough to eliminate gauge thermal expansion: Heating of the strain gauge as a result of current flow from the measuring device may be a source of significant error, since the gauge is also a temperature-sensitive element. The temperature sensitivity of a strain gauge is an obstacle to accurate mechanical strain measurements that must be considered. Fortunately, there are effective ways to deal with it.

Figure 11.13 shows two circuit arrangements that provide temperature compensation for a strain measurement. The strain gauge mounted on the test specimen experiences changes in resistance caused by temperature changes and by applied strain, whereas the compensating gauge experiences resistance changes caused only by temperature changes. As long as the compensating gauge, as shown in Figure 11.13, experiences an identical thermal environment as the measuring gauge, temperature effects will be eliminated from the circuit. To show this, consider the case when all the bridge resistances are initially equal, and the bridge is therefore balanced. If the temperature of the strain gauges now changes, their resistance will change as a result of thermal expansion, creating an apparent thermal strain. Under an applied axial load, the output of the bridge is derived from equation (11.22),

$$\delta E_o = E_i \frac{GF(\varepsilon_1 - \varepsilon_2)}{4} = E_i \frac{(GF)\varepsilon_a}{4} \tag{11.34}$$

where $\varepsilon_1 = \varepsilon_a + \varepsilon_T$ and $\varepsilon_2 = \varepsilon_T$, where ε_T refers to the apparent thermal strain. Under

Table 11.1 Common Gauge Mountings

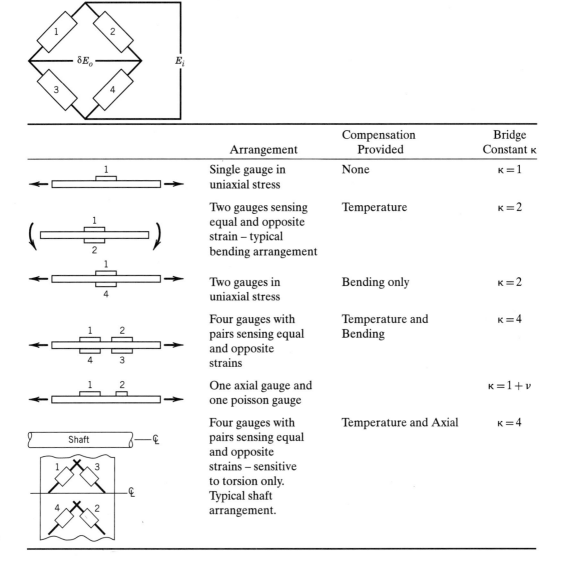

	Arrangement	Compensation Provided	Bridge Constant κ
	Single gauge in uniaxial stress	None	$\kappa = 1$
	Two gauges sensing equal and opposite strain – typical bending arrangement	Temperature	$\kappa = 2$
	Two gauges in uniaxial stress	Bending only	$\kappa = 2$
	Four gauges with pairs sensing equal and opposite strains	Temperature and Bending	$\kappa = 4$
	One axial gauge and one poisson gauge		$\kappa = 1 + \nu$
	Four gauges with pairs sensing equal and opposite strains – sensitive to torsion only. Typical shaft arrangement.	Temperature and Axial	$\kappa = 4$

this condition, the output value will not be affected by thermal effects. The result will be the same if the compensating gauge is mounted to arm 3 instead, as shown in Figure 11.13(b). Furthermore, any two active gauges mounted on adjacent bridge arms will compensate for temperature.

Since this relationship holds in general, we can state that as long as two gauges, which are mounted to a specimen or similar specimens, remain at the same temperature and are connected to adjacent arms of a Wheatstone bridge, they will provide temperature compensation for each other.

Looking back, the arrangement in Figure 11.12 will not provide temperature compensation as gauges 1 and 4 are on opposite bridge arms. However, temperature compensation could be provided for this installation by having two additional strain gauges, which are at the same temperature as gauges 1 and 4, occupy arms 2 and 3 in the bridge.

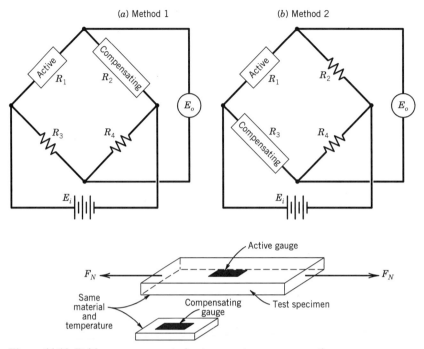

Figure 11.13 Bridge arrangements for temperature compensation.

The sensitivity of the bridge arrangement in Figure 11.13(*a*) (method 1) is

$$K_B = \frac{E_o}{\varepsilon} = E_i \frac{R_1 R_2}{(R_1 + R_2)^2} \text{GF} \tag{11.35}$$

and with $E_i = (2R_g)I_g$ and $R_g = R_1, R_2$, the sensitivity may be expressed in terms of the current flowing through the gauge (I_g) as

$$K_B = \frac{1}{2}\text{GF}\sqrt{(I_g^2 R_1)R_1} \tag{11.36}$$

Note that $(I_g)^2 R_1$ is the power dissipated in the strain gauge as a result of the bridge current. Excessive power dissipation in the gauge would cause temperature changes and introduce uncertainty into a strain measurement. These effects can be minimized by good thermal coupling between the strain gauge and the object to which it is bonded, to allow effective dissipation of thermal energy.

Consider the sensitivity of the bridge arrangement in Figure 11.13(*b*). With identical gauges at positions R_1 and R_3 and equal resistance changes for the two gauges, no change in bridge output would occur. However, the sensitivity for this arrangement is not the same as for method 1 but is given by

$$K_B = \frac{R_1/R_2}{1 + R_1/R_2}\text{GF}\sqrt{(I_g^2 R_1)R_1} \tag{11.37}$$

Here the sensitivity is the same as for a bridge having a single active gauge and without temperature compensation. However, the sensitivity depends on the choice of the fixed resistor R_2. If $R_1 = R_2$, the resulting sensitivity will be the same as for method 1. However, resistor R_2 can be chosen to provide the desired sensitivity for the circuit, within the limitations of measurement capability and allowable bridge current.

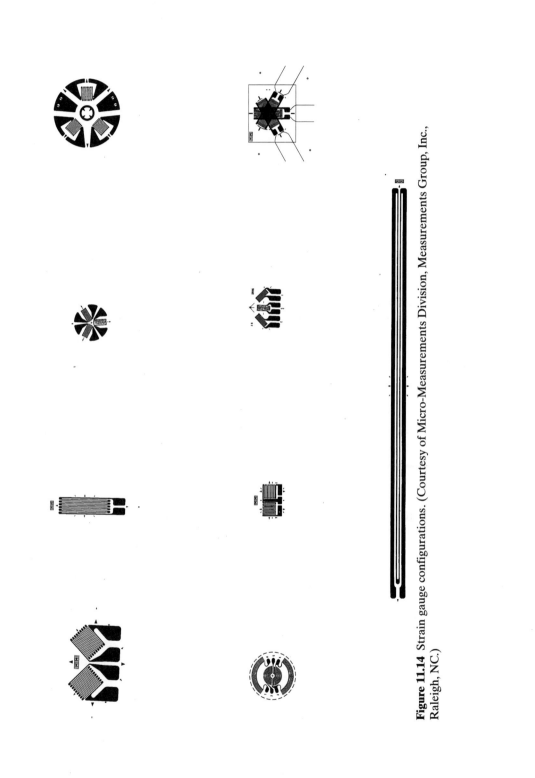

Figure 11.14 Strain gauge configurations. (Courtesy of Micro–Measurements Division, Measurements Group, Inc., Raleigh, NC.)

449

Construction and Installation

Figure 11.14 shows a variety of strain gauge configurations. Practical construction of gauges is accomplished by techniques similar to those used in printed circuit board technology, and it results in a thin-film metallic grid of conductors that form the strain gauge. The first metallic foil gauges were developed around 1950, and most commercial gauges are of this type. Standard gauge resistances are 120 and 350 Ω. Because of the accuracy of the process that produces the metal foil, a variety of patterns for the metal film are possible.

The operating assumption that leads to the definition of the gauge factor is that the change in resistance of the gauge is linear with applied strain, for a particular gauge. However, a strain gauge installed in a measurement environment will exhibit some nonlinearity. Also, in cycling between a loaded and unloaded condition, there will be some degree of hysteresis and a shift in the resistance for a state of zero strain. A typical cycle of loading and unloading is shown in Figure 11.15. The strain gauge will typically indicate lower values of strain during unloading than are measured as the load is increased. The extent of these behaviors is determined not only by the strain gauge characteristics, but also by the characteristics of the adhesive, and by the previous strains that the gauge has experienced. For properly installed gauges, the deviation from linearity should be of the order of 0.1% [3]. In contrast, first-cycle hysteresis and zero shift are difficult to predict. The effects of first-cycle hysteresis and zero shift can be minimized by cycling the strain gauge between zero strain and a value of strain above the maximum value to be measured.

In dynamic measurements of strain, the dynamic response of the strain gauge itself is generally not the limiting factor for such dynamic measurements. However, the transient response of bonded resistance strain gauges has been experimentally examined. A strain wave is made to propagate through a specimen to which a strain gauge is mounted, and the response of the gauge observed. The rise time (90%) of a bonded resistance strain gauge may be approximated as [9]

$$t_{90} \approx 0.8(L/a) + 0.5 \, \mu s \tag{11.38}$$

where L is the gauge length and a is the speed of sound in the material on which the gauge is mounted. Typical response times for gauges mounted on steel specimens are of the order of 1 μs.

Analysis of Strain Gauge Data

Strain gauges mounted on the surface of a test specimen respond only to the strains that occur at the surface of the test specimen. As such, the results from strain gauge measurements must be analyzed to determine the state of stress occurring at the strain gauge locations. The complete determination of the stress at a point on the surface of a particular test specimen will in general require the measurement of three strains at the point under consideration. The result of these measurements will yield the principal strains and allow determination of the maximum stress [3].

If more information is available concerning the expected state of stress, less than three strain gauges may be employed. When the directions of the principal axes are known in advance, only two strain measurements are necessary to calculate the maximum stress at the measured point. The multiple-element strain gauges used to measure more than one strain at a point are called strain gauge rosettes. An example of a two-element rosette is shown in Figure 11.16. For general measurements of

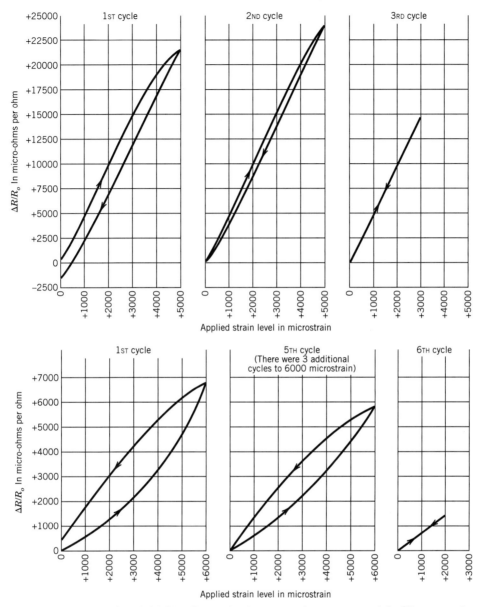

Figure 11.15 Hysteresis in initial loading cycles for two strain gauge materials. (Courtesy of Micro-Measurements Division, Measurements Group, Inc., Raleigh, NC.)

strain and stress, commercially available strain rosettes can be chosen that have a pattern of multiple-direction gauges that is compatible with the specific nature of the particular application [5]. In practice, the applicability of the single-axis strain gauge is extremely limited, and improper use can result in large errors in the measured stress.

Signal Conditioning

The most common form of signal conditioning in strain gauge–bridge circuits is to amplify the signal by using a low-noise amplifier. For an amplifier of gain G_A,

(*a*)   Single-plane type.

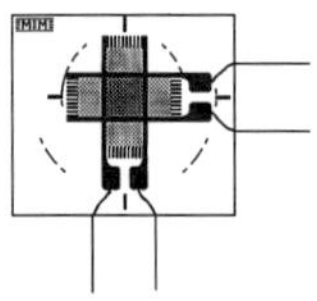

(*b*)   Stacked type.

Figure 11.16 Biaxial strain gauge rosettes.
(Courtesy of Micro-Measurements Division,
Measurements Group, Inc., Raleigh, NC.)

equation (11.25) becomes

$$\frac{\delta E_o}{E_i} = \frac{G_A \kappa \delta R/R}{4 + 2\delta R/R} = \frac{G_A \kappa (\mathrm{GF})\varepsilon}{4 + 2\mathrm{GF}\varepsilon} \approx \frac{G_A \kappa (\mathrm{GF})\varepsilon}{4} \tag{11.39}$$

The simplified form of equation (11.39) is suitable for all but those measurements that demand the highest accuracy and remains valid for values of $\delta R/R \ll 1$.

A common means of recording strain gauge–bridge circuit signals is the automated data-acquisition system. A schematic diagram of such a setup is shown in Figure 11.17. Also, a typical interface card was presented in Figure 7.26.

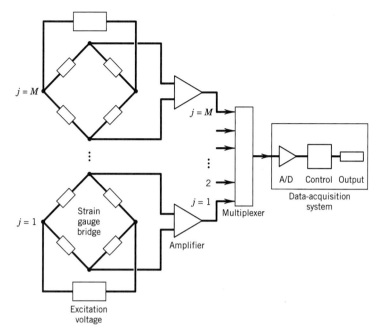

Figure 11.17 Data-acquisition and reduction system for Example 11.6.

Uncertainties in Multichannel Measurements

Uncertainties caused by an automated multichannel strain measurement system can be estimated in a system calibration. Example 11.6 describes an initial test of such a system, often called a shakedown test, and discusses the information that results from this test.

The strain developed at M different locations about a large test specimen is to be measured at each location by using a setup similar to that shown in Figure 11.17. At each location similar strain gauges are appropriately mounted and connected to a Wheatstone bridge that is powered by an external supply. The bridge deflection voltage is to be amplified and measured. A similar setup is used at all M locations. The output from each amplifier is input through an M-channel multiplexer to an automated data-acquisition system. If N (say 30) readings for each of the M setups are taken while the test specimen is maintained in a condition of uniform zero strain, what information is obtained?

KNOWN

Strain setup of Figure 11.17
M ($j = 1, 2, \ldots, M$) setups
N ($i = 1, 2, \ldots, N$) readings per setup

ASSUMPTIONS

All setups are to be operated in a similar manner. Gauges are operated in their linear regimes.

SOLUTION

Consider the data that will become available. First, each setup is exposed to a similar strain. Hence, each setup should indicate the same strain. We can calculate the pooled mean value of the $M \times N$ readings to obtain

$$\langle \bar{\varepsilon} \rangle = \frac{1}{MN} \sum_{j=1}^{M} \sum_{i=1}^{N} \varepsilon_{ij} = \frac{1}{M} \sum_{j=1}^{M} \bar{\varepsilon}_j$$

The difference between the pooled mean strain and the applied strain (zero here) is an estimate of the bias limit that can be expected from any channel during data acquisition.

Precision errors will manifest themselves through scatter in the data set. The pooled standard deviation

$$\langle s_\varepsilon \rangle = \sqrt{\frac{\sum\limits_{j=1}^{M} \sum\limits_{i=1}^{N} (\varepsilon_{ij} - \bar{\varepsilon}_j)^2}{M(N-1)}}$$

will provide a representative estimate of the precision error to be expected from any channel that is due to the data-acquisition and reduction instrumentation.

COMMENT

This test yields an estimate of the bias and precision error that is due to propagation of the elemental errors amid M setups caused by the following:

- excitation voltage errors (differences in settings: variations)
- amplifier error (differences in gain; noise)
- A/D converter, multiplexer, and conversion errors
- computer errors (noise and roundoff)
- bridge null errors
- apparent strain error during the test
- variations in gauge factors and gauge heating

The test data do not include temporal variations of the measurands or procedural variations under loading, instrument calibration errors, temperature variation effects and electrical noise induced by operation of the loading test of the specimen, dynamic effects on the gauges, including differences in creep and fatigue, or reduction curve fit errors.

11.6   OPTICAL STRAIN MEASURING TECHNIQUES

Optical methods for experimental stress analysis can provide fundamental information concerning directions and magnitudes of the stresses in parts under design loading conditions. Optical techniques have been developed for the measurement of stress and strain fields, either in models made of materials having appropriate optical properties, or through coating techniques for existing specimens. Photoelasticity takes advantage of the changes in optical properties of certain materials that occur when these materials are strained. For example, some plastics display a change in optical properties when strained that causes an incident beam of polarized light to be split into two polarized beams that travel with different speeds and that vibrate along the principal axes of stress. Since the two light beams are out of phase, they can be made to interfere; measuring the resulting light intensity yields information concerning applied stress. To implement this method, a model is constructed of an appropriate material, or a coating is applied to an existing part.

A second optical method of stress analysis is based on the development of a moiré pattern, which is an optical effect resulting from the transmission or reflection of light from two overlaid grid patterns. The fringes that result from relative displacement of the two grid patterns can be used to measure strain; each fringe corresponds to the locus of points of equal displacement.

Recent developments in strain measurement include the use of lasers and holography to very accurately determine whole field displacements for complex geometries.

Basic Characteristics of Light

To utilize optical strain measurement techniques, we must first examine some basic characteristics of light. Electromagnetic radiation, such as light, may be thought of as a transverse wave with sinusoidally oscillating electric and magnetic field vectors that are at right angles to the direction of propagation. In general, a light source emits a series of waves containing vibrations in all perpendicular planes, as illustrated in Figure 11.18. A light wave is said to be plane polarized if the transverse oscillations of the electric field are parallel to each other at all points along the direction of propagation of the wave.

Figure 11.18 illustrates the effect of a polarizing filter on an incident light wave; the transmitted light will be plane polarized, with a known direction of polarization. Complete extinction of the light beam could be achieved by introduction of a second polarizing filter, with the axis of polarization at 90° to the first filter (labeled an

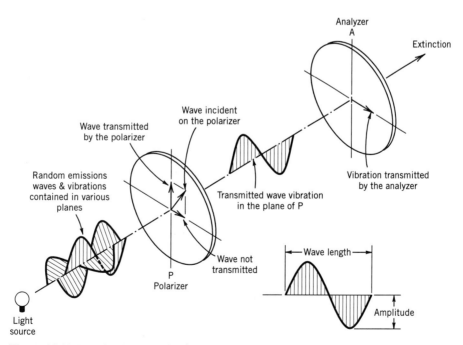

Figure 11.18 Polarization of light. (Courtesy of Measurements Group, Inc., Raleigh, NC.)

Analyzer in Figure 11.18). These behaviors of light are employed to measure direction and magnitude of strain in photoelastic materials.

Photoelastic Measurement

Photoelastic methods of stress analysis take advantage of the anisotropic optical characteristics of some materials, notably plastics, when subject to an applied load. Stress analysis may be accomplished either by constructing a model of the part to be analyzed from a material selected for its optical properties, or by coating the actual part or prototype with a photoelastic coating. If a model is constructed from a suitable plastic, the required loads for the model are significantly less than the service loads of the actual part, which reduces effort and expense in testing.

The changes in optical properties, known as artificial birefringence, that occur in certain materials subject to a load or loads was first observed by Sir David Brewster [10] in 1815. He observed that when light passes through glass that is subject to uniaxial tension, such that the stress is perpendicular to the direction of propagation of the light, the glass becomes doubly refracting, with the axes of polarization in the glass aligned with and perpendicular to the stress. Maxwell [11] and Neumann [12] first put forward the mathematical observation that the relationship between artificial birefringence and applied stress or strain is linear. These relations are known as the stress optic law.

The anisotropy that occurs in photoelastic materials results in two refracted beams of light and one reflected beam, produced for a single incident beam of appropriately polarized light. The two refracted beams propagate at different velocities through the material because of an anisotropy in the index of refraction. In an appropriately designed photoelastic (two-dimensional) model, these two refracted components of the incident light travel in the same direction and can be examined in a polariscope. The degree to which the two light waves are out of phase is related to the stress by the stress optic law:

$$\delta \propto n_x - n_y \tag{11.40}$$

where

δ = relative retardation between the two light beams
n_x, n_y = indices of refraction in the directions of the principal strains

The index of refraction changes in direct proportion to the amount the material is strained, such that

$$n_x - n_y = K(\varepsilon_x - \varepsilon_y) \tag{11.41}$$

The strain optical coefficient, K, is generally assumed to be a material property that is independent of the wavelength of the incident light. However, if the photoelastic material is strained beyond the elastic limit, this constant may become wavelength dependent, a phenomenon known as photoelastic dispersion.

Figure 11.19 shows the use of a plane polariscope to examine the strain in a photoelastic model. Plane-polarized light enters the specimen and emerges with two planes of polarization along the principal strain axes. This light beam is then passed through a polarizing filter, called the analyzer, which transmits only the component of each of the light waves that is parallel to the plane of polarization. The transmitted waves will interfere, since they are out of phase, and the resulting light intensity will be a function of the angle between the analyzer and the principal strain direction and the phase shift between the beams. The variations in strain in the specimen produce a pattern of fringes, which can be related to the strain field through the strain optic relation.

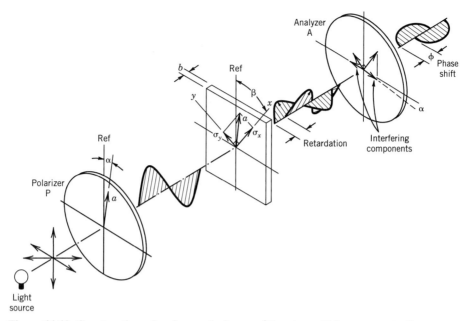

Figure 11.19 Construction of a plane polariscope. (Courtesy of Measurements Group, Inc., Raleigh, NC.)

When a photoelastic model is observed in a plane polariscope, a series of fringes is observed. The complete extinction of light occurs at locations where the principal strain directions coincide with the axes of the analyzer or where either the strain is zero or $\varepsilon_x - \varepsilon_y = 0$. These fringes are termed "isoclinics," and they are used to determine the principal strain directions at all points in the photoelastic model. Figure 11.20 shows the isoclinics in a ring subject to a compression load (as shown in the figure). A reference direction is selected along the horizontal compression load and labeled 0°. For each measurement angle, one of the principal strains at a point on an isoclinic is parallel to the specified angle and the other is perpendicular. For the 0° isoclinics, the principal strains are oriented at 0 and 90°.

Using the fact that the direction of the principal axes is known at a free surface and the fact that the shear stress is zero on a free surface, the magnitude of the stress on the boundary can be determined. The primary applications for photoelasticity, especially in a historical sense, have been in the study of stress concentrations around holes or reentrant corners. In these cases, the maximum stress is at the boundary and corresponds to one of the principal stresses. This maximum stress can be obtained directly by the optical method, since the shear stress is zero on the boundary.

Optical methods provide information about the strain and stress at every point in the object being examined, in contrast to a strain gauge, which supplies information about the strain at a single location on the object. The optical methods provide the possibility of identifying stress concentration locations and may allow for an improvement in design or guide detailed measurements with strain gauges.

Moiré Methods

A moiré pattern results from two overlaid, relatively dense patterns that are displaced relative to each other. This observable optical effect occurs, for example, in color printing, where patterns of dots form an image. If the printing is slightly out of register,

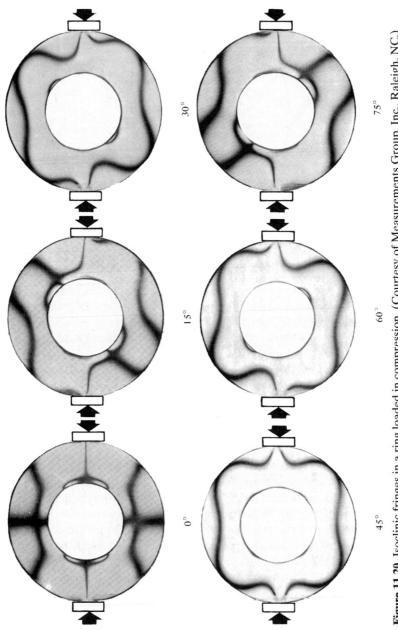

Figure 11.20 Isoclinic fringes in a ring loaded in compression. (Courtesy of Measurements Group, Inc., Raleigh, NC.)

0° 15° 30° 45° 60° 75°

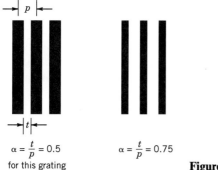

$$\alpha = \frac{t}{p} = 0.5$$
for this grating

$$\alpha = \frac{t}{p} = 0.75$$

Figure 11.21 Moiré gratings.

a moiré pattern will result. Another common example is the striking shimmering effect that occurs with some patterned clothing on television. This effect results when the size of the pattern in the fabric is essentially the same as the resolution of the television image.

In experimental mechanics, moiré patterns are used to measure surface displacements, typically in a model constructed specifically for this purpose. The technique uses two gratings, or patterns of parallel lines spaced equally apart. Figure 11.21 shows two line gratings. There are two important properties of line gratings for moiré techniques. The pitch is defined as the distance between the centers of adjacent lines in the grating, and for typical gratings has a value of from 1 to 40 lines/mm. The second characteristic of gratings is the ratio of the open, transparent area of the grating to the total area, or, for a line grating, the ratio of the distance between adjacent lines to the center-to-center distance, as illustrated in Figure 11.21. Clearly, a greater density of lines per unit width allows a greater sensitivity of strain measurement; however, as line densities increase, coherent light is required for practical measurement.

To determine strain using the moiré technique, a grating is fixed directly to the surface to be studied. This can be accomplished through photoengraving, cementing film copies of a grating to the surface, or interferometric techniques. The master or reference grating is next placed in contact with the surface, forming a reference for determining the relative displacements under loaded conditions. A series of fringes results when the gratings are displaced relative to each other; the bright fringes are the loci of points where the projected displacements of the surface are integer multiples of the pitch. The technique then is two dimensional, providing information concerning the projection of the displacements into the plane of the master grating. Once a fringe pattern is recorded, data-reduction techniques are employed to determine the stress and strain field. Graphical techniques exist that allow the strain components in two orthogonal directions to be determined. Further information on moiré techniques and additional references may be found in the review article by Sciammarella [13].

Recently, techniques such as moiré-fringe multiplication have greatly increased the sensitivity of moiré techniques, with possible reference grating frequencies of 2400 lines/mm. Moiré interferometry is an extension of moiré-fringe multiplication that uses coherent light and has sensitivities of the order of 0.4 μm/fringe [14]. A reflective grating is applied to the specimen, which experiences deformation under load conditions. The technique provides whole field readings of in-plane strain, with a four-beam optical arrangement currently in use [15].

11.7 SUMMARY

Experimental stress analysis can be accomplished through several practical techniques, including electrical resistance, photoelastic, and moiré strain measurement techniques. Each of these methods yields information concerning the surface strains for a test specimen. The design and selection of an appropriate strain measurement system begins with the choice of a measurement technique.

Optical methods are useful in the initial determination of a stress field for complex geometries, and the determination of whole-field information in model studies. Such whole-field methods provide the basis for design and establish information necessary to make detailed local strain measurements.

The bonded electrical resistance strain gauge provides a versatile means of measuring strain at a specific location on a test specimen. Strain gauge selection involves the specification of strain gauge material, the backing or carrier material, and the adhesive used to bond the strain gauge to the test specimen, as well as the total electrical resistance of the gauge. Other considerations include the orientation and pattern for a strain gauge rosette, and the temperature limit and maximum allowable elongation. In addition, for electrical resistance strain gauges appropriate arrangement of the gauges in a bridge circuit can provide temperature compensation and elimination of specific components of strain.

The techniques for strain measurement described in this chapter provide the basis for determining surface strains for a test specimen. Although the focus here has been the measurement techniques, the proper placement of strain gauges and the interpretation of the measured results requires further analysis. The inference of load-carrying capability and safety for a particular component is a result of the overall experimental program.

REFERENCES

1. Hibbeler, R. C., *Mechanics of Materials*, 3rd ed, Prentice-Hall, Upper Saddle River, NJ, 1997.
2. Timoshenko, S. P., and Goodier, J. M., *Theory of Elasticity*, Engineering Society Monographs, McGraw-Hill, New York, 1970.
3. Dally, J. W., and Riley, W. F., *Experimental Stress Analysis*, 3rd ed., McGraw-Hill, New York, 1991.
4. Thomson, W. (Lord Kelvin), On the electrodynamic qualities of metals, *Philosophical Transactions of the Royal Society (London)*, 146:649–751, 1856.
5. Micro-Measurements Division, Measurements Group, Inc., *Strain Gauge Selection: Criteria, Procedures, Recommendations*, Technical Note 505-1, Raleigh, NC, 1989.
6. Micro-Measurements Division, Measurements Group, Inc., *Strain Gauge Technical Data*, Catalog 500, Part B, and TN-509, *Errors Due to Transverse Sensitivity in Strain Gauges*, Raleigh, NC, 1982.
7. Kulite Semiconductor Products, Inc, Bulletin KSG-5E, *Semiconductor Strain Gauges*, Leonia, NJ.
8. Weymouth, L. J., Starr, J. E., and Dorsey, J., Bonded resistance strain gauges, *Experimental Mechanics*, 6(4):19A, 1966.
9. Oi, K., Transient response of bonded strain gauges, *Experimental Mechanics*, 6(9):463, 1966.
10. Brewster, D., On the effects of simple pressure in producing that species of crystallization which forms two oppositely polarized images and exhibits the complementary colours by polarized light, *Philosophical Transactions A (GB)*, 105:60, 1815.
11. Maxwell, J. C., On the equilibrium of elastic solids, *Transactions of the Royal Society*, 20(Part I):87, 1853.
12. Neumann, F. E., Uber Gesetze der Doppelbrechung des Lichtes in comprimierten oder ungleichformig Erwamten unkrystallischen Korpern, *Abh. Akad. Wiss. Berlin*, Part II: 1, 1841 (in German).
13. Sciammarella, C. A., The Moiré method—A review, *Experimental Mechanics*, 22(11):418, 1982.
14. Post, D., Moiré interferometry at VPI & SU, *Experimental Mechanics*, 23(2):203, 1983.
15. Post, D., Moiré interferometry for deformation and strain studies, *Optical Engineering*, 24(4):663, 1985.

NOMENCLATURE

b	width $[l]$	δR	resistance change $[\Omega]$
c	speed of sound $[l\,t]$	M	bending moment $[m\,l^2\,t^{-2}]$
h	height $[l]$	R	electrical resistance $[\Omega]$
n_i	index of refraction in the direction of the principal strain in the i direction	T	temperature $[°]$
		T_0	reference temperature $[°]$
$(u_d)_x$	design-stage uncertainty in the variable x	α	Moiré grating width to spacing parameter
A_c	cross-sectional area $[l^2]$	γ_{xy}	shear strain in the xy plane $[m\,l^{-1}\,t^{-2}]$
D	diameter $[l]$	δ	relative retardation between two light beams in a photoelastic material
E_m	modulus of elasticity $[m\,l^{-1}\,t^{-2}]$		
E_i	input voltage $[V]$	ε_a	axial strain
E_o	output voltage $[V]$	ε_i	strain in the i coordinate direction (i.e., x direction)
e_l	strain gauge lateral sensitivity error as a percentage of axial strain		
		ε_l	lateral or transverse strain
F_N	force normal to A_c $[m\,l\,t^{-2}]$	κ	bridge constant
G	shear modulus $[m\,t^{-2}]$	ν_p	Poisson ratio
GF	gauge factor	ν_{po}	Poisson ratio for gauge factor calibration test specimen
K	strain optical coefficient		
K_B	bridge sensitivity $[V]$	π_1	piezoresistance coefficient $[t^2\,l\,m]$
K_l	strain gauge lateral sensitivity	ρ_e	electrical resistivity $[\Omega\,l]$
L	length $[l]$	σ	stress $[m\,l^{-1}\,t^{-2}]$
δE_o	voltage change $[V]$	σ_a	axial stress $[m\,l^{-1}\,t^{-2}]$
δL	change in length $[l]$	τ_{xy}	shear stress in the xy plane $[m\,l^{-1}\,t^{-2}]$

PROBLEMS

11.1 Calculate the change in length of a steel rod ($E_m = 30 \times 10^6$ psi) having a circular cross section, a length of 10 in., and a diameter of 1/4 in. The rod supports a weight of 40 lb$_m$ in such a way that a state of uniaxial tension is created in the rod.

11.2 Calculate the change in length of a steel rod ($E_m = 20 \times 10^{10}$ Pa) that has a length of 0.3 m and a diameter of 5 mm. The rod supports a mass of 50 kg in a standard gravitation field in such a way that a state of uniaxial tension is created in the rod.

11.3 An electrical coil is made by winding copper wire around a core. What is the resistance of 20,000 turns of 16-gauge wire (0.051 in. diameter) at an average radius of 2.0 in.?

11.4 Compare the resistance of a volume of $\pi \times 10^{-5}$ m^3 of aluminum wire having a diameter of 2 mm, with the same volume of aluminum formed into 1-mm-diameter wire. (The resistivity of aluminum is 2.66×10^{-8} Ω m.)

11.5 A conductor made of nickel ($\rho_c = 6.8 \times 10^{-8}$ Ω m) has a rectangular cross section 5×2 mm and is 5 m long. Determine the total resistance of this conductor. Calculate the diameter of a 5-m-long copper wire having a circular cross section that yields the same total resistance.

11.6 Consider a Wheatstone bridge circuit having all resistances equal to 100 Ω. The resistance R_1 is a strain gauge that cannot sustain a power dissipation of more than 0.25 W. What is the maximum applied voltage that can be used for this bridge circuit? At this level of bridge excitation, what is the bridge sensitivity?

11.7 A resistance strain gauge with $R = 120\,\Omega$ and a gauge factor of 2 is placed in an equal-arm Wheatstone bridge in which all the resistances are equal to $120\,\Omega$. If the maximum gauge current is to be 0.05 A, what is the maximum allowable bridge excitation voltage?

11.8 A strain gauge having a nominal resistance of 350 Ω and a gauge factor of 1.8 is mounted in an equal-arm bridge, which is balanced at a zero applied strain condition. The gauge is mounted on a 1-cm^2 aluminum rod, having $E_m = 70$ GPa. The gauge senses axial strain. The bridge output is 1 mV for a bridge input of 5 V. What is the applied load, assuming the rod is in uniaxial tension?

11.9 Consider a structural member subject to loads that produce both axial and bending stresses, as shown in Figure 11.12. Two strain gauges are to be mounted on the member and connected in a Wheatstone bridge in such a way that the bridge output indicates the axial component of strain only (the installation is bending compensated). Show that the installation of the gauges shown in Figure 11.12 will not be sensitive to bending.

11.10 A steel beam member ($\nu_p = 0.3$) is subjected to simple axial tensile loading. One strain gauge aligned with the axial load is mounted on the top and center of the beam. A second gauge is similarly mounted on the bottom of the beam. If the gauges are connected as arms 1 and 4 in a Wheatstone bridge (Figure 11.11), determine the bridge constant for this installation. Is the measurement system temperature compensated? (Clearly explain why or why not). If $\delta E_o = 10\,\mu V$ and $E_i = 10$ V, determine the axial and transverse strains. The gauge factor for each gauge is 2, and all resistances are initially equal to 120 Ω.

11.11 An axial strain gauge and a transverse strain gauge are mounted to the top surface of a steel beam that experiences a uniaxial stress of 2222 psi. The gauges are connected to arms 1 and 2 (Figure 11.11) of a Wheatstone bridge. With a purely axial load applied, determine the bridge constant for the measurement system. If $\delta E_o = 250\,\mu V$ and $E_i = 10$ V, estimate the average gauge factor of the strain gauges. For this material Poisson's ratio is 0.3 and the modulus of elasticity is 29.4×10^6 psi.

11.12 A strain gauge is mounted on a steel cantilever beam of rectangular cross section. The gauge is connected in a Wheatstone bridge; initially $R_{gauge} = R_2 = R_3 = R_4 = 120\,\Omega$. A gauge resistance change of 0.1 Ω is measured for the loading condition and gauge orientation shown in Figure 11.22. If the gauge factor is $2.05 \pm 1\%$ (95%) estimate the strain. Suppose the uncertainty in each resistor value is 1% (95%). Estimate an uncertainty in the measured strain caused by the uncertainties in the bridge resistances and gauge factor. Assume that the bridge operates in a null mode, which is detected by a galvanometer. Also assume reasonable values for other necessary uncertainties and parameters, such as input voltage or galvanometer sensitivity.

11.13 Two strain gauges are mounted so that they sense axial strain on a steel member in uniaxial tension. The 120-Ω gauges form two legs of a Wheatstone bridge and are mounted on opposite arms. For a bridge excitation voltage of 4 V and a bridge output voltage of 120 μV under load, estimate the strain in the member. What is the resistance change experienced by each gauge? The gauge factor for each of the strain gauges is 2 and E_m for steel is 29×10^6 psi.

11.14 A rectangular bar is instrumented with strain gauges and subjected to a state of uniaxial tension. The bar has a cross-sectional area of 2 in.^{2}, and the bar is 12 in. long. The two strain gauges are mounted such that one senses the axial strain, while the other senses the lateral strain. For an axial load of 1500 lb, the axial strain is measured as 1500 $\mu\varepsilon$ (μin./in.), and the lateral gauge indicates a strain of $-465\,\mu\varepsilon$. Determine the modulus of elasticity and Poisson's ratio for this material.

11.15 A round member having a cross-sectional area of 3 cm^2 experiences an axial load of 10 kN. Two strain gauges are mounted on the member, one measuring an axial strain of 600 $\mu\varepsilon$ (μin./in.), and the other measuring a lateral strain of $-163\,\mu\varepsilon$. Determine the modulus of elasticity and Poisson's ratio for this material.

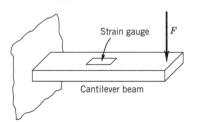

Figure 11.22 Loading for Problem 11.12.

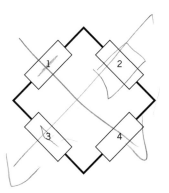

Figure 11.23 Bridge arrangement for Problem 11.19.

11.16 Show that the use of a dummy gauge together with a single active gauge compensates for temperature but not for bending. Consider the case in which the active gauge is subjected to axial loading with minimal bending and both gauges experience the same temperature.

11.17 Show that four strain gauges mounted to a shaft such that the gauge pairs measure equal and opposite strain can be used to measure torsional twist (as suggested in Table 11.1). Show that this method compensates for axial and bending strains and for temperature.

11.18 A galvanometer having an internal impedance of $100\,\Omega \pm 0.5\%$ and a 1-μA resolution is used to indicate strain in a strain gauge Wheatstone bridge circuit arrangement. A single gauge, R_1, is aligned on a steel member in uniaxial tension, so as to sense axial strain. The bridge is composed of two fixed resistors ($R_2 = R_3 = 120\,\Omega \pm 1\%$) and a null adjust potentiometer (R_p known to $\pm 1\%$). With no applied load R_p must be set to 120.07 Ω to balance the bridge. Under load the galvanometer indicates a current flow of 2 μA. Estimate the strain and the uncertainty in this measured value. The gauge factor is 2 and the input voltage is 4 V.

11.19 For each of the bridge configurations shown in Figure 11.23, determine the bridge constant and show reasoning. Assume that all of the active gauges are identical, and that all of the fixed resistances are equal.

Bridge Arrangement	Description
(a) 1: Active gauge 2–4: Fixed resistors	Single gauge in uniaxial tension.
(b) 1: Active gauge 2: Poisson gauge 3, 4: Fixed resistors	Two active gauges in uniaxial stress field. Gauge 1 aligned with maximum axial stress: gauge 2 lateral.
(c) 1: Active gauge 3: Active gauge 2, 4: Fixed resistors	Equal and opposite strains applied to the active gauges (bending compensation).
(d) Four active gauges	Gauges 1 and 4 aligned with uniaxial stress; gauges 2 and 3 transverse.
(e) Four active gauges	Gauges 1 and 2 subject to equal and opposite strains; gauges 3 and 4 subject to the same equal but opposite strains.

11.20 A bathroom scale uses four internal strain gauges to measure the displacement of its diaphragm as a means for determining load. Four active gauges are used in a bridge circuit. The gauge factor is 2.0, and gauge resistance is 120 Ω for each gauge. If an applied load to the diaphragm causes a compression strain on R_1 and R_4 of 20 $\mu\varepsilon$ while gauges R_2 and R_3 experience a tensile strain of 20 $\mu\varepsilon$, estimate the bridge deflection voltage. The supply voltage is 9 V. Refer to Figure 11.23 for gauge positions.

11.21 Suppose in Problem 11.20 that the lead wires of two gauges (R_2 and R_4) are accidentally interchanged on the assembly line such that a compression strain is now sensed on R_1 and R_2 of 20 $\mu\varepsilon$ while gauges R_4 and R_3 experience a tensile strain of 20 $\mu\varepsilon$. Should this really matter? Estimate the bridge deflection voltage. Refer to Figure 11.23 for gauge positions.

Consider measuring strains associated with a thin-walled pressure vessel for Problems 11.22–11.24. The pressure vessel is constructed of steel and has a circular cross section. The tangential and longitudinal stresses in the wall are

$$\sigma_t = pD/2t \qquad \sigma_l = pD/4t$$

where

σ_t = tangential stress
σ_l = longitudinal stress
p = pressure
t = wall thickness

11.22 A single-strain gauge is mounted on the surface of a thin-walled pressure vessel having a diameter of 1 m. For a strain gauge oriented along the tangential stress direction, determine the percentage error in tangential strain measurement caused by lateral sensitivity as a function of pressure and vessel wall thickness. The lateral sensitivity is 0.03.

11.23 Consider a Wheatstone bridge circuit having all fixed resistances equal to 100 Ω and with a strain gauge located at the R_1 position, which has a value of 100 Ω under conditions of zero strain. The strain gauge is mounted so as to sense longitudinal strain for a thin-walled pressure vessel made of steel having a wall thickness of 2 cm and a diameter of 2 m. The strain gauge has a gauge factor of 2 and cannot sustain a power dissipation of more than 0.25 W. What is the maximum static sensitivity that can be achieved with this proposed measurement system? Is this static sensitivity constant with input pressure? If not, under what conditions would it be reasonable to assume a constant static sensitivity? The static sensitivity should be expressed in units of V/kPa.

11.24 Design a Wheatstone bridge measurement system to measure the tangential strain in the wall of the pressure vessel just described, and develop a reasonable estimate of the resulting uncertainty. You may assume that the strain does not vary with time and that the bridge will be operated in a balanced mode. Your selection of specified values for the input voltage, the fixed resistances in the bridge and the galvanometer sensitivity, and their associated uncertainties will allow completion of this design.

11.25 A steel cantilever beam is fixed at one end and free to move at the other. A load F of 980 N is applied to the free end. Four axially aligned strain gauges (GF = 2) are mounted to the beam a distance L from the applied load, two on the upper surface, R_1 and R_4, and two on the lower surface, R_3 and R_2. The bridge deflection output is passed through an amplifier (Gain, $G_A = 1000$) and measured. For a cantilever, the relation between applied load and strain is

$$F = \frac{2E_m I \varepsilon}{Lt}$$

where I is the beam moment of inertia ($= bt^3/12$), t is the beam thickness, and b is the beam width. Estimate the measured output for the applied load if $L = 0.1$ m, $b = 0.03$ m, $t = 0.01$ m, and the bridge excitation voltage is 5 V. $E_m = 200$ GPa.

11.26 A cantilever beam is to be used as a scale. The beam, made of 2024-T4 aluminum, is 21 cm long, 0.4 cm thick, and 2 cm wide. The scale load of between 0 and 200 g is to be concentrated at a point along the beam centerline 20 cm from its fixed end. Strain gauges are to be used to measure beam deflection and mounted to a Wheatstone bridge to provide an electrical signal that is proportional to load. Either 1/4-, 1/2-, or full-bridge gauge arrangements can be used. Design the sensor arrangement and its location on the beam. Specify appropriate signal conditioning to excite the bridge and to measure its output on a data-acquisition system using a ± 5 V, 12-bit A/D converter. Will the system achieve a 4% uncertainty at the design stage? Specifications and system choices as follows are stated at 95% confidence levels.

Sensors: one, two, or four, axial gauges at 120 $\Omega \pm 0.2\%$ (selectable). Gauge factor of
    $2.0 \pm 1\%$
Bridge excitation: 1, 3, 5 V $\pm 0.5\%$ (selectable)
Bridge null: within 5 μV
Signal conditioning:
    Amplifier gain: 1×, 10×, 100×, 1000 × $\pm 1\%$ (selectable)
    Low-pass filter: f_c at 0.5, 5, 50, 500 Hz @ -12 dB/octave (selectable)
Data-acquisition system:
    Conversion errors: <0.1% reading
    Sample rate: 1, 10, 100, 1000 Hz (selectable)

Chapter 12

Metrology, Motion, Force, and Power Measurements

12.1 INTRODUCTION

In this chapter, an introduction to common techniques for dimensional measurements, and the measurement of mass, force, torque, and power will be presented. Measurement methods for displacement, linear and angular velocity, and vibration and acceleration will be described. The measurement of vibration or shock will be discussed in the context of a specific instrument design based on a second-order dynamic system, the seismic instrument.

Forces ranging over many orders of magnitude are measured in situations ranging from the laboratory to a busy highway. Techniques for force measurement will be discussed, which are used in the primary ranges of interest for engineering applications. Mechanical power is largely transmitted through rotating shafts. The measurement of shaft power and torque will be presented, primarily as applied in dynamometers.

12.2 DIMENSIONAL MEASUREMENTS: METROLOGY

The term "metrology" implies the science of weights and measures, referring primarily to the measurements of lengths, angles, and weight, but including other measurements essential to engineering practice such as the establishment of a flat, plane reference surface. Length is one of the most common specifications and measurements. Measurements of length may span a range extending from the distance to the Great Nebula in Andromeda, which is 2×10^{22} m, to the radius of a hydrogen atom, 5×10^{-11} m. The tremendously difficult task of making accurate measurements on these scales requires highly specialized measurement techniques and systems that are not applicable to most engineering measurements. We limit the scope of the present discussion to common dimensional measurements, such as would be necessary in design and manufacturing.

Historical Perspective

Several concepts based on accurate dimensional measurement that are now generally accepted as commonplace are actually quite recent. We certainly expect to be able to purchase standard threaded bolts and nuts that are completely interchangeable, and even to be able to purchase precision parts for automobiles, such as pistons and camshafts, which are finished to such exacting standards that they are completely interchangeable. In fact, mass production requires the ability to accurately measure specified dimensions. The techniques and instruments for accurate dimensional measurements have not always existed in manufacturing; they could only be achieved through the adoption of common standards for length and the development of sufficiently accurate measurement techniques and instruments.

The need for accurately reproducible weights and measures has been recognized since ancient times. For example, the ancient Egyptians maintained a primary standard measuring stick that defined the unit known as the cubit. More modern development of accurate measuring instruments and machine tools in American industry can be traced to the early 19th century. The Brown and Sharpe company was founded in 1833, and by 1851 it was manufacturing steel rules, scales, and vernier calipers [1]. Industrial practice in this period was based upon tailoring individual parts to fit each item produced. The firearms industry was the first to attempt to produce interchangeable parts, demonstrating completely interchangeable parts for rifles around 1850. The desire to manufacture parts to specified tolerances, and thus eliminate manually fitting parts to a specific item, prompted the development of standards for dimensional measurements. These standards were first established within a particular company, but it soon became apparent that universal standards were needed.

The gauge block is the major innovation that allowed a dimensional standard to exist in a practical sense on every shop floor. Gauge blocks were developed in the 1890s in Sweden by Carl Edvard Johansson (1864–1934). He developed the concept of using precisely dimensioned blocks in combination to produce a wide range of dimensions. His original concept entailed 102 blocks that would allow measurement of 20,000 distinct dimensions through combinations of these gauge blocks.

Principles of Linear Measurement

Accurate measurements of length can be accomplished only through comparison with a standard. Local standards for the measurement of length, such as line standards, end standards, and gauge blocks, are essential to manufacturing. Two marks on a dimensionally stable material define a *line standard*. The lines are produced by a diamond tool to a width as small as 0.002 mm. The original SI[1] length standard was such a line standard. The distance between marks is established by using interferometric methods. The length of an *end standard* is the distance between its flat, parallel end faces.

Gauge blocks are the most often used length standard for machining processes, and they are a necessary tool for dimensional quality control in the manufacture of interchangeable parts. Figure 12.1 shows a set of gauge blocks. The process whereby a length standard is established involves a unique property of highly polished, extremely flat surfaces on which there is a thin lubricant film. When such surfaces are slid together with a slight contact pressure, the surfaces adhere with a significant force; this procedure for combining gauge blocks is called wringing. By combining gauge blocks, a range of lengths can be produced; for example, a set of 81 blocks can yield a range of lengths from 0.100 to 12.000 in. in increments of 0.001 in. Thus, a set of gauge blocks forms a useful and accurate local standard for length measurement. Table 12.1 lists the federally established accuracy standards and designations for gauge blocks.

---

[1] The acronym SI is derived from the French *Le Systéme International d'Unités.*

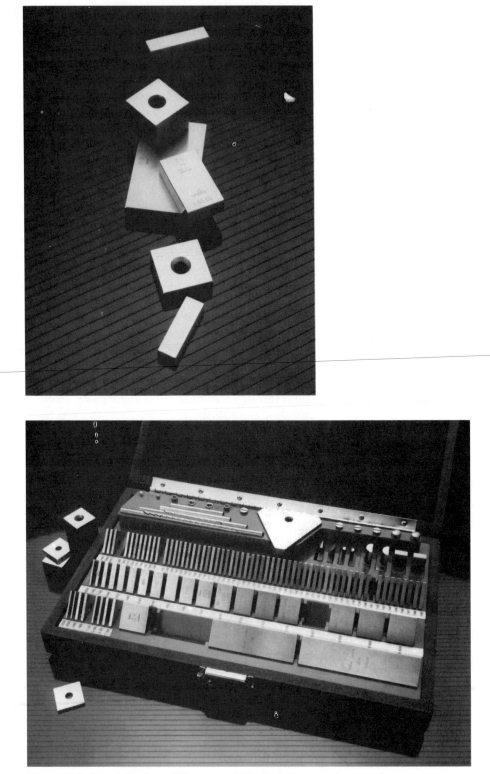

Figure 12.1 Gauge block set. (Courtesy of The L. S. Starrett Company.)

Table 12.1 US Accuracy Grades for Gauge Blocks

US Accuracy Grade	Accuracy [in.]	Accuracy [mm]
0.5	0.000001	0.00003
1	0.000002	0.00005
2	+0.000004	+0.0001
	−0.000002	−0.00005
3	+0.000006	+0.00015
	−0.000002	−0.00005

Hand measuring tools form the basis for sufficiently accurate measurements of length for most engineering applications. The most common instruments for the measurement of length are undoubtedly the ruler and measuring tape. A measuring tape can be used to measure distances of the order of 100 ft with total uncertainties as low as 0.05%.

The sources of error in a ruler or tape include resolution errors and bias errors caused by thermal expansion; in addition, error in a measurement using a tape may arise from improper tension in the tape. Typically, bias errors in the scale markings are negligible compared to the resolution errors in reading the scale.

EXAMPLE 12.1

A steel ruler 1 ft long (30.48 cm) is used to measure a length of 8 in. (20.32 cm). The ruler is graduated in 1/16 (0.159 cm) in. increments and was manufactured in a controlled temperature environment of 77°F (25°C). At this reference temperature, the ruler has negligible bias error. Determine the design-stage uncertainty in a reading of 8 in. if the measurement is taken in a 100°F (37.8°C) environment, but not corrected for thermal expansion.

KNOWN

A ruler referenced to 77°F is used in a 100°F environment. The coefficient of thermal expansion of the steel is 9.9×10^{-6} in./in. °F.

FIND

The uncertainty in the measured length.

SOLUTION

The uncertainty analysis will include two contributions: a contribution from thermal expansion, and a contribution from interpolation errors. The interpolation uncertainty, u_o, is estimated in the usual manner as ±1/2 the least division, which in this case is 1/32 in. The instrument uncertainty, u_c, contains only one elemental error and is found from the equation

$$\Delta L = (L)(C_\alpha)(\Delta T) \tag{12.1}$$

where

ΔL = change in length caused by thermal expansion
L = length

C_α = coefficient of thermal expansion
ΔT = change in temperature

This relationship yields

$$\Delta L = (8\ \text{in.})(9.9 \times 10^{-6}\ \text{in./in.} \cdot {}^\circ\text{F})(100 - 77^\circ\text{F}) = 1.82 \times 10^{-3}\ \text{in.}\ (0.046\ \text{mm})$$

The uncertainty contributions may be combined as

$$u_d = \sqrt{(u_0)^2 + (u_c)^2} = \sqrt{(1/32)^2 + (1.82 \times 10^{-3})^2} = 0.0313\ \text{in.}$$

which yields a design-stage uncertainty of 0.031 in. (0.787 mm).

COMMENTS

The uncertainty of 0.031 in. results almost entirely from the interpolation error of the ruler. Thermal expansion associated with a temperature change of 23°F is negligible for this application. However, the error is approximately 0.002 in. (0.051 mm) over the 8-in. measured distance. In some applications, such an error would be totally unacceptable. Environmental controls for temperature are a common practice in machine tool manufacturing and precision machining facilities.

Other measurement tools are designed for specific kinds of measurements, depending on the required accuracy and the specific geometry being measured. Vernier calipers, as shown in Figure 12.2, allow precise measurement of both inside and outside dimensions. Consider the construction of a Vernier scale, named for the French mathematician Pierre Vernier (1580–1673), who invented this method for increased resolution in measurements in the early 17th century. The basic principle of a Vernier scale is illustrated in Figure 12.3. The upper stationary scale is divided into equal lengths of one unit each. Suppose it is desired to further resolve the measured distance to the nearest 1/4 unit. The movable Vernier scale would be graduated in four equal divisions of length 3/4 of a unit. As the Vernier scale moves to the right, say between 2 and 3, at 2 1/4 the 1/4 mark would exactly line up with the number 3 above it. Reading the Vernier scale is then accomplished by locating the line on

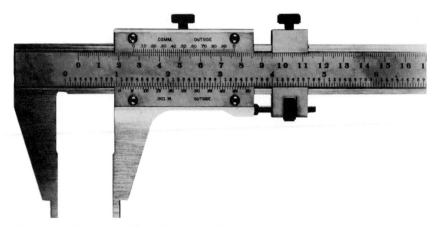

Figure 12.2 Vernier caliper. (Courtesy of The L. S. Starrett Company.)

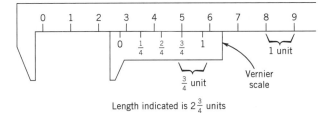

Length indicated is $2\frac{3}{4}$ units

Figure 12.3 Operating principle for a Vernier scale.

the Vernier that is most closely aligned with a line on the main scale. In this case the reading is 2 3/4. Figure 12.4 shows a Vernier caliper and its associated reading.

A micrometer is commonly used for accurate measurements of a part that is small enough to be located between the frame and the spindle of the micrometer, as shown in Figure 12.5. The micrometer spindle moves by virtue of a precise screw thread, typically 40 threads/in. for U.S. customary units system micrometers. Equipped with a Vernier scale, such a micrometer has a resolution of 0.0001 in. Further information on errors in micrometer measurements and calibration techniques may be found in [2].

Dial indicators measure distance of travel of a spindle; the construction of a typical dial indicator is shown in Figure 12.6. Dial indicators are most useful in setup work for machining, such as alignment of machine components, and in inspection of uniformly dimensioned parts. Dial indicators can be used as depth gauges and hole gauges and may be employed in a variety of caliper gauges. The principle employed in a dial indicator provides the means of output for a variety of specialized measurement devices.

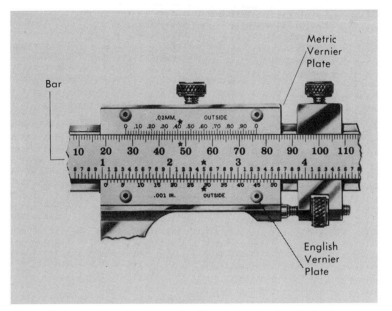

Figure 12.4 Reading a Vernier scale. The metric reading is 27.42 mm. (Courtesy of The L. S. Starrett Company.)

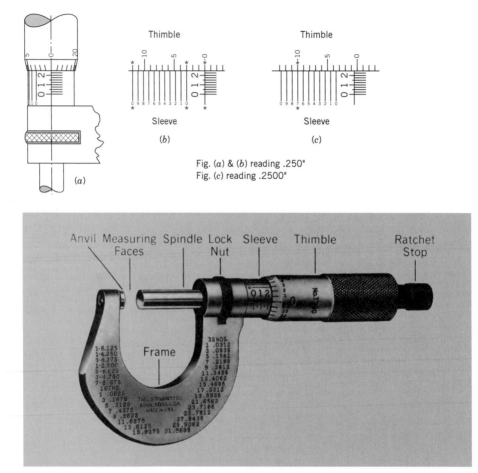

Fig. (*a*) & (*b*) reading .250"
Fig. (*c*) reading .2500"

Figure 12.5 Micrometer construction. (Courtesy of The L. S. Starrett Company.)

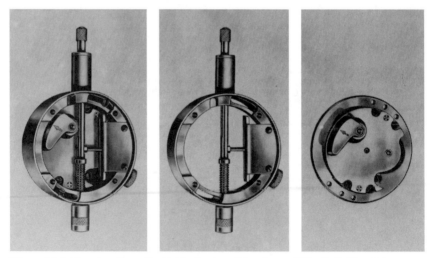

Figure 12.6 Construction of a dial indicator. (Courtesy of The L. S. Starrett Company.)

Optical Methods

The primary standard for length is currently based on the standard for time, such that 1 m is the length of the path traveled by light in a vacuum in 1/299,792,458 of a second. However, until 1982 the standard for length was based on a specific number of wavelengths of light having a particular wavelength, specifically 1 m equaled 1,650,763.73 wavelengths of the orange-red light of a krypton-86 lamp.

One of the most accurate means of measuring length currently is through the use of an interferometer [3]. The history of the interferometer includes the progress of the standard for length measurement from a single prototype line standard to a universal light wave standard. The meter was originally defined by a platinum-iridium bar, called the International Prototype Meter, from late in the 18th century to well into the 20th century. The physicist Albert A. Michelson (1852–1931) designed an interferometer for length comparisons between the primary standard meter bar and copies, which were to be used as laboratory standards at locations remote from the primary standard. The discovery that these metal bars changed in length over time led to the adoption of a light wave standard for length in 1960. In France in 1799 the definition of the meter based on terrestrial distances was accurate to one part in 100,000. The next standard, the prototype platinum-iridium bar, was accurate to 1 part in 10^6, and current standards are accurate to approximately 1 part in 10^8.

Figure 12.7 shows a schematic diagram of an interferometer as first employed by Michelson. A light ray from the source is divided into two beams, one of which is incident upon mirror M_A and the other upon mirror M_B. These waves form an interference pattern that is visible at the location of the observer. As the mirror M_B is moved, interference fringes will pass the field of view; the number of interference fringes passing the field of view is a measure of the distance that M_B has moved. Thus, it is important to note that the interferometer measures the change in distance between the two mirrors, not the absolute distance to the movable mirror. Length measurements made in this manner are extremely accurate, since a large number of

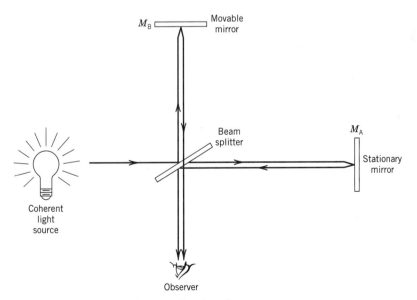

Figure 12.7 Schematic diagram of an interferometer.

fringes are counted. Each fringe corresponds to a movement of the mirror of one-half of a wavelength. The laser has made it possible to use an interferometer to measure significant distances with resolution into the μin. range [4].

Optical means for length measurement also include measuring microscopes, which allow the measurement of small dimensions, typically less than 2 mm, by direct comparison with a scale that is visible through the reticle. A length measurement may also be accomplished by moving a telescope with a very precise, fine-thread screw that is calibrated to yield length measurements. A hairline in the telescope optics allows greater resolution in length measurements than can be accomplished by using some other techniques, and it allows for remote measurement. Such a device is called a cathetometer. Cathetometers consist of a vertical post mounted on a tripod that may be adjusted so that the post is oriented vertically. A telescope is mounted on a rack and pinion, such that the telescope can be moved vertically along the post. Using a scale on the post, and a Vernier scale that moves with the telescope, the vertical position of the telescope may be determined. Accuracies to ±0.05 mm or ±0.001 in. are possible. One application of a cathetometer is to remotely read a liquid level in a manometer.

12.3   DISPLACEMENT MEASUREMENTS

The transducers and physical principles used to measure displacement are highly dependent upon the particular application; therefore, only the most common methods will be discussed. The measurement of displacement involves the determination of the relative motion of two points, of which one is usually fixed. The special subject of the measurement of the small displacements that occur in engineering materials under load was described in Chapter 11.

Potentiometers

A potentiometer[2] or variable electrical resistance transducer is depicted in Figure 12.8. The transducer is composed of a sliding contact and a winding. The winding is made of many turns of wire, wrapped around a nonconducting substrate. Output signals from such a device can be realized by imposing a known voltage across the total resistance of the winding and by measuring the output voltage, which is proportional to the fraction of the distance the contact point has moved along the winding. Potentiometers can also be configured in a rotary form, with numerous total revolutions of the contact possible in a helical arrangement. The output from the sliding contact as it moves along the winding is actually discrete, as illustrated in Figure 12.8; the resolution is limited by the number of turns per unit distance. The loading errors associated with voltage-dividing circuits, discussed in Chapter 6, should be considered in choosing a measuring device for the output voltage.

Linear Variable Differential Transformers

The LVDT, as shown in Figure 12.9, produces an ac output with an amplitude that is proportional to the displacement of a movable core. The waveform of the output from the LVDT is sinusoidal; the amplitude of the sine wave is proportional to

---

[2]The potentiometer–transducer should not be confused with the potentiometer–instrument. Although both are based on voltage-divider principles, the latter measures emf as is described in Chapter 6.

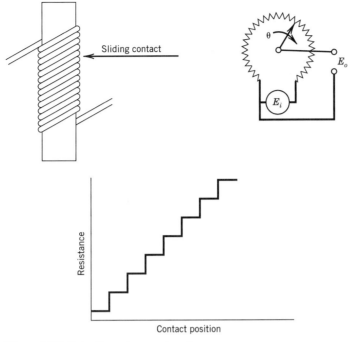

Figure 12.8 Potentiometer construction.

the displacement of the core for a limited range of core motion. The movement of the core causes a mutual inductance in the secondary coils for an ac voltage applied to the primary coil. In 1831, the English physicist Michael Faraday (1791–1867) demonstrated that a current could be induced in a conductor by a changing magnetic field. An interesting account of the development of the transformer may be found in [5].

Recall that for two coils in close proximity, a change in the current in one coil will induce an emf in the second coil according to Faraday's law. The application of this inductance principle to the measurement of distance begins by applying an ac voltage to the primary coil of the LVDT. The two secondary coils are connected in a series circuit, such that when the iron core is centered between the two secondary coils the output voltage amplitude is zero (see Figure 12.9). Motion of the magnetic core changes the mutual inductance of the coils, which causes a different emf to be induced in each of the two secondary coils. Over the range of operation, the output amplitude is essentially linear with core displacement, as first noted in a U.S. patent by G. B. Hoadley in 1940 [6].

The output of a differential transformer is illustrated in Figure 12.10. Over a specific range of core motion the output is essentially linear. Beyond this linear range, the output amplitude will rise in a nonlinear manner to a maximum, and eventually fall to zero. The output voltages on either side of the zero displacement position are 180° out of phase. With appropriate circuitry it is possible to determine positive or negative displacement of the core. However, note that because of harmonic distortion in the supply voltage and because the two secondary coils are not identical, the output voltage with the coil centered is not zero but instead reaches a minimum. Resolution of an LVDT strongly depends on the resolution of the measurement system used to determine its output. Resolutions in the microinch range can be accomplished.

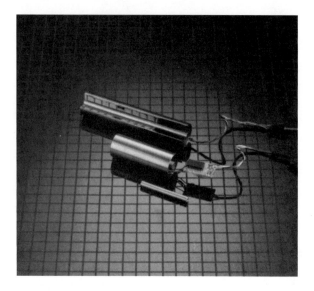

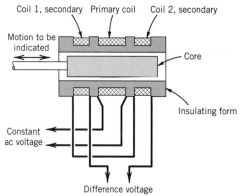

Figure 12.9 Construction of an LVDT.
(Courtesy of Schaevitz Engineering; from [6].)

The differential voltage output of an LVDT, as shown in Figure 12.10, may be analyzed by assuming that the magnetic field strengths are uniform along the axis of the coils, neglecting end effects, and limiting the analysis to the case in which the core does not move beyond the ends of the coils [6]. Under these conditions the differential voltage may be expressed in terms of the core displacement. The sensitivity of the LVDT in the linear range is a function of the number of turns in the primary and secondary coils, the root-mean-square (rms) current in the primary coil,

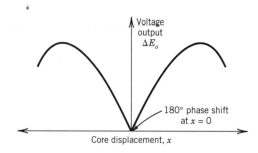

Figure 12.10 LVDT output as a function of core position.

and the physical size of the LVDT. In practical applications a range of core motion over which nonlinearity is negligible is specified as the operating range.

Excitation Voltage and Frequency

The dynamic response of an LVDT is directly related to the frequency of the applied ac voltage, since the output voltage of the secondary coil is induced by the variation of the magnetic field induced by the primary coil. For this reason the excitation voltage should have a frequency at least 10 times the maximum frequency in the measured input. An LVDT can be designed to operate with input frequencies ranging from 60 Hz up to 25 kHz (for specialized applications, frequencies in the megahertz range can be used). A constant current source is preferable for an LVDT, to limit temperature effects.

The appropriate means of measuring and recording the output signal from an LVDT and the ac frequency applied to the primary coil should be chosen based on the highest frequencies present in the input signal to the LVDT. For example, for static measurements and signals having frequency contents much lower than the excitation frequency of the primary coil, an ac voltmeter may be an appropriate choice for measuring the output signal. In this case, it is likely that the frequency response of the measuring system would be limited by the averaging effects of the ac voltmeter. For higher-frequency signals, the output signal can be sampled at a sufficiently high frequency by using a computer data-acquisition system to allow signal processing for a variety of purposes, including quality assurance or control.

The measurement of distance by using an LVDT is accomplished with an assembly known as an LVDT gauge head. Such devices are widely used in machine tools and various types of gauging equipment. Control applications will have similar transducer designs. The basic construction is shown in Figure 12.11, which can yield instrument errors as low as 0.05% and repeatability of 0.0001 mm.

Angular displacement can also be measured by using inductance techniques employing a rotary variable differential transformer (RVDT). The output curve of an RVDT and a typical construction are shown in Figure 12.12, where the linear output range is approximately $\pm 40°$.

12.4   MEASUREMENT OF MASS

Newton's second law states that the fundamental quantity force is proportional to the product of the mass and acceleration. In the U.S. customary and SI systems of units, the defined quantities are as follows.

Mass	lb_m	kg
Length	foot	m
Time	second	s
Force	lb (or lb_f)	

In the U.S. system, the units of force and mass are related by a definition: 1 lb_m exerts a force of 1 lb in a standard gravitational field. Newton's second law, based on this definition, is

$$1\ lb = \frac{(1\ lb_m)(32.174\ ft/s^2)}{g_c} \tag{12.2}$$

Thus, g_c is seen to have a value of 32.174 ft lb_m/lb s^2. However, in the SI system

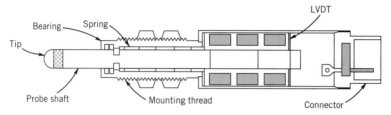

(a) Cross-section of typical LVDT gauge head.

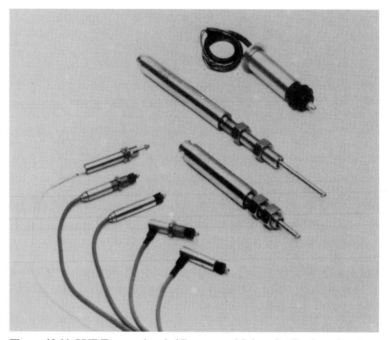

Figure 12.11 LVDT gauge head. (Courtesy of Schaevitz Engineering; from [6].)

g_c is defined as unity (1 kg m/s^2 N), and force, the newton, becomes a derived unit. In the U.S. customary system the unit of force is the defined quantity, and g_c must be derived from Newton's second law.

In the mid-17th century, Newton postulated the law of universal gravitation. This statement of the proportionality between mass and gravitational force allows the measurement of mass by comparison with the known gravitational force on a standard mass (or through the direct measurement of gravitational force under a known gravitational acceleration). The instrument for determining an unknown mass by comparison with a standard mass is known as an analytical balance. This null balance device compares the moment that results from the force exerted by a known mass in a gravitational field with the moment created by an unknown mass. Such a balance is shown in Figure 12.13.

To achieve accurate comparisons, such a balance must be properly designed and utilized. The sensitivity of the balance increases with increasing length of the balance arms, L, but decreases with increasing weight of the balance arm. The sensitivity is also approximately inversely proportional to the vertical distance from the point P to the center of gravity of the arm–pointer assembly, which must be below the point P for a stable system.

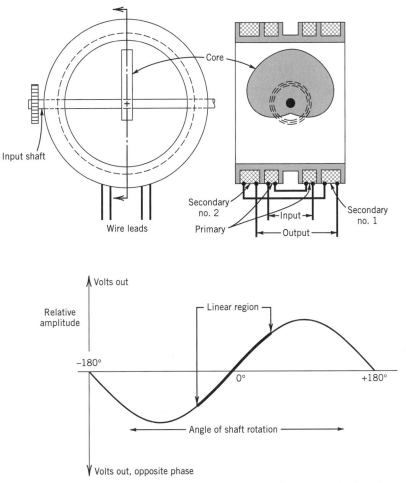

Figure 12.12 Rotary variable differential transformer. (Courtesy of Schaevitz Engineering; from [6].)

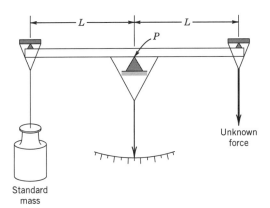

Figure 12.13 Construction of an analytical balance.

EXAMPLE 12.2

The mass of a Teflon object is to be determined through direct comparison with a standard weight on an analytical balance. The density of the Teflon is 2200 kg/m³, and the density of the material from which the standard weights are constructed is 7900 kg/m³. Determine the error that would occur if the mass of the Teflon was measured without accounting for buoyancy forces. The comparison is made in air at 25°C and 1 atm, and a standard mass of 0.4 kg is required to balance the unknown mass of Teflon.

KNOWN

An unknown mass of Teflon is in equilibrium with a standard mass, made of a material of known density, on an analytical balance.

FIND

The effect of buoyancy forces on the measurement, expressed as the error resulting from ignoring buoyancy.

SOLUTION

A free-body diagram of the standard mass and the Teflon object results in three forces acting on each mass: the weight, W, the supporting force from the balance, F_{bal}, and the buoyancy force, F_{air}, as shown in Figure 12.14. The volume of the standard mass can be determined from

$$V_s = \frac{M_s}{\rho_s} = \frac{0.4 \text{ kg}}{7900 \text{ kg m}^3} = 5.063 \times 10^{-5} \text{ m}^3$$

From the ideal gas equation of state, the density of air at 25°C and 1 atm is calculated as 1.181 kg/m³. The buoyancy force on the standard mass is then

$$F_{air} = V_s \rho_{air} g = (5.063 \times 10^{-5} \text{ m}^3)(1.81 \text{ kg/m}^3)(9.8 \text{ m/s}^2)$$
$$= 5.8602 \times 10^{-4} \text{ N}$$

which yields the force that the standard mass exerts on the balance as 3.919 N (which corresponds to an effective mass of 0.39994 kg).

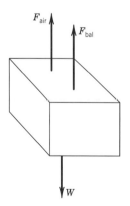

F_{air}

F_{bal}

W

Figure 12.14 Free-body diagram for Example 12.2.

For the Teflon, the value of F_{bal} must be the same, if the balance is in static equilibrium, which requires

$$(F_{bal})_{Teflon} = (F_{bal})_{standard} = 3.919 \text{ N}$$

and by examining a free-body diagram of the Teflon or standard mass,

$$F_{bal} = W - F_{air}$$

which implies

$$\rho_{Teflon} V_{Teflon} g - \rho_{air} V_{Teflon} g = F_{bal}$$

$$3.919 \text{ N} = V_{Teflon} g(\rho_{Teflon} - \rho_{air})$$

and yields for the volume of the Teflon

$$V_{Teflon} = \frac{3.919 \text{ N}}{(9.8 \text{ kg/m}^3)(2200 - 1.181) \text{ kg/m}^3} = 1.819 \times 10^{-4} \text{ m}^3$$

Thus, the mass of the Teflon is 0.400155 kg, or a difference of 0.039% between the corrected value of the mass and 0.4 kg. The analysis of the forces on the two masses can be combined to yield the general expression to correct for buoyancy forces as

$$m_u = \rho_u/(\rho_u - \rho_{air})[1 - (\rho_{air}/\rho_s)] \tag{12.3}$$

where

m_u = the unknown mass
ρ_u = density of the unknown mass
ρ_{air} = density of air at the local conditions
ρ_s = density of the standard mass

Various adaptations of the analytical balance to eliminate the need for standard masses have been devised. Figure 12.15 shows an unequal arm balance and a schematic diagram of a pendulum scale. As previously stated, a balance compares the moment created by a standard mass to the moment created by an unknown mass. To achieve an equilibrium condition, any change in the system that changes the moment can be utilized. The moment about the pivot point for a balance may be expressed as

$$\mathbf{M} = \mathbf{r} \times \mathbf{F} \tag{12.4}$$

where

$\mathbf{r}$ = radius to the pivot (vector)
$\mathbf{F}$ = force created by the mass
$\mathbf{M}$ = moment

The unequal arm balance changes the moment applied at the pivot by changing the position of the mass along the beam; the output is determined by a scale on the beam. For the pendulum scale, the vector cross product is affected by the change in the angle between the radius and the line of action of the force. The output of the pendulum scale is the angle, α. The cam changes the inherently nonlinear relationship between the unknown force and the angle into a linear one. For the pendulum scale, summing

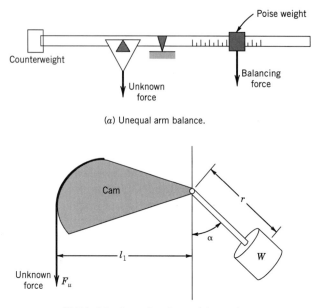

(a) Unequal arm balance.

(b) Principle of operation of a pendulum scale.

Figure 12.15 Unequal arm balance and pendulum scale.

moments about the pivot yields

$$F_u l_1 = Wr \sin \alpha \tag{12.5}$$

which may be directly solved for F_u.

12.5   MEASUREMENT OF ACCELERATION AND VIBRATION

The measurement of acceleration is required for a variety of purposes, ranging from machine design to guidance systems. Because of the range of applications for acceleration and vibration measurements, there exists a wide variety of transducers and measurement techniques, each associated with a particular application. In this chapter we address some fundamental aspects of these measurements along with some common applications.

Displacement, velocity, or acceleration measurements are also referred to as shock or vibration measurements, depending on the waveform of the forcing function that causes the acceleration. A forcing function that is periodic in nature generally results in accelerations that are analyzed as vibrations. In contrast, a force input having a short duration and a large amplitude would be classified a shock load.

The fundamental aspects of acceleration, velocity, and displacement measurements can be discerned through examination of the most basic device for measuring acceleration and velocity, a seismic transducer.

Seismic Transducer

A seismic transducer consists of three basic elements, as shown in Figure 12.16; a spring–mass–damper system, a protective housing, and an appropriate output

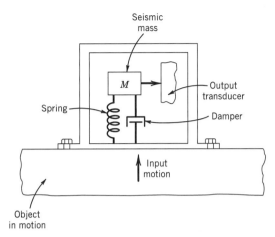

Figure 12.16 Seismic transducer.

transducer. Through the appropriate design of the characteristics of this spring–mass–damper system, the output is a direct indication of either displacement or acceleration. To accomplish a specific measurement, this basic seismic transducer is rigidly attached to the object experiencing the motion that is to be measured.

Consider the case in which the output transducer senses the position of the seismic mass; a variety of transducers could serve this function. Under some conditions, the displacement of the seismic mass serves as a direct measure of the acceleration of the housing, and the object to which it is attached. To illustrate the relation between the relative displacement of the seismic mass and acceleration, consider the case in which the input to the seismic instrument is a constant acceleration. The response of the instrument is illustrated in Figure 12.17. At steady-state conditions, under this constant acceleration, the mass will be at rest with respect to the housing. The spring deflects an amount proportional to the force required to accelerate the seismic mass, and since the mass is known, Newton's second law yields the corresponding acceleration. The relationship between a constant acceleration and the displacement of the seismic mass is linear for a linear spring (where $F = kx$).

We wish to measure not only constant accelerations, but also complex acceleration waveforms. Recall from Chapter 2 that a complex waveform can be represented

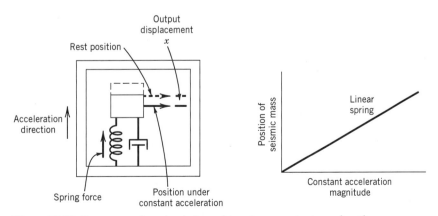

Figure 12.17 Response of a seismic transducer to a constant acceleration.

as a series of sine or cosine functions, and that by analyzing a measuring system response to a periodic waveform input we can discern the allowable range of frequency inputs. Consider an input to our seismic instrument such that the displacement of the housing is a sine wave, $y_h = A \sin \omega t$, and the absolute value of the resulting acceleration of the housing is $A\omega^2 \sin \omega t$. If we consider a free-body diagram of the seismic mass, the spring force and the damping force must balance the inertial force for the mass. (Notice that gravitational effects could play an important role in the analysis of this instrument, for instance, if the instrument were installed in an aircraft.) The spring force and damping force are proportional to the relative displacement and velocity between the housing and the mass, whereas the inertial force is dependent only on the absolute acceleration of the seismic mass.

Since

$$y_m = y_h + y_r \tag{12.6}$$

Newton's second law may be expressed

$$m\frac{d^2 y_m}{dt^2} + c\frac{dy_r}{dt} + ky_r = 0 \tag{12.7}$$

Substituting equation (12.6) in equation (12.7) yields

$$m\left(\frac{d^2 y_h}{dt^2} + \frac{d^2 y_r}{dt^2}\right) + c\frac{dy_r}{dt} + ky_r = 0 \tag{12.8}$$

But we know that $y_h = A\sin \omega t$ and

$$\frac{d^2 y_h}{dt^2} = -A\omega^2 \sin \omega t \tag{12.9}$$

Thus,

$$m\frac{d^2 y_r}{dt^2} + c\frac{dy_r}{dt} + ky_r = mA\omega^2 \sin \omega t \tag{12.10}$$

This equation is identical in form to equation (3.12). As in the development for a second-order system response, we will examine the steady-state solution to this governing equation. The transducer will sense the relative motion between the seismic mass and the instrument housing. Thus, for the instrument to be effective, the value of y_r must provide indication of the desired output.

For the input function $y_h = A \sin \omega t$, the steady-state solution for y_r is

$$(y_r)_{\text{steady}} = \frac{(1/\omega_n^2)\, A\omega^2 \cos(\omega t - \phi)}{\{[1 - (\omega/\omega_n)^2]^2 + [2\zeta(\omega/\omega_n)]^2\}^{1/2}} \tag{12.11}$$

where

$$\omega_n = \sqrt{\frac{k}{m}} \qquad \zeta = \frac{c}{2(km)^{1/2}} \qquad \phi = \tan^{-1}\frac{2\zeta(\omega/\omega_n)}{1 - (\omega/\omega_n)^2} \tag{12.12}$$

The characteristics of this seismic instrument can now be discerned by examining equations (12.11) and (12.12). The natural frequency and damping will be fixed for

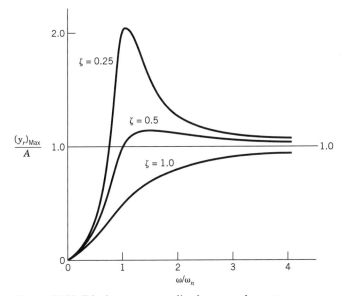

Figure 12.18 Displacement amplitude at steady state as a
function of input frequency for a seismic transducer.

a particular design. We wish to examine the motion of the seismic mass and the
resulting output for a range of input frequencies.

For vibration measurements, it is desired to measure the amplitude of the displacements associated with the vibrations; thus, the desired behavior of the seismic
instrument would be to have an output that gave a direct indication of y_h. For this
to occur, the seismic mass should remain essentially stationary in an absolute frame
of reference, and the housing and output transducer should move with the vibrating object. To determine the conditions under which this behavior would occur,
the amplitude of y_r at steady state can be examined. The ratio of the maximum
amplitude of the output divided by the equivalent static output in the present case
would be $(y_r)_{max}/A$. For vibration measurements, this ratio should have a value of 1.
Figure 12.18 shows $(y_r)_{max}/A$ as a function of the ratio of the input frequency to the
natural frequency. Clearly, as the input frequency increases, the output amplitude, y_r,
approaches the input amplitude, A, as desired. Thus, a seismic instrument that is to
be used as a vibrometer should have a natural frequency smaller than the expected
input frequency. Damping ratios near 0.7 are common for such an instrument.

EXAMPLE 12.3

A seismic instrument like the one shown in Figure 12.16 is to be used to measure a
periodic vibration having an amplitude of 0.05 in. and a frequency of 15 Hz.

1. Specify an appropriate combination of natural frequency and damping ratio
   such that the amplitude error in the output is less than 5%.
2. What spring constant and damping coefficient would yield these values of
   natural frequency and damping ratio?

3. Determine the phase lag for the output signal. Would the phase lag change if the input frequency were changed?

KNOWN

Input function $y_h = 0.05 \sin 30\pi t$

FIND

Values of ω_n, ζ, k, m, and c to yield a measurement with less than 5% magnitude error. Examine the phase response of the system.

SOLUTION

Numerous combinations of the mass, spring constant, and damping coefficient would yield a workable design. Let's choose $m = 0.05$ lb$_m$ and $\zeta = 0.07$. We know that

$$\omega_n = \sqrt{k/m} \qquad c_c = 2\sqrt{km}$$

for a spring–mass–damper system. With $\zeta = c/c_c = 0.7$, the damping coefficient, c, is found as

$$c = 2(0.7)\sqrt{km}$$

We can now examine the values of $(y_r)_{max}/A$ for $\zeta = 0.7$. The results are shown in tabular form:

$\dfrac{\omega}{\omega_n}$	$\dfrac{(y_r)_{max}}{A}$
10	1.000
8	1.000
6	1.000
4	0.999
3	0.996
2	0.975
1.7	0.951

Acceptable behavior is achieved for values of $\omega/\omega_n \geq 1.7$ for $\zeta = 0.7$, and an acceptable maximum value of the natural frequency is

$$\omega_n = \omega/1.7 = 30\pi/1.7 = 55.4 \text{ rad/s}$$

Since[3]

$$\omega_n = \sqrt{k/m}$$

we find that with $m = 0.05$ lb$_m$, the value of k is 4.8 lb/ft. Then the value of the damping coefficient is found as

$$c = 2(0.7)\sqrt{km} = 0.12$$

---

[3]Note that if k has units of lb/ft, and mass is in units of lb$_m$, g_c is required in the expression for ω_n.

The phase behavior for this system is given by equation (12.12) and results in the phase response tabulated as

$\dfrac{\omega}{\omega_n}$	$\phi[°]$
10	172
8	169.9
6	166.5
4	159.5
3	152.3
2	137.0
1.7	128.5

COMMENTS

The design of such an instrument would have other constraints that would have to be considered in the choice of design parameters. The design would be influenced by such factors as size, cost, and operating environment. The phase behavior determined for this combination of design parameters does not result in a linear relationship between phase shift and frequency. As such, the possibility of distortion of complex waveform inputs does exist.

If it is desired to measure acceleration, the behavior of the seismic mass must be quite different. The amplitude of the acceleration input signal is $A\omega^2$. To have the output value y_r represent the acceleration, it is clear from equation (12.11) that the value of $y_r/A(\omega/\omega_n)^2$ must be a constant over the design range of input frequencies. If this is true, the output will be proportional to the acceleration. The amplitude of $y_r/A(\omega/\omega_n)^2$ can be expressed

$$\frac{(y_r)_{\text{steady}}}{A(\omega/\omega_n)^2} = \frac{\cos(\omega t - \phi)}{\{[1 - (\omega/\omega_n)^2]^2 + [2\zeta(\omega/\omega_n)]^2\}^{1/2}} \tag{12.13}$$

and

$$M(\omega) = \frac{1}{\{[1 - (\omega/\omega_n)^2]^2 + [2\zeta(\omega/\omega_n)]^2\}^{1/2}} \tag{12.14}$$

where $M(\omega)$ is the magnitude ratio as defined by equation (3.21). The magnitude ratio is plotted as a function of input frequency and damping ratio in Figure 3.16.

For the desired behavior to be achieved, it is clear that the magnitude ratio should be unity. Over a range of input frequency ratios from 0 to 0.4, as determined from Figure 3.16, the magnitude ratio is approximately 1.0. Typically, in an accelerometer the damping ratio is designed to be near 0.7, so that the phase shift is linear with frequency, and distortion is minimized.

In summary, the seismic instrument can be designed so that the output can be interpreted in terms of either the input displacement or the input acceleration. Acceleration measurements may be integrated to yield velocity information; the differentiation of displacement data to determine velocity or acceleration introduces significantly more difficulties than the integration process.

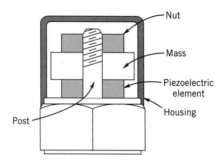

Figure 12.19 Basic piezoelectric accelerometer.

Transducers for Shock and Vibration Measurement

In general, the destructive forces generated by vibration and shock are best quantified through the measurement of acceleration. Although a variety of accelerometers are available, strain gauge and piezoelectric transducers are widely employed for the measurement of shock and vibration [7].

A piezoelectric accelerometer employs the principles of a seismic transducer through the use of a piezoelectric element to provide a portion of the spring force. Figure 12.19 illustrates one basic construction of a piezoelectric accelerometer. A preload is applied to the piezoelectric element simply by tightening the nut that holds the mass and piezoelectric element in place. Upward or downward motion of the housing will change the compressive forces in the piezoelectric element, resulting in an appropriate output signal. Instruments are available with a range of frequency response from 0.03 to 10,000 Hz. Depending on the piezoelectric material used in the transducer construction, the static sensitivity can range from 1 to 100 mV/g. Steady accelerations cannot be effectively measured with such a piezoelectric transducer.

Strain gauge accelerometers are generally constructed by using a mass supported by a flexure member, with the strain gauge sensing the deflection that results from an acceleration of the mass. The frequency response and range of acceleration of these instruments are related, such that instruments designed for higher accelerations have a wider bandwidth, but significantly lower static sensitivity. Table 12.2 provides typical performance characteristics for two strain gauge accelerometers that employ semiconductor strain gauges.

Many other accelerometer designs are available, including potentiometric, reluctive, or accelerometers that use closed-loop servo systems to provide a high output level. Piezoelectric transducers have the highest frequency response and range of

Table 12.2 Representative Performance Characteristics for Piezoresistive Accelerometers[a]

Characteristic	25-g Range	2500-g Range
Sensitivity [mV/g]	50	0.1
Resonance frequency [Hz]	2700	30,000
Damping ratio	0.4–0.7	0.03
Resistance [Ω]	1500	500

[a] Adapted from reference [7].

acceleration, but they have relatively lower sensitivity. Amplification can overcome this drawback to some degree. Semiconductor strain gauge transducers have a lower frequency response limit than piezoelectric, but they can be used to measure steady accelerations. Other transducers generally have lower frequency response behaviors compared to piezoelectric and strain gauge units.

12.6   VELOCITY MEASUREMENTS

Linear and angular velocity measurements utilize a variety of approaches ranging from radar and laser systems for speed measurement to mechanical counters to provide indication of a shaft rotational speed. For many applications, the sensors employed provide a scalar output of speed. However, sensors and methods exist that can provide indication of both speed and direction, when properly employed. Here we will consider techniques for the measurement of linear and angular speed.

Displacement, velocity, and acceleration measurements are made with respect to some frame of reference. Consider the case of a game of billiards in a moving railway car. Observers on the ground and on the train would assign different velocity vectors to the balls during play. The velocity vectors would differ by the relative velocity of the two observers. Simply differentiating the vector velocity equation, however, shows that the accelerations of the balls are the same in all reference frames moving relative to one another with constant relative velocity.

Linear Velocity Measurements

As previously discussed, measurements of velocity require a frame of reference. On one hand, the velocity of a conveyor belt might be measured relative to the floor of the building where it is housed. On the other hand, advantages may be realized in a control system if the velocity of a robotic arm, which "picks" parts from this same conveyor belt, is measured relative to the moving conveyor. In the present discussion, it is assumed that velocity is measured relative to a ground state, which is generally defined by the mounting point of a transducer.

Consider the measurement of velocity relative to a fixed frame of reference. Typically, if the measurement of linear velocity is to be made on a continuous basis, an equivalent angular rotational speed is measured, and the data are analyzed in such a way so as to produce a measured linear velocity. For example, a speedometer on an automobile provides a continuous record of the speed of the car, but the output is derived from measuring the rotational speed of the driveshaft or transmission.

Velocity from Displacement or Acceleration

Velocity can, in general, be directly measured by mechanical means only over very short times or small displacements, as a result of limitations in transducers. However, if the displacement of a rigid body is measured at identifiable time intervals, the velocity can be determined through differentiation of the time-dependent displacement. Alternatively, if acceleration is measured the velocity may be determined from integration of the acceleration signal. The following example demonstrates the effect of integration and differentiation on the uncertainty of velocities computed from acceleration or displacement.

Table 12.3 Specifications and Uncertainty Analyses for Displacement and Acceleration

Measured Variable	Functional Form	Full-Scale Output Range
Displacement (cm)	$y(t) = 20 \sin 2t$	0–10 V
Acceleration (cm/s^2)	$y''(t) = -\dfrac{20}{4} \sin 2t$	−5 to 5 V

Uncertainty Values for Displacement and Acceleration	
Measured Variable	Uncertainty
Displacement	Accuracy: 1% full scale $= \pm 0.2$ cm A/D 8 bit (0.04 V) $= \pm 0.08$ cm Total uncertainty $= \pm 0.22$ cm
Acceleration	Accuracy: 1% full scale $= \pm 0.05$ cm/s^2 A/D 8 bit (0.04 V) $= \pm 0.04$ cm/s^2 Total uncertainty $= \pm 0.064$ cm/s^2

EXAMPLE 12.4

Our goal is to assess the merits of measuring velocity through the integration of an acceleration signal as compared to differentiating a displacement signal. The following conditions are assumed to apply.

For both $y(t)$ and $y''(t)$ the data-acquisition system and transducers may be assumed to have the following characteristics: 8-bit A/D resolution and 1% accuracy for the measured variable (acceleration or displacement). The sensor outputs and uncertainties are described in Table 12.3. Note that the uncertainties are derived directly from the A/D resolution error and accuracy. Assume that the signals for both acceleration and displacement are sampled at 10 Hz and that numerical techniques are used to differentiate or integrate the resulting signals.

SOLUTION

Consider first determining the velocity through differentiation of the displacement signal. Displacement is measured digitally by the data acquisition, with a digitized value of displacement recorded at time intervals δt. The velocity at any time $n \, \delta t$ can be approximated as

$$v(t) = y'(t) = \frac{y_{n+1} - y_n}{\delta t} \qquad (12.15)$$

where

$y_{n+1} =$ the $(n+1)$ measurement of displacement, at time $(n+1)\delta t$
$\quad y_n =$ the nth measurement of displacement, at time $n \, \delta t$
$\; v(t) =$ velocity at time t

If the signal is sampled at 10 Hz, δt is 0.1 s. For the present, we will assume that the uncertainty in time is negligible, so that the uncertainty in v, u_v, can be expressed as

$$u_v = \left\{ \left[\frac{\partial v}{\partial y_{n+1}} u_{y_{n+1}} \right] + \left[\frac{\partial v}{\partial y_n} u_{y_n} \right]^2 \right\}^{1/2} \qquad (12.16)$$

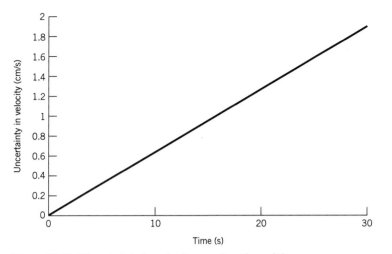

Figure 12.20 Uncertainty in velocity as a function of time.

The uncertainties in the measured displacements, y_n and y_{n+1}, will be equal, and are listed in Table 12.3. Substituting these values for uncertainty in equation (12.16) yields an uncertainty of ± 3 cm/s in the velocity measurement. Notice that this uncertainty magnitude is not a function of time or the measured velocity. This corresponds to a minimum uncertainty of 30% in the velocity measurement.

To determine velocity from acceleration, the measured values of accleration must be integrated. Since we have a digital signal, the integration can be accomplished numerically as

$$v(t) = y'(t) = \sum_i y_i'' \delta t \tag{12.17}$$

Assuming the uncertainty in time is negligible, the uncertainty in velocity at any time t is simply

$$u_v = u_{y''} t \tag{12.18}$$

Clearly, the integration process tends to accumulate error as the calculation of velocity proceeds in time, as illustrated in Figure 12.20.

COMMENT

This example can be used to illustrate several useful principles for data analysis. Consider the effects of adding noise, or a degree of precision error, to the measured data. In general, errors of this type tend to be minimized through a process of integration and amplified through differentiation. Differentiation tends to be extremely sensitive to low amplitude, high-frequency noise, which can create very large errors in derivatives, especially at high sampling frequencies. Appropriate filtering or smoothing techniques are generally effective at reducing errors associated with noise in derivatives in cases in which the noise has low amplitude.

Integration tends to eliminate low amplitude, high-frequency noise from signals. However, unless a means is devised to prevent accumulation of error during integration, this advantage of integrating may be outweighed by error accumulation. In the present problem, for example, since the velocity is periodic, the integration could possibly be "reset" at each zero of the velocity, and the error accumulation eliminated.

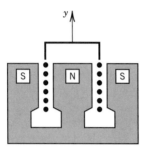

Figure 12.21 Moving coil transducer.

Moving Coil Transducers

Moving coil transducers take advantage of the voltage generated when a conductor experiences a displacement in a magnetic field. This is the same phenomenon used to generate electric power in generators and alternators. An illustration of a moving coil velocity pickup is provided in Figure 12.21. Recall that a current carrying conductor experiences a force in a magnetic field, and that a force is required to move a conductor in a magnetic field. For the latter case, an emf is induced in the conductor. Consider the case in which the magnetic field strength is at right angles to the conductor. The induced emf is given by

$$\text{emf} = \pi B D_c \, l \, N \frac{dy}{dt} \tag{12.19}$$

where

$$B = \text{magnetic field strength}$$
$$D_c = \text{coil diameter}$$
$$\text{emf} = \text{induced electromotive force}$$
$$N = \text{number of turns in coil}$$
$$dy/dt = \text{velocity of coil linear motion}$$
$$l = \text{coil length}$$

A moving coil transducer is appropriate for vibration applications in which the velocities of small amplitude motions are measured. The output voltage is proportional to the coil velocity, and the output polarity indicates the velocity direction. Static sensitivities of the order of 2 V s/m are typical. Moving coil transducers find application in seismic measurements as well as in vibration applications.

Angular Velocity Measurements

The measurement of angular velocity finds a wide range of applications, including such familiar examples as speedometers on automobiles. We will consider a variety of applications and measurement techniques.

Mechanical Measurement Techniques

Mechanical means of measuring angular velocity or rotational speed were developed primarily to provide feedback for control of engines and steam turbines. Mechanical governors and centrifugal tachometers [8] operate on the principle illustrated in Figure 12.22. Here the centripetal acceleration of the flyball masses result in a steady-state displacement of the spring, which provides a control signal or is a direct

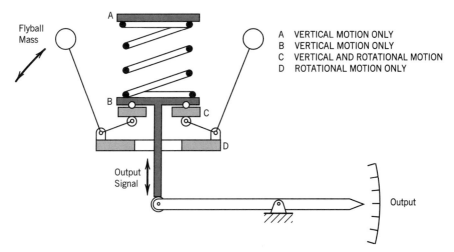

Figure 12.22 Mechanical angular velocity sensor.

indication of rotational speed. For this arrangement, the spring force is proportional to the square of the angular velocity.

A variety of angular velocity measurement techniques exist. Figure 12.23 shows a schematic diagram of a measurement system which allows the total number of revolutions and the total time to be used to determine angular velocity.

Stroboscopic Angular Velocity Measurements

A stroboscopic light source provides high-intensity flashes of light, which can be caused to occur at a precise frequency. A stroboscope is illustrated in Figure 12.24. Stroboscopes permit the intermittent observation of a periodic motion in a manner that appears to stop or slow the motion. Advances in electronics in the mid-1930s allowed the development of a stroboscope with a very well-defined flashing rate, and led to the use of stroboscopic tachometers. Figure 12.25 illustrates the use of a strobe to measure rotational speed. A timing mark on the rotating object is illuminated

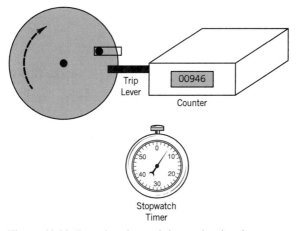

Figure 12.23 Rotational speed determination from independent time and rotation measurements.

Figure 12.24 Stroboscope. (Courtesy of Mill Devices Co., a division of A. B. Carter Inc.)

with the strobe, and the strobe frequency is adjusted such that the mark appears to remain motionless, as shown in Figure 12.25 (*a*). Thus, the highest synchronous speed is the actual rotational speed. At this speed, the output value is available from a calibration of the stroboscopic lamp flash frequency, with accuracies to less than 0.1%. Clearly, at rotational speeds higher than can be tracked by the human eye, the mark would appear motionless for integer multiples of the actual rotational speed and for integral submultiples, as illustrated in Figures 12.25(*b*)–12.25(*e*).

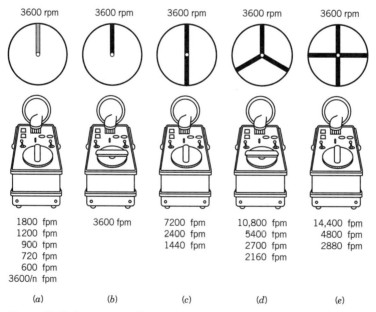

Figure 12.25 Images resulting from harmonic and subharmonic flashing rates for stroboscopic angular speed measurement. (Courtesy of Mill Devices Co., a division of A. B. Carter Inc.)

The synchronization of images at flashing rates other than the actual speed requires some practical approaches to ensuring the accurate determination of speed. Spurious images can easily result for symmetrical objects, and some asymmetric marking is necessary to prevent misinterpretation of stroboscopic data. To distinguish the actual speed from a submultiple, the flashing rate can be decreased until another single synchronous image appears. If this flashing rate corresponds to one-half the original rate, then the original rate is the actual speed. If it does not occur at one-half the original value, then the original value is a submultiple.

The upper limit of the flash rate of the strobe does not limit the ability of the stroboscope to measure rotational speed. For high speeds, synchronization can be achieved N times, with ω_1 representing the maximum achievable synchronization speed, and $\omega_2, \omega_3, \ldots$ representing successively lower synchronization speeds. The measured rotational speed is then calculated as

$$\omega = \frac{\omega_1 \omega_N (N-1)}{\omega_1 - \omega_N} \tag{12.20}$$

Electromagnetic Techniques

Several measurement techniques for rotational velocity utilize transducers that generate electrical signals, which are indicative of angular velocity. One of the most basic is illustrated in Figure 12.26. This transducer consists of a toothed wheel and a magnetic pickup, which consists of a magnet and a coil. As the toothed wheel rotates, an emf is induced in the coil as a result of changes in the magnetic field. As each ferromagnetic tooth passes the pickup, the reluctance of the magnetic circuit changes in time, yielding a voltage in the coil given by

$$E = C_B N_t \omega \sin N_t \omega t \tag{12.21}$$

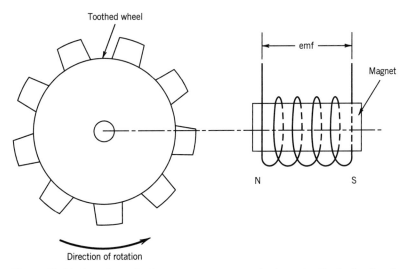

Figure 12.26 Angular velocity measurement employing a toothed wheel and magnetic pickup.

where

E = output voltage
C_B = proportionality constant
N_t = number of teeth
ω = angular velocity of wheel

The angular velocity can be found either from the amplitude or the frequency of the output signal. The voltage amplitude signal is susceptible to noise and loading errors. Thus, less error is introduced if the frequency is used to determine the angular velocity; typically, some means of counting the pulses electronically is employed. This frequency information can be transmitted digitally for recording, which eliminates the noise and loading error problems associated with voltage signals.

12.7 FORCE MEASUREMENT

The measurement of force is most familiar as the process of weighing, ranging from weighing micrograms of a medicine to weighing trucks on the highway. Force is a quantity derived from the fundamental dimensions mass, length, and time. Standards and units of measure for these quantities are defined in Chapter 1. The most common techniques for force measurement will be described in this section.

Load Cells

"Load cell" is a term used to describe a transducer that generates a voltage signal as a result of an applied force, usually along a particular direction. Such force transducers often consist of an elastic member and a deflection sensor. A technology overview for such devices is provided in [8]. These deflection sensors may employ changes in capacitance, resistance, or the piezoelectric effect to sense deflection. Consider first load cells, which are designed by using a linearly elastic member instrumented with strain gauges.

Strain Gauge Load Cells

Strain gauge load cells are most often constructed of a metal and have a shape such that the range of forces to be measured results in a measurable output voltage over the desired operating range. The shape of the linearly elastic member is designed to meet the following goals: (1) provide an appropriate range of force measuring capability with necessary accuracy and (2) provide sensitivity to forces in a particular direction, and have low sensitivity to force components in other directions. A variety of designs of linearly elastic load cells are shown in Figure 12.27. In general, load cells may be characterized as beam-type load cells, proving rings, or columnar-type designs. Beam-type load cells may be characterized as bending beam load cells or shear beam load cells.

A bending beam load cell, as shown in Figure 12.28, is configured such that the sensing element of the load cell functions as a cantilever beam. Strain gauges are mounted on the top and bottom of the beam to measure normal or bending stresses. Figure 12.28 provides qualitative indication of the shear and normal stress distributions in a cantilever beam. In the linear elastic range of the load cell, the bending stresses are linearly related to the applied load.

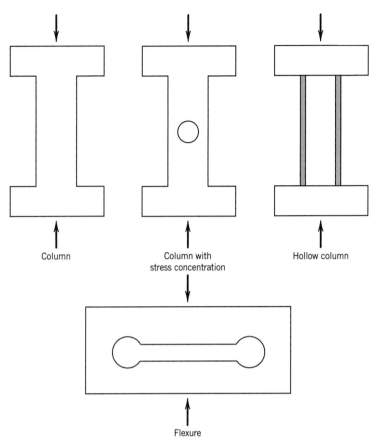

Figure 12.27 Elastic load cell designs.

In a shear beam load cell the beam cross section is that of an I-beam. The resulting shear stress in the web is nearly constant, allowing placement of a strain gauge essentially anywhere on the web with reasonable accuracy. Such a load cell is illustrated schematically in Figure 12.29, along with the shear stress distribution in the beam. In general, bending beam load cells are less costly because of their construction; however, the shear beam load cells have several advantages, including lower creep and faster response times. Typical load cells for industrial applications are illustrated in Figure 12.30.

Piezoelectric Load Cells

Piezoelectric materials are characterized by their ability to develop a charge when subject to a mechanical strain. The most common piezoelectric material is single-crystal quartz. The basic principle of transduction, which occurs in a piezoelectric element, may best be thought of as a charge generator and a capacitor. The frequency response of piezoelectric transducers is very high, since the frequency response is determined primarily by the size and material properties of the quartz crystal. The modulus of elasticity of quartz is approximately 85 GPa, yielding load cells with typical static sensitivities ranging from 0.05 to 10 mV/N, and frequency response up to 15,000 Hz. A typical piezoelectric load cell construction is shown in Figure 12.31.

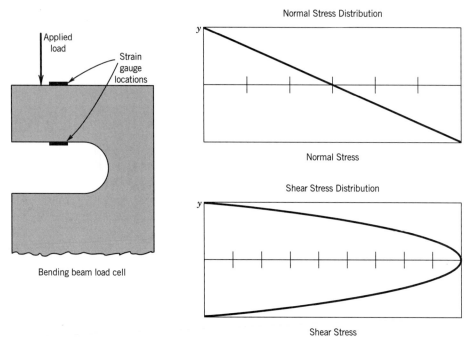

Figure 12.28 Bending beam load cell and stress distributions.

Proving Ring

A ring-type load cell can be employed as a local force standard. Such a ring-type load cell, as shown in Figure 12.32, is often employed in the calibration of materials testing machines because of the high degree of precision and accuracy possible with this arrangement of transducer and sensor. If the sensor is approximated as a circular right cylinder, the relationship between applied force and deflection is given by

$$\delta y = \left(\frac{\pi}{2} - \frac{4}{\pi} \right) \frac{F_n D^3}{16 EI} \tag{12.22}$$

where

δy = deflection along the applied force
F_n = applied force

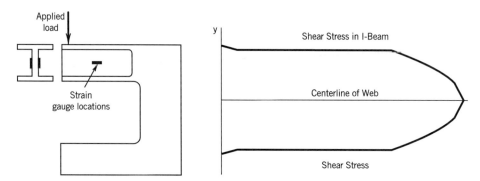

Figure 12.29 Shear beam load cell and shear stress distribution.

Figure 12.30 Typical load cells. (Courtesy of Transducer Techniques, Inc.)

D = diameter
E = modulus of elasticity
I = moment of inertia

The application of the proving ring involves measuring the deflection of the proving ring in the direction of the applied force. Typical methods for this displacement

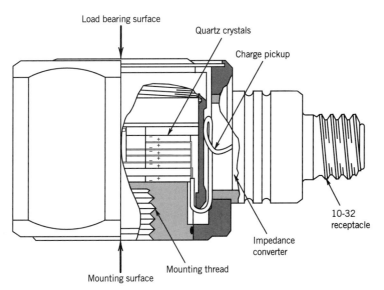

Figure 12.31 Piezoelectric load cell design. (Courtesy of The Kistler Instrument Co.)

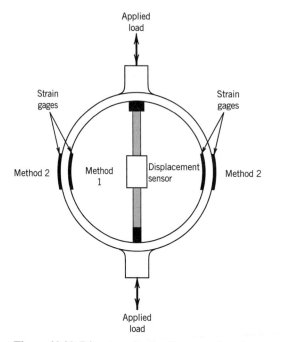

Figure 12.32   Ring-type load cell, or proving ring.

measurement include displacement transducers, which measure overall displacement, and strain gauges. These methods are illustrated in Figure 12.32.

12.8   TORQUE MEASUREMENTS

Torque and mechanical power measurements are often associated with the energy conversion processes that serve to provide mechanical and electrical power to our industrial world. Such energy conversion processes are largely characterized by the mechanical transmission of power produced by prime movers such as internal combustion engines. From automobiles to turbine-generator sets, mechanical power transmission occurs through a torque acting through a rotating shaft.

The measurement of torque is important in a variety of applications, including sizing of load-carrying shafts. This measurement is also a crucial aspect of the measurement of shaft power, such as in an engine dynamometer. Strain-gauge-based torque cells are constructed in a manner similar to load cells, in which a torsional strain in an elastic element is sensed by strain gauges appropriately placed on the elastic element. Figure 12.33 shows a circular shaft instrumented with strain gauges for the purpose of measuring torque, and a commercially available torque sensor.

Consider the stresses created in a shaft subject to a torque, T. The maximum shearing stress in a circular shaft occurs on the surface and may be calculated from the torsion formula[4]

$$\tau_{max} = TR_0/J \qquad (12.23)$$

where

τ_{max} = maximum shearing stress

---

[4]Coulomb developed the torsion formula in 1775 in connection with electrical instruments.

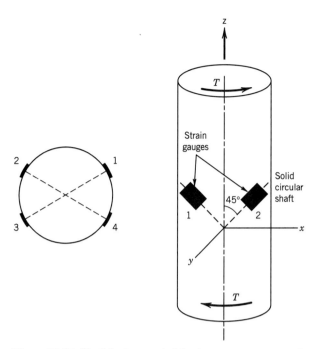

Figure 12.33 Shaft instrumented for torque measurement.

T = applied torque
J = polar moment of inertia ($\pi R_0^4/2$ for a solid circular shaft)

For a shaft in pure torsion, there are no normal stresses, σ_x, σ_y, or σ_z. The principal stresses lie along a line, which makes a 45° angle with the axis of the shaft, as illustrated in Figure 12.33, and have a value equal to τ_{max}. Strains that occur along the curve labeled A–A' are opposite in sign from those that occur along B–B'. These locations allow placement of four active strain gauges in a Wheatstone bridge arrangement, and the direct measurement of torque in terms of bridge output voltage.

Measurement of Torque on Rotating Shafts

A significant challenge to measuring the torque being transmitted by an engine or other prime mover is a result of the shaft rotation. Two primary means of transmitting data from a rotating shaft are typically employed. The traditional method of transmitting dc voltage signals from rotating equipment uses *slip rings*. Slip rings are composed of a rotating electrically conducting cylindrical conductor, which is rigidly attached to a rotating member. An electrical circuit is formed through sliding contact with a stationary brush. Variations in contact resistance in the sliding contact cause some degree of signal degradation and noise. Higher signal fidelity is achieved using mercury slip rings, although this technique is most often employed in research applications.

Telemetry can also be used to transmit signals from sensors on a rotating shaft. Such telemetry systems often employ radio frequencies and operate in a manner similar to cordless telephones or wireless microphones. Digital encoding of data before telemetry provides a reduced level of noise and improved signal fidelity.

Table 12.4 Shaft Power, Torque and Speed Relationships

	SI	U.S. Customary
Shaft power, P	$P = \omega T$	$P = \frac{2\pi n T}{550}$
Power	P (W)	P (hp)
Rotational speed	ω (rad/s)	n (rev/sec)
Torque	T (N m)	T (ft lb)

12.9 MECHANICAL POWER MEASUREMENTS

Almost universally, prime movers such as internal combustion (IC) engines and gas turbines convert chemical energy in a fuel to thermodynamic work transmitted by a shaft to the end use. In automotive applications, the pistons create a torque on the crankshaft, which is ultimately transmitted to the driving wheels. In each case, the power is transmitted through a mechanical coupling. This section is concerned with the measurement of such mechanical power transmission.

Rotational Speed, Torque, and Shaft Power

Shaft power is related to rotational speed and torque as

$$\vec{P_s} = \vec{\omega} \times \vec{T} \qquad (12.24)$$

where P_s is the shaft power, ω is the rotational velocity vector, and T the torque vector. In general, the orientation of the torque and rotational velocity vectors are such that the equation may be written in scalar form as

$$P_s = \omega T \qquad (12.25)$$

Table 12.4 provides a summary of useful equations related to shaft power, torque, and speed as employed in mechanical measurements. Historically, a device called a Prony brake was used to measure shaft power. A typical Prony brake arrangement is shown in Figure 12.34. Consider using the Prony brake to measure power output for an IC engine. The Prony brake serves to provide a well-defined load for the engine, with the power output of the engine dissipated as thermal energy in the braking

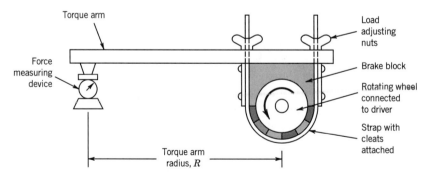

Figure 12.34 Prony brake. (Courtesy of the American Society of Mechanical Engineers, New York. Reprinted from PTC 19.7-1980 [10].)

material. By adjusting the load, the power output over a range of speeds and throttle settings can be realized. The power is measured by recording the torque acting on the torque arm, and the rotational speed of the engine. Clearly, this device is limited in speed and power, but it does serve to demonstrate the operating principles of power measurement and is historically significant as the first technique for measuring power.

Cradled Dynamometers

A Prony brake is an example of an absorbing dynamometer. The term "dynamometer" refers to a device that absorbs and measures the power output of a prime mover. Prime movers are large mechanical power-producing devices such as gasoline or diesel engines or gas turbines. Several methods of energy dissipation are utilized in various ranges of power, but the measurement techniques are governed by the same underlying principles. Thus, we will consider first the measurement of power, and then discuss means for dissipating the sometimes large amounts of power generated by prime movers.

The cradled dynamometer measures mechanical power by measuring the rotational speed of the shaft, which transmits the power, and the reaction torque required to prevent movement of the stationary part of the prime mover. This reaction torque is impressively illustrated by so-called wheelies by motorcycle riders. A cradled dynamometer is supported in bearings, which are called trunion bearings, such that the reaction torque is transmitted to a torque or force measuring device. The state-of-the-art dynamometer shown in Figure 12.35 is designed for emissions testing, with power absorption ratings above 200 hp and a top speed of 120 mph.

In principle, the operation of the dynamometer involves the steady-state measurement of the load F_r created by the reaction torque and the measurement of shaft speed. From equation (12.25) the transmitted shaft power can be calculated directly.

The ASME Performance Test Code [10] provides guidelines for the measurement of shaft power. According to this Performance Test Code, overall uncertainty in the

Figure 12.35 Dynamometer. (Courtesy of Burke E. Porter Machinery Co., Grand Rapids, MI.)

measurement of shaft power by a cradled dynamometer results from (a) trunnion bearing friction; (b) force measurement uncertainty (F_r); (c) moment arm length uncertainty (L_r); (d) static unbalance of dynamometer; and (e) uncertainty in rotational speed measurement.

A means of supplying a controllable load to the prime mover and dissipating the energy absorbed in the dynamometer is an integral part of the design of any dynamometer. Several techniques are described for providing an appropriate load.

Eddy current dynamometers: A direct current field coil and a rotor allow shaft power to be dissipated by eddy currents in the stator winding. The resulting conversion to thermal energy by joulian heating of the eddy currents necessitates some cooling be supplied, typically using cooling water.

Alternating current and dc generators: Cradled ac and dc machines are employed as power absorbing elements in dynamometers. The ac applications require variable frequency capabilities to allow a wide range of power and speed measurements. The power produced in such dynamometers may be dissipated as thermal energy using resistive loads.

Waterbrake dynamometers: A waterbrake dynamometer employs fluid friction and momentum transport to create a means of energy dissipation. Two representative designs are provided in Figure 12.36. The viscous shear-type brake is useful for high rotational speeds, and the agitator type unit is used over a range of speeds and loads. Waterbrakes may be employed for applications up to 10,000 hp (7450 kW). The load absorbed by waterbrakes can be adjusted by using water level and flow rates in the brake.

12.10   SUMMARY

The design and fabrication of any finished product requires the ability to ensure that the product meets design specifications; in part, this requires accurate dimensional measurements. Engineering measurements of size are accomplished by using a variety of methods and instruments, with the most basic tools, such as calipers and micrometers, providing sufficient accuracy for most applications. Gauge blocks serve as a local standard for length and are widely used in quality control for machining.

The measurement of displacement is used in construction of sensors for a variety of measured quantities. The most common means of displacement measurement include LVDTs and potentiometric methods. The seismic instrument provides a means for measuring acceleration and vibration, through appropriately chosen parameters for the spring–mass–damper system, which forms the sensor stage for the seismic instrument.

This chapter has provided a brief introduction to basic measurements associated with length and displacement. However, many applications and measurement techniques exist for the establishment of the flatness of surfaces, surface finish, curvature, and so on, all of which are related to dimensional measurement. Further information is found in [1, 2].

The measurement of linear velocity is most often accomplished by the measurement of an angular velocity coupled with a data-reduction process. Automobile speedometers are an example of such a measurement. Shaft power measurements are a key element in testing prime movers, such as IC engines or gas turbines. In this chapter, methods of measuring torque and rotational speed are shown to be sufficient for determining shaft power.

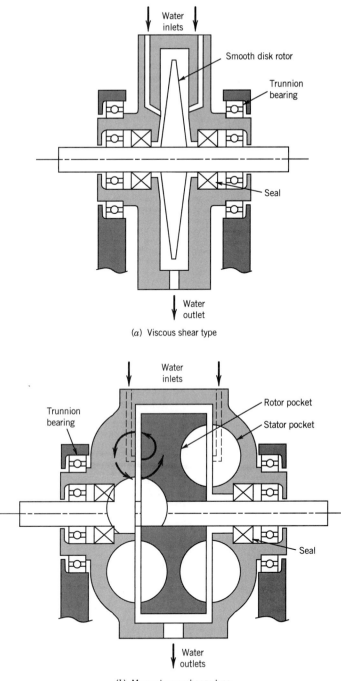

(a) Viscous shear type

(b) Momentum exchange type

Figure 12.36 Waterbrake dynamometers. (Courtesy of the American Society of Mechanical Engineers, New York. Reprinted from PTC 19.7-1980 [10].)

REFERENCES

1. Moore, W. R., *Foundations of Mechanical Accuracy*, The Moore Special Tool Company, Bridgeport, CT, 1970.
2. Sharp, K. W. B., *Practical Engineering Metrology*, Pitman, London, 1970.
3. Pontius, P. E., *Measurement Assurance Program—A Case Study: Length Measurements*, Part 1, *Long Gage Blocks (5 to 20 in.)*, National Bureau of Standards monograph 149, 1975.
4. Dukes, J. N., and Gordon, G. B., A two-hundred-foot yardstick with graduations every microinch, *Hewlett Packard Journal*, 1970; reprinted in *Inventions of Opportunity: Matching Technology with Market Needs*, Hewlett Packard Co., Palo Alto, CA, 1983.
5. Coltman, J. W., The transformer, *Scientific American*, January: 86, 1988.
6. Herceg, E. E., *Schaevitz Handbook of Measurement and Control*, Schaevitz Engineering, Pennsauken, NJ, 1976.
7. Bredin, H., Measuring shock and vibration, *Mechanical Engineering*, February: 30, 1983.
8. Measurement of Rotary Speed, ASME Performance Test Codes, ANSI/ASME PTC 19.13–1961, The American Society of Mechanical Engineers, New York, 1961.
9. Gindy, S. S., Force and torque measurement, a technology overview, *Experimental Techniques*, 9: 28, 1985.
10. Measurement of Shaft Power, ASME Performance Test Codes, ANSI/ASME PTC 19.7–1980, The American Society of Mechanical Engineers, New York, 1980.

NOMENCLATURE

a	acceleration $[l\,t^2]$		v	linear velocity $[l\,t^{-1}]$
c	damping coefficient $[m\,t]$		y	displacement $[l]$
c_c	critical damping coefficient $[m\,t]$		y_h	displacement of housing, seismic instrument $[l]$
g	acceleration of gravity $[l\,t^{-2}]$		y_m	displacement of seismic mass $[l]$
k	spring constant $[m\,t^{-2}]$		y_r	relative displacement between seismic mass and housing in a seismic instrument $[l]$
m	mass $[m]$			
m_s	standard mass $[m]$			
m_u	unknown mass, equal arm balance $[m]$		δy	deflection $[l]$
$\mathbf{r}$	radius, vector $[l]$		A	amplitude
t	time $[t]$		B	magnetic field strength
δt	time interval for data sampling $[t]$		C_α	coefficient of thermal expansion $[°]$
D_c	coil diameter for magnetic pickup $[l]$		T	temperature $[°]$
E_i	input voltage $[V]$		T	torque $[m\,l^2\,t^{-2}]$
E_o	output voltage $[V]$		ΔT	change in temperature $[°]$
F	force, vector $[m\,l\,t^{-2}]$		V	volume $[l^3]$
F_{air}	buoyancy force of air $[m\,l\,t^{-2}]$		V_s	volume of standard mass $[l^3]$
F_{bal}	force exerted by equal arm balance $[m\,l\,t^{-2}]$		W	weight $[m\,l\,t^{-2}]$
F_n	applied force $[m\,l\,t^{-2}]$		α	angle
F_u	unknown force $[m\,l\,t^{-2}]$		ζ	damping ratio
J	polar moment of inertia $[l^4]$		θ	angle
L	length $[l]$		ρ_s	density of standard mass $[m\,l^{-3}]$
ΔL	change in length $[l]$		ρ_{air}	density of air $[m\,l^{-3}]$
M	moment, vector $[m\,l^2\,t^{-2}]$		τ_{max}	maximum shearing stress $[m\,t^2\,l^{-1}]$
P	power $[m\,l^2\,t]$		ϕ	phase angle
R_o	outer radius $[l]$		ω	rotational speed $[t^{-1}]$
u	uncertainty		ω_n	natural frequency $[t^{-1}]$

PROBLEMS

12.1 Identify techniques for the measurement of lengths in the following size ranges, and specify the expected uncertainties: (a) 0.1–1 m, (b) 20–100 Å, (c) 10^6 miles, (d) 1 km, and (e) 100 light-years.

12.2 A steel tape that is graduated in 1/16-in. (1.59 mm) increments is used to measure 2 × 6-in. (5 × 15 cm) planking for constructing a deck. Identify sources of error in the measurement. If the planks could be cut to the measured size with negligible error, within what range of values would you expect the length of a single plank to fall with 95% confidence? What errors are introduced in the cutting process?

12.3 Provide a list of possible error sources for use in uncertainty analysis for the following instruments having the specified range: (a) ruler (12 in. or 30 mm), (b) micrometer (2 in. or 5 cm), (c) interferometer (1 in. or 2.5 cm), and (d) calipers (4 in. or 10 cm).

12.4 A micrometer has a least division of 0.0001 in. (0.0025 mm) and is calibrated by using two gauge blocks (wrung together) that form a value of 0.1501 in. (3.8125 mm) with a stated accuracy for each block of +4 and −2 μin. (+101 and −50 μm). If the micrometer is subsequently used to measure 0.1501 in. (3.8125 mm), estimate the design-stage uncertainty in the measured length, assuming only these elemental errors.

12.5 Consider a linear potentiometer as shown in Figure 12.8. The potentiometer consists of 0.1-mm copper wire ($\rho_e = 1.7 \times 10^{-8}$ Ω-m) wrapped around a core to form a total resistance of 1 kΩ. The sliding contact surface area is very small.

1. Estimate the range of displacement that could be measured with this potentiometer for a 1.5-cm core.

2. The circuit shown in Figure 12.37 is used to record position. On a single plot, show the loading error in an indicated displacement as a function of displacement, over the range found in step 1, for values of the meter resistance, R_m, of 1, 10, and 100 kΩ. For practical meters, would the loading error be significant?

12.6 Determine the design-stage uncertainty in the measured force, F_u, for a pendulum scale having the following parameters and associated uncertainties:

$l_1 = 25$ cm ± 0.5 mm
$r = 12$ cm ± 0.5 mm
$W = 350 \pm 0.1$ g
$F_u = 100$ g (nominally)

All uncertainty values represent total uncertainties for the indicated variables. The uncertainty in the angle α may be assumed to be ± 0.02 rad.

12.7 A seismic instrument, as shown in Figure 12.16, is used to measure a vibration given by

$$y = 0.2 \cos 10t + 0.3 \cos 20t$$

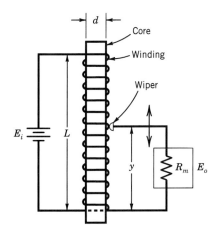

Figure 12.37 Circuit for Problem 12.5.

where

$y =$ displacement in inches
$t =$ time in seconds

The seismic instrument is to have a damping ratio of 0.7, and a spring constant of 1.2 lb/ft.

1. Select a combination of seismic mass and damping coefficient to yield less than a 10% amplitude error in measuring the input signal.

2. Describe the phase response of the system, either in a plot or tabular form. Under what conditions would the output signal experience significant distortion?

12.8 A seismic instrument has a natural frequency of 20 Hz and a damping ratio of 0.65. Determine the maximum input frequency for a vibration such that the amplitude error in the indicated displacement is less than 5%.

12.9 Determine the bandwidth for an accelerometer having a seismic mass of 0.2 g and a spring constant of 20,000 N/m, with very low damping. Discuss the advantages of a high natural frequency and a low damping ratio. Piezoelectric sensors are well suited for the construction of accelerometers, since they possess these characteristics.

12.10 In Example 12.4, integration is identified as a method for reducing the effects of noise in a signal. Discuss how a moving average can be used to reduce the effects of noise in a velocity measurement, through the integration of an acceleration signal. Discuss the effects of averaging time on the elimination of noise. Assume that the noise has significant amplitude, but higher frequency content than the velocity being measured.

12.11 Consider a moving coil transducer having a coil diameter D_c of 0.8 cm and a coil length of 2 cm. The nominal range of velocities to be measured is 1–10 cm/s. The resulting emf will be measured by a PC-based data-acquisition system with an 8-bit A/D converter and a range of -1 to 1 V. The accuracy is 0.1% full scale. Plot the number of turns as a function of magnetic field strength to provide an accuracy of 1% in the resulting velocity measurement.

12.12 Consider measuring a rotational speed using a stroboscope. The rotational speed is higher than the flash rate of the stroboscope. The stroboscope is observed to synchronize at 10,000, 18,000, and 22,000 flashes/s. Determine the rotational speed.

12.13 Design a proving ring load cell appropriate to serve as a laboratory calibration standard in the range 250–1000 N. The proving ring material is steel.

12.14 Power transmitted through the drive shaft of a car results in a rotational speed of 1800 rpm with a power transmission of 40 hp. Determine the torque that the driveshaft must support.

Appendix A

A Guide for Technical Writing

The usual way to document test results is through a technical report. Depending on the intended audience, reports usually take one of three formats: (1) executive summary, (2) lab (brief) report, or (3) formal report (technical paper).

The executive summary is usually a one- or two-page brief stating the reasons for the tests conducted, important results, and substantial conclusions: Objectives, Results, and Conclusions. This format is intended for upper-level management, who will plan corporate decisions around the input from the report. Brevity with clarity of purpose and clearly directed conclusions are crucial.

Laboratory reports are internal reports accessed by members of your engineering group. These reports provide progress information or document useful information in a brief format that stresses results in detail. These reports should include the reasons for the tests, the test plan, and the methods and test conditions used, with an emphasis on the discussion of the results and conclusions: Objectives, Approach, Results and Discussion and Conclusion. Data in the form of tables and, graphs are cited in the report and attached, including an appendix containing useful raw data.

Formal technical reports are intended for an audience outside of the immediate group. They are written at a level and style and with detail that is appropriate for the receiving audience. Accordingly, these reports should include or cite sufficient background information so that the reader can understand the purpose of the tests, the manner in which the tests were analyzed, and the results of the tests. A formal report must present enough information for the reader to follow the logic of the tests and to interpret the test results. Data should be presented only in the form of well-prepared tables and graphs cited from discussions within the report. A report ends with a concise conclusion.

Each company, agency, or test laboratory has its own format for these reports. However, we provide sample guidelines for the preparation of a formal technical report as follows. Further information concerning technical reporting may be found in numerous guides for technical writing.[1]

---

[1]Further information on the various aspects of technical writing may be found in Houp, K. W., and Pearsall, T. E., *Reporting Technical Information*, 6th ed., Macmillan, New York, 1988, or in Tichy, H. J., with Foudrinier, S., *Effective Writing for Engineers, Managers, and Scientists*, 2d ed., Wiley Interscience, New York, 1988.

A.1   A GUIDE FOR TECHNICAL WRITING[2]

Competent engineers are able to communicate their ideas in both oral and written formats to both technical and nontechnical audiences. This appendix provides a brief guide to constructing a written technical report, primarily for a technical audience. The format of this appendix is representative of most technical reporting formats. The major headings, such as Abstract and Introduction, are typical but are not meant to be exclusive. We provide this guide to show you the generally expected organization of a technical report. Different report formats will use fewer headings and more or less detail.

Abstract

Effective technical communication abilities are important in a technologically advancing society. The purpose of this document is to serve as a format example for a technical report and to provide specific procedures and ideas for generating sound technical reports. Guidelines for preparing each section of a report and detailed ideas for the presentation of results in plots and tables are provided.

Introduction

In a 1980 study, the U.S. Department of Education and the National Science Foundation concluded that the majority of Americans are moving toward "virtual scientific and technological illiteracy" [1]. The technical person's ability to communicate effectively with both technical and nontechnical segments of society is essential for addressing the myriads of technological problems and decisions facing our society. In fact, career advancement in any profession is largely based on how well a person communicates. It is especially important for engineers to develop effective technical writing skills that enable them to convey the results and significance of their work to a variety of audiences. The primary form of writing for practicing engineers and engineering managers (as well as researchers) is the technical report. In general, this is directed toward a technically knowledgable audience. The purpose of this guide is to provide reasonable ideas and suggestions for producing clear and concise technical manuscripts.

Specific Writing Steps

The following steps are recommended for effective technical writing.

1. Collect all of your data, organize your thoughts, and sketch figures and tables. The scope of the material to be presented and the particular audience for whom you are writing should be kept in mind while deciding what to present.
2. Arrange your materials in the order in which you plan to present them.
3. Establish a thesis that adequately covers your material, and create a detailed outline (Note: A thesis is a document that tests a hypothesis or idea against supporting tests, leading to a conclusion). Then expand your outline to at least the level at which topics and some sentences have been formulated. The

---

[2] Adapted with permission from Henry M. H., and Lonsdale, H. K., The researcher's writing guide, *Journal of Membrane Science*, 13:101–107, 1983.

use of this outline will help you avoid making a chronological or "stream of consciousness" presentation; the order in which you did the work and the order in which you report it will rarely match.

4. Before you begin writing, study the format or style guide for the specific publication.

5. Don't expect to create a perfect product the first time through. Write a first draft, where you concentrate on organization and documenting your ideas for the particular audience for which you are writing. Write a first draft as rapidly as possible to take advantage of your continuity of thought. The draft should be revised and polished later.

6. Read the draft and, as necessary for presentation, clarity, and accuracy, rewrite it. Be careful to make sure that your writing clearly says what you intended; don't read your intentions into whatever is on the paper. As a last step, proofread the manuscript. Reading and proofing should not be attempted at the same time. Few writers can successfully perform both functions simultaneously. Remember: We cannot read your mind, just your report—put what you want to say in writing.

Writing Technical Papers

The key to effectively communicating the results of your work is to have a clear understanding of your current knowledge. Forget all previous misconceptions, mistakes, bad data, and so on. Do not report them. Instead, take a fresh look at the results and organize your thoughts to present the work in the best way. There are several supporting blocks for constructing a technical report: written text, figures and tables, and, if needed, appendixes. Each figure and table must be described and explicitly referred to in the text. An appendix has to be cited within the text. The following guidelines for preparing each section of a technical report provide the basics necessary for sound reporting procedures.

The Abstract—A Summary of the Entire Report

An Abstract is a complete, concise distillation of the full report. Although first in the report, it is always written last. It provides a brief (one sentence) introduction to the subject, a statement of the problem, highlights of the results (quantitative, if possible), and the major conclusion. It must stand alone without citing figures or tables. A concise, clear approach is essential. Most abstracts are short and rarely exceed 200 words. An executive summary is an extended abstract of up to 500 words and may include an essential figure.

The Introduction—Why Did You Do What You Did?

An Introduction generally identifies the subject of the report, provides the necessary background information, including appropriate literature review, and provides the reader with a clear rationale for the work described. It states the hypothesis or concept tested. The introduction does not contain results and generally does not contain equations. The use of figures and tables should be limited in the Introduction.

Analysis—What Does Theory Have to Say?

An Analysis section describes a proposed theory or a descriptive model, if available. It does not contain results, nor should extreme mathematical details be provided. Sufficient detail (mathematical or otherwise) should be provided for the reader to clearly understand the physical assumptions associated with a theory or model.

Experimental Program—What Did You Measure and How?

The Experimental Program section is intended to describe how experimental results were obtained. Provide an overview of the approach, test facilities, validations, and range of measurements. As a rule of thumb, provide just sufficient detail to allow the experiment to be conducted by someone else. Do not give instructions or commands to the reader; rather report what was done. If a list of equipment is included in the report, it should be a table in the body of the report, or it should be placed in an appendix. Uncertainty analysis information can be described either here or in the Results section, or both. In cases in which both an analysis and experiment are described, these two sections of the report should complement and support each other. The relationship of the analysis to the experiment should be clearly stated. If a numerical simulation was performed, it might be described under a separate heading, such as Numerical Model, using guidelines similar to those under the Experimental Program.

Results and Discussion—So What Did You Find?

Here you present and discuss your test results and tie them back to your original objectives or hypothesis. Data must be interpreted to be useful. This transforms raw data into useful results. When presenting your results, remember that even though you are usually writing to an experienced technical audience, what may be clear to you may not be obvious to the reader. Assuming too much knowledge can be a big mistake, so explain your results even if it seems unnecessary. If you can't figure them out, say so: "The mechanism is unclear and we are continuing to examine this phenomenon." Often the most important vehicles for the clear presentation of results are figures and tables. All of the figures and tables should be numbered and have descriptive titles. Column heads in tables should accurately describe the data that appear in those columns. Each table and figure must be explicitly and individually cited and described in the text of the Results section. Since you have spent significant time in preparing the plots and tables, you are intimately familiar with their trends and implications; the reader needs your insight to understand the results as well as you.

As a guideline, Figure A.1 provides an example of an appropriately prepared plot. Each plot must have a figure number and a descriptive caption, and clearly labeled axes. A few hints for creating effective plots are as follows.

1. The independent parameter is always plotted on the x axis; the dependent parameter is always plotted on the y axis.
2. Try to use at least four tics or increments for each coordinate, but fewer than 10. Multiples of 1, 2, or 5 are good increments. Avoid axis increments that make interpolation unnecessarily difficult. It is hard to justify strange increments such as 0-7-14. . . .

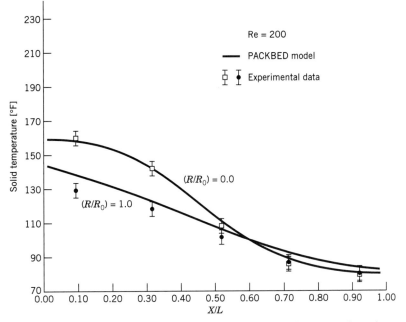

Figure A.1. Comparison of experimentally measured and computed results for axial temperature distributions for spheres of radius R in a thermal energy storage bed of radius R_0.

3. In drawing smooth curves through experimental data points, try to follow these rules: (a) always show the data points as symbols, such as open or filled squares, circles, or triangles; (b) do not extend a curve beyond the ends of the data points. If for some valid reason you find it necessary to extend the line, use a dashed line to indicate that it is some sort of extrapolation. If appropriate, indicate the curve-fit equations; (c) if you are certain that $y = 0$ when $x = 0$, put the curve through the origin. If you are uncertain, don't use the origin; stop the line slightly below the lowest point. Some judgment is required here.

4. Use a minimum variety of symbols for your data. Always include a symbol legend.

5. To compare experimentally determined data with theoretical predictions, show the experimental points as symbols and the theory as a smooth curve. Similarly, show numerical predictions as a smooth curve. Clear labeling is essential for this kind of plot.

6. Whenever possible, label curves individually, even if it means using arrows to match the lines with the labels. Use a curve legend in your figure only as a last resort.

7. The figure should contain enough information to stand alone, especially if there's a chance it will later be used by itself (e.g., as a slide). However, each figure must be discussed in the text.

8. Always provide axes quantities and units. In labeling axes, try to use the written name for the quantity plotted, and then present its units in parentheses. For example, use "time (seconds)" instead of merely "seconds."

Table A.1 Characteristics of a Thermistor
Anemometer in a Uniform Flow Field

Velocity [ft/s]	E[V]
0.467	3.137
0.950	3.240
2.13	3.617
3.20	3.811
3.33	3.876
4.25	3.985
5.00	4.141
6.67	4.299
8.33	4.484
10.0	4.635
12.0	4.780

9. Position the data on the graph (i.e., adjust the scales) so that the curves are not bunched near the top, bottom, or one of the sides. Don't run the scale up to 100% if your data only go up to 38%. The only time white space serves a valid function in a graph is when you are comparing it to another graph that makes appropriate use of the full scale.

10. Uncertainty limits should be indicated for a number of measured points on a given plot by using interval bands about the mean value, as shown in Figure A.1.

The visual impact of a plot conveys considerable information about the relationship between the plotted dependent and independent variables. Data-presentation or spreadsheet software packages are readily available that are fast and easy to use for the construction and manipulation of good quality plots. If hand drawn, plots should be constructed by using a straight edge, french curve, label maker, and symbol maker. Several forms of axis plotting formats are discussed in Chapter 1.

Table A.1 is an example of results presented in a tabular form. Each table must be explicitly described in the text of the Results section, and each table must have a title.

Conclusions—What Do I Now Know?

The Conclusions section is where you should concisely restate your answer to the question, "What do I know now?" It must support or refute your hypothesis. It is not a place to offer new facts, nor should it contain another rendition of experimental results or rationale. In a short summary, restate why the work was done and how it was done, and provide a conclusion to the work. An appropriate conclusion might be "The temperature measuring system calibrated in this study was found to indicate the correct temperature over the range 30–250°F with no more than a ±1°F uncertainty at 95% confidence." It would not normally be useful or appropriate to conclude, "The temperature measuring system was tested and worked well." Conclusions should be clear and concise statements of the important findings of a particular study; most conclusions require some quantitative aspect to be useful.

Appendices

Appendices are places to place superfluous but possibly useful information. They should stand on their own and should not provide information critical to the report—that information should be in the main body of the report. Uncertain whether information should be in an Appendix? Ask yourself: If the reader did not read the Appendix, would the report be sufficient? It should be!

References

The references cited in a formal list should be available to the reader and described in sufficient detail for the reader to obtain the source with a reasonable effort. References for this document provide a format guide.

Writing Tips

1. Accuracy is important, but so is consistency. Define all nonstandard terms the first time they are used and stick to those terms and definitions throughout all writing on that subject. Err on the side of clarity if you must err, so that if a particular construction is questionable, add the extra words that make it longer but guarantee its clarity.
2. Don't overdo significant figures. This is one of the surest ways of convincing the astute reader that you are an amateur. How many figures can you reproduce for a given measurement? Use that number.
3. Avoid the use of contractions and possessives and jargon. On occasion, jargon serves a useful function—one of neatly describing or labeling an otherwise troublesome concept or process. Still, use jargon only when your audience will understand it and a simple substitution doesn't exist.
4. Technical writing is often in the third person to focus attention on the subject matter at hand. The active voice is preferred where possible, but choose the style that suits your writing best. The important thing is to communicate effectively.

Conclusions

Technical authors must be cognizant of primary goals in presenting the results of their writing. This report has outlined the essential features germane to technical report writing and is a format example. An outline for each section of the report was stated as being the essential starting point for effective writing. A polished and professional product results only through careful and persevering revision of the first and subsequent drafts and by using effective figures and tables.

REFERENCES

1. Naisbitt, J., *Megatrends*, Warner Books, New York, 1984. (See also, U.S. report fears most Americans will become scientific illiterates, *New York Times*, Oct. 23, 1984.)
2. Henry, M. H., and Lonsdale, H. K., The researcher's writing guide, *Journal of Membrane Science*, 13:101–107, 1983.

Appendix B

Property Data
and Conversion Factors

Table B.1 Properties of Pure Metals and Selected Alloys

	Density [kg/m³]	Modulus of Elasticity [GPA]	Coefficient of Thermal Expansion [10⁻⁶m/m-K]	Thermal Conductivity [W/m-K]	Electrical Resistivity [10⁻⁶ Ω-cm]
Pure Metals					
Aluminum	2 698.9	62	23.6	247	2.655
Beryllium	1 848	275	11.6	190	4.0
Chromium	7 190	248	6.2	67	13.0
Copper	8 930	125	16.5	398	1.673
Gold	19 302	78	14.2	317.9	2.01
Iron	7 870	208.2	15.0	80	9.7
Lead	11 350	12.4	26.5	33.6	20.6
Magnesium	1 738	40	25.2	418	4.45
Molybdenum	10 220	312	5.0	142	8.0
Nickel	8 902	207	13.3	82.9	6.84
Palladium	12 020	—	11.76	70	10.8
Platinum	21 450	130.2	9.1	71.1	10.6
Rhodium	12 410	293	8.3	150.0	4.51
Silicon	2 330	112.7	5.0	83.68	1×10^5
Silver	10 490	71	19.0	428	1.47
Tin	5 765	41.6	20.0	60	11.0
Titanium	4 507	99.2	8.41	11.4	42.0
Zinc	7 133	74.4	15.0	113	5.9
Alloys					
Aluminum (2024, T6)	2 770	72.4	22.9	151	4.5
Brass (C36000)	8 500	97	20.5	115	6.6
Brass (C86500)	8 300	105	21.6	87	8.3
Bronze (C90700)	8 770	105	18	71	1.5
Constantan annealed (55% Cu 45% Ni)	8 920	—	—	19	44.1
Steel (AISI 1010)	7 832	200	12.6	60.2	20
Stainless Steel (Type 316)	8 238	190	—	14.7	—

Source: Compiled from *Metals Handbook* 9th ed., American Society for Metals, Metals Park, OH, 1978, and other sources.

Table B.2 Thermophysical Properties of Selected Metallic Solids

Composition	Melting point [K]	Properties at 300 K				Properties at various temperatures [K]							
		ρ [kg/m³]	c_p [J/kg·K]	k [W/m·K]	$\alpha \times 10^4$ [m²/s]	k[W/m·K]				c_p[J/kg·K]			
						100	200	400	600	100	200	400	600
Aluminum													
Pure	933	2702	903	237	97.1	302	237	240	231	482	796	949	1033
Alloy 2024-T6 (4.5% Cu, 1.5% Mg, 0.6% Mn)	755	2770	875	177	73.0	65	163	186	186	473	787	925	1042
Alloy 195, cast (4.5% Cu)	—	2790	883	168	68.2	—	—	174	185	—	—	—	—
Chromium	2118	7160	449	93.7	29.1	159	111	90.9	80.7	192	384	484	542
Copper													
Pure	1358	8933	385	401	117	482	413	393	379	252	356	397	417
Commercial bronze (90% Cu, 10% Al)	1293	8800	420	52	14	—	42	52	59	—	785	460	545
Phosphor gear bronze (89% Cu, 11% Sn)	1104	8780	355	54	17	—	41	65	74	—	—	—	—
Cartridge brass (70% Cu, 30% Zn)	1188	8530	380	110	33.9	75	95	137	149	—	360	395	425
Constantan (55% Cu, 45% Ni)	1493	8920	384	23	6.71	17	19	—	–	237	362	—	—
Iron													
Pure	1810	7870	447	80.2	23.1	134	94.0	69.5	54.7	216	384	490	574
Armco (99.75%)	—	7870	447	72.7	20.7	95.6	80.6	65.7	53.1	215	384	490	574
Carbon steels													
Plain carbon (Mn ≤ 1%, Si ≤ 0.1%)	—	7854	434	60.5	17.7	—	—	56.7	48.0	—	—	487	559
AISI 1010	—	7832	434	63.9	18.8	—	—	58.7	48.8	—	—	487	559
Carbon-silicon (Mn ≤ 1%, 0.1% < Si ≤ 0.6%)	—	7817	446	51.9	14.9	—	—	49.8	44.0	—	—	501	582

(Continued)

Table B.2 (*Continued*)

Composition	Melting point [K]	Properties at 300 K				Properties at various temperatures [K]							
		ρ [kg/m³]	c_p [J/kg·K]	k [W/m·K]	$\alpha \times 10^4$ [m²/s]	k[W/m·K]				c_p[J/kg·K]			
						100	200	400	600	100	200	400	600
Carbon–manganese–silicon (1% < Mn ≤ 1.65%, 0.1% < Si ≤ 0.6%)	—	8131	434	41.0	11.6	—	—	42.2	39.7	—	—	487	559
Chromium (low) steels 1/2 Cr-1/4 Mo-Si (0.18% C, 0.65% Cr, 0.23% Mo, 0.6% Si)	—	7822	444	37.7	10.9	—	—	38.2	36.7	—	—	492	575
1 Cr–1/2 Mo (0.16% C, 1% Cr, 0.54% Mo, 0.39% Si)	—	7858	442	42.3	12.2	—	—	42.0	39.1	—	—	492	575
1 Cr–V (0.2% C, 1.02% Cr, 0.15% V)	—	7836	443	48.9	14.1	—	—	46.8	42.1	—	—	492	575
Stainless steels													
AISI 302	—	8055	480	15.1	3.91	—	—	17.3	20.0	—	—	512	559
AISI 304	1670	7900	477	14.9	3.95	9.2	12.6	16.6	19.8	272	402	515	557
AISI 316	—	8238	468	13.4	3.48	—	—	15.2	18.3	—	—	504	550
AISI 347	—	7978	480	14.2	3.71	—	—	15.8	18.9	—	—	513	559
Lead	601	11340	129	35.3	24.1	39.7	36.7	34.0	31.4	118	125	132	142
Magnesium	923	1740	1024	156	87.6	169	159	153	149	649	934	1074	1170
Molybdenum	2894	10240	251	138	53.7	179	143	134	126	141	224	261	275
Nickel													
Pure	1728	8900	444	90.7	23.0	164	107	80.2	65.6	232	383	485	592
Nichrome (80% Ni, 20% Cr)	1672	8400	420	12	3.4	—	—	14	16	—	—	480	525
Inconel X—750 (73% Ni, 15% Cr, 6.7% Fe)	1665	8510	439	11.7	3.1	8.7	10.3	13.5	17.0	—	372	473	510

Table B.3 Thermophysical Properties of Saturated Water (Liquid)

T [K]	ρ [kg/m^3]	c_p [kJ/kg·K]	$\mu \times 10^6$ [N·s/m^2]	k [W/m·K]	Pr	$\beta \times 10^6$ [K^{-1}]
273.15	1000	4.217	1750	0.569	12.97	−68.05
275.0	1000	4.211	1652	0.574	12.12	−32.74
280	1000	4.198	1422	0.582	10.26	46.04
285	1000	4.189	1225	0.590	8.70	114.1
290	999	4.184	1080	0.598	7.56	174.0
295	998	4.181	959	0.606	6.62	227.5
300	997	4.179	855	0.613	5.83	276.1
305	995	4.178	769	0.620	5.18	320.6
310	993	4.178	695	0.628	4.62	361.9
315	991	4.179	631	0.634	4.16	400.4
320	989	4.180	577	0.640	3.77	436.7
325	987	4.182	528	0.645	3.42	471.2
330	984	4.184	489	0.650	3.15	504.0
335	982	4.186	453	0.656	2.89	535.5
340	979	4.188	420	0.660	2.66	566.0
345	977	4.191	389	0.664	2.46	595.4
350	974	4.195	365	0.668	2.29	624.2
355	971	4.199	343	0.671	2.15	652.3
360	967	4.203	324	0.674	2.02	679.9
365	963	4.209	306	0.677	1.90	707.1
370	961	4.214	289	0.679	1.79	728.7
373.15	958	4.217	279	0.680	1.73	750.1
400	937	4.256	217	0.688	1.34	896
450	890	4.40	152	0.678	0.99	
500	831	4.66	118	0.642	0.86	
550	756	5.24	97	0.580	0.88	
600	649	7.00	81	0.497	1.14	
647.3	315	00	45	0.238	00	

Formulas for interpolation (T = absolute temperature)

$$f(T) = A + BT + CT^2 + DT^3$$

$f(T)$	A	B	C	D	Standard deviation, σ
		$273.15 < T < 373.15$ K			
ρ	766.17	1.80396	-3.4589×10^{-3}		0.5868
c_p	5.6158	-9.0277×10^{-3}	14.177×10^{-6}		4.142×10^{-3}
k	−0.4806	5.84704×10^{-3}	-0.733188×10^{-5}		0.481×10^{-3}
		$273.15 < T < 320$ K			
$\mu \times 10^6$	0.239179×10^6	-2.23748×10^3	7.03318	-7.40993×10^{-3}	4.0534×10^{-6}
$\beta \times 10^6$	-57.2544×10^3	530.421	−1.64882	1.73329×10^{-3}	1.1498×10^{-6}
		$320 < T < 373.15$ K			
$\mu \times 10^6$	35.6602×10^3	−272.757	0.707777	-0.618833×10^{-3}	1.0194×10^{-6}
$\beta \times 10^6$	-11.1377×10^3	84.0903	−0.208544	0.183714×10^{-3}	1.2651×10^{-6}

Source: From F. P. Incropera and D. P. DeWitt. *Fundamentals of Heat and Mass Transfer*, Wiley, New York, 1985.

Table B.4 Thermophysical Properties of Air

T [K]	ρ [kg/m^3]	c_p [kJ/kg·K]	$\mu \times 10^7$ [N·s/m^2]	$\nu \times 10^6$ [m^2/s]	$k \times 10^3$ [W/m·K]	$\alpha \times 10^5$ [m^2/s]	Pr
200	1.7458	1.007	132.5	7.590	18.1	10.3	0.737
250	1.3947	1.006	159.6	11.44	22.3	15.9	0.720
300	1.1614	1.007	184.6	15.89	26.3	22.5	0.707
350	0.9950	1.009	208.2	20.92	30.0	29.9	0.700
400	0.8711	1.014	230.1	26.41	33.8	38.3	0.690
450	0.7740	1.021	250.7	32.39	37.3	47.2	0.686
500	0.6964	1.030	270.1	38.79	40.7	56.7	0.684
550	0.6329	1.040	288.4	45.57	43.9	66.7	0.683
600	0.5804	1.051	305.8	52.69	46.9	76.9	0.685
650	0.5356	1.063	322.5	60.21	49.7	87.3	0.690
700	0.4975	1.075	338.8	68.10	52.4	98.0	0.695
750	0.4643	1.087	354.6	76.37	54.9	109.	0.702
800	0.4354	1.099	369.8	84.93	57.3	120.	0.709
850	0.4097	1.110	384.3	93.80	59.6	131.	0.716
900	0.3868	1.121	398.1	102.9	62.0	143.	0.720
950	0.3666	1.131	411.3	112.2	64.3	155.	0.723
1000	0.3482	1.141	424.4	121.9	66.7	168.	0.726

Formulas for Interpolation (T = absolute temperature)

$$\rho = \frac{348.59}{T}(\sigma = 9 \times 10^{-4})$$

$$f(T) = A + BT + CT^2 + DT^3$$

$f(T)$	A	B	C	D	Standard deviation, σ
c_p	1.0507	-3.465×10^{-4}	8.388×10^{-7}	-3.848×10^{-10}	4×10^{-4}
$\mu \times 10^7$	13.554	0.6738	-3.808×10^{-4}	1.183×10^{-7}	0.4192
$k \times 10^3$	-2.450	0.1130	-6.287×10^{-5}	1.891×10^{-8}	0.1198
$\alpha \times 10^8$	-11.064	7.04×10^{-2}	1.528×10^{-4}	-4.476×10^{-8}	0.4417
Pr	0.8650	-8.488×10^{-4}	1.234×10^{-6}	-5.232×10^{-10}	1.623×10^{-3}

Source: From F. P. Incropera and D. P. DeWitt. *Fundamentals of Heat and Mass Transfer*, Wiley, New York, 1985.

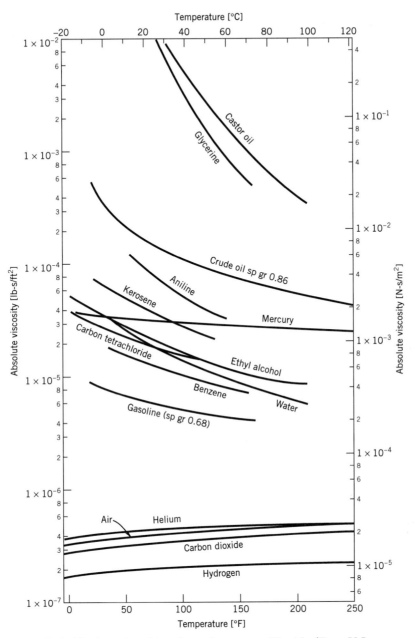

Figure B.1 Absolute viscosities of certain gases and liquids. (From V. L. Streeter and E. B. Wylie, *Fluid Mechanics*, 8th ed., McGraw-Hill, New York, 1985.)

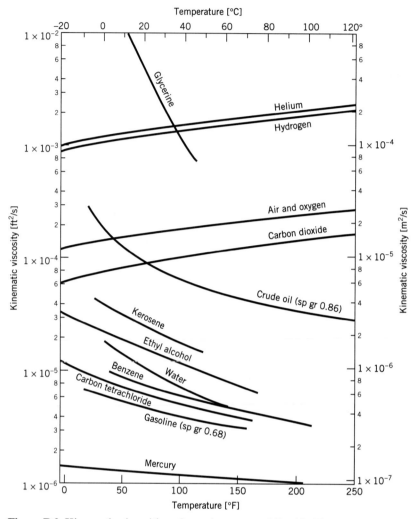

Figure B.2 Kinematic viscosities of certain gases and liquids. The gases are at standard pressure. (From V. L. Steeter and E. B. Wylie, *Fluid Mechanics*, 8th ed., McGraw-Hill, New York, 1985.)

GLOSSARY OF NEW TERMS

A/D converter   A device, which converts an analog voltage into a digital number.

absolute temperature scale   A temperature scale referenced to the absolute zero of temperature; equivalent to thermodynamic temperature.

accelerometer   An instrument for measuring acceleration.

accuracy   The closeness with which a measuring system indicates the actual value.

advanced-stage uncertainty   The uncertainty due to measurement system, measurement procedure, and measurand variation effects.

alias frequency   A frequency appearing in a discrete data set which is not in the original signal.

apparent strain   A false output from a strain gauge as a result of input from extraneous variables, such as temperature.

band-pass filter   A device which allows a range of frequencies in a signal to pass unaltered, while frequencies higher and lower than certain thresholds are attenuated.

Bessel filter   Filter design having a linear phase shift over its frequency passband.

bias error   The constant offset between the average indicated value and the actual value measured.

bimetallic thermometer   A temperature measuring device that utilizes differential thermal expansion of two materials.

bit   The smallest unit in binary representation.

bit number   The number of bits used to represent a word in a digital device.

bonded resistance strain gauge   A strain sensor that exhibits an electrical resistance change indicative of strain. The sensor is bonded to the member in which strain is being measured in such a way that the gauge experiences the same strain conditions.

bridge constant   The ratio of the actual bridge voltage to the output of a single strain gauge sensing the maximum strain.

bus   The main computer line used to transmit information.

Butterworth filter   Filter design having a flat magnitude ratio over its passband.

byte   A grouping of 8 bits.

calibration   The act of applying a known input to a system to observe the system output.

calibration curve   Plot of the input versus the output signals.

central tendency   Tendency of normal data scatter to group about a central (mean) or most probable value.

charge amplifier   Converts a high impedance charge into a voltage.

common-mode voltage   Voltage difference between two ground points.

concomitant method   An alternate method to assess a value of a variable.

control   A means to set and maintain a value during a test.

conversion error   The errors associated with converting an analog value into a digital number.

conversion time   The duration associated with converting an analog value into a digital number.

d'Arsonval movement   The basis for most analog electrical meters, which senses the flow of current through torque on a current carrying loop.

D/A converter   A device that converts a digital number into an analog voltage.

damping ratio   A measure of system damping—a measure of a system's ability to absorb or dissipate energy.

data acquisition board   A plug-in board for a PC that contains a minimum of devices required for data acquisition.

data acquisition system   A system that quantifies and stores data.

dependent variable   A variable whose value depends on the value of one or more other variables.

design-stage uncertainty   The uncertainty due to instrument error and instrument resolution.

deterministic signal   A signal that is predictable in time or in space, such as a sine wave or ramp function.

differential-ended connection   Dual wire connection scheme that measures the difference between the two signals without regard to ground.

direct memory access   Transfer protocol that allows a direct transmission path between a device and computer memory.

dynamic pressure   Difference between the total and static pressures at any point.

dynamometer   Device for measuring shaft power.

earth ground   A ground path whose voltage level is zero.

end standard   Length standard described the length between two flat square ends of a bar.

error   Difference between the value indicated by a measurement system and the actual value measured.

error fraction   A measure of the time-dependent error in the time response of a first-order system.

extraneous variable   A variable that is not or cannot be controlled during a measurement but that affects the measured value.

first-order system   A system characterized as having time-dependent storage or dissipative ability but having no inertia.

fixed point temperature   Temperatures specified through a reproducible physical behavior, such as a melting or boiling point.

flip-flop   A switching circuit that toggles either on or off on command. A bi-stable multivibrator.

frequency bandwidth   The range of frequencies in which the magnitude ratio remains within 3 db of unity.

frequency response   The output signal amplitude versus input frequency relation characteristic to a measurement system.

full-scale output (FSO)   The arithmetic difference between the end points (range) of a calibration expressed in units of output variable.

galvanometer   Instrument that is sensitive to current flow through torque on a current loop; typically used to sense balanced conditions for zero current flow.

gauge blocks   Working length standard for machining and calibration.

gauge factor   Property of a strain gauge, which relates changes in electrical resistance to strain.

ground   Signal return path to earth.

ground loop   A signal circuit formed by grounding a signal at multiple ground points of differing potentials.

handshake   An interface procedure to control data flow between different devices.

high-pass filter   A device that allows frequencies in a signal to pass unaltered, while frequencies lower than a certain threshold are attenuated.

Hooke's law   The fundamental linear relation between stress and strain. The proportionality constant is the modulus of elasticity.

hysteresis   The difference in the indicated value for any particular input when that input is approached in an increasing input direction versus when approached in a decreasing input direction.

immersion errors   Temperature measurement errors, which result from differences between sensor temperature and the temperature being measured.

impedance matching   Ensures that loading errors are kept to an acceptable level; requires assessing signal levels and instrument characteristics.

independent variable   A variable whose value can be changed directly as opposed to a dependent variable.

indicated value   The value displayed or indicated by the measurement system.

input   Process information sensed by the measurement system.

interference   Extraneous effect that imposes a deterministic trend on the measured signal.

interferometer   Device for measuring length, which uses interference patterns of light sources; provides very high resolution.

interrupt   A signal to initiate a different procedure. Analogous to raising one's hand.

junction   For thermocouples, an electrical connection, which measures temperature.

least squares regression   Analysis in curve-fitting that minimizes the sum of the squares of the deviations between the data set and the predicted polynomial curve fit.

line standard   Standard for length made by scribing two marks on a dimensionally stable material.

linearity   The closeness of a calibration curve to a straight line.

load cell   Sensor for measuring force or load; a variety of principles may be employed.

loading errors   Errors that arise in measurements of all types due to the energy extracted from the system being measured; in electrical voltage measurements, loading errors could result from the flow of a current.

low-pass filter   Device that allows low frequencies in a signal to pass unaltered, while frequencies higher than a certain threshold are atttenuated.

Linear Variable Differential Transformer (LVDT)   A sensor/transducer that provides an emf output as a function of core displacement.

magnitude ratio   The ratio of output amplitude to the input amplitude of a dynamic signal.

mean value   The central tendency of a data set. The preferred or most probable value.

measured variable   A variable whose value is measured; also known as the measurand.

metrology   The science of weights and measures.

Moiré patterns   An optical effect resulting from overlaying two dense grids, which are displaced slightly in space from one another.

monostable   One-shot device that toggles on and then off state on a single command.

moving coil transducer   A sensor that uses a displacement of a conductor in a magnetic field to sense velocity.

multiplexer   A switch having multiple input lines but only one output line.

multivibrator   Switching circuit that toggles between on and off in response to a command.

natural frequency   The frequency of the free oscillations of an undamped system. A system property.

noise   An extraneous effect that imposes random variations on the measured signal.

Nyquist frequency   One-half of the sample rate.

ohmmeter   An instrument for measuring resistance.

operating conditions   Conditions of the test, including all equipment settings, sensor locations, and environmental conditions.

optical pyrometer   Nonintrusive temperature measurement device. Measurement is made through an optical comparison.

oscilloscope   An instrument for measuring a time-varying voltage and displaying a trace; very high frequency signals can be examined. Oscilloscopes are the primary instruments for examining electrical signals.

output   The value indicated by the measurement system.

overall error   The square root of the sum of the squares (RSS) of all known errors that could affect a measuring system.

parallel communication   Communication that transmits information in simultaneous groups of bits.

parameter   The value defined by some functional relationship between variables related to the measurement.

passband   Range of frequencies over which the signal amplitude is attenuated less than -3 dB.

Peltier effect   Describes the reversible conversion of energy from electrical to thermal at a junction of dissimilar materials through which a current flows.

phase shift   A shift or offset between the angle of the input signal and the corresponding measured signal.

photoelastic   Describes a change in optical properties with strain.

Poisson's ratio   Ratio of lateral strain to axial strain.

pooled   Statistics determined from separate data sets that are grouped, such as replications.

potentiometer   Refers to a variable resistor, or to an instrument for measuring small voltages with high accuracy.

potentiometer transducer   Variable electrical resistance transducer for measurement of length or rotation angle.

precision (instrument)   Ultimate performance claim of instrument repeatability based on multiple tests conducted at multiple labs and on multiple units.

precision error   Statistical measure of the variation of the measured value during repeated measurements.

precision interval   Interval evaluated from the data scatter that defines the probable range of a statistical value.

primary standard   The definitive value of a unit.

Prony brake   A historically significant dynamometer design, which uses mechanical friction to dissipate and measure power.

proving ring   An elastic load cell, which may be used as a local calibration standard.

pyranometer   Optical instrument used to measure irradiation on a plane.

quantization   Process of converting analog value into a digital number.

quantization error   An error brought on by the resolution of an A/D converter.

radiation shield   A device which reduces radiative heat transfer to a temperature sensor, to reduce insertion errors associated with radiation.

randomization methods   Methods used to break up the interference effects from extraneous variables.

random variable   A measured variable sampled at random.

range   The lower to upper limits of an instrument or test.

recovery errors   Temperature measurement error in a flowing gas stream; important in high-speed flows.

repeatability   Precision claim based on multiple tests within a given lab on a single unit.

repetition   Repeated measurements during the same test.

replication   The duplication of a test under similar operating conditions.

reproducibility   Precision claim based on multiple tests performed in different labs on a single unit.

resistivity   Material property describing the electrical resistance of a material.

resolution   The smallest detectable change in measured value as indicated by the measuring system.

resonance frequency   The frequency at which the magnitude ratio reaches a maximum value greater than unity.

ringing frequency   The frequency of the free oscillations of a damped system. A function of the natural frequency and damping ratio.

rise time   The time required for a first-order system to respond to 90% of a step change in signal.

rms value   Root-mean-square value; derived from power dissipation by an ac current.

roll-off slope   Decay rate associated with a filter.

RTD   Resistance temperature detector; senses temperature through changes in electrical resistance of a conductor.

sample rate   The rate or frequency, at which data points are acquired. Reciprocal of sample time increment.

sample statistics   Statistics generated from a finite number of measurements of a measured variable.

sample time increment   The time interval between two successive data measurements. Reciprocal of sample rate.

saturation error   Error due to an input value exceeding the device maximum.

second-order system   A system whose behavior includes time-dependent inertia.

Seebeck effect   Source of open circuit emf in thermocouple circuits.

seismic transducer   Instrument for measuring displacement in time, velocity, or acceleration; based on a spring-mass-damper system.

sensitivity   The rate of change of a variable (y) relative to a change in some other variable (*x*), for example, dy/dx.

sensitivity index   The derivative between a resultant and a measured variable ($\partial R/\partial x$) that describes the change in the resultant due to a change in the variable.

sensor   The portion of the measurement system that senses or responds directly to the process variable being measured.

serial communication   Communication that transmits information one bit at a time.

settling time   The time required for a second order system to settle to within $\pm 10\%$ of the final value of a step change in input value.

shield   Metal foil or braid that is connected to a ground and surrounds a signal path.

shielded twisted pair   Describes electrical wiring, which reduces mutual induction between conductors through twisting the conductors around each other, and which reduces noise through shielding.

single-ended connection   Two-wire connection scheme that measures signal relative to ground.

signal-to-noise ratio   Ratio of signal power to noise power.

span   The difference between the maximum and minimum values of operating range of an instrument.

standard   The known value or basis of a calibration.

standard deviation   Probable deviation (variation) of data scatter about a mean.

standard error of fit   Probable deviation of data scatter about a curve fit line.

static pressure   Pressure sensed by a fluid particle as it moves with the flow.

static sensitivity   The rate of change of the output signal relative to a change in a static input signal. The slope of the static calibration curve at a point. (See *sensitivity*.)

steady response   A portion of the time-dependent system response, which either remains constant or repeats its waveform with time.

strain   Elongation per unit length of a member subject to an applied force.

stress   Internal force per unit area. These internal forces maintain in equilibrium the applied external forces.

stroboscope   High-intensity source of light, which can be made to flash at a precise rate, for a precise duration. It is used to measure the frequency of rotation.

temperature scale   Establishes a universal means of assigning a quantitative value to temperatures.

thermistor   Temperature-sensitive semiconductor resistor.

thermocouple   Junction of two dissimilar conductors used to measure temperature.

thermoelectric   A type of device that indicates a thermally induced emf.

thermopile   A multiple junction thermocouple circuit, designed to measure a particular temperature or temperature difference.

Thomson effect   Describes the creation of an emf through a temperature difference in a homogeneous conductor.

time constant   System property defining the time required for a first-order system to respond to 63.2% of a step input.

time delay   The delay or lag between an applied input signal and the measured output signal.

time response   The complete time-dependent system response.

total pressure   Pressure sensed by a fluid particle when brought to rest in an isentropic manner. It is the sum of the static and dynamic pressures. Also called *stagnation pressure*.

total sample period   Duration of the measured signal represented by the data set.

transducer   The portion of measurement system, which transforms the sensed information into a different form. Also loosely refers to a device that houses the sensor, transducer and, often, signal conditioning stages of a measurement system.

transient response   Portion of the time-dependent system response that decays to zero with time.

true rms   Indicates that measuring system or data reduction technique can correctly provide rms value for a nonsinusoidal signal.

true value   The actual or exact value of the measured variable.

TTL (true-transistor-logic)   Switched signal that toggles between a high (e.g. 5 V) and low (e.g. 0 V) state.

turndown   Ratio of the highest flow rate to the lowest flow rate a flow meter can measure.

uncertainty   An estimate of the range of a possible error or errors. An estimate of the probable error in a reported value.

uncertainty analysis   A process of identifying the errors in a measurement and quantifying their effects.

USB   Universal Serial Bus.

Vernier calipers   Tool for measuring both inside and outside dimensions.

Vernier scale   Allows increased resolution in reading a length scale.

VOM   Acronym for volt-ohm meter.

VTVM Acronym for vacuum tube volt meter.

Wheatstone bridge Electrical circuit for measuring resistance with high precision; can be used to measure static or dynamic signals.

word A collection of bits used to represent a number.

zero drift A shift away from zero output under a zero input value condition.

zero-order system A system whose behavior is independent of the time-dependent characteristics of storage or inertia.

zero-order uncertainty Uncertainty due only to a measurement system's resolution errors.

Index